For Jami

Table of Contents

Introduction: What is Data Science?...12

1. Experimental Design 14

　　Experimental Research 15

　　Observational Research 16

　　Types of Variables...17

　　Exploring Variable Types with R...17

　　Exploring Variable Types with Python 19

2. Descriptive Statistics 20

　　Measures of Central Tendency 20

　　Frequency Distribution 21

　　Probability Distribution 22

　　Central Limit Theorem 24

　　Standard Error..25

　　Confidence Intervals 25

　　p-Values 26

　　Types of Probability Distributions..27

　　Multivariate Descriptive Statistics 30

　　Descriptive Statistics with R...30

　　Descriptive Statistics with Python..33

3. Statistical Modeling...37

　　Test Statistics 38

　　Hypothesis Tests 38

　　When to Use Hypothesis Tests 39

　　How to Choose a Hypothesis Test 40

　　Type I and Type II Error..41

　　Parametric Tests and Standard Assumptions 42

　　Testing the Normality Assumption 43

　　The Importance of Sample Size 44

　　Testing the Homogeneity of Variance Assumption 44

　　Testing the Other Assumptions 45

　　Hypothesis Tests and Validating Parametric Assumptions with R..........................45

　　Hypothesis Tests and Validating Parametric Assumptions with Python 51

4. Data Cleaning and Preparation 58

    Dealing with Null or Missing Data 58

    Detecting Outliers 62

    Dealing with Outliers 68

    Transforming Data 68

    Data Cleaning with R 71

    Data Cleaning with Python 76

5. Correlation 83

    Covariance 83

    Pearson's Correlation Coefficient = Standardized Covariance 83

    Spearman's Rank Correlation Coefficient 85

    Kendall's Tau Correlation Coefficient 86

    Point-Biserial and Biserial Correlations 87

    Partial Correlation 88

    Comparing Correlations 89

    Correlations with R 90

    Correlations with Python 90

6. Linear Regression 92

    Goodness of Fit: R2 and the F-test 93

    Assessing Individual Predictors: t-test and p-values 94

    Interaction Terms 95

    Preventing Overfitting: AIC, BIC, and Adjusted R2 Measures of Fit 96

    Clearing Up Confusion with the SSM, SSR, and SST, and Other Measures of Fit 98

    Bias-Variance Tradeoff 101

    Linear Regression Assumptions 103

    Dealing with Categorical Predictors 106

    Sample Size, Training/Testing Split, and Outliers 107

    Linear Regression with R 107

    Linear Regression with Python 114

7. Optimization 123

    First and Second Derivative Tests 124

    Ordinary Least Squares (OLS) 126

    Weighted Least Squares (WLS) 127

Generalized Least Squares (GLS) 129

Partial Least Squares (PLS) 129

Maximum Likelihood Estimation (MLE) 129

Gradient Descent 130

Stochastic Gradient Descent 131

AdaGrad, AdaDelta, RMSprop, and Adam 134

Simulated Annealing 135

Genetic Algorithm 136

Particle Swarm 137

Optimization with Python 139

8. Hyperparameter Optimization 145

Grid Search and Random Search 145

Bayesian Optimization 146

Tree-Structured Parzen Estimation 149

Hyperparameter Optimization with R 151

Hyperparameter Optimization with Python 155

9. Logistic Regression 161

Goodness of Fit: Log-Likelihood Ratio 162

Other Measures of Fit: R, AIC, BIC, Confusion Matrix, ROC Curve 164

Residual Diagnostics 166

Assessing Individual Predictors: Odds Ratio, Wald Statistic, Likelihood Ratio Test 167

Logistic Regression Assumptions 168

Complete Separation 169

Probit Regression 169

Logistic Regression with R 169

Logistic Regression with Python 175

10. Regression for Count or Integer Variables 180

Poisson Regression 180

Negative Binomial Regression 181

Zero Inflation 182

Count and Integer Regression with R 182

Count and Integer Regression with Python 183

11. Regression with Censored Data 185

Tobit Model 185

Truncated Regression 185

Interval Regression 186

Censored Data Regression with R 186

12. Generalized Linear Models: t-Test, ANOVA, ANCOVA, Chi-Square Test, and Log-Linear Analysis 188

t-Test: Comparing 2 Means 188

ANOVA: Comparing Several Means 190

Post Hoc Analysis 194

ANCOVA: Comparing Several Means with Covariates 197

Chi-Square Test: Comparing 2 Groups of Categories 198

Loglinear Analysis: Comparing Several Groups of Categories 199

GLMs with R 201

GLMs with Python 208

13. Factorial ANOVA 214

Sphericity 215

Factorial ANOVA with R 216

Factorial ANOVA with Python 217

14. Nonparametric Tests for Comparing Group Means 219

Wilcoxon Rank-Sum Test and Mann-Whitney Test 219

Wilcoxon Signed-Rank Test 220

Kruskal-Wallis Test 221

Jonckheere-Terpstra Test 221

Friedman's ANOVA 221

Nonparametric Tests with R 222

Nonparametric Tests with Python 223

15. Discriminant Function Analysis for Classification 225

Discriminant Function Analysis Assumptions 226

Linear Discriminant Analysis 227

Quadratic Discriminant Analysis 229

LDA and QDA Classification with R 229

LDA and QDA Classification with Python 232

16. Dimension Reduction 236

Feature Selection Methods 236

Feature Extraction Methods 236

Eigendecomposition 237

Singular Value Decomposition 239

Principal Component Analysis 240

Multiple Correspondence Analysis 245

Factor Analysis 249

Factor Analysis of Mixed Data 251

Linear Discriminant Analysis 251

Dimension Reduction with R 253

Dimension Reduction with Python 259

17. Nonlinear Dimension Reduction and Manifold Learning 265

Multidimensional Scaling 265

Isomap 267

Locally Linear Embedding 270

Modified Locally Linear Embedding 273

Stochastic Neighbor Embedding 273

t-Distributed Stochastic Neighbor Embedding 275

Nonlinear Dimension Reduction with Python 276

18. Multivariate Analysis of Variance (MANOVA) 281

Discriminant Function Variates 282

Test Statistics 282

Choosing a Test Statistic 284

MANOVA Assumptions 284

Univariate ANOVA Tests after MANOVA 285

MANOVA with R 285

19. Regularized Regression 287

Ridge Regression 287

Least Absolute Shrinkage and Selection Operation Regression 288

Elastic Net 289

Regularized Regression with R 289

Regularized Regression with Python 292

20. Nonlinear Regression and Classification 296

Polynomial Regression 296

Isotonic Regression 297

Exponential and Logarithmic Regression 297

Smoothing Functions 298

Generalized Additive Models 299

Regression and Classification Trees 302

K-Nearest Neighbors 305

Data Reduction with KNN 309

Nonlinear Regression and Classification with R 309

Nonlinear Regression and Classification with Python 313

21. Multilevel Models 321

Random Effects and Fixed Effects Models 322

When to Use Multilevel Models 323

Multilevel Models with R 324

22. Resampling 328

Bootstrapping 328

Jackknifing 329

Cross Validation 329

When to Use Bootstrapping vs Cross Validation 331

Resampling to Deal with Class Imbalance..331

Resampling with R..332

Resampling with Python 334

23. Probability and Bayesian Statistics...338

Bayesian vs Frequentist Statistics 338

Probability 338

Bayes' Theorem 345

Introduction to Graph Theory..348

Bayesian Networks...350

Markov Networks..352

Bayesian Regression...356

Classification with Naïve Bayes...358

Bayesian Networks and Classification with R 359

Bayesian Regression and Classification with Python 362

24. Markov Models 374

Hidden Markov Models 374

Kalman Filter 377

Markov Chain Monte Carlo Algorithms 380

Metropolis-Hastings Algorithm 381

Gibbs Sampling 383

Markov Models with Python 384

25. Time Series Analysis 393

Serial Correlation/Autocorrelation 394

Time Series Decomposition 396

Random Walks 397

Autoregressive Models and Moving Averages 398

Stochastic Volatility 404

Quantitative Trading: Mean Reversion and Cointegration 406

Anomaly Detection 409

Benford's Law 411

Time Series Segmentation: Change Point Detection and Regime Change 413

Time Series Analysis with R 416

Time Series Analysis with Python 423

26. Signal Processing 433

Basics of Wave Physics 433

Frequency vs Time Domain 435

Fourier Transform 436

Fourier Transform in More Detail 441

Wavelet Transform 446

Spectral Density 447

Spectrum versus Cepstrum 448

Filters... 450

Signal Processing with Python 451

27. The Data Analytics Process 461

Defining the Problem 461

Gathering Data 462

Exploratory Data Analysis 464

Data Modeling... 465

Interpreting and Communicating the Results...466

Putting Models into Production...467

Maintaining Models...468

Wrapping Up..468

References and Bibliography 469

Acknowledgements...480

About the Author...481

Introduction: What is Data Science?

Data science is an interdisciplinary field that combines knowledge of mathematics, statistics, computer science, information science, communication studies, and typically some kind of domain knowledge. It involves harvesting data, extracting meaning from it, and using the insights to solve problems.

There are countless applications for data science. It can be used for everything from diagnosing diseases to modeling risk, predicting product sales, segmenting customers, classifying images, recognizing speech, and optimizing decision making under uncertainty. The application of data science requires data, which can be collected through experimentation and observation.

The word "science" is important because it distinguishes data science from data pseudo-science. Data science requires the full rigor of the scientific method, which means abandoning all pre-conceived notions about problems and testing all assumptions. When pre-conceived notions are not abandoned, but used as a benchmark for which to compare the results of data analysis, the dangerous territory of data pseudo-science is entered. In this territory, it is likely that any analysis will only reinforce pre-conceived assumptions, and data will be selectively mined to support the assumptions. Data pseudo-science leads to erroneous conclusions.

In this book, we will explore data science through the lens of statistics. We will explore best practices to avoid data pseudo-science, learn how to describe the relationships between variables, and apply statistical models to make predictions and classify data. By the end of the book, the reader will have general knowledge of statistics and how to apply it with R and Python. While most of the concepts are explained in a way that builds ideas from the ground up, the reader is expected to be familiar with the basis of computer programming for the R and Python sections. This book provides the logical background for the concepts presented, with minimal math. Equations are presented only when they enhance the reader's understanding of the topics, and there are no mathematical proofs of the concepts. The purpose of this book is to give a lay person a conceptual understanding of statistics and the ability to apply the concepts presented with R and Python. The reader is expected to have an understanding of basic mathematics and algebra. Knowledge of linear algebra is required only from chapter 16 until the end of the book, and an understanding of geometry would be useful for chapter 17. Knowledge of physics, as it pertains to waves, would be helpful although not required, for chapter 26.

This book is the compilation and organization of years of the author's own notes about the subject. Therefore, the chapters are ordered by increasing complexity, such that the concepts in later chapters build upon the concepts from earlier chapters. This should be advantageous for beginners or intermediate practitioners of data science, as the explanations do not require the reader to make inferential leaps in order to understand the basic ideas. The book is by no means comprehensive of data science or statistics, but it does cover an enormous amount of material. It will likely be the most useful for practitioners of data science and engineers, rather than researchers and academics. The sections where the concepts are applied in R and Python give useful tips for dealing with the messiness of data in the real world, where statistics do not always come out cleanly like they do in sterile academic exercises.

!!!!!! IMPORTANT !!!!!

All of the code shown in this book can be found in the GitHub repository located here: https://github.com/nlinc1905/dsilt-stats-code. Additionally, an Ubuntu Virtual Machine has been set up with all of the R packages and Python libraries required to run the code, as well as a copy of the GitHub Repository. Downloading this Virtual Machine (VM) will allow the reader to run code out of the box in an environment that is invariant to library and software updates, ensuring that it will always run without errors. The VM can run on any VM software, such as Oracle's VirtualBox, which is free to download and works well with Microsoft Windows. It requires a minimum partition of 11 GB. The VM can be download here: http://www.mediafire.com/file/za5a18lr5kroak8/DSILT_Ubuntu_18.04.ova/file.

1. Experimental Design

The goal of statistics is to describe random variables, and their relationships with other random variables, by performing research. Research can be conducted through experimentation or observation. Experimental research involves the manipulation of one variable, while all other variables are controlled or randomized, in order to determine the impact of manipulating the one variable. Observational research involves observing phenomenon in natural or uncontrolled settings. Experimental research is usually preferable, especially in fields like psychology and medicine, because it can be used to identify causal relationships. Observational research cannot identify causality, because the experimenter cannot control any of the variables. Observational research is more common in fields like econometrics and natural sciences.

The variables whose relationships we seek to describe with statistics are categorized into independent and dependent variables. Independent variables, or predictor variables, are not influenced by any other variables. They are typically represented by x in simple linear equations such as y = x + 2. Independent variables influence dependent variables. Dependent variables, or outcome variables, are determined by the independent variables. They are typically represented by y in simple linear equations such as y = x + 2.

Take a look at the tabular data from the Iris flower dataset in figure 1.1 (Anderson, 1936). Each column or field is a variable. Whether or not a variable is independent or dependent depends on the research question to be answered. For example, if the question is how to determine flower species, then the species variable would be dependent on the other variables. Another possible question is how to predict the sepal length, given species and the other variables. In this case, sepal length would be the dependent variable, and species would be independent. With statistics, it is possible to describe the relationships between these variables through the use of a mathematical equation, or model.

Sepal.Length	Sepal.Width	Petal.Length	Petal.Width	Species
5.1	3.5	1.4	0.2	setosa
4.9	3.0	1.4	0.2	setosa
4.7	3.2	1.3	0.2	setosa

Figure 1.1

A statistical model is just another phrase for an equation. An equation that describes the relationships between variables is called a model because, like any model, it is an imprecise simplification of a more complex relationship. Every model in statistics contains some small measure of error, typically represented by the variable epsilon (ε). A reasonable question is why should models be used at all if they are imprecise? The answer is that models can be useful for describing the relationships between variables with fairly high accuracy, accepting that they will be incorrect once in a while due to the fact that they simplify highly complex relationships. For example, consider how engineers who want to build a bridge first model the bridge to ensure that it could withstand a variety of conditions. The model is a synthetic or small scale version of what the bridge would actually be, but it is useful to determining how the bridge would stand up in the real world. There is no way the model can test every possible set of environmental conditions, but it works *nearly* every time. Similarly, statistical models that simplify the relationships between variables can be accurate most of the time. The trick is to find the model that works the best, or the equation that best describes the relationship between dependent and independent variables, and that is what the bulk of this book is about.

Experimental Research

The two types of research are experimental and observational. In an experiment, the goal is to manipulate an independent variable (the predictor) and observe its effect on a dependent variable (the outcome). Experiments must be reliable and valid. That means they must be repeatable, generate the same results every time they are repeated, and the measurements taken for the experiment must actually measure what they intend to. For example, it would not be accurate to count the number of participants in an experiment the first time and estimate it the second time, because each measurement would be fundamentally different. Experiments must be carried out the exact same way every time they are repeated.

Experimental research is inspired by the **scientific method**. The steps to the scientific method are as follows:

1. An observation about the world leads to a question
2. A question is answered by an educated guess, or hypothesis
3. An experiment is designed to test the hypothesis
4. An experiment requires variables to be measured, and data is collected to measure those variables
5. The collected data is analyzed to either confirm or reject the hypothesis
6. If the hypothesis is rejected, it is revised and tested again
7. If the hypothesis is confirmed, then the data is fit to a model that can explain a phenomenon
8. Models are the foundations of theory

When data is collected during an experiment, there is a possibility that some measurements could be influenced by outside factors. The outside factors could cause variability in a measurement. Good experiments employ techniques to reduce variance in measurement. For example, if an experiment requires measuring a group of people's heart rates, outside factors could influence heart rate. Some people could be nervous about being in a laboratory setting, or others might have taken the stairs instead of the elevator so their heart rates are faster. While it may not always be possible to control every possible variable, experiments can be designed in ways that minimize outside effects. There are 2 types of experimental designs, and each minimizes the outside effects in different ways.

Independent measures design is an experimental design in which groups of different people are asked to perform the same task, with only 1 difference between the ways by which they are asked to complete the task. For example, to test the effect of showering on body odor after gym use, we could have 1 group shower after using the gym and another group not shower. The group who did not shower would be the control group, because nothing was done to try to alter their body odor. The group who showered would be the test group (also called the treatment group), because we are interested in the effect of the shower on body odor. In this experiment, shower would be the independent variable and body odor would be the dependent variable.

The drawback to independent measures design experiments is that they do not account for innate differences between the participants. For example, some people in our body odor experiment could have naturally have more body odor than others, regardless of whether or not they showered. If differences between participants could be an issue, then it may be better to use a repeated measures design.

Repeated measures design is a design in which the same group of people is asked to perform the same task, with only 1 difference between the ways by which they are asked to complete the task. As an example, think back to our body odor experiment. Instead of having 1 group shower after using the gym and the other group not shower, we could have the same group of people shower after using the gym on 1 day, and then not shower after using the gym on another day. The same group of people would be both the control group and the test group. In this experiment, shower would again be the independent variable, and body odor would be the dependent variable.

Repeated measures design experiments reduce unmeasured error that could influence independent design experiments, however there could still be outside factors that cannot be controlled even by using the same participants in each group. For example, people who eat spicy food before using the gym could smell differently because they of the food they ate, rather than because of gym use or showering. One way to reduce other outside factors is to randomly split the group of people in half. On day one, half of the participants would shower after using the gym and the other half would not. On the second day, the half that showered before *would not* shower and the half that did not shower before *would* shower. Splitting the same group of people and alternating which half is the control improves the reliability of the experiment.

Observational Research

The two types of research are experimental and observational. When it is not possible to manipulate or control any of the independent variables, observational research is the only option. For example, if we wanted to test the effect of cocaine usage on body odor, it would not be possible to administer doses of cocaine to participants. Instead, we could conduct a survey or look at what happens in naturally occurring examples. The 2 most common types of observational research are cross sectional research and longitudinal research.

Longitudinal research, also called panel research, involves repeated observations of the same variables or features over a period of time. The data used for longitudinal research are called panel data. Panel data describe several entities over the research time period. Sometimes only a single entity is described however, and when this happens the result is a special case of panel data called time series data. So panel data is essentially a collection of time series data where the same features are observed, at the same points in time, and throughout the same time span. An example of time series data collected through longitudinal research would be 1 month of daily opening and closing stock prices for a single stock. An example of panel data collected through longitudinal research would be 1 month of daily opening and closing stock prices for multiple stocks.

Cross sectional research involves observing a subset of a population at some random point or cross section in time. The data collected from cross sectional research is a special case panel data in which the features of 1 or many entities are observed at 1 point in time. Surveys that are carried out once are examples of cross sectional research. Cross sectional research cannot identify causal relationships, only correlations.

It is important to know the differences between the types of research and the types of data that result from them, because they will determine how the data can be analyzed and modeled.

Types of Variables

Statistical research involves 2 types of variables: dependent and independent. Variables can also be called features, attributes, dimensions, fields, or columns. The last two synonyms presume that the data is in tabular format like the data in figure 1.1, which is not always true, so it is more broadly applicable to refer to variables as features, attributes, or dimensions.

Variables can be measured at different levels, and the level of measurement often determines the type of analysis that can be performed. It is therefore important to know the level of measurement of all the variables in a dataset before performing any analysis. There are 2 broad categories of variables: categorical and continuous.

Categorical variables consist of entities divided into categories. Categorical variables can be further divided into 2 sub-types: binary and nominal. If a categorical variable has only 2 categories, like true or false, then it is binary. If a categorical variable has more than 2 categories, like the colors of the rainbow, then it is nominal. Ordinal variables are a sub-type of nominal variables that have categories with a defined order, such the letter grades on tests. Figure 1.2 shows the breakdown of categorical variables.

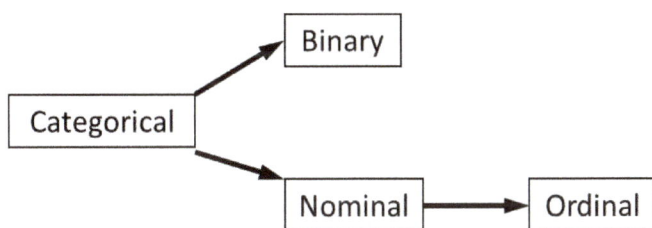

Figure 1.2

Continuous variables consist of entities with distinct numerical values. In other words, they are quantifiable variables. Continuous variables can be further divided into 2 sub-types: interval and ratio. If a continuous variable has equal intervals between each value, then it is an interval variable. Temperature in degrees Celsius is an example of an interval variable because the difference between 50 and 60 is the same as the difference between 60 and 70. If a continuous variable has both equal intervals and proportionality between values, then it is a ratio variable. If there is none of a variable when its value is 0, then it is a ratio level variable. For example, if temperature is measured in Kelvins, 0 means there is no temperature. But if temperature is measured in Celsius, 0 does not mean there is no temperature, so temperature in degrees Celsius is not a ratio variable whereas temperature in Kelvins is.

Exploring Variable Types with R

The R and Python sections of this book assume that the reader has installed R and Python and has basic knowledge of them.

Suppose we wanted to test the hypothesis that first class passengers on the Titanic were less likely to have died than second or third class passengers. We could not set up an experiment to test the hypothesis, so we would need to perform observational research, and collect historic data about deaths and passenger class. This would be cross sectional data because it would consist of data about several individuals at one point in time.

Kaggle is a website that features machine learning and predictive modeling competitions. One of their introductory datasets includes the class, age, sex, ticket fare, and other features of Titanic passengers (Kaggle, 2018). It can be used to test the hypothesis that first class passengers were more likely to have survived than the other classes. The dataset can be downloaded from Kaggle's Titanic competition page: https://www.kaggle.com/c/titanic

Kaggle has a training dataset, which is used to train models, and a test dataset, which is used to test the models on unseen data. Using different datasets for training and testing is standard practice in data science, because it ensures that the model can hold up to new data. A model that can only explain the relationship between variables in 1 dataset would not be useful. It would be like a model bridge that can only endure sunny, calm weather. We will look deeper into the reasoning behind using different datasets for training and testing in a later chapter.

After downloading the training and test datasets and selecting a working directory for R, they can be read into R as data frames. A data frame is an object that represents a data table. It consists of a list of vectors that are all the same length. In other words, a data frame is a table of features and their values for some number of observations. The syntax for reading a csv file into R as a data frame is below.

```
train <- read.csv("train.csv", header=T)
test <- read.csv("test.csv", header=T)
```

Let us explore the training data. The first 6 lines can be read by using the head function.

```
head(train)
```

The first variable, PassengerId, is used to uniquely identify each passenger. By calling the str function, it is possible to see that it is an integer type. Its range is from 1 to 891.

```
str(train$PassengerId)
range(train$PassengerId)
```

Based on that information about PassengerId, it appears to be a continuous, interval variable. However, it does not make sense to have intervals between passengers. The interval between passenger 1 and passenger 2 has no meaning. It would make sense to categorize passengers though. A unique identifier puts every passenger into his or her own category. So PassengerId is really a categorical variable, or more specifically, a nominal variable. The unique identifiers in most datasets are nominal variables.

By calling the str and range function on the second variable, Survived, it becomes clear that it is a binary variable. Doing the same for the third variable, Pclass, shows that it is an integer with a range from 1 to 3. Pclass is not a continuous variable because the interval between classes is subjective. Pclass is used to categorize passengers, so it must be categorical. It is not binary, so it must be nominal, but the classes can be ordered, so Pclass is actually an ordinal variable.

The same process can be used to determine the variable type of each feature. Name is a nominal variable, and Age is an interval variable. SibSp and Parch represent the number of siblings/spouses on board, and the number of parents/children on board, respectively. They are both ratio variables because they have all the properties of interval variables and when they equal 0, they mean that there are none of them. Ticket is nominal, Fare is ratio, Cabin is ordinal, and Embarked is nominal but could be transformed into an ordinal variable if we consider the path of the Titanic and rank embarkation port oldest to most recent.

Exploring Variable Types with Python

The R and Python sections of this book assume that the reader has installed R and Python and has basic knowledge of them.

Suppose we wanted to test the hypothesis that first class passengers on the Titanic were less likely to have died than second or third class passengers. We could not set up an experiment to test the hypothesis, so we would need to perform observational research, and collect historic data about deaths and passenger class. This would be cross sectional data because it would consist of data about several individuals at one point in time.

Kaggle is a website that features machine learning and predictive modeling competitions. One of their introductory datasets includes the class, age, sex, ticket fare, and other features of Titanic passengers (Kaggle, 2018). It can be used to test the hypothesis that first class passengers were more likely to have survived than the other classes. The dataset can be downloaded from Kaggle's Titanic competition page: https://www.kaggle.com/c/titanic

Kaggle has a training dataset, which is used to train models, and a test dataset, which is used to test the models on unseen data. Using different datasets for training and testing is standard practice in data science, because it ensures that the model can hold up to new data. A model that can only explain the relationship between variables in 1 dataset would not be useful. It would be like a model bridge that can only endure sunny, calm weather. We will look deeper into the reasoning behind using different datasets for training and testing in a later chapter.

After downloading the training and test datasets and selecting a working directory for Python, they can be read into Python as data frames through the help of the Pandas library. A data frame is an object that represents a data table. It consists of a list of vectors that are all the same length. In other words, a data frame is a table of features and their values for some number of observations. The Pandas library makes manipulating tabular data very easy and somewhat similar to the way it is done in R. Pandas is not a native Python library, so it must be installed separately.

```
import pandas as pd
train = pd.read_csv("train.csv", sep=",")
test = pd.read_csv("test.csv", sep=",")
```

Let's explore the training data. The first 6 lines can be read by using the head function.

```
train.head(n=6)
```

Notice that Python begins counting at 0, so the first 6 rows end with row = 5. The first variable, PassengerId, is used to uniquely identify each passenger. By requesting the dtype (data type) attribute with dot notation, it is possible to see that it is an integer type. Its range is from 1 to 891.

```
train.PassengerId.dtype
min(train.PassengerId)
max(train.PassengerId)
```

By following this procedure for each variable, it is possible to discern the variable types, just like we did in the section for R. A shortcut is to use the info function, which is similar to the str function in R.

```
train.info()
```

2. Descriptive Statistics

A dataset is a collection of information. The simplest way to quantitatively describe a dataset is through descriptive statistics. Descriptive statistics, also called summary statistics, are numbers that summarize one or more features of the dataset. One of the most common examples of a descriptive statistic is the number of observations in a dataset. If the data is in tabular format, then the number of observations is just the number of rows. Each observation is a sample, so the number of observations in a dataset is also called the sample size. Three more common descriptive statistics are the mean (average), median (middle value), and mode (most frequent value). These statistics describe the center values of variables and are called measures of central tendency.

Measures of Central Tendency

The **mean** of a variable is the sum of all its values divided by the number of values. For example, the mean of the set of numbers {5, 7, 4, 4, 9, 3, 1} is 33/7 or 4.7143. The **median** of a variable is the middle number when the variable's values are sorted from least to greatest. For example, the median of the previous set of numbers is 5. When there are an even number of values in a set, the median is the mean of the middle two. The **mode** of a variable is the most common value. For example, the mode of the previous set of numbers is 4, because 4 appears the most frequently. When a set of numbers has two values that occur most frequently, then both are the mode. Sets with 2 modes are called bimodal. Sets could have more than 2 modes as well. Sets with more than 2 modes are called multimodal.

The median of a variable is also its 50th percentile. A **percentile** is the value below which a certain percentage of observations fall. So if the median is the 50th percentile, then 50% of the observations will be less than the median. Dividing a variable into some number of percentiles such that the interval between each percentile is the same results in a quantile. A **quantile** shows the percentiles for certain divisions of a variable. For example, a quantile that divides a variable into 4 equal intervals of percentiles (25th, 50th, and 75th) is called a quartile.

Recall from chapter 1 that only certain types of analysis can be performed on certain types of variables. The type of variable determines which measure of central tendency is the best to use. For example, the mode is the only measure that can be applied to nominal variables. The chart below shows which measures of central tendency can be applied to each type of variable, and the best measures to use for each type of variable are in green text.

	Categorical			Continuous	
	Binary	Nominal	Ordinal	Interval	Ratio
Mean	No	No	Maybe	Yes*	Yes*
Median	No	No	Yes	Yes*	Yes*
Mode	Yes	Yes	Yes	Yes	Yes

Figure 2.1

*The mean is only the best measure of central tendency if the distribution is not skewed. If the distribution is skewed, then the best measure would be the median.

Notice that ordinal data has no single best measure. If an ordinal scale like letter grades in a class is used, then the mode is appropriate. However the letter grades could be replaced by numbers such that A = 1, B = 2, etc., and then the median and mean would be possible. Taking the mean of numbered ordinal data might not make sense though. For example, if the mean was found to be 1.5, would it be considered to be an A or a B? As long as there were an odd number of categories, the median could work. The mean could be applied to other ordinal scales though. If the ordinal scale had a range of 1 to 5 measuring a scale of strongly disagree to strongly agree (a.k.a. a likert scale), then the mean would make more sense even if it were in between two values. While there are purists who think that the mean should never be applied to ordinal data, in practice it is up to one's own judgement.

Frequency Distribution

The mean, median, and mode are referred to as measures of central tendency because they measure the center of a distribution of values. A distribution of values can be summarized by the number of times each value occurs. This is called a frequency distribution, and it is usually displayed as a histogram. Suppose that the set of numbers {5, 7, 4, 4, 9, 3, 1} represents the number of students in a classroom whose favorite colors are red, orange, yellow, green, blue, indigo, and violet, respectively. The frequency distribution of favorite colors would look like figure 2.2.

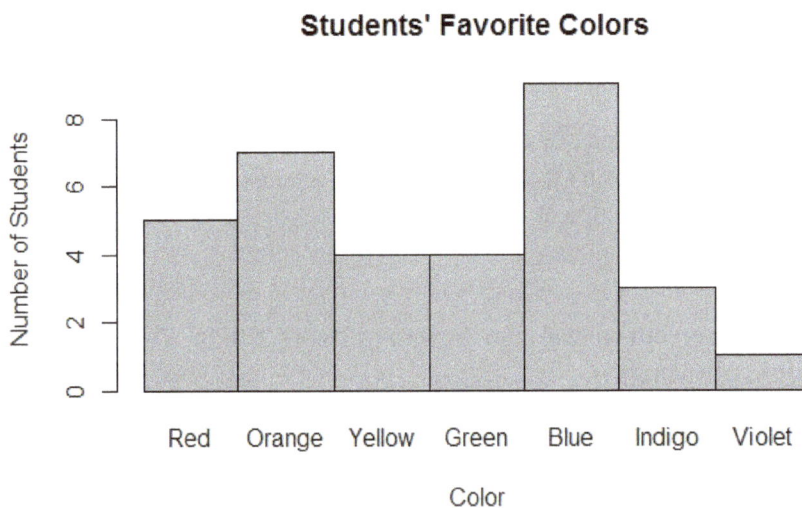

Figure 2.2

While the frequency distribution shows how often a variable equals a certain value, another way of looking at it is to ask how likely a variable equals a certain value. For example, 5 out of 33 students in figure 2.2 said that their favorite color was red. So the probability of red being a student's favorite color is 5/33, or about 15.2%. Looking at a distribution in terms of probability instead of frequency is called a probability distribution.

Before defining the probability distribution, it is important to note that probability is always measured from 0 to 1, or from 0% to 100%. This is because if all of the observations were in the same category, then the probability of any randomly chosen observation belonging to that category would be the number of observations in that category divided by total observations, which would just be a number divided by itself. Any number divided by itself is 1. Conversely, if none of the observations belonged to a particular category, then the probability of any randomly chosen observation belonging to that category would be the number of observations in that category (0) divided by the total observations. Zero divided by any number is always 0.

Probability Distribution

A probability distribution is the distribution of values in terms of how likely they are to occur. While the likelihood of a categorical variable assuming some value can be calculated by dividing the frequency of that value by the total of all possible values for all categories, the likelihood of a continuous variable being equal to some value must be calculated by finding the area under the probability distribution curve. Finding the area under a curve requires integral calculus, but fortunately, statisticians have built tables showing the areas under the most common probability distribution curves. A few of the most common distributions are the normal distribution, Student's t-distribution, and the chi-squared distribution. Of these, the normal distribution is by far the most common (see figure 2.4). Any distribution can be converted to a normal distribution by calculating the **z-score**. The practice of taking the z-score of a distribution of numbers is called **standardization**, because it translates the numbers so that they fit the standard normal distribution. The z-score formula is shown in figure 2.3.

$$Z_x = \frac{x - \mu}{\sigma}$$

Figure 2.3

Mu in the z-score formula in figure 2.3 represents the population mean, x represents the observation value, and lowercase sigma represents the population standard deviation. As an example, let's calculate the z-score for the number 5 in the set of numbers {5, 7, 4, 4, 9, 3, 1}:

$Z_5 = (5 - 5.2857) / 2.628 = -0.1087$

The standard deviation was provided for now, but we will learn how to calculate it soon. The z-score falls on the x-axis of the standard normal distribution.

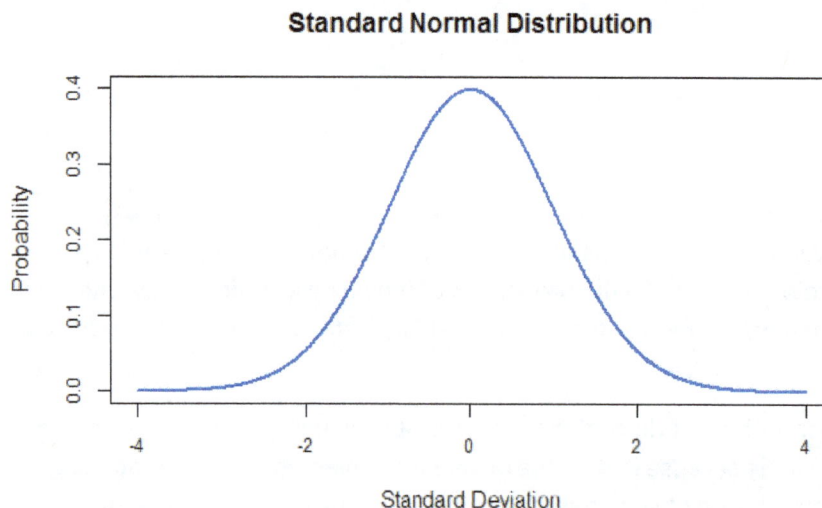

Figure 2.4

The standard normal distribution, colloquially referred to as the bell curve, is centered on mean 0 and has a standard deviation of 1. We will define standard deviation in a moment, but for now look at the y-axis of the normal distribution in figure 2.4. The y-axis is probability. The area under the entire bell curve is equal to the total probability of 1, because it accounts for the entire range of possible values in the distribution. Figure 2.5 shows how the cumulative sum of probabilities in the normal distribution starts at 0 and goes to 1. Figure 2.5 is called the **cumulative distribution function**. It shows what would happen if we totaled all of the probabilities in the bell curve up to each standard deviation. For example, the cumulative probability up to a standard deviation of 0 is 0.5, because 50% of the observations in normally distributed data are less than or equal to the mean. That means that the normal distribution is perfectly symmetrical.

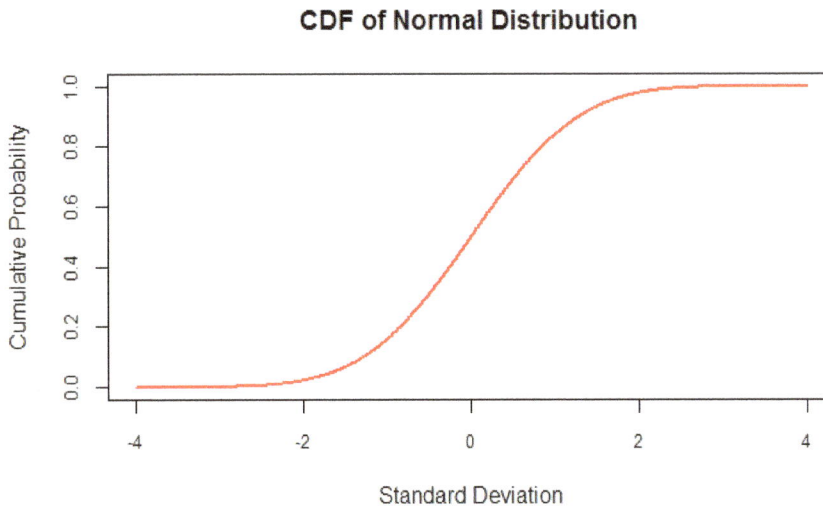

CDF of Normal Distribution

Figure 2.5

By slicing the bell curve at certain standard deviations, it is possible to see the probability of a randomly picking an observation less than a standardized value, more than the standardized value, or between two standardized values. Slicing the bell curve is the same as drawing a vertical line at any standard deviation in the cumulative distribution function in figure 2.5. If we wanted to know the probability of finding an observation 2 or more standard deviations greater than the mean, we could either calculate the area under the bell curve by reading the standard normal statistical table or by finding where the line SD = 2 intersects the cumulative distribution function, and subtracting that value from 1. Similarly, if we wanted to know the percentage of observations that lied within 1 SD on either side of the mean, we could use the normal statistical table to find the answer to be 68.26%, or used the CDF and find the difference between the cumulative probability at SD = 1 and SD = -1. In fact, for any bell curve, 68.26% of the observations will lie within 1 SD of the mean, 95.44% of the observations will lie within 2 SD of the mean, and 99.73% of the observations will lie within 3 SD of the mean (Glen, 2018).

Standard deviation measures the average distance from an observation to the mean in standardized units. It is mathematically defined as the square root of the variance. Standard deviation is represented by lowercase sigma. The **variance** of a variable is a measure of the variable's dispersion. It is a descriptive statistic that summarizes how widely spread the values are from the mean. Variance is defined mathematically as the average of the squared differences between the each value and the mean.

Standard Deviation	Variance
$$\sigma = \sqrt{\frac{1}{N}\sum_{i=1}^{N}(x_i - \mu)^2}$$	$$\sigma^2 = \frac{1}{N}\sum_{i=1}^{N}(x_i - \mu)^2$$

Figure 2.6

The variance is represented by lowercase sigma squared, N represents the number of observations, mu is the population mean, and x sub i is the value of the ith observation. The differences between each value and the mean are squared to get rid of negative numbers and to amplify the impact of larger differences. Sets of numbers with outliers, or values that lie very far from the mean, will therefore have larger variances. If a set of numbers has a large variance, then it means that the values vary greatly from the mean. The same is true for standard deviation. Thus, the mean is typically reported with either the variance or standard deviation so that anyone reading the summary statistics is aware of how well the mean describes the center of the distribution.

Central Limit Theorem

So far we have looked at standard deviation and variance in terms of the population mean. The population is the entire set of possible values that a variable can take. Often in statistics, data is collected for samples however, because it is not possible to observe the entire population. For example, political surveys are carried out on smaller samples of the population, because surveyors cannot possibly ask every registered voter about their political preferences. This presents a problem with the equations used to calculate variance and standard deviation, because they rely on the mean. The same mean can be calculated for very different sets of numbers. For example, the sets {5, 10, 15} and {1, 12, 17} both have a mean of 10. If we were to pick a third set of 3 numbers that have an average of 10, then we have 2 degrees of freedom, because the first two numbers we choose can be any we like, but our choices for the first two eliminate our ability to choose the third number. Whatever that third number is, it must result in a mean of 10. For example, if we pick 3 and 7 as the first two numbers, then the only possible choice for the third number is 20. The ability to vary the numbers used to calculate the mean is called the **degree of freedom**.

As an analogy, think of a soccer team. There are 11 positions for players to take on the soccer field. As positions are filled, there are fewer options for the remaining players. Finally, only 1 position remains open so the next player must fill that position. So there are only 10 degrees of freedom on a soccer team, and 1 spot that cannot vary. The same logic applies to all possible averages. The degrees of freedom will always be the number of observations minus one. Thus, the variance of a sample is the sum of squared difference of each observation from the mean, divided by the numbers of observations minus 1, as in figure 2.7.

Sample Standard Deviation	Sample Variance
$s = \sqrt{\dfrac{1}{N-1}\sum_{i=1}^{N}(x_i - \bar{x})^2}$	$s^2 = \dfrac{1}{N-1}\sum_{i=1}^{N}(x_i - \bar{x})^2$

Figure 2.7

In figure 2.7, x bar is the sample mean. The sample standard deviation is represented by a lowercase s.

Samples are drawn from a population. The means of each sample form a frequency distribution. As the sample size increases, the closer the distribution of sample means gets to the true population mean. So even if the population mean cannot be directly measured, it can be inferred. This property is called the Central Limit Theorem.

The Central Limit Theorem states that as a sample gets larger, the distribution of sample means gets closer to the population mean. This allows the population parameters (mean, variance, and standard deviation) to be estimated. Since the Central Limit Theorem allows us to estimate how likely a statistical model can generalize to the population, it is critical to the application of inferential statistics. Fortunately, the theorem holds regardless of the shape of the underlying population distribution, because it is standardized. A large sample for the purpose of the Central Limit Theorem is one that has more than 30 observations (Mordkoff, 2000/2011/2016).

Standard Error

Due to the fact that population parameters like mean, variance, and standard deviation are estimated, there is a chance that the parameters are inaccurate. The **standard error** measures the accuracy the population parameter estimates. Usually the standard error is referred to in reference to the population mean (SEM), but all parameters have a standard error. As the sample size increases, parameters converge on their true population values and the standard error approaches 0. So the lower the standard error, the better the estimate of the parameter.

The variation in a sampling distribution will always be less than or equal to the variation in an individual sample, so the standard error of the mean (SEM) will always be less than or equal to a sample standard deviation. SEM indicates how far a sample mean is likely to be from the population mean. It is calculated by the equation in figure 2.8.

$$SEM = \frac{s}{\sqrt{N}}$$

Figure 2.8

In the equation in figure 2.8, s is the sample standard deviation and N is the number of observations in the sample.

Confidence Intervals

The standard error of the mean is one way to look at the variation across sample means for a population. Another way to do it is to create a range of values in which the true population mean can lie. These ranges are called confidence intervals.

Since large sample distributions are normally distributed according to the Central Limit Theorem, then the same heuristics that apply to the normal distribution can apply to any large sample distribution (large meaning more than 30 observations). For example, 95% of the observations in a sample will lie between -1.96 and 1.96 standard deviations from the sample mean. Therefore, 1.96 is considered to be a critical value for the 95% confidence level. **Critical values** are represented by Z* and are defined as the standard deviation limits beyond which the null can be rejected in hypothesis tests. We will explore what this statement means in chapter 3, but for now, it is only necessary to know that the critical value is one of two ways to calculate the confidence interval of a population parameter. The population mean confidence intervals can be calculated as follows:

1. Calculate the confidence interval of the sample mean:
   a. Lower Limit = sample mean + (-1.96*standard error)
   b. Upper Limit = sample mean + (1.96*standard error)

c. If the interval is large, then the samples are not good representations of the population, and more sampling may be needed to better estimate the population parameters.
2. Calculate the z-score of the sample mean and compare it to the critical value:
   a. The z-score equals the sample mean minus the population mean, divided by the population standard deviation: $Z = \frac{\bar{x} - \mu}{\sigma}$
   b. Determine whether or not the absolute value of the z-score is > the critical value: Is $|Z| > 1.96$?
   c. If it is, then the samples are not good representations of the population, and more sampling may be needed to better estimate the population parameters

Confidence intervals can be also used to calculate the range of other parameter estimates of a population. If the confidence intervals for 2 samples believed to be drawn from the same population do not overlap, then the samples are likely from different populations. However, at the 95% confidence level, there is also a 5% chance that one of the samples did not contain the true population parameter. If 2 samples are drawn from experimental research, then either there is something in the experiment setup that is causing one of the samples to be different, or the samples were not drawn correctly. In either case, experimental design should be revisited if samples drawn from an experiment do not have overlapping confidence intervals.

Confidence intervals are not limited to the normal distribution: they can be calculated for other distributions as well. For example, calculating the confidence intervals for Student's t-distribution requires replacing the z-score with the t-statistic.

When a population is described by its mean, the mean is typically reported with a 90%, 95%, or 99% confidence interval. The mean + z-score gives the upper confidence interval and the mean – z-score gives the lower interval. For example, if a shoe company is manufacturing shoes that must be precisely 9 inches in length, then a confidence interval can be constructed around the mean of the population to serve as a quality control measure. Shoes with lengths outside the confidence interval fail the quality control test and are destroyed. The z confidence interval is hardly ever used in the real world however, because it require knowing the population standard deviation. It is much more common to use the standard deviation of a sample, and when sample standard deviation replaces population standard deviation in the z-score formula, the result is the t-statistic. We will explore the t-statistic and the t-distribution shortly.

p-Values
The p-value is another way of looking at variation in a probability distribution. The "p" in p-value stands for probability. It is defined as the probability of obtaining a result from a distribution that is more extreme than was observed. For example, if the confidence interval is 95%, then the p-value would be the likelihood of obtaining a result in the 5% of the distribution that is not included in the confidence interval (see the red area in figure 2.9).

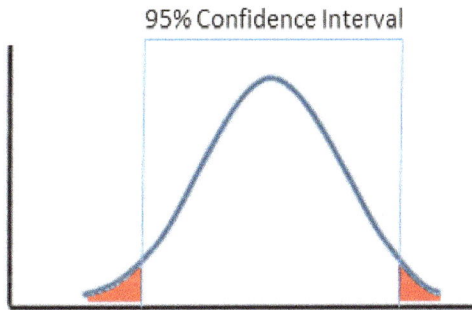

Figure 2.9

If the p-value is small, then the samples are likely a good representation of the population. Conversely, larger p-values suggest that the samples are likely not a good representation of the population. As we will see in chapter 3, test statistic p-values can be used to determine whether or not to reject the null hypothesis in hypothesis tests, because if the test statistic's p-value is smaller than a certain level of significance, then it is not likely that the sample is representative of the population. The threshold for determining the size of the p-value is denoted as alpha (α), and it is up to the experimenter to choose, depending on the level of confidence that the experimenter wants to use. The most common values for alpha are 0.05 and 0.01, meaning that p-values less than or equal to those numbers are low, and anything greater than those values are high (Lu & Belitskaya-Levy, 2015). For example, a p-value less than 0.05 would mean that there is less than a 5% chance that an extreme outcome would occur.

p-values are often criticized for being misused or overused in research, but they are indeed very useful. No better alternative to the p-value has been found and garnered enough consensus to replace it (Lu & Belitskaya-Levy, 2015).

Types of Probability Distributions

Probability distributions can be described by their skew and kurtosis. Skewness measures the asymmetry of a distribution. Skew can be positive, meaning the distribution is skewed toward positive standard deviations and has a longer right tail, or negative, meaning the distribution is skewed toward negative standard deviations and has a longer left tail. Figure 2.10 shows skewness.

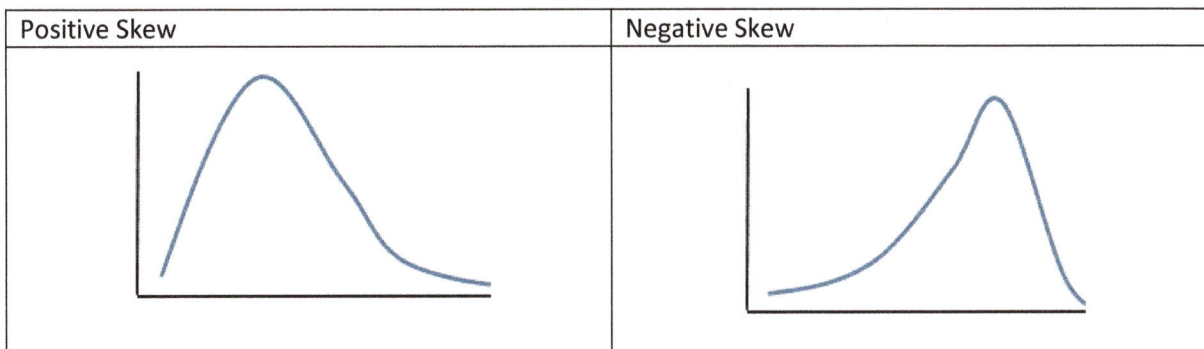

Positive Skew	Negative Skew

Figure 2.10

Kurtosis measures the fatness of the tails of a distribution. Distributions with many outliers (observations that are very far from the mean) or high variance have fatter tails and are said to be platykurtic. Distributions with fewer or no outliers, or with low variance, have thinner tails and are said to be leptokurtic. The normal distribution has no kurtosis and is said to be mesokurtic.   Kurtosis is commonly measured in relation to the normal distribution, which has a kurtosis measure of 3. When kurtosis is measured as compared to the normal distribution, it is called excess kurtosis, or Pearson's Kurtosis, and it is calculated as raw kurtosis - 3. Figure 2.11 shows excess kurtosis measures, which we will refer to as simply "kurtosis" from here on.

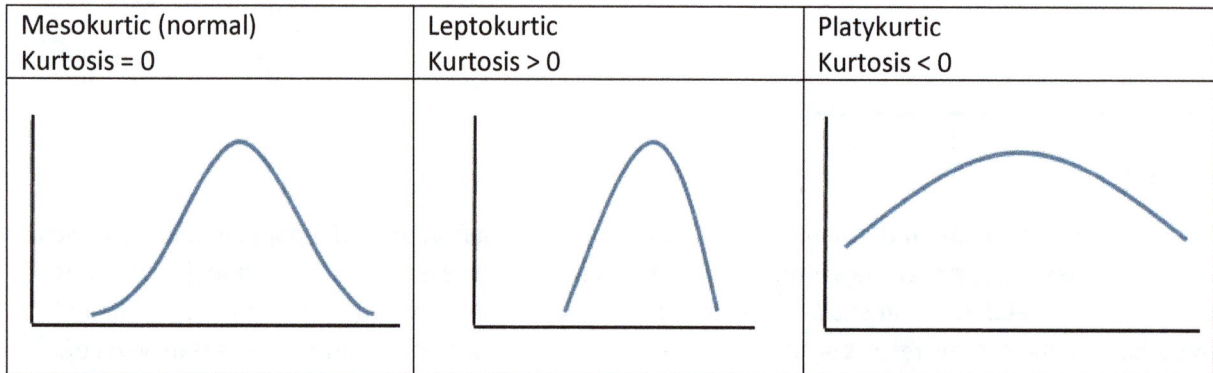

Mesokurtic (normal) Kurtosis = 0	Leptokurtic Kurtosis > 0	Platykurtic Kurtosis < 0

Figure 2.11

Student's t-distribution is more platykurtic than the normal distribution. The t-distribution is like a small sample version of the normal distribution. If the normal distribution describes a population, then the t-distribution describes a small sample of that population. The larger the sample size, the closer the t-distribution becomes to the normal distribution, thus, the t-distribution varies depending on sample size. Smaller samples have more platykurtic distributions. The t-distribution is approximately normal for large samples (samples with more than 30 observations). Recall how z-scores are calculated using the population mean in the denominator (look back to the confidence intervals section for a refresher). In real world situations, the population standard deviation is hardly ever known, so the sample standard deviation is used instead. When the sample standard deviation, s, is used in the denominator instead of the population standard deviation, the result is the t-statistic instead of the z-score. Instead of z-scores, the t-distribution relies on the t-statistic for hypothesis testing. The degrees of freedom for a t-distribution is 1 less than the number of observations. Figure 2.12 shows the t-distribution with 2 degrees of freedom (green) in comparison to the normal distribution (blue).

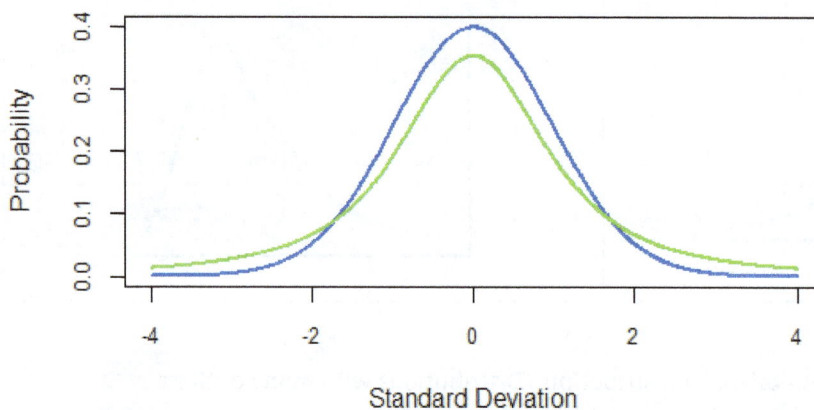

Figure 2.12

For some historical trivia, Student was actually a pseudonym for the statistician William Sealy Gossett. Gossett developed the t-distribution and the t-test while working for Guinness in 1908 to help improve the quality and consistency of Guinness beer (Coladarci, 2010). The t-statistic and t-distribution were developed specifically to account for the higher sampling error of small samples (small samples have higher error in the mean and standard deviation than large samples). Using the t-test, Guinness was able to ensure that its beer had the right mixture of ingredients and alcohol content by testing only a small sample of its products. Gossett was allowed to publish his work only if he used a pseudonym, hence we all study Student's t-test.

If k samples are drawn from the z-scores of the normal distribution, squared, and then the squares are summed, the resulting samples form a chi-square distribution. The **chi-square distribution** is defined as the distribution formed by the sum of k squared standard normal variables. Take a look at the normal distribution again, and notice that most of the observations lie at 0 standard units (the mean is 0). There are very few large observations at the tails. Given that squaring small numbers less than 1 produces even smaller numbers, we should expect that the chi-square distribution reflect a large probability of small numbers and a small probability of large numbers. This is exactly what the chi-square distribution displays, as shown in figure 2.13.

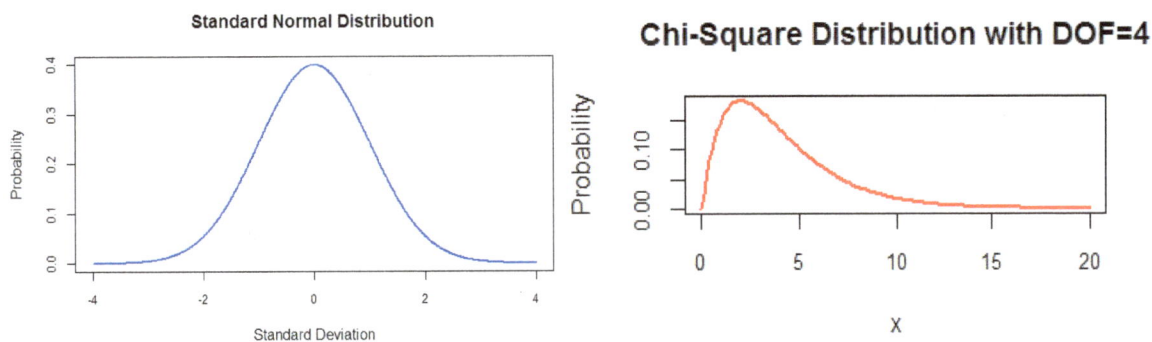

Figure 2.13

As more samples are drawn from the normal distribution, or in other words as k increases, the more similar the chi-square distribution becomes to the normal distribution.

The chi-square distribution is an approximation of a multinomial distribution, which is useful for testing for relationships between two categorical variables, as we will see later. Instead of z-scores, the t-distribution relies on the chi-square statistic for hypothesis testing.

Probability distributions can be constructed for both continuous and discrete random variables. A function of the probability distribution gives the likelihood that a sample from the distribution equals a certain value. The terms below define the types of functions for each type of variable:

- **Probability Density Function (PDF)** – A function that gives the probability that a sample from the distribution of a continuous random variable equals some value.
- **Probability Mass Function (PMF)** – A function that gives the probability that a discrete random variable equals some value. If the frequencies of a histogram were converted to probabilities, then the function that represented the probability of any given value would be the probability

mass function. This is different from the probability density function because it is for discrete variables, whereas the probability density function is for continuous variables.

- **Cumulative Distribution Function (CDF)** – The CDF was defined previously, but it is important to note that this function applies to both continuous and discrete random variables.

The general term "probability distribution" encompasses all of the terms above. The PDF, PMF, and CDF all draw from probability distributions.

Multivariate Descriptive Statistics

All of the descriptive statistics explained hitherto are univariate: they summarize single variables. While the remainder of this book deals with bivariate statistics (describing the relationship between 2 variables) or multivariate statistics (describing the relationships between multiple variables), multivariate statistics require statistical inference. **Statistical inference** is the process of analyzing the underlying distribution of a variable so that population parameters can be estimated and the relationships between multiple variables can be described. Descriptive statistics do not require statistical inference though. One type of multivariate descriptive analysis that does not require statistical inference is cross-tabulation or contingency table analysis.

A **cross-tabulation** or contingency table displays the frequencies of multiple variables. It can display the frequency distributions as raw numbers, percentages, or both. Cross-tabulation can often reveal the basic picture of relationships between variables, although they should never be relied on exclusively to determine the relationships. The process of verifying assumptions and considering the variable type and level of measurement should be followed with rigor. An example of a cross-tabulation of the number of cylinders by car model from the automobile miles per gallon dataset (a.k.a. the Motor Trends car dataset) is shown in figure 2.14 (Quinlan, 1993).

	4	6	8
AMC Javelin	0	0	1
Cadillac Fleetwood	0	0	1
Camaro Z28	0	0	1
Chrysler Imperial	0	0	1
Datsun 710	1	0	0
Dodge Challenger	0	0	1
Duster 360	0	0	1
Ferrari Dino	0	1	0
Fiat 128	1	0	0
Fiat X1-9	1	0	0

Figure 2.14

Cross-tabulations that can be filtered, sorted, or display averages, ranks or frequencies at different levels of aggregation are commonly referred to as **pivot tables** in the business world. Most enterprise spreadsheet and database management tools have pivot table features built in to allow users to drag and drop variables for summarization.

Descriptive Statistics with R

Let's look at some descriptive statistics for the variables in the Titanic dataset that we began to explore in chapter 1. For simplicity, let's focus on the continuous numeric variable Fare, which is the ticket price that passengers paid. R has built in functions for the sample size, mean, median, variance, and standard deviation, but no built in function for the mode. So we will first build a function to calculate the mode of a variable:

```
Mode <- function(x) {
  ux <- unique(x)
  ux[which.max(tabulate(match(x, ux)))]
}
```

Our mode function takes an input x, de-duplicates the input through the unique() function, and stores the list of unique values as ux. The mode function returns the value from the unique list ux that has the greatest frequency in the input list x. The frequency is calculated by the tabulate() function.

Now let's look at the basic summary statistics about Fare.

```
length(train$Fare)    #Number of observations
mean(train$Fare)      #Mean
median(train$Fare)    #Median
Mode(train$Fare)      #Mode
var(train$Fare)       #Sample Variance
sd(train$Fare)        #Sample Standard Deviation
```

It is cumbersome to write all of these lines of code for the descriptive statistics of individual variables. Fortunately, the pastecs package has a function to display them all. The pastecs package must first be installed if it has not already been. All R packages are installed the same way:

```
install.packages("pastecs")
```

If the pastecs package is already installed, it can be loaded by calling the library() function. All R packages are loaded the same way:

```
library(pastecs)
```

Now the stat.desc() function can be called to display the summary statistics for Fare. By default, the function includes information like the minimum and maximum values, the number of null values, and other measures. They can be omitted by setting the basic option equal to false (F). The pastecs summary statistics also show the 95% confidence interval and standard error of the mean.

```
stat.desc(train$Fare, basic=F)
```

The standard error of the mean could also be computed by creating a custom function for it.

```
SEM <- function(x) {
  sd(x)/sqrt(length(x))
}
```

The default functions for variance and standard deviation, as well as the stat.desc() function from the pastecs package, provide the sample variance and standard deviation, which means that they are average over the total number of observations N-1. If we want to see the population variance and population standard deviation, we will need to write custom functions:

```
popVar <- function(x) {
  N <- length(x)
  var(x) * ((N-1)/N)
}
```

```
popSD <- function(x) {
  N <- length(x)
  sd(x) * sqrt((N-1)/N)
}

popVar(train$Fare)   #Population Variance
popSD(train$Fare)    #Population Standard Deviation
```

To view the quartiles (or any other quantiles) of the variable, use the quantile() function. By entering a question mark in front of the empty function, it is possible to see helpful information about how to use the function and what the possible input parameters are. To view the quartiles, the probs argument is set to increment at 0.25. The summary function also shows the quartiles.

```
?quantile
quantile(train$Fare, probs=seq(0, 1, 0.25))
summary(train$Fare)
```

Now let's build a frequency distribution for the passenger class variable. Note that the Pclass variable must be aggregated by each category by wrapping it in the table() function, otherwise the bar plot will show messy, non-aggregated data.

```
barplot(table(train$Pclass), main="Passengers by Class", xlab="Class", ylab="Number
of Passengers", space=0)
```

Before we examine the distribution of fare, it may be helpful to see how to draw the normal distribution for comparison. A simple way to draw it is to generate a sequence of numbers from -4 to 4 that increases by 0.01, and then build the normal distribution around that sequence, with mean 0 and standard deviation 1. For more information, look at what the dnorm function does: ?dnorm

```
x <- seq(-4 ,4, 0.01)
dens <- dnorm(x, 0, 1)
plot(x, dens, col="blue", xlab="Standard Deviation", ylab="Density", type="l", lwd=2,
cex=2, main="Standard Normal Distribution", cex.axis=.8)
```

To view the probability density of fare, we need to aggregate fare in a table and then pass it into the density function. We can store the density of fare as densf, and then plot it.

```
densf <- density(table(train$Fare))
plot(densf, xlab="Fare", ylab="Probability", type="l", lwd=2, cex=2,
main="Probability Density of Fare", cex.axis=.8)
```

The probability density that this code produces shows the likelihood of different fares on a scaled x-axis. Notice that the mean of fare is about 32. The probability density does not appear to have a mean of 32, but that is because the fare prices have been automatically scaled by R to show the very low likelihood of the extreme outliers that exist in the distribution of fare prices. The distribution of fare is very positively skewed because it has a long right tail. It definitely does not appear to be a normal distribution. That means that the median is the better measure of central tendency. The abnormal distribution also has implications for any kind of statistical inference used to describe the relationship between fare and other variables, but we will dive into that in a later chapter.

We can look at multivariate descriptive statistics through cross-tabulation. Basic cross-tabulations can be performed by using the table() function. For example, let's look at the number of passengers by class.

```
table(train$Pclass)
```

It is also possible to cross-tabulate many variables, for example, the number of passengers by class for each embarkation port.

```
table(train$Embarked, train$Pclass)
```

The gmodels package has a very clean cross-tabulation function with many more features than the basic table() function.

```
library(gmodels)
CrossTable(train$Pclass, train$Embarked, prop.chisq=F)
```

This table shows the count, row percentage, column percentage, and total percentage of passengers for each combination of class and embarkation port. From this table, it is possible to see that port Q, Queenstown, consisted of mostly third class passengers (93.5%), and that most passengers boarded at port S, Southampton (72.3% of all passengers).

Descriptive Statistics with Python

Let's look at some descriptive statistics for the variables in the Titanic dataset that we began to explore in chapter 1. For simplicity, let's focus on the continuous numeric variable Fare, which is the ticket price that passengers paid. Python's statistics library has functions for the mean, median, mode, variance, standard deviation, population variance, and population standard deviation. The scipy library has a function to calculate the standard error of the mean. We must first import the libraries, taking care to name them differently, and then we can access all of the descriptive statistics functions.

```
import statistics as stats
from scipy import stats as scistats

len(train.Fare)             #Number of observations
stats.mean(train.Fare)      #Mean
stats.median(train.Fare)    #Median
stats.mode(train.Fare)      #Mode
stats.variance(train.Fare)  #Sample Variance
stats.stdev(train.Fare)     #Sample Standard Deviation
stats.pvariance(train.Fare) #Population Variance
stats.pstdev(train.Fare)    #Population Standard Deviation
scistats.sem(train.Fare)    #Standard Error of the Mean
```

While there is no function to print all of these descriptive statistics in a nicely formatted way, it is trivial to define one:

```
def summaryStats(x):
    return print('\n',
                'Observations: ' + str(len(x)) + '\n',
                'Mean: ' + str(stats.mean(x)) + '\n',
                'Median: ' + str(stats.median(x)) + '\n',
                'Mode: ' + str(stats.mode(x)) + '\n',
                'Variance: ' + str(stats.variance(x)) + '\n',
```

```
                        'Std Dev: ' + str(stats.stdev(x)) + '\n',
                        'Pop Variance: ' + str(stats.pvariance(x)) + '\n',
                        'Pop Std Dev: ' + str(stats.pstdev(x)) + '\n',
                        'SEM: ' + str(scistats.sem(x)) + '\n',
                        )
```

```
summaryStats(train.Fare)
```

Every Pandas data frame has a quantiles method, which takes the desired percentiles as arguments. We can calculate the quartiles (quantiles 0.25, 0.5, and 0.75) for every column in the data frame like so:

```
train.quantile(q=[0.25, 0.5, 0.75])
```

The matplotlib pyplot library enables frequency distributions to be plotted. We will use it to plot the frequency distribution of passengers by class. Frequency distributions are plotted by the histogram function, as they are technically histograms, so the terms frequency distribution and histogram are used interchangeably here.

```
import matplotlib.pyplot as plt
```

Plot options can be specified before plotting. They are options so it is not necessary to use them, but the settings below show nice and clean histograms.

```
plt.figure(figsize=(12, 9))            #Specify plot dimensions
ax = plt.subplot(111)                  #Access a specific plot
ax.spines["top"].set_visible(False)    #Remove top frame
ax.spines["right"].set_visible(False)  #Remove right frame
ax.get_xaxis().tick_bottom()           #x-axis ticks on bottom
ax.get_yaxis().tick_left()             #y-axis ticks on left
```

The plot title and axes labels are added next:

```
plt.title("Histogram of Passenger Class")
plt.xlabel("Class")
plt.ylabel("Number of Passengers")
```

The number of tick marks on the x-axis can be specified optionally as well. I added this line after drawing the plot for the first time because the x-axis was showing tick marks at 1.5 and 2.5, which did not make sense for passenger class. By setting the tick mark range between the minimum and maximum values, and the interval between ticks to 1.0, even categories are shown on the x-axis. This option requires the numpy library to be imported.

```
import numpy as np
plt.xticks(np.arange(min(train.Pclass), max(train.Pclass)+1, 1.0))
```

Pyplot's histogram function has a bin option to specify how many categories to split the data into. Since we know that there are 3 passenger classes, we can specify a bin size of 3. However, a better way to specify bin size is to pretend we do not know how many passenger classes there are. That way we can write code that will calculate the number of categories to bin, if a situation ever arises when we truly do not know the it. Numpy's unique() function can help accomplish this:

```
b = len(np.unique(train.Pclass))
```

Finally, the histogram is plotted. Pyplot requires the show() function to be called in order for the plot to appear on the screen. Otherwise it will remain hidden.

```
plt.hist(train.Pclass, color="#3F5D70", bins=b)
plt.show()
```

Before we examine the distribution of fare, it may be helpful to see how to draw the normal distribution for comparison. A simple way to draw it is to generate a sequence of numbers from -4 to 4 that increases by 0.01, and then build the normal distribution around that sequence, with mean 0 and standard deviation 1. For more information, look at what the norm.pdf function does: help(scistats.norm.pdf). Note that the plot options reset themselves after plt.show() is called.

```
x = np.arange(-4, 4, 0.01)
plt.title("Normal Distribution")
plt.xlabel("Standard Deviation")
plt.ylabel("Probability")
plt.plot(x, scistats.norm.pdf(x, 0, 1))
plt.show()
```

Viewing the probability density is more challenging in Python than in R. To view the probability density of fare, we need to perform Gaussian kernel density estimation. Gaussian kernel density estimation is a non-parametric method to estimate probability density of a variable. We will define non-parametric methods in the next chapter, but for now we will just accept that it estimates the probability density. The x-axis range must also be defined as the continuous range from the minimum value of fare to the maximum value with a step interval of 1. We could then plot the probability density and allow Python to automatically select a covariance smoothing factor. Alternatively, we could specify our own smoothing factor. The smoothing factor removes the jagged peaks in the plot. To see what this means, uncomment the two lines that are commented below, and test covariance factors of 0.005 and 0.4 to see how the larger covariance factor smooths the plot more. Unlike R, Python does not automatically scale the x-axis.

```
dens = scistats.gaussian_kde(train.Fare)
dist_range = np.arange(min(train.Fare), max(train.Fare), 1)
#dens.covariance_factor = lambda : 0.4
#dens._compute_covariance()
plt.title("Probability Density of Fare")
plt.plot(dist_range, dens(dist_range))
plt.show()
```

The distribution of fare is very positively skewed because it has a long right tail. It definitely does not appear to be a normal distribution. That means that the median is the better measure of central tendency. The abnormal distribution also has implications for any kind of statistical inference used to describe the relationship between fare and other variables, but we will dive into that in a later chapter.

We can look at multivariate descriptive statistics through cross-tabulation. Basic cross-tabulations can be performed by using the crosstab() function from the pandas library. For example, let's look at the number of passengers by class.

```
pd.crosstab(len(train), train.Pclass,
            rownames=["Number of Passengers"])
```

It is also possible to cross-tabulate many variables, for example, the number of passengers by class for each embarkation port.

```
pd.crosstab(train.Pclass, train.Embarked,
            rownames=["Number of Passengers"])
```

Unfortunately there is no easy way to display row, column, and cell percentages in Python cross tabulation, so they must be calculated one at a time in separate tables, using the apply function. Use the parameter axis=1 to tell the apply function to do something with the rows, and axis=0 to tell apply to do something with the columns. For the cell percentages (the percentage of passengers in a class for a particular embarkation port), divide r by the total number of passengers (the total number of observations in the datasets). For row percentages (the percentage of passengers in each class from each embarkation port), divide r by the sum of the row and specify axis=1. For column percentages (the percentage of passengers from each embarkation port in each class), divide r by the sum of the column and specify axis=0.

```
ct = pd.crosstab(train.Pclass, train.Embarked,
            rownames=["Perc of Passengers by Class"])
print(ct.apply(lambda r: r/len(train), axis=1)) #Perc by cell
print(ct.apply(lambda r: r/sum(r), axis=1))     #Perc by row
print(ct.apply(lambda r: r/sum(r), axis=0))     #Perc by col
```

These tables the count, row percentage, column percentage, and cell percentage of passengers for each combination of class and embarkation port. From these tables, it is possible to see that port Q, Queenstown, consisted of mostly third class passengers (93.5%).

3. Statistical Modeling

In chapter 1, a statistical model was defined as an equation that describes the relationships between variables. A model or equation is no more than a set of assumptions about the variables whose relationship it describes. A logical question might be to wonder what a relationship between variables actually is. In statistics, a relationship refers to the probability distributions of random variables. That means that an equation like y = x is a model that describes the probability distribution of y in terms of the probability distribution of x. The fact that models are always in terms of probability distributions separates statistics from regular mathematics.

Looking at a variable in terms of its probability distribution requires statistical inference, which involves estimating the population parameters of the variable based on a smaller sample. We saw in chapter 2 that estimating the population parameters of a variable is made possible by the Central Limit Theorem. Therefore, to tie everything together, we can say that statistical inference is the process of examining the relationships between random variables in terms of a set of assumptions about those variables that are described by a statistical model. This chapter explains what assumptions statistical models describe, and how the models are interpreted.

Chapter 1 explained how statistical models are imprecise simplifications of more complex relationships, which means that models always contain some small measure of error that is usually represented by the variable epsilon (ε). The goal of statistical modeling is to find the model that best describes the relationship between dependent and independent variables. In order to determine how good a model is, there need to be ways to measure its fit. Measures of accuracy, like confidence intervals and p-values, show the likelihood of a model being the true estimate of the population parameters, while measures like **goodness of fit** show how accurately the model reflects the sample data. Both accuracy and goodness of fit are required to adequately evaluate a statistical model.

Let's revisit the mean, variance, and standard deviation. The mean of a variable is a rudimentary statistical model. To designate how accurately the mean describes the variable's distribution, we need some measure of the differences between the mean and the observations. This measure would indicate how well the model (in this case the mean) fits the variable. One such measure is the sum of the deviations from the mean, or in other words the sum of the differences from each observation to the mean. The problem with the sum of deviations is that it could equal 0, because the deviations could cancel one another out. Squaring each deviation rectifies this potential problem. By squaring the deviations, observations that are farther from the mean (the outliers) are given more influence in the evaluation of the mean's fit. Another consequence of squaring the deviations is that the results are always positive numbers, which means that the more observations there are in a dataset, the higher the sum of squared deviations will be. The sum of squared deviations will make the mean appear to be a poor model for larger samples. To account for sample size, the sum of squared deviations can be divided by the number of observations in the sample to get the average squared distance from the mean that each observed value lies. This is the exact definition of the variance. So the variance is a measure of how well the mean fits a variable's distribution of values.

The disadvantage to using variance to measure the mean's fit is that it is a squared measure of distance. So the obvious solution is to take the square root of the variance to get the average distance from the mean that each observation lies, which results in the standard deviation. A low standard deviation means that the mean is a good fit of the data.

One way of thinking about statistical models is to think of them in terms of outcome = model + error. So if the mean is 3, and a random observation has a value of 1, then the error is 1-3 = -2. The best model will have the smallest error, because the difference between the model's predicted outcome and the true income will be minimized.

Test Statistics

Test statistics are measurements used to determine how good a model is. There are many types of test statistics, such as t, F, and chi-squared for example. They are all calculated a bit differently, but they all represent the same thing: the ratio of the variation in the data explained by a model (the effect) to the variation that is not explained by the model (the error).

$$\text{test statistic} = \frac{variation\ explained\ by\ model}{variation\ NOT\ explained\ by\ model} = \frac{effect}{error}$$

Figure 3.1

The higher the test statistic, the better the model. In other words, the more variation in the data that is explained by the model, the more likely it is that the null hypothesis can be rejected. Although strong models increase the likelihood that the null hypothesis can be rejected, neither the null nor the alternative hypothesis can ever be completely rejected, or even completely accepted. That is because there is always a small chance that the model could be wrong. For example, consider the following syllogism:

    a. If a person plays basketball then they are most likely do not play in the NBA. This statement is true because there are many people who play basketball, but only a few people who play in the NBA.
    b. LeBron James plays basketball.
    c. Therefore, LeBron James most likely does not play in the NBA.

While the logic in the syllogism would be correct most of the time if basketball players are selected at random to test whether or not they play in the NBA, it is incorrect if the randomly chosen player is LeBron James.

Hypothesis Tests

Recall from chapter 1 that a hypothesis is an educated guess about what could be true. All hypotheses must be tested before they can be verified as true. Hypothesis tests test for relationships between variables by considering two competing hypotheses:

1. The **null hypothesis** states that there is no relationship between the variables being tested.
2. The **alternative hypothesis** states that there is some kind of relationship between the variables being tested.

It is only possible to confirm or reject the null in hypothesis testing, and even then it is never possible to be 100% certain of the results. It is also never possible to prove the alternative hypothesis is true, even when the null can be rejected (Glen, 2018).

Hypothesis tests can be 1 directional (one-tailed) or 2 directional (two-tailed). The direction of a test simply refers to which side of a distribution the test looks for an effect on.

Two-tailed tests look for a relationship between variables in either direction (variable x causes y to decrease, increase, or both). P-values in statistical tables are almost always for two-tailed tests. Two-tailed tests are also the default for most statistical software packages.

One-tailed tests look for a relationship between variables in one direction (variable x causes y to only increase or only decrease). One-tailed tests have a greater ability to spot relationships in a particular direction, because the significance level is focused on one tail, rather than being split between each end, as shown in figure 3.2.

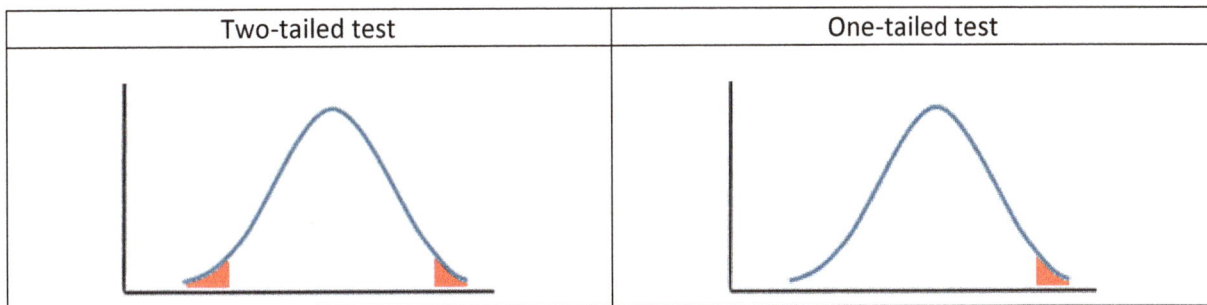

Two-tailed test	One-tailed test

Figure 3.2

When to Use Hypothesis Tests

If the goal is to look for a relationship between variables in one direction, it is best practice to perform a two-tailed test first, before proceeding to a one-tailed test (Field, Miles, & Field, 2012). This eliminates that possibility that an effect in the opposite direction could be overlooked if only a one-tailed test is performed. As an example, suppose we want to test the hypothesis that sleepiness is a side-effect of taking an antihistamine. A two-tailed test would look for the two effects: one effect is that an antihistamine makes people more tired than they were before taking it, and the other effect is that the antihistimine makes people more alert than they were before taking it. Suppose however that we wanted to test whether the antihistimine were less effective than some other drug at extinguishing an allergic reaction. In this case, one of the directions would be that the antihistimine is more effective than the other drug, but since we only care about whether or not it is less effective, it would be ok to only perform a one-tailed test. One-tailed tests are better than two-tailed tests at detecting small effects in one direction. One-tailed tests are also commonly used for testing for effects in skewed distributions with only 1 tail, such as the chi-square distribution.

As an example of performing a hypothesis test, let's suppose we want to know whether a coin that is being flipped is biased toward heads. Since we're looking for an effect in 1 direction, a one-tailed test seems appropriate. The null hypothesis would be that the coin is not biased. The alternative hypothesis would be that the coin is biased toward heads. What would happen if the coin were flipped several times and landed on tails every time? Since the one-tailed test was looking for a bias toward heads, it would conclude that there was no significance in the results, and the null could not be rejected. A two-tailed test however, would look for an effect in either direction (the coin is biased for either heads or tails). So if the coin landed on either all heads or all tails, both outcomes would be significant and the null (that the coin was not biased) could be rejected.

How to Choose a Hypothesis Test

The four most common test statistics used in hypothesis tests are the Z-statistic, the t-statistic, the F-statistic, and the Chi-square statistic. Note that it is also common to see these statistics referred to as scores (z-score, t-score, etc.) Each of these test statistics are used for different purposes. The chart below summarizes when each test should be used.

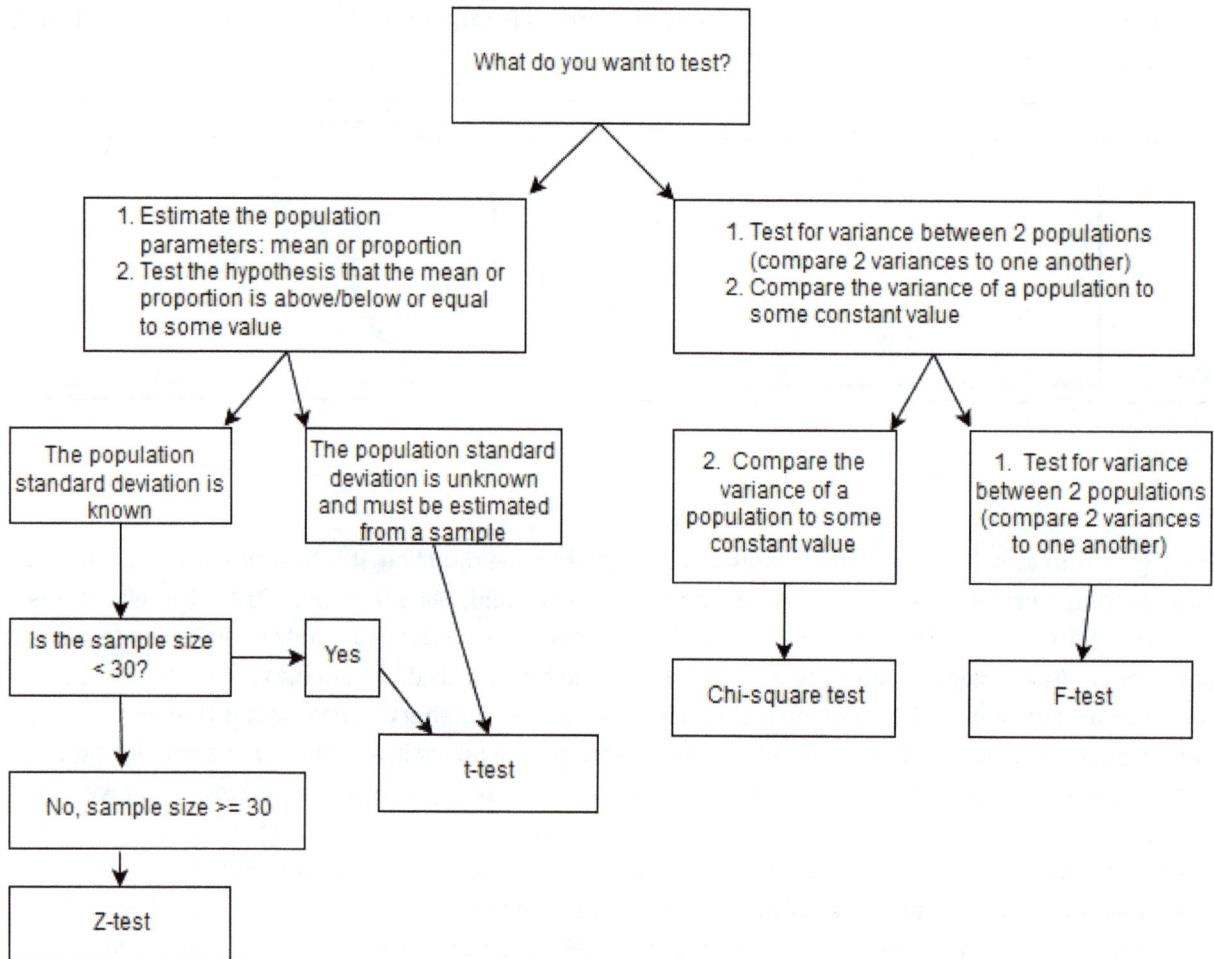

Figure 3.3

These are the equations to calculate the test statistic for each type of test:

Z-statistic	t-statistic
$z = \dfrac{\bar{x} - \mu_0}{\dfrac{\sigma}{\sqrt{n}}}$	$t = \dfrac{\bar{x} - \mu_0}{\dfrac{s}{\sqrt{n}}}$
F-statistic	Chi-square statistic
$F = \dfrac{s_1^2}{s_2^2}, \quad s_1^2 > s_2^2$	$\chi^2 = \sum \dfrac{(observed - expected)^2}{expected}$

Figure 3.4

In the equations in figure 3.4, x bar is the sample mean, mu subscript 0 is the population mean or the value to be tested, lowercase sigma is the population standard deviation, s is the sample standard deviation, n is the number of observations in the sample, and s squared is the sample variance (the subscripts indicate which sample). Note that the sample with the larger variance is the numerator for the F-test, because the F distribution is not symmetrical and putting the larger variance in the numerator makes it so that only the right tail critical value needs to be found.

Once the test statistic has been calculated, the null hypothesis can be accepted or rejected by comparing the test statistic's p-value to a desired level of significance, or by calculating a critical value (or confidence interval of critical values for two-tailed tests) to compare the test statistic to. The critical value is usually found in statistics classes by looking at tables of critical values for the test, for a desired level of significance. In summary, the null hypothesis can be rejected by using either of the two methods below:

- If the test statistic's p-value is below the desired level of significance, then the null hypothesis can be rejected.

OR

- For two-tailed tests, the null can be rejected if the test statistic is outside of the confidence interval.
- For one-tailed tests, whether the test statistic should be above or below the critical value depends on which tail it is being compared to. For the upper tail, the null is rejected when the test statistic is >= the critical value. For the lower tail, the null is rejected when the test statistic is <= the critical value.

Type I and Type II Error

Whenever predictions or classifications are made in data science and statistics, there is always error. **Type I error** occurs when the dependent variable is determined to have an effect on the independent variable, but it really does not. Type I errors are called false positives. **Type II error** occurs when the dependent variable is determined not to have an effect on the independent variable, but it really does. Type II errors are called false negatives. Error is typically summarized in a **confusion matrix**. Figure 3.3 shows the confusion matrix for a hypothesis test.

Confusion Matrix			
	True state of null hypothesis		
Determination about null hypothesis		Null is True and Valid	Null is False and Invalid
	Reject the Null	Type I Error (false positive)	Correct (true positive)
	Cannot Reject the Null	Correct (true negative)	Type II Error (false negative)

Figure 3.5

The probability of type I error is called **alpha level probability**. At a 95% confidence interval, there is a 5% probability of type I error, so the alpha level probability is 5%. The probability of type II error is called **beta level probability**. Although there is no steadfast rule for what constitutes an acceptable beta level probability, it is common practice to use 20%. So in a sample of 100 observations in which an effect is present, it should be expected that no more than 20 of them are missed.

There is a tradeoff between the alpha and beta level probabilities. Increasing one decreases the other, but it is unknown by how much (Field, Miles, & Field, 2012). The two probabilities of error cannot be compared mathematically because they rely on opposing assumptions: type 1 error requires no effect to be present, but type 2 error does require some effect. Due to the fact that the probabilities of error cannot be weighed beforehand, they have to be adjusted by trial and error.

Statistical power is the likelihood that a model will detect an effect from an experiment when there is truly an effect present. It can also be defined as the likelihood of rejecting a false null hypothesis. A model's statistical power is the level of confidence after subtracting beta level probability. For example, if the beta level probability is 20%, or 0.2, then the model's statistical power is $1 - 0.2$, or 80%.

While statistical power is the likelihood that a model will detect an effect, it is also useful to know the size of the effect. The **effect size** is defined as the strength and direction of the relationship between variables. Effect size is important in determining the practicality of an experiment. For example, if the effect is small, the relationship between variables could be statistically significant but not practically significant. That is, the effect could be so slight that the new drug whose effect is being tested would not be worth mass producing, for example. Statisticians have provided general guidelines for choosing sample sizes that are large enough to detect different effect sizes. For an alpha level probability of 0.05 and a beta level probability of 0.8, 783 observations are needed to detect small effect sizes ($r = 0.1$), 85 observations are needed to detect medium effect sizes ($r = 0.3$), and 28 observations are needed for large effect sizes ($r = 0.5$) (Field, Miles, & Field, 2012).

Parametric Tests and Standard Assumptions

Recall that statistical inference involves estimating the parameters of a population. When data is drawn from a distribution, the assumption is that its parameters (or defining properties), like mean and standard deviation, have known characteristics. For example, two parameters of the normal distribution are that it has a mean of 0 and a standard deviation of 1. The inferences and tests applied to statistical models that rely on the assumption that the data is drawn from a distribution are called **parametric tests**. Parametric tests have four general assumptions:

1. The sampling distribution is normally distributed. Notice that this does not mean that the observations in a single sample must be normally distributed, but that the sampling distribution is. Most large samples can meet this assumption by the Central Limit Theorem.
2. The variance in the dependent variable is homogenous throughout the dataset. This means that every sample comes from a population with the same variance, or dispersion around the mean. The condition in which the variance is homogenous is called **homoskedasticity**. The opposite condition, in which the variance changes throughout the dataset, is called **heteroskedasticity**.
3. The dependent variable is continuous. In other words, it is at least measured at the interval or ratio level.
4. Observations are independent of one another. This means that in repeated measure designs, individual participants or entities should be different from one another. This assumption also implies that the errors should not be correlated in regression analysis, as we will see later.

Nonparametric tests do not require these assumptions, because they do not assume that the data comes from an underlying distribution. A reasonable question might be why to use parametric tests at all if they require certain assumptions to be made? For large samples (more than 30 observations), parametric and nonparametric tests perform nearly the same. So if the sample size is large enough, either type of test could be applied. The differences between the tests are amplified when the sample size is small however. When the sample size is small and a parametric test is applied to data that does not meet the parametric assumptions, the Central Limit Theorem may not hold, so the p-values will be unreliable. Similarly, when the sample size is small and a nonparametric test is applied to data that meets the parametric assumptions, the p-values will be high and the test will lack statistical power. Therefore, it is important to choose the right type of test: parametric tests are not robust but nonparametric tests are weak.

Before we examine each of the standard assumptions in detail, let's take a look at a rudimentary statistical model: the mean. The mean is a good baseline to compare other models to. The variance and standard deviation are measures of how well the mean fits a dataset as a model. Recall from chapter 2 that the mean is not the best measure of central tendency when the distribution of a variable is skewed. When a distribution is skewed, the median is the better measure. In general, since measures of central tendency aim to represent typical values of a random variable, they are in a sense, rudimentary statistical models against which other models can be compared.

Testing the Normality Assumption

When testing for relationships between variables, it is best to first decide whether to use parametric tests or nonparametric tests. Assuming that the sample is either sufficiently large for the Central Limit Theorem to hold, or that the data comes from some underlying distribution, parametric tests can be carried out.

The first assumption is that the sampling distribution is normally distributed. Since it is not possible to view the sampling distribution, the Central Limit Theorem must be relied on to infer that it is normal. Using the Central Limit Theorem to infer normality requires comparing the sample to the normal distribution.

One way to check for normality is to view a histogram or probability density plot of the dependent variable. If the plot looks similar to a normal distribution, then normality can be assumed. Note that testing for normality is not even necessary if the dependent variable is categorical. A categorical variable would not be normally distributed and the third assumption (that the data is continuous) would also be violated.

While plotting the dependent variable as a histogram or probability density is quick and easy, it is not the most reliable method. A better method is to create a **Q-Q Plot**. The Q-Q Plot shows the quantiles of 1 variable plotted against the quantiles of another. In the case of a normality test, the Q-Q plot shows the cumulative values in the data against the cumulative probabilities of the normal distribution.

1. If the Q-Q Plot follows a straight diagonal line, then the distribution is normal. Note that the plot does not have to be perfectly straight as long as it is pretty close. If in doubt, do multiple kinds of normality tests.
2. If the plot has an S shape, then the distribution is skewed.
3. If the plot sags below or floats above a straight diagonal line, then the distribution has kurtosis.

The Q-Q Plot may be more reliable than a histogram or probability density plot, but it is still a plot that must be visually inspected, and thus it is open to interpretation. The **Jarque-Bera Test** and **Shapiro-Wilk Test** are two quantitative tests for normality that are easier to interpret. These tests are essentially calculating the same test statistic, albeit through different formulas. They both result in a number between 0 and 1 (the test statistic), which is compared to an alpha level probability for a desired level of confidence, such as 0.05 for the 5% level. If the test statistic is less than the alpha level probability, then the null can be rejected and the sample is likely not drawn from a normal distribution. Since these two tests can be biased for large samples, it is best practice to use them in conjunction with a Q-Q plot.

Other quantitative tests for normality include looking at the skewness and kurtosis of the distribution of the dependent variable. If the distribution is normal, then skewness and kurtosis should both equal 0. Skewness and kurtosis can be converted to z-scores and compared to a desired alpha level probability. If the z-score is less than the alpha level probability, then the null can be rejected and the sample is likely not drawn from a normal distribution. Like the Jarque-Bera and Shapiro-Wilk tests, skewness and kurtosis tests can be biased for large samples.

The Importance of Sample Size

In the previous section, the Jarque-Bera, Shapiro-Wilk, skewness, and kurtosis tests were all said to be biased for large samples. Let's look at what that means. In general, the larger a sample is, the more amplified smaller effects will appear to be. For example, there are more cases of human immunodeficiency virus (HIV) in Los Angeles than in Des Moines. However, this does not mean that Los Angeles is a less healthy place to live. Los Angeles has more people infected with HIV simply because it has a larger population than Des Moines. The ability of large samples to inflate the significance and size of small effects is important to keep in mind when conducting experiments and hypothesis tests.

One way to counter the tendency of effect exaggeration in large samples is to increase the alpha level probability to a higher percentage (such as to 99%). This technique works particularly well for samples with 30-200 observations. For samples with more than 200 observations, visual inspections like the probability density plot and Q-Q plot are generally more reliable than test statistics, because the test statistics are more likely to suggest non-normality even for small deviations from the normal distribution (Field, Miles, & Field, 2012).

The magnification of small effects in large samples is often overlooked in statistics. It is important to keep in mind because it has implications beyond checking for normality. Hypothesis tests that do not show significant effects with small samples could completely reverse and show significant effects with larger samples. It is up to the researcher to determine if the effect size and p-value are of enough practical significance to be of interest.

Testing the Homogeneity of Variance Assumption

The second standard assumption for parametric tests is homogeneity of variance: every sample is assumed to have the same spread around the mean. There are two common ways to test this assumption: Levene's Tests and the variance ratio. Of the two, Levene's test is biased for large samples, so it is useful to try both tests when the sample size is large (Field, Miles, & Field, 2012).

Levene's Test compares the spread around each mean or median between groups. In other words, it compares the distances between each group's observations and their centers. When applied to testing homogeneity of variance, groups are defined as different sub-samples of the data. The sub-samples can be defined by grouping by any categorical independent variable in the dataset, but is properly used to compare the variances between the control group and treatment group (or groups). Recall from chapter 1 that experiments involve a control group and a treatment group (or groups). With observational research, there may not always be a clearly defined control group. One example is fiscal quarters: any of the 4 quarters could be a control group. When this happens, it is fine to just pick any one group to act as the control. Levene's Test determines whether or not the variance is the same between groups or samples, which either confirms or denies the homogeneity of variance assumption about the data. Performing this test will result in an F-statistic and p-value for the F-statistic. If the p-value for the F-statistic is less than a desired alpha level probability, such as 0.05, then the variances are not homogenous and the assumption is violated.

The **variance ratio**, also called Hartley's F Max, is the ratio of the variance between the group or sample with the largest variance to the group or sample with the smallest. Performing this test will result in an F-statistic (called the variance ratio) and p-value for the F-statistic. The variance ratio or F-statistic must be compared to a table of F max critical values, such as this one: http://archive.bio.ed.ac.uk/jdeacon/statistics/table8.htm, for a given number of degrees of freedom, and the number of variances being compared. For this test, the degrees of freedom is the number of observations in the group minus one, and the number of variances being compared is simply the number of groups or samples. If the variance ratio is less than or equal to the critical value, then the variance is homogenous and the assumption is validated, but if it is greater, then the variance is not homogenous and the assumption of homogeneity of variance is violated.

Testing the Other Assumptions

The third assumption of parametric tests, that the data is continuous or at least measured at the interval or ratio level, does not require any testing. It is fairly self-evident by looking at the values of a variable whether or not it meets this assumption. The fourth assumption of parametric tests, that the observations are independent of one another, is usually covered by the experimental design. For example, if none of the observations are duplicate measurements of the same participant in a study, then the independence assumption holds. For observational research, or time series analysis in particular, the independence assumption is tested by determining whether or not there are correlations between observations occurring at different times. We will test this assumption with time series data in a later chapter.

Hypothesis Tests and Validating Parametric Assumptions with R

Let's perform some hypothesis tests with the variable passenger age from the Titanic dataset used in the first two chapters. For the purpose of these examples, age will serve as the dependent variable in a statistical model. An interesting question might be whether the mean age of Titanic passengers differed from the mean age of the UK population. According to the Daily Mail, the average age for people living in the UK was 34 in 1974 (McTague, 2015). Unfortunately, that is as far back as the data goes. So we will use 34 as the age to test. Therefore, the null hypothesis will be that the average age of Titanic passengers is not significantly different from 34. The alternative hypothesis will be that the average age of Titanic passengers is significantly different from 34. Since the age could differ from 34 in either direction (higher or lower), ad two-tailed test should be used. Furthermore, the test should be a t-test because the population standard deviation is unknown. Refer back to figure 3.3 to trace the logic used to determine that a t-test should be performed.

Before calculating the t-statistic, combine the training and test sets, taking care to remove the survived column from the training set, and removing the rows with missing age values. It is bad practice to combine the training and test sets for hypothesis testing, because it is possible that the average age is vastly different between the training and test sets. Maybe the data was split unevenly along certain features. But since the dataset documentation puts our fears about the dataset being divided unevenly to rest, we will proceed with combining the training and test sets.

```
t_all <- rbind(train[,-2], test)
t_all <- t_all[!is.na(t_all$Age),]
```

Refer back to figure 3.4 for the formula to calculate the t-statistic.

```
xbar <- mean(t_all$Age)        #Sample mean of passenger age
mu <- 34                       #The hypothesized mean age
s <- sd(t_all$Age)             #Sample standard deviation of passenger age
n <- nrow(t_all)               #The sample size
t <- (xbar-mu)/(s/sqrt(n))     #The t-test statistic
```

Recall that there are two ways to determine whether the t-statistic means the null should be accepted or rejected. One way is to calculate a critical value (or confidence interval of critical values for a two-tailed test) for a desired level of significance. A two-tailed test with 95% confidence has a 5% likelihood of type I error split between the two tails of the distribution. The qt function in R calculates the t-distribution for a vector of probabilities (the alpha value) and a number of degrees of freedom. Since the 5% error is split between the 2 tails of the t-distribution, the vector of probabilities is (1-alpha)/2:

```
alpha <- 0.05                            #Level of significance
tdist.half.alpha <- qt(1-alpha/2, df=n-1)
c(-tdist.half.alpha, tdist.half.alpha) #The confidence interval for alpha
t
```

Looking at the t-statistic, it is clear that it lies outside of the range of the confidence interval, meaning the null hypothesis, that the mean age of Titanic passengers is not significantly different from the mean age of the UK population in 1974 (34 years old), can be rejected. An alternative way to determine whether or not the null should be rejected is by calculating the p-value of the test statistic. In statistics classes, this is usually done by looking up the critical value in a table of critical values, but R can calculate the exact critical value:

```
pval <- 2*pt(t, df=n-1)
pval
```

Since the p-value of the t-statistic is < 0.05, then the null hypothesis can be rejected with 95% confidence, leaving a 5% chance of type I error (the significance level).

It should be evident that performing hypothesis tests in R is somewhat tedious. To eliminate the need to examine each type of test ad nauseam, we can define functions to perform them:

```
#Function to perform z-tests
ztest <- function(data, mu, sig, alpha=0.05, tails='two') {
  #data is a numeric vector
  #mu is the estimated population value to compare against
  #sig is the population standard deviation
  #alpha is the significance level, defaults to 0.05
```

```r
  #tails is the number of tails for the test, defaults to 'two', other options are
'upper' and 'lower'
  xbar <- mean(data)
  n <- length(data)
  z <- (xbar-mu)/(sig/sqrt(n))
  if (tails=='two') {
    zdist.half.alpha <- qnorm(1-alpha/2)
    pval <- 2*pnorm(z)
    print(paste('Confidence Interval:', -zdist.half.alpha, ',', zdist.half.alpha))
    print(paste('z-statistic:', z))
    print(paste('z-statistic p-value:', pval))
  } else if (tails=='lower') {
    zdist.alpha <- qnorm(1-alpha)
    pval <- pnorm(z)
    print(paste('Critical Value:', -zdist.alpha))
    print(paste('z-statistic:', z))
    print(paste('z-statistic p-value:', pval))
  } else if (tails=='upper') {
    zdist.alpha <- qnorm(1-alpha)
    pval <- pnorm(z, lower.tail=F)
    print(paste('Critical Value:', zdist.alpha))
    print(paste('z-statistic:', z))
    print(paste('z-statistic p-value:', pval))
  } else {
    return (message('Error: invalid tails argument'))
  }
}
ztest.prop <- function(data, criterion, p0, alpha=0.05, tails='two') {
  #data is a numeric vector
  #criterion is a numeric vector of the number of samples that meet some criterion (a
subset of data)
  #p0 is the estimated proportion to compare against
  #alpha is the significance level, defaults to 0.05
  #tails is the number of tails for the test, defaults to 'two', other options are
'upper' and 'lower'
  pbar <- length(criterion)/length(data)
  n <- length(data)
  z <- (pbar-p0)/sqrt(p0*(1-p0)/n)
  if (tails=='two') {
    zdist.half.alpha <- qnorm(1-alpha/2)
    pval <- 2*pnorm(z, lower.tail=F)
    print(paste('Confidence Interval:', -zdist.half.alpha, ',', zdist.half.alpha))
    print(paste('z-statistic:', z))
    print(paste('z-statistic p-value:', pval))
  } else if (tails=='lower') {
    zdist.alpha <- qnorm(1-alpha)
    pval <- pnorm(z)
    print(paste('Critical Value:', -zdist.alpha))
    print(paste('z-statistic:', z))
    print(paste('z-statistic p-value:', pval))
  } else if (tails=='upper') {
    zdist.alpha <- qnorm(1-alpha)
    pval <- 2*pnorm(z, lower.tail=F)
    print(paste('Critical Value:', zdist.alpha))
    print(paste('z-statistic:', z))
```

```
      print(paste('z-statistic p-value:', pval))
    } else {
      return (message('Error: invalid tails argument'))
    }
}
#Function to perform t-tests
ttest <- function(data, mu, alpha=0.05, tails='two') {
  #data is a numeric vector
  #mu is the estimated population value to compare against
  #alpha is the significance level, defaults to 0.05
  #tails is the number of tails for the test, defaults to 'two', other options are
'upper' and 'lower'
  data <- data[!is.null(data)]
  xbar <- mean(data)
  s <- sd(data)
  n <- length(data)
  t <- (xbar-mu)/(s/sqrt(n))
  if (tails=='two') {
    tdist.half.alpha <- qt(1-alpha/2, df=n-1)
    pval <- 2*pt(t, df=n-1)
    print(paste('Confidence Interval:', -tdist.half.alpha, ',', tdist.half.alpha))
    print(paste('t-statistic:', t))
    print(paste('t-statistic p-value:', pval))
  } else if (tails=='lower') {
    tdist.alpha <- qt(1-alpha, df=n-1)
    pval <- pt(t, df=n-1)
    print(paste('Critical Value:', -tdist.alpha))
    print(paste('t-statistic:', t))
    print(paste('t-statistic p-value:', pval))
  } else if (tails=='upper') {
    tdist.alpha <- qt(1-alpha, df=n-1)
    pval <- pt(t, df=n-1, lower.tail=F)
    print(paste('Critical Value:', tdist.alpha))
    print(paste('t-statistic:', t))
    print(paste('t-statistic p-value:', pval))
  } else {
    return (message('Error: invalid tails argument'))
  }
}
```

Now let us try a few examples. Can we reject the null that the mean age of Titanic passengers is at least 34 years old? This requires a one-tailed (lower) t-test, so the null can only be rejected if the t-statistic <= critical value and its p-value is < 0.05.

```
ttest(t_all$Age, 34, tails='lower')
```

Yes, the t-statistic is <= critical value and its p-value is < 0.05, so the null can be rejected with 95% confidence, leaving a 5% chance of type I error. Can we reject the null that the mean age of Titanic passengers is no more than 34 years old? This requires a one-tailed (upper) t-test, so the null can only be rejected if the t-statistic >= critical value and its p-value is < 0.05.

```
ttest(t_all$Age, 34, tails='upper')
```

No, we cannot reject the null because the t-statistic < critical value and its p-value is > 0.05. The examples so far have focused on comparing the mean to a value. Let's try an example of comparing proportions of passengers, using the sex variable. We would expect the ratio of males to females to be 50:50. Can we reject the null that passengers are evenly split between male and female? This requires a two-tailed z-test, because it is known that the ratio of males to females in the general population is 50:50, our sample size (number of Titanic passengers with known gender) > 30, and we want to test for a difference in either direction. We will take a shortcut and use the function we created to calculate everything automatically, but first we should make sure the data is clean.

```
t_all <- rbind(train[,-2], test)
t_all <- t_all[!is.na(t_all$Sex),]
```

```
ztest.prop(t_all$Sex, t_all[t_all$Sex=='male',which(colnames(t_all) %in% 'Sex')],
p0=0.5)
```

The ztest.prop function takes the variable with all observations as the first argument, and the subset of the variable that meets some criterion as the second argument. Judging by the results of the test, we can reject the null that the proportion of male to female passengers is 50:50, however the test gives no indication of which direction the ratio leans. A simple proportion table will show us that the ratio heavily favors males.

```
prop.table(table(t_all$Sex))
```

After viewing the proportion table, it would be reasonable to wonder why a statistical test was needed at all to determine that the ratio of males to females was not 50:50. The reason a statistical test was needed was that the hypothesis was comparing the ratio of males to females in the sample (Titanic passengers) to the ratio of males to females in the general population. While the ratios could be different, the sample could have been drawn from the population. The test showed that the ratios were significantly different to the extent that it was possible to reject the null, meaning there was evidence to suggest that the Titanic passengers were a fundamentally different group of people than a random sample of the general population.

Let's move on to testing the standard assumptions of parametric tests. All of the hypothesis tests so far have relied on these assumptions. The first assumption is that the dependent variable is normally distributed. If we consider age to be the dependent variable, we can test for normality visually by drawing a distribution of values:

```
t_all <- rbind(train[,-2], test)
t_all <- t_all[!is.na(t_all$Age),]
plot(table(t_all$Age))           #Histogram
plot(density(table(t_all$Age))) #Smoothed density plot
```

A Q-Q Plot can also be drawn:

```
qqnorm(t_all$Age)
qqline(t_all$Age, col=2)  #Optional trend line in red
```

Finally, the Shapiro-Wilk and Jarque-Bera tests can be applied:

```
shapiro.test(t_all$Age)
library(tseries)  #Required to run Jarque-Bera
```

```
jarque.bera.test(t_all$Age)
```

In every case, it is clear to see that Titanic passenger age is not normally distributed. The W-statistic from the Shapiro-Wilk test is close to 1, the p-values for both Shapiro-Wilk and Jarque-Bera tests were < 0.05, the Q-Q Plot had tails at each end that curved away from a straight line, and the histogram was off kilter, which all point to non-normality.

The next assumption is homogeneity of variance. If we again consider age to be the dependent variable, we can use Levene's test for homogeneity of variance. Since Levene's test checks for homogeneity of variance between groups, we will need to split age into subgroups by some other variable. Sex and passenger class are two categorical variables in the dataset that would be convenient to group by. Let us use sex.

```
leveneTest(t_all$Age, t_all$Sex)
```

Since the p-value of the resulting F-statistic is > 0.05, the assumption of homogeneity of variance holds. What happens when passenger class is used as the grouping variable?

```
leveneTest(t_all$Age, as.factor(t_all$Pclass))
```

Now the assumption of homogeneity of variance is violated! So the variance of passenger age is homogeneous between the sexes, but varies between passenger classes. This suggests that the distributions of ages among passenger classes are not the same. If we wish to include both sex and passenger class in a model, then we should also consider the interaction between these two variables. Levene's test handles the interaction terms automatically. We will define what the interaction terms are in a later chapter.

```
leveneTest(t_all$Age ~ t_all$Sex*as.factor(t_all$Pclass))
```

Notice that the interaction violates the homogeneity of variance assumption, so the variance of passenger age is inconsistent across sex and passenger class. Another test to verify the homogeneity of variance assumption is the variance ratio, a.k.a. Hartley's F Max. Although the SuppDists package contains functions to calculate Harley's F Max, there is no straightforward way of doing it, so let's write our own function:

```
library(SuppDists)
hartleys_f_max <- function(num_variable, group_variable) {
  #num_variable is a numeric variable to compare variances
  #group_variable is the variable with the groups to compare the variance
  group_variances <- tapply(num_variable, as.factor(group_variable), var)
  f <- max(group_variances)/min(group_variances)
  pval <- pmaxFratio(f, df=length(num_variable)-1,
                     k=length(levels(as.factor(group_variable))),
                     lower.tail=F)
  print(paste('F-statistic:', f))
  print(paste('p-value:', pval))
}
```

Applying this test to sex and passenger class results in the same conclusions about homogeneity of variance as Levene's test: age variance between sexes is homogeneous but age variance between passenger classes is not.

```
hartleys_f_max(t_all$Age, t_all$Sex)
hartleys_f_max(t_all$Age, t_all$Pclass)
```

There are two more standard assumptions that must be verified, but they only require looking at the data itself.

```
head(t_all)
```

The third standard assumption is that the data is continuous, meaning it is measured at the interval or ratio level. Looking at the age variable, it is obvious that age is measured at the ratio level, which satisfies the assumption. As long as we are building a model to predict passenger age, we will not need to worry about the third assumption being violated. If we were to build a model to predict whether or not a passenger survived the voyage of the Titanic, we would be in trouble however. The survived variable is categorical, which means that the third assumption is violated. Therefore, we could not use parametric tests to build the model to predict survival. We will discuss non-parametric tests in a later chapter.

The fourth standard assumption is that the observations are independent of one another. We can be confident that there is no serial dependence among observations in the Titanic dataset for any of the variables.

Hypothesis Tests and Validating Parametric Assumptions with Python

Let's perform some hypothesis tests with the variable passenger age from the Titanic dataset used in the first two chapters. For the purpose of these examples, age will serve as the dependent variable in a statistical model. An interesting question might be whether the mean age of Titanic passengers differed from the mean age of the UK population. According to the Daily Mail, the average age for people living in the UK was 34 in 1974 (McTague, 2015). Unfortunately, that is as far back as the data goes. So we will use 34 as the age to test. Therefore, the null hypothesis will be that the average age of Titanic passengers is not significantly different from 34. The alternative hypothesis will be that the average age of Titanic passengers is significantly different from 34. Since the age could differ from 34 in either direction (higher or lower), ad two-tailed test should be used. Furthermore, the test should be a t-test because the population standard deviation is unknown. Refer back to figure 3.3 to trace the logic used to determine that a t-test should be performed.

Before calculating the t-statistic, import the libraries needed for this chapter and the data.

```
import pandas as pd
import statistics as stats
import math
from scipy import stats as scistats
import matplotlib.pyplot as plt
import seaborn as sns
from statsmodels.api import qqplot
from sklearn.preprocessing import LabelEncoder

train = pd.read_csv("train.csv", sep=",")
test = pd.read_csv("test.csv", sep=",")
```

Combine the training and test sets, taking care to remove the survived column from the training set, and removing the rows with missing age values. It is bad practice to combine the training and test sets for hypothesis testing, because it is possible that the average age is vastly different between the training and test sets. Maybe the data was split unevenly along certain features. But since the dataset documentation puts our fears about the dataset being divided unevenly to rest, we will proceed with combining the training and test sets.

```
t_all = pd.concat([train.ix[:, train.columns != 'Survived'], test])
t_all = t_all.dropna(subset=['Age'])
```

Refer back to figure 3.4 for the formula to calculate the t-statistic.

```
xbar = stats.mean(t_all['Age']) #Sample mean of passenger age
mu = 34                         #The hypothesized mean age
s = stats.stdev(t_all['Age'])   #Sample standard deviation of passenger age
n = t_all.shape[0]              #The sample size
t = (xbar-mu)/(s/math.sqrt(n))  #The t-test statistic
```

Recall that there are two ways to determine whether the t-statistic means the null should be accepted or rejected. One way is to calculate a critical value (or confidence interval of critical values for a two-tailed test) for a desired level of significance. A two-tailed test with 95% confidence has a 5% likelihood of type I error split between the two tails of the distribution. The ppf method from the scipy.stats t-distribution object calculates the t-distribution for a vector of probabilities (the alpha value) and a number of degrees of freedom. Since the 5% error is split between the 2 tails of the t-distribution, the vector of probabilities is (1-alpha)/2:

```
alpha = 0.05                              #Level of significance
tdist_half_alpha = scistats.t.ppf(1-alpha/2, df=n-1)
print([-tdist_half_alpha, tdist_half_alpha]) #The confidence interval for alpha
print(t)
```

Looking at the t-statistic, it is clear that it lies outside of the range of the confidence interval, meaning the null hypothesis, that the mean age of Titanic passengers is not significantly different from the mean age of the UK population in 1974 (34 years old), can be rejected. An alternative way to determine whether or not the null should be rejected is by calculating the p-value of the test statistic. In statistics classes, this is usually done by looking up the critical value in a table of critical values, but we can calculate the exact critical value:

```
pval = 2*scistats.t.cdf(t, df=n-1)
print(pval)
```

Since the p-value of the t-statistic is < 0.05, then the null hypothesis can be rejected with 95% confidence, leaving a 5% chance of type I error (the significance level).

It should be evident that performing hypothesis tests in Python is somewhat tedious. To eliminate the need to examine each type of test ad nauseam, we can define functions to perform them:

```
#Function to perform z-tests
def ztest(data, mu, sig, alpha=0.05, tails='two'):
    #data is a numeric vector
    #mu is the estimated population value to compare against
    #sig is the population standard deviation
    #alpha is the significance level, defaults to 0.05
```

```python
    #tails is the number of tails for the test, defaults to 'two', other options are
'upper' and 'lower'
    xbar = stats.mean(data)
    n = len(data)
    z = (xbar-mu)/(sig/math.sqrt(n))
    if (tails=='two'):
        zdist_half_alpha = scistats.norm.ppf(1-alpha/2)
        pval = 2*scistats.norm.cdf(z)
        print('Confidence Interval:', -zdist_half_alpha, ',', zdist_half_alpha)
        print('z-statistic:', z)
        print('z-statistic p-value:', pval)
    elif (tails=='lower'):
        zdist_alpha = scistats.norm.ppf(1-alpha)
        pval = scistats.norm.cdf(z)
        print('Critical Value:', -zdist_alpha)
        print('z-statistic:', z)
        print('z-statistic p-value:', pval)
    elif (tails=='upper'):
        zdist_alpha = scistats.norm.ppf(1-alpha)
        pval = scistats.norm.sf(z)
        print('Critical Value:', zdist_alpha)
        print('z-statistic:', z)
        print('z-statistic p-value:', pval)
    else:
        return print('Error: invalid tails argument')
def ztest_prop(data, criterion, p0, alpha=0.05, tails='two'):
    #data is a numeric vector
    #criterion is a numeric vector of the number of samples that meet some criterion
(a subset of data)
    #p0 is the estimated proportion to compare against
    #alpha is the significance level, defaults to 0.05
    #tails is the number of tails for the test, defaults to 'two', other options are
'upper' and 'lower'
    pbar = len(criterion)/len(data)
    n = len(data)
    z = (pbar-p0)/math.sqrt(p0*(1-p0)/n)
    if (tails=='two'):
        zdist_half_alpha = scistats.norm.ppf(1-alpha/2)
        pval = 2*scistats.norm.sf(z)
        print('Confidence Interval:', -zdist_half_alpha, ',', zdist_half_alpha)
        print('z-statistic:', z)
        print('z-statistic p-value:', pval)
    elif (tails=='lower'):
        zdist_alpha = scistats.norm.ppf(1-alpha)
        pval = scistats.norm.cdf(z)
        print('Critical Value:', -zdist_alpha)
        print('z-statistic:', z)
        print('z-statistic p-value:', pval)
    elif (tails=='upper'):
        zdist_alpha = scistats.norm.ppf(1-alpha)
        pval = 2*scistats.norm.sf(z)
        print('Critical Value:', zdist_alpha)
        print('z-statistic:', z)
        print('z-statistic p-value:', pval)
    else:
```

```
            return print('Error: invalid tails argument')
#Function to perform t-tests
def ttest(data, mu, alpha=0.05, tails='two'):
    #data is a numeric vector
    #mu is the estimated population value to compare against
    #alpha is the significance level, defaults to 0.05
    #tails is the number of tails for the test, defaults to 'two', other options are
'upper' and 'lower'
    data = data.dropna()
    xbar = stats.mean(data)
    s = stats.stdev(data)
    n = len(data)
    t = (xbar-mu)/(s/math.sqrt(n))
    if (tails=='two'):
        tdist_half_alpha = scistats.t.ppf(1-alpha/2, df=n-1)
        pval = 2*scistats.t.cdf(t, df=n-1)
        print('Confidence Interval:', -tdist_half_alpha, ',', tdist_half_alpha)
        print('t-statistic:', t)
        print('t-statistic p-value:', pval)
    elif (tails=='lower'):
        tdist_alpha = scistats.t.ppf(1-alpha, df=n-1)
        pval = scistats.t.cdf(t, df=n-1)
        print('Critical Value:', -tdist_alpha)
        print('t-statistic:', t)
        print('t-statistic p-value:', pval)
    elif (tails=='upper'):
        tdist_alpha = scistats.t.ppf(1-alpha, df=n-1)
        pval = scistats.t.sf(t, df=n-1)
        print('Critical Value:', tdist_alpha)
        print('t-statistic:', t)
        print('t-statistic p-value:', pval)
    else:
        return print('Error: invalid tails argument')
```

Now let us try a few examples. Can we reject the null that the mean age of Titanic passengers is at least 34 years old? This requires a one-tailed (lower) t-test, so the null can only be rejected if the t-statistic <= critical value and its p-value is < 0.05.

```
ttest(t_all['Age'], 34, tails='lower')
```

Yes, the t-statistic is <= critical value and its p-value is < 0.05, so the null can be rejected with 95% confidence, leaving a 5% chance of type I error. Can we reject the null that the mean age of Titanic passengers is no more than 34 years old? This requires a one-tailed (upper) t-test, so the null can only be rejected if the t-statistic >= critical value and its p-value is < 0.05.

```
ttest(t_all['Age'], 34, tails='upper')
```

No, we cannot reject the null because the t-statistic < critical value and its p-value is > 0.05. The examples so far have focused on comparing the mean to a value. Let's try an example of comparing proportions of passengers, using the sex variable. We would expect the ratio of males to females to be 50:50. Can we reject the null that passengers are evenly split between male and female? This requires a two-tailed z-test, because it is known that the ratio of males to females in the general population is 50:50, our sample size (number of Titanic passengers with known gender) > 30, and we want to test for a difference in either direction. We will take a shortcut and use the function we created to calculate everything automatically, but first we should make sure the data is clean.

```
t_all = pd.concat([train.ix[:, train.columns != 'Survived'], test])
t_all = t_all.dropna(subset=['Sex'])
```

```
ztest_prop(t_all['Sex'], t_all[t_all['Sex']=='male'].filter(['Sex'], axis=1), p0=0.5)
```

The ztest_prop function takes the variable with all observations as the first argument, and the subset of the variable that meets some criterion as the second argument. Judging by the results of the test, we can reject the null that the proportion of male to female passengers is 50:50, however the test gives no indication of which direction the ratio leans. A simple proportion table will show us that the ratio heavily favors males.

```
print(pd.crosstab(t_all["Sex"], columns="Prop")/(pd.crosstab(t_all["Sex"],
columns="Prop").sum()))
```

After viewing the proportion table, it would be reasonable to wonder why a statistical test was needed at all to determine that the ratio of males to females was not 50:50. The reason a statistical test was needed was that the hypothesis was comparing the ratio of males to females in the sample (Titanic passengers) to the ratio of males to females in the general population. While the ratios could be different, the sample could have been drawn from the population. The test showed that the ratios were significantly different to the extent that it was possible to reject the null, meaning there was evidence to suggest that the Titanic passengers were a fundamentally different group of people than a random sample of the general population.

Let's move on to testing the standard assumptions of parametric tests. All of the hypothesis tests so far have relied on these assumptions. The first assumption is that the dependent variable is normally distributed. If we consider age to be the dependent variable, we can test for normality visually by drawing a distribution of values:

```
t_all = pd.concat([train.ix[:, train.columns != 'Survived'], test])
t_all = t_all.dropna(subset=['Age'])
plt.hist(t_all['Age'], bins='auto')   #Histogram
plt.show()
sns.kdeplot(np.array(t_all['Age']))   #Smoothed density plot
plt.show()
```

A Q-Q Plot can also be drawn:

```
qqplot(t_all['Age'])
plt.show()
```

Finally, the Shapiro-Wilk and Jarque-Bera tests can be applied:

```
print(scistats.shapiro(t_all['Age']))
print(scistats.jarque_bera(t_all['Age']))
```

In every case, it is clear to see that Titanic passenger age is not normally distributed. The W-statistic from the Shapiro-Wilk test is close to 1, the p-values for both Shapiro-Wilk and Jarque-Bera tests were < 0.05, the Q-Q Plot had tails at each end that curved away from a straight line, and the histogram was off kilter, which all point to non-normality.

The next assumption is homogeneity of variance. If we again consider age to be the dependent variable, we can use Levene's test for homogeneity of variance. Since Levene's test checks for homogeneity of variance between groups, we will need to split age into subgroups by some other variable. Sex and passenger class are two categorical variables in the dataset that would be convenient to group by. Let us use sex. Note that the sex variable has a type of "object" because it is a character string. So before we can use it in Levene's test, we need to label encode it. We will formally define label encoding in a later chapter.

```
le = LabelEncoder()
t_all['Sex_Encoded'] = le.fit_transform(np.array(t_all['Sex']))
```

The Levene function from the scipy package requires each group to be passed as a separate argument, which means we must group the variable that we wish to test before it can be tested. The function below takes a numeric variable to be grouped, and an arbitrary number of grouping variables. It splits the numeric variable into groups and returns the result of running Levene's test on every combination of the groups.

```
def levenes_test(num_variable, *group_variables, center='median'):
    temp = list(num_variable.groupby(group_variables))
    temp = [temp[i][1] for i,v in enumerate(temp)]
    return scistats.levene(*temp, center=center)
```

Now we are ready to test the homogeneity of variance assumption for passenger age across males and females.

```
print(levenes_test(t_all['Age'], t_all['Sex_Encoded']))
```

Since the p-value of the resulting F-statistic is > 0.05, the assumption of homogeneity of variance holds. What happens when passenger class is used as the grouping variable?

```
print(levenes_test(t_all['Age'], t_all['Pclass']))
```

Now the assumption of homogeneity of variance is violated! So the variance of passenger age is homogeneous between the sexes, but varies between passenger classes. This suggests that the distributions of ages among passenger classes are not the same. Now we will try grouping age by both sex and passenger class.

```
print(levenes_test(t_all['Age'], t_all['Sex_Encoded'], t_all['Pclass']))
```

The homogeneity of variance assumption is violated, so the variance of passenger age is inconsistent across the combination of sex and passenger class. Another test to verify the homogeneity of variance assumption is the variance ratio, a.k.a. Hartley's F Max. Unfortunately, there is no way to carry out a hypothesis test using Hartley's F Max in Python. We can define a function to calculate the F Max statistic, but we will need to manually compare the result to an F Max statistical table.

```
def hartleys_f_max(num_variable, group_variable):
    #num_variable is a numeric variable to compare variances
    #group_variable is the variable with the groups to compare the variance
    group_variances = num_variable.groupby(group_variable).apply(stats.variance)
    f = max(group_variances)/min(group_variances)
    print('F-max statistic:', f)
```

```
    print('Degrees of Freedom:', len(num_variable)-1, ', k:',
len(np.unique(group_variable)))
    print('Compare the F-max statistic for n-1 DoF and k to the F-max table here:
\n',
          'http://archive.bio.ed.ac.uk/jdeacon/statistics/table8.htm \n',
          'and if the F-max statistic is smaller than the critical value in the
table, then the variance is homogenous.')
```

Applying this test to sex and passenger class results in slightly different conclusions about homogeneity of variance than Levene's test: neither age variance between sexes nor age variance between passenger classes is homogenous. Levene's test suggested that age variance between sexes was homogenous. It is up to the researcher which results to trust, although the p-value for Hartley's F Max was only slightly above the critical value for the age variance between sexes test, so it might be ok to let it slide since Levene's test pointed towards homogeneity.

```
hartleys_f_max(t_all['Age'], t_all['Sex'])
hartleys_f_max(t_all['Age'], t_all['Pclass'])
```

There are two more standard assumptions that must be verified, but they only require looking at the data itself.

```
t_all.head()
```

The third standard assumption is that the data is continuous, meaning it is measured at the interval or ratio level. Looking at the age variable, it is obvious that age is measured at the ratio level, which satisfies the assumption. As long as we are building a model to predict passenger age, we will not need to worry about the third assumption being violated. If we were to build a model to predict whether or not a passenger survived the voyage of the Titanic, we would be in trouble however. The survived variable is categorical, which means that the third assumption is violated. Therefore, we could not use parametric tests to build the model to predict survival. We will discuss non-parametric tests in a later chapter.

The fourth standard assumption is that the observations are independent of one another. We can be confident that there is no serial dependence among observations in the Titanic dataset for any of the variables.

4. Data Cleaning and Preparation

Before advancing any further with data analysis, it is worth taking time to think about data cleaning and preparation, which can also be called data pre-processing. Data is rarely perfectly clean except for in classroom settings. Real world data is messy. It has missing values and outliers, and it can be in forms that make it difficult for analysis. Although the majority of this book deals with data analysis, data cleaning and preparation usually takes between 80% of the time in any real world analysis (Press, 2016). In this chapter, we will look at ways to clean data and prepare it for analysis. We will not look at ways to rearrange data; a process called data wrangling or data munging, as that is too broad a topic to fit into one chapter. Arguably, data cleaning and preparation is too large for one chapter as well, but we will look at the most common techniques to handle a wide range of situations, keeping in mind that they are by no means comprehensive.

Dealing with Null or Missing Data

Missing data is the most common problem with data, and unfortunately it is also one of the hardest to work around. Sometimes it is possible to simply fill in missing data with the correct value. For example, if daily average temperatures were collected from a meteorology website, and a few of the days were missing, one could easily find the missing values from another source. If finding the missing values is not possible, then the first action that should be taken is to find out how many values are missing. If the percentage of missing values is very high, it may be worth considering the removal of the entire feature. There are 3 general ways to handle missing values. The best method depends on the problem that is trying to be solved with the data, the reason there are missing values, and the researcher's own intuition.

1. Remove an entire feature or column – If there are many missing values for a single feature, while other features do not have as many missing values, then the feature could be deleted entirely. This assumes of course that the feature is not of critical importance to the problem, like if the feature were the only independent variable.
2. Listwise deletion (remove certain rows of the dataset) – If there are specific observations that are missing values, especially in multiple features, removing the individual rows might be the best option. When the dataset is small, this can be hard to do however, as the information loss could be enough to prevent statistical inference. In the worst case scenario, listwise deletion can bias statistical analysis, such as when the data is missing systematically. One example of when listwise deletion could bias statistical analysis is survey data collected about a sensitive topic, and people of a certain demographic are reluctant to answer the questions.
3. Impute the missing values – Missing values can be imputed by using other features to build a model to predict what the missing values might be, or by using interpolation to estimate the missing values, or by filling in the missing values with a measure of central tendency.

Before deciding which method to use to handle missing values, it is important to first try to determine if there is a reason the values are missing. For example, if the data was collected from an experiment, and some participants did not complete the experiment, then the missing values are not random. When missing values are not missing at random, any analysis performed on the data will be biased. It is usually good to use imputation when values are not missing at random, as some methods of imputation are robust to data that is not missing at random.

Imputation is a useful technique to fill in missing values with approximations. There are several methods of imputation. The easiest is just to fill in the missing values with the mean. This will ensure that the mean of the univariate distribution of the feature does not change. The drawback to this method is that it weakens any possible correlations between the feature whose values are mean substituted and others. The same is true for replacing missing values of a categorical feature with the mode.

Another way to impute missing values is to predict them by building a regression or other model. We will look at regression methods in the coming chapters, but other models, such as decision trees and machine learning techniques are beyond the scope of this book. The way this technique could work would be to treat the feature with the missing values as the dependent variable and use the other features as independent variables. The fitted model could then be used to predict the missing values. The drawback to imputation using regression is that the predicted values that are used to replace the original missing values fit perfectly on the regression line. The drawback to imputation using a machine learning is that the replacement values will increase the risk of overfitting subsequent models, because models built using the replacement values will have a high variance.

As long as the missing data is missing at random, then **single random imputation** can be performed. To perform random imputation, missing values are sampled from the range of existing values. The samples can come from a pre-defined distribution, typically a normal distribution.

Similar to single random imputation, another technique is **multiple imputation**. Multiple imputation is useful in cases when multiple variables are missing values, although it can be used even if a single variable is missing values. The logic behind multiple imputation is that repeated random samples provide a good estimate of the missing data, assuming the missing values come from a distribution. Multiple imputation is carried out in 3 steps:

1. Either use a model to impute the missing values, or draw a sample of values from the range of the non-missing values. Use either the predicted values or samples to fill in the missing values. Repeat this process m times so that there are m vectors. Note that m can be chosen arbitrarily by the researcher, but a good rule of thumb is that it should be at least 20, and it should increase with the number of observations in the dataset (Buuren, 2010).
2. Carry out the statistical analysis as planned, but do it m times for each new dataset. The result will be m models. For example, if the plan was to regress y on x, and x had missing values that were imputed in step 1, then m regression models would result.
3. The parameter estimates of the m models are pooled by calculating their mean, variance, and confidence interval.

There is currently only one type of imputation for categorical data: likelihood encoding, a.k.a. target encoding. **Likelihood encoding**, or target encoding, is encoding using a measure of central tendency. Categorical features can be encoded as the measure of central tendency of the dependent variable for each category. The measurer of central tendency is the mean for regression problems and the mode for classification problems. For example, suppose the dependent variable y is continuous and feature x has categories A, B, and C. The categories of x can be encoded as the mean of y for category A, the mean of y for category B, and the mean of y for category C. While this would work, it may bias any model built off of the encoded data. So a better approach is to use cross validation (10 or 20 fold would be ideal) to create an averaged mean for each category (Dorogush, Ershov, & Gulin, 2017). This approach is similar to using cross validation to tune any model, and since the mean is a rudimentary model, using cross validation to tune it is no different. The same process can be used if the dependent variable y is categorical, but instead of the mean, use the mode or the most frequent class of y for each category of x, over k folds. The class with the highest frequency for category c over k folds is the one that the category c of x is encoded as. Cross validation will be explained in more detail in a later chapter. Aside from using likelihood encoding, missing categorical data could be predicted with another model.

Suppose we want to derive new features based on calculations off of features that are missing values. We could impute the missing values first, and then calculate the new features, or we could calculate the new features and leave missing values where we cannot calculate anything. Either method could work, but some research has suggested that leaving the missing values as they are, and imputing them for each feature independently results in more accurate models (White, Royston, & Wood, 2011). This approach treats the any derived feature as "just another variable". As an example, let's suppose we have average daily temperatures for a particular city, and we want to calculate the average weekly temperature. If some days are missing average temperatures, then we could calculate the average weekly temperatures with the data available, and impute the missing values for both average daily and average weekly temperatures.

Imputation is useful for randomly missing data, but what if the data follows a general trend and the missing value is somewhere in the middle? For example, consider the following data:

Age in Years	Height in Inches
14	62
15	66
16	
17	70

Figure 4.1

If we performed single random imputation, we would be sampling values from 62 to 70, but if we end up with an average of 65, we know it cannot be right because that would mean the individual shrank an inch between the ages of 15 and 16. In this case, interpolation would be a better way to estimate the missing value. **Interpolation** can be done by averaging the values around the missing value, fitting a linear regression to the data, or by fitting a spline to the data. The spline method is usually the most accurate. A spline is a piecewise function defined by polynomials. **Cubic splines**, or splines defined by cubic polynomials, are perhaps the most common. The advantage to using cubic splines for interpolation instead of linear regression is that the result will have a better fit, and thus a lower error.

A reasonable question might be to wonder why use a cubic spline, instead of a 4th degree polynomial, 5th degree, or higher still. The problem with using polynomial functions instead of splines is that they can vary greatly in between points. Figure 4.2 shows one possible cubic polynomial fit versus a cubic spline. Notice how the cubic spline fits the data more smoothly. The spline fits better because it is a piecewise cubic curve with a continuous second derivative. The second derivative defines the curvature at each point. If the curvature is 0 at each end of the function, or in other words if the second derivative is 0 at each end, then the cubic spline is said to be relaxed.

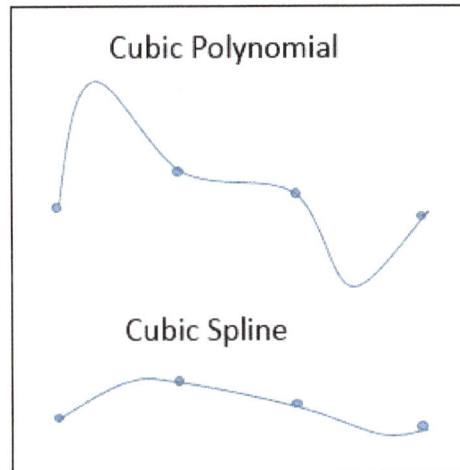

Figure 4.2

Similar to interpolation, there is a method called extrapolation, which is used when the data must be projected beyond the current range of values. Extrapolation is used to project population growth curves for example. The most basic method of extrapolation is to fit a line to the end of the data. Linear extrapolation can be good when the points being extrapolated are not far outside the range of available values, and when the values can easily be fitted by a line. For example, linear extrapolation would work well for the dataset in figure 4.3. The height at age 17 could be extrapolated and a reasonable estimation would be 70 inches.

Age in Years	Height in Inches
14	64
15	66
16	68
17	

Figure 4.3

There is always a chance that the individual whose height is being measured stopped growing at age 16, and the extrapolated height of 70 inches could be inaccurate. The possibility for error is represented by the error term in a linear equation, but since we usually do not have a value to compare against, it is impossible to calculate the true error for an extrapolated value.

In the coming chapters, we will learn about linear regression. When looking back on this chapter, a reasonable question might be what the difference is between linear regression and linear extrapolation or linear interpolation. The answer lies in what each technique is trying to accomplish. Linear regression attempts to minimize a cost function, such as the sum of squared errors between predicted and actual values. Linear interpolation or extrapolation tries to find the function that fits a predefined line passing through exact points. Interpolation and extrapolation assume that the points follow a curve exactly, whereas regression produces an estimate of possible values. It is very rare that data will follow a function perfectly, so it is not a good idea to extrapolate new values for a dependent variable for example, rather than using linear regression to predict the new values.

Censoring is a special case of missing data. It occurs when data is clipped in some way. For example, if a researcher is measuring people's weights, and the scale being used has a maximum capacity of 300 pounds, then anyone weighing more than 300 pounds would be clipped. In the data, people weighing more than 300 pounds might be listed as having a weight of 300 pounds. Without knowing the details of how each feature in a dataset was collected, it is difficult to know for certain whether censoring has occurred. However, inspection of the univariate distribution of a feature gives clues. In figure 4.4, there is evidence of censoring at 300 pounds.

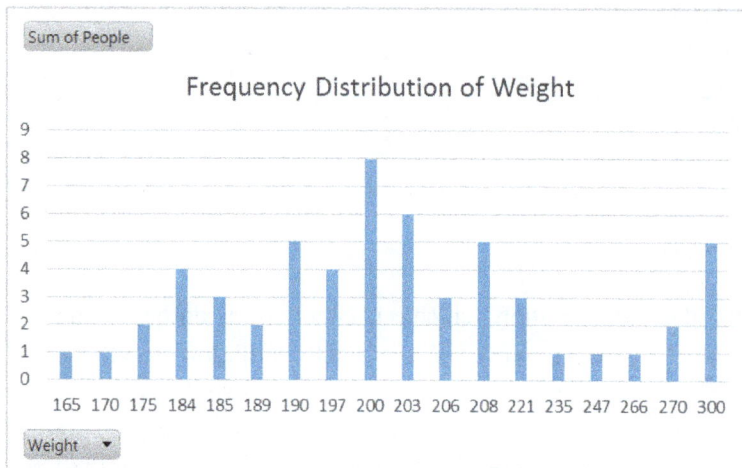

Figure 4.4

When censoring is a problem, the researcher can choose which method to use to handle the data. A common technique is to use a robust estimation method, such as **maximum likelihood estimation** (MLE). MLE is also good at handling missing values. We will look at MLE in a later chapter.

Detecting Outliers

Outliers are observations that are very far from others in the dataset. For example, in the vector of numbers [2, 5, 6, 3, 4, 2, 5, 1, 6, 3, 2, 4, 87], 87 is clearly very far away from the other numbers, so it would be considered an outlier. Outliers add variation to a dataset, so they increase both the variance and standard deviation. They may also skew the univariate distribution of a dataset. This could cause a violation in the parametric assumption that dependent variable is normally distributed. Therefore, it is important to always inspect data for outliers.

A quantitative test to detect outliers is the **interquartile range test** (IQR test). Recall from chapter 2 that a quantile results from dividing a variable into some number of percentiles such that the interval between each percentile is the same. A quantile shows the percentiles for certain divisions of a variable. For example, a quantile that divides a variable into 4 equal intervals of percentiles (25th, 50th, and 75th) is called a quartile. The **interquartile range** (IQR) is the distance between the 25th and 75th percentiles, or the first and third quartiles. The consensus among statisticians is that extreme values that lie further than 1.5 times the IQR from the first and third quartiles (or 25th and 75th percentiles) are considered outliers, while points the lie further than 3 times the IQR from the first and third quartiles are extreme outliers (Glen, 2018a). The equations to find the boundaries beyond which points are considered outliers by the IQR test are:

$$25\text{th percentile} - (1.5*(75\text{th percentile} - 25\text{th percentile}))$$

$$75^{th} \text{ percentile} + (1.5*(75^{th} \text{ percentile} - 25^{th} \text{ percentile}))$$

Figure 4.5

If 1.5 were changed to 3 in the equations in figure 4.5, then the result would be the range for extreme outliers. As an example, let's apply the IQR test to the vector [2, 5, 6, 3, 4, 2, 5, 1, 6, 3, 2, 4, 87]. The first quartile is 2 and the third quartile is 5, so the difference between them is 3. Multiplying that number by 1.5 gives 4.5, which means the range outside of which points can be considered outliers is [-2.5, 9.5] (found by calculating: 2-4.5 = -2.5 and 5+4.5=9.5). So 87 is an outlier according to the IQR test. If we use 3 instead of 1.5 then the range outside of which points can be considered extreme outliers is [-7, 14] and 87 is an extreme outlier.

The simplest way to look for outliers is to look at a box plot. A boxplot like the one in figure 4.5 has an inner fence at 1.5*IQR. Some boxplots (a.k.a. box and whisker plots) extend the arms out to the minimum and maximum values of the dataset, which means that they do not show any outliers. If we want to use the boxplot to visually inspect data for outliers, it would only be necessary to extend the arms out to the inner fences at 1.5*IQR.

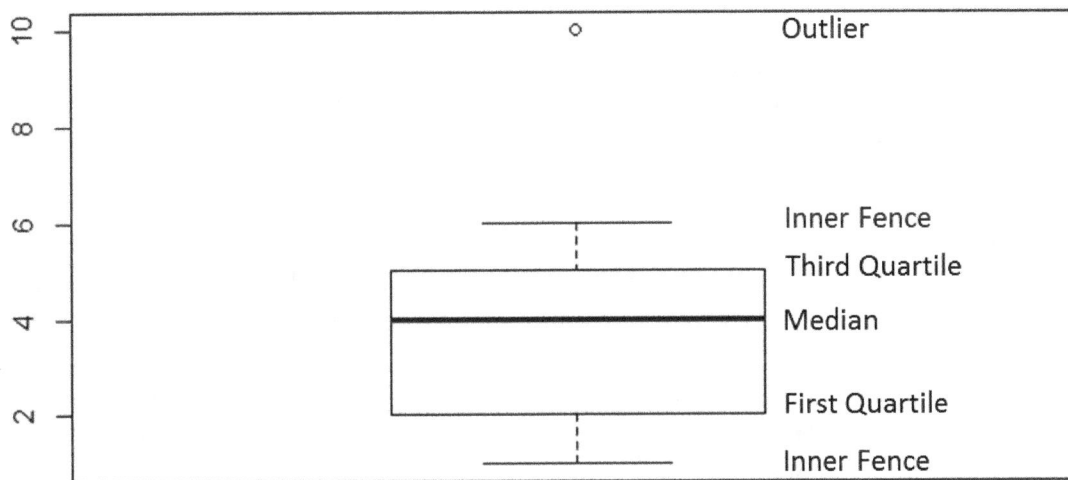

Figure 4.6

It should be noted that the IQR test works for distributions that are approximately normal. If the distribution is known for certain to be normal, then 1.35 can be used as the coefficient of the IQR instead of 1.5., otherwise the data may have to be transformed in such a way to make it normal. We will examine transformations later in this chapter. Tests for normality should be performed before using the IQR test.

While the IQR test can be used to find many outliers at once, **Grubb's test** (a.k.a. extreme studentized deviate test) is designed to find single outliers (Grubbs, 1969). Like the IQR test, Grubb's test only works for normally distributed data, so a test for normality should be performed before applying Grubb's test. Grubb's test looks for the largest absolute deviation from the sample mean, in terms of standard deviations. It is a hypothesis test with the null hypothesis being that there are no outliers in the data, and the alternative hypothesis being that there is at least one outlier. The test statistic is:

$$G = \frac{max_{i=1,...,N}|Y_i - \bar{Y}|}{s}$$

Figure 4.7

In figure 4.7, Y bar is the sample mean, s is the sample standard deviation, and Y sub I is a single value in the feature vector. To find the test statistic, G, we must find the maximum absolute deviation from the sample mean by testing every value in the feature vector. Once this maximum value is found it is divided by the sample standard deviation, s.

Grubb's is a two-sided test and the null can be rejected at significance level alpha if the test statistic, G, is:

$$G > \frac{N-1}{\sqrt{N}}\sqrt{\frac{t^2_{\frac{\alpha}{(2N)},N-2}}{N-2+t^2_{\frac{\alpha}{2/N},N-2}}}$$

Figure 4.8

In figure 4.8, N is the number of values in the feature, and $t^2_{\frac{\alpha}{(2N)},N-2}$ is the upper critical value of the t-distribution with N-2 degrees of freedom, and a significance level of alpha / 2N. It is because the t-distribution is the basis for comparison that this test gets its name; the extreme studentized deviate test. For ease of comparison, a table of critical values for Grubb's test can be found here: http://www.sediment.uni-goettingen.de/staff/dunkl/software/pep-grubbs.pdf.

The **generalized extreme studentized deviate test** is an iterative version of Grubb's test that expands the test to look for multiple outliers (NIST/SEMATECH e-Handbook of Statistical Methods, 2018b). The iterative version of the test is repeated until no more outliers are found. The drawback to using this test is that unless we have a pre-conceived notion of how many outliers we expect to find, the test may be too aggressive and suggest that a feature has more outliers that it may truly have. This feature of the generalized extreme studentized deviate test brings out an important point about outliers in general: they are subjective. An outlier to one researcher might be normal to another. Interpretation of possible outliers depends on the researcher's intuition, the nature of the data, and the type of problem being solved.

The **modified Z-score test** is a way of looking for point that are outliers with respect to the measure of central tendency and dispersion around this measure (NIST/SEMATECH e-Handbook of Statistical Methods, 2018a). The reason it is called a modified Z-score test is because there is a standard Z-score test, but it is infrequently used because it is not robust (the mean and standard deviation are skewed by outliers), and because of its weakness with small datasets (it has been shown that the standard Z-score test fails to identify any outliers in datasets with 12 or fewer observations). The modified Z-score test is more robust because it uses the median as the measure of central tendency instead of the mean, which as we saw in chapter 2 is more robust to outliers than the mean. The modified Z-score test also looks at the median absolute deviation instead of standard deviation. It can be calculated as in figure 4.9.

$$M_i = \frac{0.6745(Y_i - \tilde{Y})}{median_{i=1...N}(|Y_i - \tilde{Y}|)}$$

Figure 4.9

In figure 4.9, Y sub I is the value of a single observation in the feature vector, and Y tilde is the median of the feature vector, and N is the number of observations. Figure 4.9 shows that the median of all absolute deviations must be calculated first, and then it can be used in the denominator. If the absolute value of the modified Z-score, M sub I, is greater than 3.5, then it can be considered an outlier.

So far all of the tests we have seen apply only to numeric data. The **Z-score test** can be applied to categorical data as well. Detecting outliers in categorical data requires that we know the expected frequency of each category. Then we can compare the observed frequencies to the expected frequencies. The Z-score test compares the expected ratio of a category's frequency to the overall number of observations, to the observed ratios of a category's frequency to the number of observations. These ratios are just the probabilities of each category, so the Z-score compares the expected probability of a category to the observed probability. The Z-score test must be applied to one category at a time. The equation is:

$$Z_c = \frac{|\hat{p} - p| - \frac{1}{2N}}{\sqrt{\frac{p*(1-p)}{N}}}$$

Figure 4.10

In figure 4.10, p hat is the actual proportion of observations in a class to total observations, p is the expected proportion, and N is the number of observations. Unless there is some known distribution that the frequencies in each class should follow, the expected proportion could assume that each category should have equal observations. The numerator contains a continuity correction term 1/2N. This continuity correction term should only be included in the calculation when it is smaller than the first term, the absolute difference between the expected and observed proportions. The continuity correction term is needed because the Z-score test is approximating a discrete function with a continuous one (it's approximately a frequency distribution with a probability distribution). As shown in figure 4.11, the discrete function being approximated is the proportion of observations in each category. The continuous function (normal distribution) does not exactly match the discrete one, as shown by the green highlights, so a slight correction is needed.

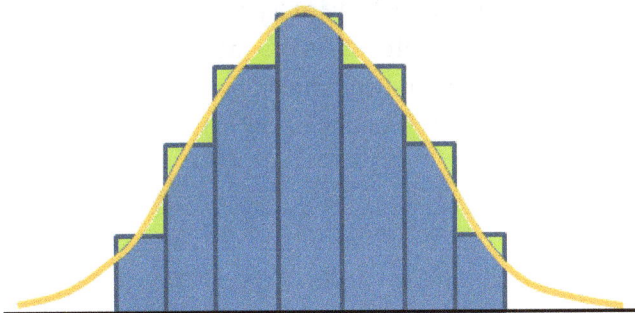

Figure 4.11

After the test statistic has been calculated, it can be compared to the critical value for a desired level of significance. For example, at the 95% level, if the absolute value of the Z-score > 1.96, then the category has outliers.

The Z-score test shows which categories have outliers, but it does not specify which observations are the outliers. The drawback to using the Z-score test for outliers in categorical data is that it is biased for large samples (NIST/SEMATECH e-Handbook of Statistical Methods, 2018a). Recall from chapter 3 that large samples exaggerate the statistical significance of even small effects. One possible work around is to use a high alpha level, such as the 99% level. Another outlier test for categorical data that is robust to sample size is the **mean absolute deviation** (MAD). This is similar to the median absolute deviation that we saw in the modified Z-score test for outliers in numeric data, except that the mean is used as the measure of central tendency instead of the median. If we think of the expected frequency of each category to be equal, then figure 4.12 shows how MAD summarizes the fit of the observed values of a categorical variable to the expected values.

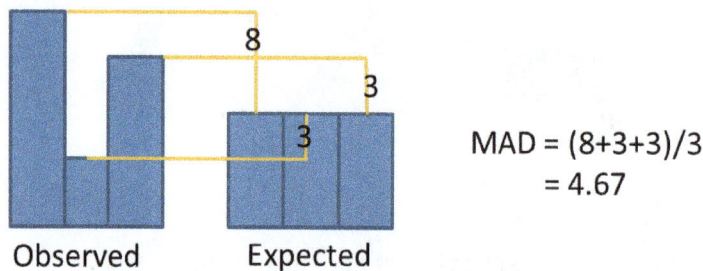

MAD = (8+3+3)/3
= 4.67

Observed   Expected

Figure 4.12

It should be clear by now that unlike the Z-score and chi-square tests, the MAD only indicates whether a feature has outliers. MAD does not tell which observations in particular are the outliers, nor which specific categories have the outliers. The equation is:

$$MAD = \frac{\sum_{i=1}^{K} |\hat{p} - p|}{K}$$

Figure 4.13

In figure 4.13, K is the number of categories, p hat is the expected probability of a particular category, and p is the actual probability of that category. The probability of a category is the ratio of observations in that category to the total number of observations. The MAD test statistic should be compared to a critical value, however there is no table of critical values for MAD. The values below can serve as general guidelines (Nigrini, 2012).

Distribution Type	Conformity	Lower MAD Critical Value	Upper MAD Critical Value
Highly Skewed, Some Categories with Many More Observations than the Rest	Close	NA	0.006
	Acceptable	0.006	0.012
	Marginal	0.012	0.015
	Non-Conformity	0.015	NA
Skewed, Some Categories with More Observations than the Rest	Close	NA	0.004
	Acceptable	0.004	0.008
	Marginal	0.008	0.012
	Non-Conformity	0.012	NA
Moderately Skewed	Close	NA	0.0012
	Acceptable	0.0012	0.0018
	Marginal	0.0018	0.0022
	Non-Conformity	0.0022	NA
Non-Skewed, Even Distribution Across Categories	Close	NA	0.00036
	Acceptable	0.00036	0.00044
	Marginal	0.00044	0.0005
	Non-Conformity	0.0005	NA

Figure 4.14

Another way of looking for outliers in categorical data is to plot a histogram of frequencies in each category. Although visual inspection is always open to interpretation, a category with many more or much fewer observations than the rest can be considered an outlier. Alternatively, a box plot of a numeric feature across the categories of a categorical feature could reveal outliers. This is called **bivariate outlier detection**. An example would be plotting height in inches by age, assuming we treat age as a categorical variable. If a 12 year-old person were 72 inches tall, they would be an outlier. The outlier in this case would show up in bivariate analysis, whereas it may not have appeared as an outlier if height were only examined in univariate analysis.

Another bivariate method of detecting outliers is to compare the Pearson correlation coefficient to Spearman's rho or Kendall's tau, although unlike box plots by category, correlations only apply to numeric variables. If the correlations are unusually different from one another (they should be nearly the same value), then there may be outliers. If there is evidence of outliers, a scatterplot between the two variables would show them. This method works only for numeric data. We will look at the Pearson correlation coefficient, Spearman's rho, and Kendall's tau in the next chapter. There are also multivariate methods that search the entire feature space for outliers, meaning they search for outliers in all of the features at the same time, but those methods are outside the scope of this book.

There is one method of outlier detection that is specific to outliers in linear regression called **Cook's Distance** (D). Cook's Distance tests the effect of a single outlier on a least squares regression, by looking at what would happen if the outlier were removed (Glen, 2018b). In general, a D value > 1 may be cause for concern. Some statistics texts have suggested using a D value of 4/n instead, where n is the sample size. While Cook's D can detect outliers affecting a linear regression model, it does not mean that the outliers should be removed – that requires investigating what caused the outlier, as we will soon see.

Dealing with Outliers

Now that we know how to detect outliers, we can look at techniques to handle them. Like missing data, it is important to first try to determine what is causing the outliers, if possible. Outliers could be caused by researcher error, such as recording the wrong value or adding a zero. There could also be systematic variance in the way the data is measured that was not controlled for in the experiment, like a participant who is having a bad day and responds to survey questions very differently because of their mood. Outliers may be able to be corrected if there is a known reason they exist.

One way to deal with outliers, especially if there are very few of them in a dataset, is to simply remove them. This is usually the best course of action if it is possible to prove that the outlier was a mistake, like in the aforementioned example of the survey participant having a bad day or the research adding a zero to a recorded value.

Another way to deal with outliers is to just change the values of the outliers to something else to make them conform better. Changing an outlier might feel unethical, like the data is being fudged, but if the outlier is significant enough to fundamentally alter a statistical model, then changing the outlier is the lesser evil: it is better to change a single observation (or a few observations) than draw an entirely wrong conclusion because the outlier was given too much weight. There are 3 common ways to change an outlier value:

1. **Use the next highest or lowest value plus one** – find the most extreme value that is not the outlier, increase it slightly, and use that value in place of the outlier.
2. **Convert back from a Z-score** – since a Z-score of 3.29 or more is an outlier, then it is possible to rearrange the Z-score equation to see which value would give 3.29. The general equation to do this is: X = (Z * standard deviation) + mean. The outlier's new value would become the mean plus 3.29 times the standard deviation (Z is replaced by 3.29 since 3.29 is the boundary for outliers in terms of Z-score).
3. Use the mean plus 2 standard deviations – instead of using 3 times the standard deviation like in option 2 above, use 2 standard deviations.

Another way to deal with outliers is to leave them as they are and use a robust estimation method, such as maximum likelihood estimation (MLE). We will look at MLE in a later chapter.

Transforming Data

Data must often be transformed to be used in models. Some models can only handle numeric data for example. Transformation can also treat outliers, although transformations meant to diminish the impact of outliers should only be considered after the other methods previously described. The reason transformation is such a drastic way to treat outliers is that in order to preserve the relationships between variables, any transformation done to one feature must be done to all of them. This can change the hypothesis being tested. For example, comparing squared numbers is much different than comparing normal numbers. There is a possibility that transforming the data to correct the distribution of one variable will introduce skew into another. The most complicated part of data transformation is that it is mostly a task of trial and error to see which transformations work best. There are some general guidelines however:

1. **Log Transformation** – can fix skewed data or data with heterogeneous variances. This squashes the right tail of the distribution. Since it is not possible to take the log of 0 or negative numbers, if these values exist in the data, a constant must be added to the values first in order to make all the numbers positive. Then the log transformation can be applied.

2. **Square Root Transformation** – can fix skewed data or data with heterogeneous variances. This is another way to squash the right tail of the distribution, because larger numbers are brought closer to the center. Since it is not possible to take the square root of negative numbers, if these values exist in the data, a constant must be added to the values first in order to make all the numbers positive. Then the square root transformation can be applied.

3. **Reciprocal Transformation** – can fix skewed data or data with heterogeneous variances. This flips a dataset around so that larger scores become small and smaller scores become large. Because of this, it is common to subtract each value from the highest score before taking the reciprocal. That will keep smaller scores below larger scores. This transformation will have a limit at 0, because 1 divided by larger numbers approaches 0.

4. **Reverse Score Transformation** – can fix negatively skewed data. To apply this transformation, subtract each score from the highest score in the dataset. The data will later have to be transformed back, after statistical analysis is applied.

Data transformation is not only a way to handle outliers, it is also a technique to rectify violations in the standard parametric assumptions. As an example, standardization and normalization are two ways to scale data so that they are normally distributed.

Standardization, a.k.a. Z-score normalization, scales data to have a mean 0 and standard deviation 1. In other words, it converts the data to a normal distribution through the calculation of Z-scores. Recall from chapter 2 that the Z-score formula is:

$$Z = \frac{x - \mu}{\sigma}$$

Figure 4.15

In figure 4.15, x is some value to be converted, mu is the population mean, and sigma is the population standard deviation. Standardization allows features of different scales to be directly compared. For example, if age and annual income were standardized, then they could be compared more easily. Without standardization, it would be impossible to compare age and annual income because they are orders of magnitude in difference, and they are measured in different units.

Normalization, a.k.a. min-max normalization, scales data so that the minimum value is 0 and the maximum value is 1. The equation is:

$$X_{norm} = \frac{X - X_{min}}{X_{max} - X_{min}}$$

Figure 4.16

In figure 4.16, x is some value to be converted, x min is the minimum value of the variable, and x max is the maximum value of the variable. Normalization bounds the data to a definite range, which results in a smaller standard deviation. This can inadvertently suppress the effect of outliers. For this reason, standardization is usually preferred to normalization for algorithms that seek to maximize explained variance like principal component analysis, as we will see in a later chapter. If outliers are important, use standardization. If outliers should not impact the model, use normalization. Normalization is preferable in situations like image processing, where pixel values can fall between 0 and 255 red, green, and blue (RGB) values.

Another type of transformation is long to wide transformation and vice versa. Most tabular data is natively in wide format, where there is a column for each variable. Long format is more common in time series data, such as when there is a row for each variable. Figure 4.17 shows the difference between the two. The data is the same.

Long Format

Week Nbr	Day	Average Temperature
1	Sunday	68.4
1	Monday	69.1
1	Tuesday	70.5
1	Wednesday	67.3
1	Thursday	69.2
1	Friday	70.2
1	Saturday	73.4

Wide Format

Week Nbr	Sunday	Monday	Tuesday	Wednesday	Thursday	Friday	Saturday
1	68.4	69.1	70.5	67.3	69.2	70.2	73.4

Figure 4.17

There is no formula to re-arrange data format, but we see how to do it programmatically in the sections for R and Python in this chapter.

Categorical variables can be encoded so that they become integers. One way to do this is to one-hot encode them. **One-hot encoding** turns every unique category of a categorical variable into a separate column in the dataset. The values of the new columns are either 1 or 0. The value is 1 if the observations falls into that category. A similar technique is dummy encoding. **Dummy encoding** follows the same procedure as one-hot encoding, except that one category is chosen as the baseline. Whereas a categorical variable with k categories is encoded into k columns using one-hot encoding, the same variable would be encoded into k-1 columns using dummy encoding. The advantage to dummy encoding is that it prevents multicollinearity, which is important for linear regression as we will see in a later chapter.

Categorical variables can also be encoded by label encoding. **Label encoding** converts nominal variables into ordinal variables by turning each category into an integer, in order. Figure 4.18 shows an example of label encoding. Label encoded variables can be used in models that only accept numeric variables, as long as the distance between the categories is preserved. In other words, if the transformed variable is measured at the interval level, it is acceptable to use it in a model that only accepts numeric variables. For favorite colors, as shown in figure 4.18, this would not work, because the distance between red and green is undefined. If the categories were grade levels in school however, it would work.

Original Variable: Favorite Color	Label Encoded Variable: Favorite Color
green	1
green	1
red	2
yellow	3
blue	4
red	2

Figure 4.18

Data Cleaning with R

The Titanic dataset that we have been using so far is messy. Now that we have explored ways to clean messy data, we can clean up the Titanic dataset to prepare it for the more elaborate analysis we will do in the coming chapters. To start, we need an idea of how many null or missing values are in the dataset. We can define a function to count how many there are in each field, taking care to include nulls, NAs, NaNs, infinite values, empty strings, and NA strings in the count of missing values, as missing values can be represented in any of these ways, depending on the source of the data.

```
train <- read.csv("train.csv", header=T)
test <- read.csv("test.csv", header=T)

naCol <- function(x){
  y <- sapply(x, function(y) sum(length(which(is.na(y) | is.nan(y) | is.infinite(y) |
y=='NA' | y=='NaN' | y=='Inf' | y==''))))
  y <- data.frame('feature'=names(y), 'count.nas'=y)
  row.names(y) <- c()
  y
}
```

This function is draconian. There may be cases where we want to keep empty strings, infinite values, or even NAs in the dataset. Perhaps NAs could be assigned to a particular category. In these situations, the function above can be modified to omit the types of data that should be kept from the count of nulls. To see an example of the different forms the data can take, apply the naCol function to the data frame defined below.

```
x <- data.frame(a=c(1, NA, 3, NaN, 5, Inf, 7, ''), b=c(1,1,1,1,1,1,1,1))
x
naCol(x)
```

When this function is applied to the Titanic dataset, the age and cabin variables are found to have many missing values. Additionally, fare has 1 missing value in the test set, and there are 2 missing values for embarked in the test set.

```
naCol(train)
naCol(test)
```

Since most of the columns have no missing values, listwise deletion of certain rows is unnecessary. The cabin field is more the 50% null in both training and test sets, so it may be worth discarding the variable entirely. However, for the rows where cabin does have a value, it might be an important feature. For example, if we want to predict who survived, it is reasonable to think that cabin location played a major role in survival. Cabins in lower levels or closer to the hole that the iceberg punctured probably had lower survival rates. But it is also reasonable to assume that cabin location is closely tied to passenger class. To test this idea, we could look at the correlation between cabin and class for the rows where cabin is defined. Since correlation is tricky for categorical variables, a simpler way to see if cabin and class are related is to simply use a cross tabulation. Alternatively, since the first letter of cabin indicates the deck level, perhaps a cross tabulation of deck and class would be more revealing. It could also be argued that cabin should not be important, because it may only represent the cabins people booked, rather than where they actually stayed. Maybe all of the passengers left their cabins. If this were the case, then only the deck level would matter because passengers on lower levels would have to climb more stairs to get to safety. Let us extract the first character of the cabin feature to get deck level, and cross tabulate it with passenger class.

```
train$Deck <- substr(train$Cabin, 1, 1)
test$Deck <- substr(test$Cabin, 1, 1)

library(gmodels)
CrossTable(train$Deck, train$Pclass, prop.chisq=F)
```

The first noticeable thing is that the percentage of missing cabin values increases with class. So without any further exploration, it seems that cabin/deck and class are related. Looking down the column for first class, it appears that decks A-C and T were exclusively first class. Deck D was shared by first and second class, deck E was shared by all three classes, deck F was shared by second and third class, and deck G was exclusively third class. So class is not a perfect substitute of cabin or deck, and it is possible that there was an interaction effect between cabin/deck, class, and survived. Let's keep deck in the dataset and get rid of cabin. Since we will likely use deck in a model later on, we can fill in the missing values with a single value to represent that the data was missing.

```
train$Cabin <- NULL
test$Cabin <- NULL

unique(train$Deck)
unique(test$Deck)  #Make sure the encoded value is not an existing category
train[train$Deck=='', 'Deck'] <- 'Z'
test[test$Deck=='', 'Deck'] <- 'Z'
```

There is 1 fare price missing from the test set. Since there is only 1 missing value, it would be safe to fill it in with the mean fare price. But since fare price likely depends on class, we will need to determine the class of the passenger with the missing fare, calculate its mean over the training and test sets, and use that to fill in the value.

```
missing_fare_pclass <- test[is.na(test$Fare), 'Pclass']
```

```
test[is.na(test$Fare), 'Fare'] <-
round(mean(c(train[train$Pclass==missing_fare_pclass, 'Fare'],
test[test$Pclass==missing_fare_pclass, 'Fare']), na.rm=T), 4)
```

There are 2 missing embarkation ports in the training set. Since there are so few missing values, it would be safe to fill them in with the mode of port of embarkation. It is unlikely that port of embarkation depends on any other variables in this dataset, so we can simply fill it in with the mode over the training and test sets. Recall from chapter 2 that we need to define our own function for the mode. Note that R converts factors to integers when they are concatenated, so the fields must be treated as characters when they are concatenated in order to prevent the result of the mode from being an integer that we cannot trace back to an embarkation port. Also note that the missing value in embarked is actually an empty string, rather than an NA.

```
#Function for calculating the mode
Mode <- function(x) {
  ux <- unique(x)
  ux[which.max(tabulate(match(x, ux)))]
}

train[is.na(train$Embarked) | train$Embarked=='', 'Embarked'] <-
Mode(c(as.character(train$Embarked), as.character(test$Embarked)))
```

The last field with missing values is age. There are too many to fill them in with the mean. Before deciding what to do, we need to know if the values are missing at random. We can get a general idea by looking at the data where the values are missing. But looking at all of the columns at once is overwhelming and gives away very little insight. A better idea is to think about what might cause the variable to be missing. For this dataset, it seems logical that passengers who died might have unknown ages. Third class passengers might have unknown ages more often than first and second class. So let's create a table with these and see if there are any patterns that emerge.

```
table(train[is.na(train$Age), 'Survived'], train[is.na(train$Age), 'Pclass'])
```

It is clear that age is missing more often for passengers who did not survive, and slightly more often for third class passengers, regardless of whether or not they survived. Therefore, since the values are not exactly missing at random, we cannot use single random imputation. Instead, we could build a model for age using the remaining values. We will return to this topic after we learn more about modeling in a later chapter.

Now it is time to inspect the data for outliers. A box plot can quickly show whether any observations of the numeric variables lie outside the inner fence (1.5*IQR).

```
for (col in 1:ncol(train[,which(sapply(train, class) == 'numeric')])) {
  boxplot(train[,which(sapply(train, class) == 'numeric')][col],
          main=paste0("Box Plot for ", colnames(train[,which(sapply(train, class) ==
'numeric')])[col]))
}
```

Looking at the box plot for age, there appear to be a few outliers, but none of them are egregious. Most of the outliers are older people, and the oldest is around 80 years old. In this case, it may be fine to just leave them as they are, since all of the outlier candidates are perfectly reasonable values. The box plot for fare however shows more considerable outliers. There is one observation above 500, and the next closest value is under 300. Grubb's test confirms that the observation above 500 is an outlier:

```
library(outliers)
grubbs.test(train$Fare)
```

Since there are 3 observations with a fare price at this value, we can determine that the value is probably not a data entry error. Inspecting the rows with the high fare price show that the price was paid by three different first class customers, all embarking from the same port, and all with the same ticket number. This confirms our belief that the value is likely not a data entry error, but it also reveals a peculiarity that we need to account for: some passengers bought their tickets together. This means that the fare price for the 3 customers sharing the same ticket is likely the sum of 3 separate first class ticket prices. So we need to compute a new feature that takes group fares into consideration. To do this accurately, we need to temporarily combine the training and test sets to account for groups of passengers who were split between the datasets.

```
train$Set <- 'train'
test$Set <- 'test'
alldata <- rbind(train[,-2], test)
library(magrittr)
library(dplyr)
alldata <- alldata  %>% group_by(Fare, Ticket) %>% mutate(Group_Size = n())
alldata <- transform(alldata, Fare_Per_Person=Fare/Group_Size)
```

Now we can view the box plots of fare per person by passenger class to more accurately spot outlier fares.

```
boxplot(alldata[alldata$Pclass==1,'Fare_Per_Person'], main="Box Plot for Fare Per
Person - First Class")
boxplot(alldata[alldata$Pclass==2,'Fare_Per_Person'], main="Box Plot for Fare Per
Person - Second Class")
boxplot(alldata[alldata$Pclass==3,'Fare_Per_Person'], main="Box Plot for Fare Per
Person - Third Class")
```

A quick peek at the apparent outliers reveals a group of 4 first class passengers, who may be related, who all purchased expensive tickets, and a couple brothers in third class who paid more for their tickets than the rest of third class. None of the outliers seem unreasonable, however, there are some fare prices that are 0. Since it is unlikely that any passenger got on board for free, these zeroes are probably data entry error. Perhaps they were unknown fares that were encoded as zero. Let's change all fare prices and fares per person that equal zero to the mean for their respective classes.

```
firstclassmean <- mean(alldata[alldata$Pclass==1, 'Fare'])
firstclassmeanpp <- mean(alldata[alldata$Pclass==1, 'Fare'])
secclassmean <- mean(alldata[alldata$Pclass==2, 'Fare'])
secclassmeanpp <- mean(alldata[alldata$Pclass==2, 'Fare'])
thirdclassmean <- mean(alldata[alldata$Pclass==3, 'Fare'])
thirdclassmeanpp <- mean(alldata[alldata$Pclass==3, 'Fare'])
alldata[alldata$Fare==0 & alldata$Pclass==1, 'Fare'] <- firstclassmean
alldata[alldata$Fare_Per_Person==0 & alldata$Pclass==1, 'Fare_Per_Person'] <-
firstclassmeanpp
alldata[alldata$Fare==0 & alldata$Pclass==2, 'Fare'] <- secclassmean
```

```
alldata[alldata$Fare_Per_Person==0 & alldata$Pclass==2, 'Fare_Per_Person'] <-
secclassmeanpp
alldata[alldata$Fare==0 & alldata$Pclass==3, 'Fare'] <- thirdclassmean
alldata[alldata$Fare_Per_Person==0 & alldata$Pclass==3, 'Fare_Per_Person'] <-
thirdclassmeanpp
rm('firstclassmean' ,'firstclassmeanpp', 'secclassmean', 'secclassmeanpp',
'thirdclassmean', 'thirdclassmeanpp')
```

Now let's visually inspect the categorical features for outliers, using histograms. We will continue using the combined dataset since it will be easier to check for outliers across the whole dataset rather than checking each split separately.

```
for (col in 1:ncol(alldata[,which(sapply(alldata, class) %in% c('factor',
'integer'))])) {

  barplot(table(alldata[,which(sapply(alldata, class) %in% c('factor',
'integer'))][col]),

        main=paste0("Bar Plot for ", colnames(alldata[,which(sapply(alldata, class)
%in% c('factor', 'integer'))])[col]))

}
```

There is one family with 11 members and a few passengers with the same name. We should inspect these rows to make sure they are not invalid data.

```
alldata[alldata$Group_Size==11,]
```

```
alldata[alldata$Name %in% names(which(table(alldata$Name)>1)),]
```

Fortunately, there do not appear to be any outliers in the categorical features of this dataset, as the only items that waved red flags were found to be benign upon inspection.

The last step in the data cleaning and preparation process is to transform the data so that it can easily be passed into a model. Models do not like strings or characters, so the first thing we will do is convert them to something more usable. One-hot encoding or dummy encoding are the best options, but if there are character variables with hundreds or thousands of different values or categories, these options may be too demanding on computer memory. So to decide how to encode the character variables, we can first check to see how many distinct values they have.

```
for (col in 1:ncol(alldata[,which(sapply(alldata, class) %in% c('factor',
'character'))])) {
  print(paste0("Unique categories of ", colnames(alldata[,which(sapply(alldata,
class) %in% c('factor', 'character'))])[col], ": ",
length(unique(alldata[,which(sapply(alldata, class) %in% c('factor',
'character'))][,col]))))
}
```

Only name and ticket have so many values that they could cause memory problems. So let us one-hot encode sex, embarked, and deck. We will not do anything to set because we are using it only to determine which dataset the observations came from.

```
library(caret)
```

```
dummies <- dummyVars( " ~ .", data=alldata[,which(colnames(alldata) %in% c('Sex',
'Embarked', 'Deck'))])
dummyenc <- data.frame(predict(dummies, newdata=alldata[,which(colnames(alldata) %in%
c('Sex', 'Embarked', 'Deck'))]))
alldata <- cbind(alldata, dummyenc)
colnames(alldata) <- gsub("[.]", "_", colnames(alldata))
rm('dummies', 'dummyenc')
```

The name field can be left alone for now, because we will return to it in a later chapter to build more interesting features. The ticket field can now be label encoded.

```
alldata$Ticket <- as.integer(alldata$Ticket)
```

Since the variables have been encoded, the dataset can be reduced to the encoded and numeric columns, plus name. Passenger ID will be discarded since it is never good to include an ID field in a model. Finally, the data can be split back to the training and test sets.

```
alldata <- alldata[,-which(colnames(alldata) %in% c('PassengerId', 'Sex', 'Embarked',
'Deck', 'Ticket'))]

train_clean <- alldata[alldata$Set=='train',]
train_clean$Set <- NULL
train_clean$Survived <- train$Survived
test_clean <- alldata[alldata$Set=='test',]
test_clean$Set <- NULL
write.csv(train_clean, file="train_clean.csv", row.names=F)
write.csv(test_clean, file="test_clean.csv", row.names=F)
```

The cleaned datasets will be used in the coming chapters to build models.

Data Cleaning with Python

The Titanic dataset that we have been using so far is messy. Now that we have explored ways to clean messy data, we can clean up the Titanic dataset to prepare it for the more elaborate analysis we will do in the coming chapters. To start, we need an idea of how many null or missing values are in the dataset. We can define a function to count how many there are in each field, taking care to include nulls, NAs, NaNs, infinite values, empty strings, and NA strings in the count of missing values, as missing values can be represented in any of these ways, depending on the source of the data. Fortunately, Pandas always reads missing values as either NaN or NaT for time. Therefore, only the isnull() function is required to check for NaN or NaT. This is much easier than using several checks in R, however, the only way to check for empty strings is to compare columns read as objects (Pandas calls string or character variables objects) to the empty string, as columns that are not of the object data type cannot be compared to a string.

```
import pandas as pd
import numpy as np
import statistics as stats
import matplotlib.pyplot as plt
from outliers import smirnov_grubbs as grubbs
import seaborn as sns
from sklearn.preprocessing import LabelEncoder

train = pd.read_csv("train.csv", sep=",")
```

```
test = pd.read_csv("test.csv", sep=",")

print(train.info())

#Define a function to count the nulls in every field
def naCol(df):
    y = dict.fromkeys(df.columns)
    for idx, key in enumerate(y.keys()):
        if df.dtypes[list(y.keys())[idx]] == 'object':
            y[key] = pd.isnull(df[list(y.keys())[idx]]).sum() +
(df[list(y.keys())[idx]]=='').sum()
        else:
            y[key] = pd.isnull(df[list(y.keys())[idx]]).sum()
    print("Number of nulls by column")
    print(y)
    return y
```

When this function is applied to the Titanic dataset, the age and cabin variables are found to have many missing values. Additionally, fare has 1 missing value in the test set, and there are 2 missing values for embarked in the test set.

```
naCol(train)
naCol(test)
```

Another useful null check is a row-wise check. The column null check will show whether there may be rows with missing values in many fields, because if that were the case there would be at least a few nulls in every column. The Titanic dataset does not have this problem, but it is useful to define this function should the need to use it ever arise. Rows with more than some percentage (the function below uses a default of 50%) of the columns missing may not be useful for any kind of data analysis and could be deleted.

```
#Define a function to count the nulls by row to see if any rows have too many missing
values to be useful
def naRow(df, threshold=0.5):
    y = dict.fromkeys(df.index)
    for idx, key in enumerate(y.keys()):
        y[key] = sum(df.iloc[[idx]].isnull().sum())
    print("Rows with more than 50% null columns")
    print([r for r in y if y[r]/df.shape[1] > threshold])
    return y
```

```
naRow(train)
naRow(test)
```

Since most of the columns have no missing values, listwise deletion of certain rows is unnecessary. The cabin field is more the 50% null in both training and test sets, so it may be worth discarding the variable entirely. However, for the rows where cabin does have a value, it might be an important feature. For example, if we want to predict who survived, it is reasonable to think that cabin location played a major role in survival. Cabins in lower levels or closer to the hole that the iceberg punctured probably had lower survival rates. But it is also reasonable to assume that cabin location is closely tied to passenger class. To test this idea, we could look at the correlation between cabin and class for the rows where cabin is defined. Since correlation is tricky for categorical variables, a simpler way to see if cabin and class are related is to simply use a cross tabulation. Alternatively, since the first letter of cabin indicates the deck level, perhaps a cross tabulation of deck and class would be more revealing. It could also be argued that cabin should not be important, because it may only represent the cabins people booked, rather than where they actually stayed. Maybe all of the passengers left their cabins. If this were the case, then only the deck level would matter because passengers on lower levels would have to climb more stairs to get to safety. Let us extract the first character of the cabin feature to get deck level, and cross tabulate it with passenger class.

```
train['Deck'] = train['Cabin'].str[:1]
test['Deck'] = test['Cabin'].str[:1]

ct = pd.crosstab(train['Deck'], train['Pclass'], rownames=["Passengers on Each Deck
by Class"])
print(ct)
print(ct.apply(lambda r: r/len(train), axis=1)) #Perc by cell
print(ct.apply(lambda r: r/sum(r), axis=1))     #Perc by row
print(ct.apply(lambda r: r/sum(r), axis=0))     #Perc by col
```

The first noticeable thing is that the percentage of missing cabin values increases with class. So without any further exploration, it seems that cabin/deck and class are related. Looking down the column for first class, it appears that decks A-C and T were exclusively first class. Deck D was shared by first and second class, deck E was shared by all three classes, deck F was shared by second and third class, and deck G was exclusively third class. So class is not a perfect substitute of cabin or deck, and it is possible that there was an interaction effect between cabin/deck, class, and survived. Let's keep deck in the dataset and get rid of cabin. Since we will likely use deck in a model later on, we can fill in the missing values with a single value to represent that the data was missing.

```
del train['Cabin'], test['Cabin']

print(train.Deck.unique())
print(test.Deck.unique())  #Make sure the encoded value is not an existing category
train.Deck.fillna('Z', inplace=True)
test.Deck.fillna('Z', inplace=True)
```

There is 1 fare price missing from the test set. Since there is only 1 missing value, it would be safe to fill it in with the mean fare price. But since fare price likely depends on class, we will need to determine the class of the passenger with the missing fare, calculate its mean over the training and test sets, and use that to fill in the value.

```
missing_fare_pclass = int(test.loc[test['Fare'].isnull()]['Pclass'])
test.Fare.fillna(round(np.nanmean(pd.concat([train.loc[train['Pclass']==missing_fare_
pclass]['Fare'], test.loc[test['Pclass']==missing_fare_pclass]['Fare']],
ignore_index=True)), 4), inplace=True)
```

There are 2 missing embarkation ports in the training set. Since there are so few missing values, it would be safe to fill them in with the mode of port of embarkation. It is unlikely that port of embarkation depends on any other variables in this dataset, so we can simply fill it in with the mode over the training and test sets.

```
train.Embarked.fillna(stats.mode(pd.concat([train['Embarked'], test['Embarked']],
ignore_index=True)), inplace=True)
```

The last field with missing values is age. There are too many to fill them in with the mean. Before deciding what to do, we need to know if the values are missing at random. We can get a general idea by looking at the data where the values are missing. But looking at all of the columns at once is overwhelming and gives away very little insight. A better idea is to think about what might cause the variable to be missing. For this dataset, it seems logical that passengers who died might have unknown ages. Third class passengers might have unknown ages more often than first and second class. So let's create a table with these and see if there are any patterns that emerge.

```
print(pd.crosstab(train.loc[train.Age.isnull()]['Survived'],
train.loc[train.Age.isnull()]['Pclass'], rownames=["Rows Missing Age by Survived and
Pclass"]))
```

It is clear that age is missing more often for passengers who did not survive, and slightly more often for third class passengers, regardless of whether or not they survived. Therefore, since the values are not exactly missing at random, we cannot use single random imputation. Instead, we could build a model for age using the remaining values. We will return to this topic after we learn more about modeling in a later chapter.

Now it is time to inspect the data for outliers. A box plot can quickly show whether any observations of the numeric variables lie outside the inner fence (1.5*IQR).

```
numeric_cols = [col for col in train.columns if train[col].dtype == 'float64']
for col in numeric_cols:
    sns.boxplot(y=train[~np.isnan(train[col])][col])
    plt.title("Box Plot for " + col)
    plt.show()
```

Looking at the box plot for age, there appear to be a few outliers, but none of them are egregious. Most of the outliers are older people, and the oldest is around 80 years old. In this case, it may be fine to just leave them as they are, since all of the outlier candidates are perfectly reasonable values. The box plot for fare however shows more considerable outliers. There is one observation above 500, and the next closest value is under 300. Grubb's test can confirm whether that the observation above 500 is an outlier. Note that the version of Grubb's test from scipy is actually the iterative version of the test (the generalized extreme studentized deviates test), so it returns all outliers. To get the most extreme, take the index of the 0th position.

```
print(train.loc[train.Fare == grubbs.max_test_outliers(train['Fare'],
alpha=0.05)[0]])
```

Since there are 3 observations with a fare price at this value, we can determine that the value is probably not a data entry error. Inspecting the rows with the high fare price show that the price was paid by three different first class customers, all embarking from the same port, and all with the same ticket number. This confirms our belief that the value is likely not a data entry error, but it also reveals a peculiarity that we need to account for: some passengers bought their tickets together. This means that the fare price for the 3 customers sharing the same ticket is likely the sum of 3 separate first class ticket prices. So we need to compute a new feature that takes group fares into consideration. To do this accurately, we need to temporarily combine the training and test sets to account for groups of passengers who were split between the datasets.

```
train['Set'] = 'train'
```

```
test['Set'] = 'test'
alldata = pd.concat([train.drop(['Survived'], axis=1), test], ignore_index=True)
alldata['Group_Size'] = alldata.groupby(['Fare',
'Ticket'])['PassengerId'].transform("count")
alldata['Fare_Per_Person'] = alldata.Fare/alldata.Group_Size
```

Now we can view the box plots of fare per person by passenger class to more accurately spot outlier fares.

```
sns.boxplot(y=alldata[alldata.Pclass==1]['Fare_Per_Person'].values)
plt.title("Box Plot for Fare Per Person - First Class")
plt.show()
sns.boxplot(y=alldata[alldata.Pclass==2]['Fare_Per_Person'].values)
plt.title("Box Plot for Fare Per Person - Second Class")
plt.show()
sns.boxplot(y=alldata[alldata.Pclass==3]['Fare_Per_Person'].values)
plt.title("Box Plot for Fare Per Person - Third Class")
plt.show()

#Inspect outliers by class
print(alldata[(alldata.Fare_Per_Person > 100) & (alldata.Pclass==1)].head())
print(alldata[(alldata.Fare_Per_Person > 15) & (alldata.Pclass==3)].head())
```

A quick peek at the apparent outliers reveals a group of 4 first class passengers, who may be related, who all purchased expensive tickets, and a couple brothers in third class who paid more for their tickets than the rest of third class. None of the outliers seem unreasonable, however, there are some fare prices that are 0. Since it is unlikely that any passenger got on board for free, these zeroes are probably data entry error. Perhaps they were unknown fares that were encoded as zero. Let's change all fare prices and fares per person that equal zero to the mean for their respective classes.

```
firstclassmean = stats.mean(alldata[alldata.Pclass==1]['Fare'])
firstclassmeanpp = stats.mean(alldata[alldata.Pclass==1]['Fare_Per_Person'])
secclassmean = stats.mean(alldata[alldata.Pclass==2]['Fare'])
secclassmeanpp = stats.mean(alldata[alldata.Pclass==2]['Fare_Per_Person'])
thirdclassmean = stats.mean(alldata[alldata.Pclass==3]['Fare'])
thirdclassmeanpp = stats.mean(alldata[alldata.Pclass==3]['Fare_Per_Person'])
alldata.loc[(alldata.Fare==0) & (alldata.Pclass==1)]['Fare'].replace(0,
firstclassmean)
alldata.loc[(alldata.Fare_Per_Person==0) &
(alldata.Pclass==1)]['Fare_Per_Person'].replace(0, firstclassmeanpp)
alldata.loc[(alldata.Fare==0) & (alldata.Pclass==2)]['Fare'].replace(0, secclassmean)
alldata.loc[(alldata.Fare_Per_Person==0) &
(alldata.Pclass==2)]['Fare_Per_Person'].replace(0, secclassmeanpp)
alldata.loc[(alldata.Fare==0) & (alldata.Pclass==3)]['Fare'].replace(0,
firstclassmean)
alldata.loc[(alldata.Fare_Per_Person==0) &
(alldata.Pclass==3)]['Fare_Per_Person'].replace(0, firstclassmeanpp)
del firstclassmean ,firstclassmeanpp, secclassmean, secclassmeanpp, thirdclassmean,
thirdclassmeanpp
```

Now let's visually inspect the categorical features for outliers, using histograms. We will continue using the combined dataset since it will be easier to check for outliers across the whole dataset rather than checking each split separately.

```
categorical_cols = [col for col in alldata.columns if alldata[col].dtype == 'int64'
or alldata[col].dtype == 'object']
for col in categorical_cols:
    sns.countplot(alldata[col])
    #plt.bar([indx for indx, _ in enumerate(alldata[col])], height=alldata[col])
    plt.title("Bar Plot for " + col)
    plt.show()
```

There is one family with 11 members and a few passengers with the same name. We should inspect these rows to make sure they are not invalid data.

```
#Inspect the 11 passengers with the same ticket
print(alldata[alldata.Group_Size==11])

#Inspect passengers with same name
print(alldata.ix[alldata.groupby('Name')['Name'].transform('count')[alldata.groupby('
Name')['Name'].transform('count')>1].index])
```

Fortunately, there do not appear to be any outliers in the categorical features of this dataset, as the only items that waved red flags were found to be benign upon inspection.

The last step in the data cleaning and preparation process is to transform the data so that it can easily be passed into a model. Models do not like strings or characters, so the first thing we will do is convert them to something more usable. One-hot encoding or dummy encoding are the best options, but if there are character variables with hundreds or thousands of different values or categories, these options may be too demanding on computer memory. So to decide how to encode the character variables, we can first check to see how many distinct values they have.

```
object_cols = [col for col in alldata.columns if alldata[col].dtype == 'object']
for col in object_cols:
    print("Unique categories of " + col + ": " + str(len(alldata[col].unique())))
```

Only name and ticket have so many values that they could cause memory problems. So let us one-hot encode sex, embarked, and deck. We will not do anything to set because we are using it only to determine which dataset the observations came from.

```
#One-hot encode sex, embarked, and deck
#The commented out portion shows how to dummy encode rather than one-hot encode
sex_dummy =
pd.get_dummies(alldata['Sex']).add_prefix('Sex_')#.drop(alldata.sort_values('Sex')['S
ex'].unique()[0], axis=1)
embarked_dummy =
pd.get_dummies(alldata['Embarked']).add_prefix('Embarked_')#.drop(alldata.sort_values
('Embarked')['Embarked'].unique()[0], axis=1)
deck_dummy =
pd.get_dummies(alldata['Deck']).add_prefix('Deck')#.drop(alldata.sort_values('Deck')[
'Deck'].unique()[0], axis=1)
dummies = pd.concat([sex_dummy, embarked_dummy, deck_dummy], axis=1)
alldata = pd.concat([alldata, dummies], axis=1)
del sex_dummy, embarked_dummy, deck_dummy
```

The name field can be left alone for now, because we will return to it in a later chapter to build more interesting features. The ticket field can now be label encoded.

```
enc = LabelEncoder()
alldata['Ticket_Enc'] = enc.fit_transform(alldata['Ticket'])
```

Since the variables have been encoded, the dataset can be reduced to the encoded and numeric columns, plus name. Passenger ID will be discarded since it is never good to include an ID field in a model. Finally, the data can be split back to the training and test sets.

```
#Remove unneeded columns
alldata.drop(['PassengerId', 'Sex', 'Embarked', 'Deck', 'Ticket'], axis=1,
inplace=True)

#Split back to training and test sets and save as cleaned data
train_clean = alldata.loc[alldata.Set=='train']
train_clean.is_copy = False
train_clean.drop(['Set'], axis=1, inplace=True)
train_clean['Survived'] = train['Survived']
test_clean = alldata.loc[alldata.Set=='test']
test_clean.is_copy = False
test_clean.drop(['Set'], axis=1, inplace=True)
train_clean.to_csv('train_clean.csv', index=False)
test_clean.to_csv('test_clean.csv', index=False)
```

The cleaned datasets will be used in the coming chapters to build models.

5. Correlation

So far we have been leading up to describing the relationships between variables, and now that we have a way to clean and prepare data for analysis, we can finally start off with the simplest way to do it.

Covariance

One way to look at the relationship between variables is to see if they covary. Recall from chapter 2 that the variance of a variable is a measure of the average squared distance from the mean that each observation lies. If two variables covary, then their variances will move together across observations. This does not mean that the variances of the two variables will be the same, only that the variables will deviate from their means in the same way. So for example if the variance of one variable increases, so will the variance of the other. Covariance is a measure of how well variables covary.

The variance of a variable is the sum of squared deviations from the mean. Squaring the deviations gets rid of negative values while preserving their values relative to one another. Likewise, when we calculate covariance, we need a way to get rid of negatives and preserve relative values. This is done by multiplying the deviation of one variable by the corresponding deviation of the second; a procedure in mathematics known as calculating the cross product. So the first step to calculate covariance is to calculate the cross product deviation for each observation. This cross product is then averaged to get the covariance. Therefore, the covariance is the average of the product of deviations from the means of each variable.

$$cov(x, y) = \frac{\Sigma(x_i - \bar{x})(y_i - \bar{y})}{N - 1}$$

Figure 5.1

If the covariance is positive, then when one variable deviates from the mean, the other deviates in the same direction. If the covariance is negative, then the variables deviate in opposite directions. Covariance measure the direction of a relationship, but not the strength. The actual value of the covariance has no meaning, because it is almost always the case that the variables being compared have different units of measure. In order to compare variables on the same unit of measure, the covariance must be standardized in some way. That is the purpose of correlation.

Pearson's Correlation Coefficient = Standardized Covariance

Statistician Karl Pearson developed a way to standardize the covariance by dividing it by the cross product of the standard deviations of the variables (Pearson, 1895). This gives the covariance in terms of standard units.

$$r = \frac{cov(x, y)}{s_x s_y} = \frac{\Sigma(x_i - \bar{x})(y_i - \bar{y})}{(N - 1)s_x s_y}$$

Figure 5.2

In figure 5.2, s is the standard deviation of a variable, N is the number of observations, and x bar and y bar are the means of two variables. Pearson's correlation coefficient, often abbreviated to just "correlation coefficient", is always a value between -1 and 1.

- If the correlation coefficient is closer to -1, then as one variable deviates in one direction, the other deviates in the **opposite** direction.
- If the correlation coefficient is closer to 1, then as one variable deviates in one direction, the other deviates in the **same** direction.
- The strength of the correlation is greater towards either extreme (-1 or 1) and weaker toward 0.
  - Generally, 0.1 is weak, 0.3 is medium strength, and 0.5 and higher is strong (Pearson, 1895).

The correlation coefficient is a test statistic for the hypothesis test whose null hypothesis is that there is no correlation between the variables (correlation = 0). Although it can be interpreted directly, it is also possible to calculate the significance by looking at the Z-score or t-statistic. The correlation coefficient is not normally distributed, but it can be manipulated to have a normal sampling distribution by calculating the Z-score through the Fisher transformation formula (Glen, 2018):

$$z_r = \frac{1}{2} \ln \left(\frac{1+r}{1-r} \right)$$

Figure 5.3

In figure 5.3, r is the correlation coefficient. The resulting Z-score has a standard error of:

$$SE_z = \frac{1}{\sqrt{N-3}}$$

Figure 5.4

So the test statistic is calculated by:

$$z_{test} = \frac{z_r}{SE_z} = \frac{\frac{1}{2} \ln \left(\frac{1+r}{1-r} \right)}{\frac{1}{\sqrt{N-3}}}$$

This number can be compared to the critical values in a one-tailed normal distribution table, or it can be multiplied by 2 and compared to the critical values in a two-tailed distribution table. The Z-score test will show whether or not a correlation is statistically significant, but the t-statistic for the correlation coefficient is usually the test statistic that is reported as the default for most statistical packages.

$$t_r = \frac{r\sqrt{N-2}}{\sqrt{1-r^2}}$$

Figure 5.5

In figure 5.5, r is the correlation coefficient and N is the number of observations. The t-statistic for the correlation coefficient is calculated with N-2 degrees of freedom.

For parametric tests, Pearson's correlation coefficient requires interval data (or 1 interval and 1 categorical with only 2 categories, which is akin to a t-test as we will see later). It also requires that the variables being compared be normally distributed. If they are not, then a different correlation coefficient, like Spearman's rho or Kendall's tau is needed. These three tests are non-parametric, and we will examine them later in the chapter.

It is important to note that Pearson's correlation coefficient gives the likelihood of 2 variables trending together, and the strength of that relationship. It does not indicate that one variable causes the other to trend the same way. In fact, there is equal likelihood of each variable influencing the other, and without additional information it is impossible to know if a third variable is the one that is truly responsible. Correlation does not imply causation.

Squaring Pearson's correlation coefficient gives **R^2**, otherwise known as the **coefficient of determination**. The coefficient of determination is a measure of how much variability is shared between 2 variables. If it is multiplied by 100 to transform it into a percentage, it will show the percentage of shared variance. People who are unfamiliar with statistics often see the R^2 value and say "this percentage of the variance in y is explained by x". That statement is not accurate however. The coefficient of determination implies nothing about causality, because once again, correlation does not imply causation. As we will see in a later chapter, since R^2 measures shared variance, it shows how well the shared variance between variables can be fit by a line, and therefore, it is often used as a measure of goodness of fit for linear regression.

Spearman's Rank Correlation Coefficient

There are non-parametric correlation coefficients, and Spearman's rho is one of them. Non-parametric correlation coefficients like Spearman's rho are useful when the data violate any of the standard parametric assumptions (Field, Miles, & Field, 2012). Spearman's rho is calculated by ranking the values of the data and applying Pearson's equation to the ranked data. This means that while Pearson's correlation requires interval or ratio data, Spearman's rho can show correlations between interval, ratio, or even ordinal data. Spearman's rho works best when there are no ties in the rank of either variable (when there is only 1 observation for each unique value in each variable).

$$p = 1 - \frac{6 \sum d_i^2}{N(N^2 - 1)}$$

Figure 5.6

In figure 5.6, N is the number of observations, and d is the difference between the rank of 1 observation of 1 variable to the corresponding rank of the value of the second variable. Here is an example of how it is calculated:

Age in Years	Weight in Pounds
38	169
16	153
24	162
30	159

1. Sort the data by the first variable, age, in ascending order and then give each row a rank:

Age in Years	Weight in Pounds	Rank of Age
16	153	1

24	162	2
30	159	3
38	169	4

2. Sort the data by the second variable, weight, in ascending order and then give each row a new rank:

Age in Years	Weight in Pounds	Rank of Age	Rank of Weight
16	153	1	1
30	159	3	2
24	162	2	3
38	169	4	4

3. Calculate the difference between the ranks. This will become d sub I in the equation in figure 5.6.

Age in Years	Weight in Pounds	Rank of Age	Rank of Weight	Rank of Age – Rank of Weight
38	169	4	4	0
16	153	1	1	0
24	162	2	3	-1
30	159	3	2	1

The sum of the squared differences in ranks is 2, so referring back to the equation in figure 5.6, Spearman's rho = 1-((6*2)/4*(4^2-1)) = 1-(12/60) = 4/5 = 0.8. Like Pearson's coefficient, this indicates a strong positive correlation.

Kendall's Tau Correlation Coefficient

Kendall's tau is another non-parametric test to be used when the data violate the standard assumptions. It is similar to Spearman's in that it works well with ordinal data, as well as interval and ratio. There is evidence to suggest that Kendall's tau is actually more accurate than Spearman's coefficient (Field, Miles, & Field, 2012). Kendall's tau is preferred over Spearman's when there are ties in the ranking of variables, or many observations with the same rank. For example, when the data is ranked and two values are the same, they will have the same rank. In this case it would be better to use Kendall's Tau than Spearman's rho.

$$\tau = \frac{(number\ of\ concordant\ pairs - number\ of\ discordant\ pairs)}{\left(\frac{N(N-1)}{2} - \sum_i t_i(t_i - 1)/2\right)\left(\frac{N(N-1)}{2} - \sum_j u_j(u_j - 1)/2\right)}$$

Figure 5.7

In the equation in figure 5.7, N is the number of observations, t sub I is the number of tied values in the ith group of ties for the first variable, and u sub j is the number of tied values in the jth group of ties for the second variable. A concordant pair is a pair of variables whose ranks agree. This means that if the first variable is in ascending order, then the count of all larger ranks in the rows below the current row are concordant. For example, if row 1 were being evaluated in the upper table below, there are 3 ranks of weight that are larger than 1 in the rows below the first row, so the number of concordant pairs is 3.

Age in Years	Weight in Pounds	Rank of Age	Rank of Weight	Concordant Pairs	Discordant Pairs
16	153	1	1	3	0
24	162	2	3		
30	159	3	2		
38	169	4	4		

Now consider the second row in the table below. There is only 1 larger rank than 3 in the rows below row 2, so there is only 1 concordant pair. The other pair is discordant.

Age in Years	Weight in Pounds	Rank of Age	Rank of Weight	Concordant Pairs	Discordant Pairs
16	153	1	1	3	0
24	162	3	3	1	1
30	159	2	2		
38	169	4	4		

The final row will not have any concordant or discordant pairs, so it is always left blank. Kendall's Tau is a fairly complex procedure, but fortunately there are libraries in R and Python that can calculate it automatically, as we will soon see.

Point-Biserial and Biserial Correlations

So far we have looked at correlations between numeric variables; continuous variables measured at the interval or ratio level, and ordinal variables (ordered categorical variables). What about binary categorical variables? Binary variables are very common in data science. Point-biserial correlation measures the correlation between a continuous variable and a binary one.

$$r_{pq} = \frac{M_1 - M_0}{s}\sqrt{pq}$$

Figure 5.8

In the equation in figure 5.8, M sub 1 is the mean of the group for which the binary variable had a value of 1, M sub 0 is the mean of the group for which the binary variable had a value of 0, s is the standard deviation for the entire dataset, p is the ratio of observations in the 0 group to total observations, and q is the ratio of observations in the 1 group to total observations.

Point-biserial correlation assumes the variables are independent. This means that point-biserial correlation cannot be used to measure the effect of a binary predictor or independent variable on a continuous outcome or dependent variable, or vice versa. As we will see in the next chapter, linear regression would be the appropriate method to do that analysis.

Sometimes a binary variable has an underlying continuity. For example, pass/fail on an exam is based on an underlying score. If this is the case, and the goal is to compare such a binary variable to a continuous variable, then biserial correlation should be used instead of point-biserial.

$$r_{biserial} = \frac{(Y_1 - Y_0) * \left(\frac{pq}{Y}\right)}{\sigma_y}$$

Figure 5.9

In the equation in figure 5.9, Y sub 0 is the mean score for data pairs for the group for which the binary variable is 0, Y sub 1 is the mean score for data pairs for the group for which the binary variable is 1, sigma is the population standard deviation for the entire dataset, p is the ratio of observations in the 0 group to total observations, and q is the ratio of observations in the 1 group to total observations.

Partial Correlation

So far we have been ignoring the elephant in the room: what happens if multiple variables are correlated with one another? If x, y, and z are all correlated to one another individually, how does the correlation between y and z effect the correlation between x and y, for example? Imagine 3 variables represented by circles, as in figure 5.10. The circles overlap where the variance is shared between the variables (that's the definition of correlation). It's possible for all 3 circles to overlap in the same spot. Therefore, if we want to describe the relationship between x and y in terms of correlation (shared variance), we would have to exclude the variance that they both share with z (the white space in the middle).

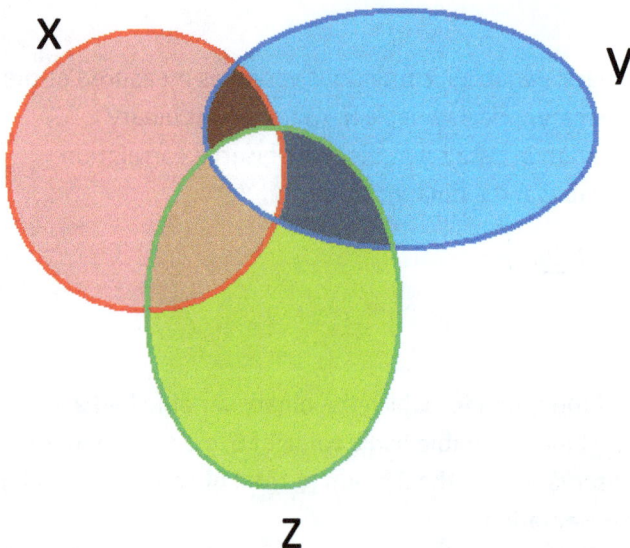

Figure 5.10

Partial correlation is the relation between 2 variables while all other variables are controlled. It is used to exclusively look at the relationship between 2 variables when all of the variables are correlated with one another.

Typically in statistics we wish to examine the relationship between an outcome variable and one or many predictors. Using the concept of correlation, we can show how much of the outcome's variance is shared with any one of the predictors. We have just seen with partial correlation that in order to accurately describe shared variance between 2 variables, the variance shared with all other variables must be controlled for. One way to control for the effects of other variables is to set them equal to 0. Doing this allows us to look at the effect of 1 variable on another when all other variables are ruled out. This concept is called **semi-partial correlation** and it is the basis for linear regression. Semi-partial correlation measures the relationship between variables when the effects of all other variables on only 1 of the variables in the relationship are controlled for. So if we look to figure 5.10, we have x, y, and z, and if we want to know the effect of x on z, we would control for the effect of y on z, but not x on y. Figure 5.11 shows the semi-partial correlation just described. Essentially we would be building an equation: $z = \alpha + \beta x + \beta y + \varepsilon$ as we will see in the following chapter. In summation, semi-partial correlation is useful for explaining the variance in an outcome variable based on a set of predictors.

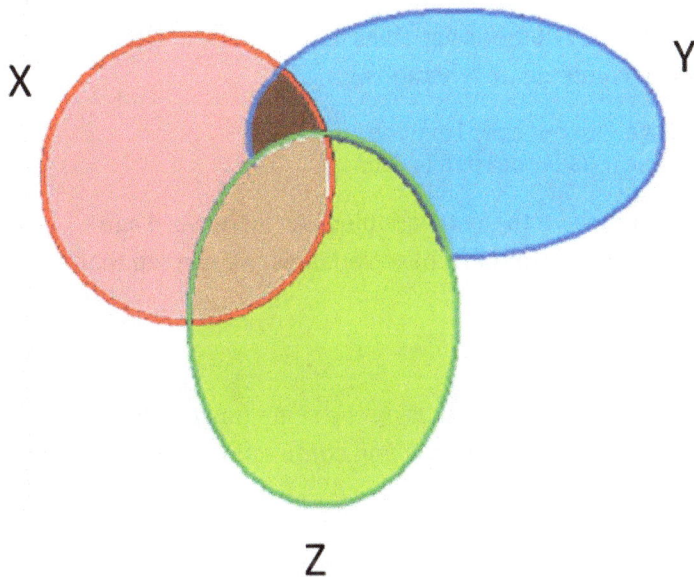

Figure 5.11

Comparing Correlations

If two correlations are calculated independently, they cannot be directly compared because they could be measuring shared variance on completely different scales. The Z-score can once again create a standardized scale that enables the comparison of independent correlations or correlations between independent groups. Refer back to figure 5.3 for the Fisher transformation Z-score formula. Once two correlations have been converted to Z-scores, they can be entered into the formula below to come up with a test statistic to look up in the normal distribution tables for a 1 or 2 tailed test.

$$z_{Difference} = \frac{z_{r_1} - z_{r_2}}{\sqrt{\frac{1}{N_1 - 3} + \frac{1}{N_2 - 3}}}$$

Figure 5.12

If the test statistic is greater than the critical value for the type of test desired, then the null hypothesis that the correlations are the same can be rejected.

Correlations with R

Correlations are commonly calculated during the data exploration phase of a data science project, usually right after the data has been cleaned. Since we have a nice clean dataset from the work in chapter 4, we can use it to calculate correlations.

```
train <- read.csv("train_clean.csv", header=T)
test <- read.csv("test_clean.csv", header=T)
train$Set <- 'train'
test$Set <- 'test'
alldata <- rbind(train[,-which(colnames(train) %in% 'Survived')], test)
```

Let us start by looking at the covariance and Pearson correlation between fare per person and age. This is not a very useful comparison, but these are the only numeric variables in this dataset. Note that age has missing values, so we must tell R to only use complete observations.

```
cov(alldata$Fare_Per_Person, alldata$Age, use='complete.obs')
cor(alldata$Fare_Per_Person, alldata$Age, use='complete.obs')
```

There is a very small positive linear relationship between the two variables. An increase in age correlates with an increase in fare per person. If we scatter plot these variables, we can see that a very small relationship exists.

```
plot(alldata$Fare_Per_Person, alldata$Age)
```

Spearman's rank and Kendall's Tau correlations can be computed using the same function in R. The default is Pearson's coefficient, but the method argument of the cov and cor functions can also be set to "kendall" or "spearman".

As an example of point biserial correlation, we can look at the correlation between age and deck A. It might be tempting to try to look at the correlation between age and survival, or fare and deck A, but remember that point biserial correlation assumes that the variables are independent. The result of a correlation between dependent variables would be spurious.

```
library(ltm)
biserial.cor(train$Age, train$DeckA, use='complete.obs')
```

Looking at this relationship in the training set, we can see that there is a small negative correlation between age and deck A. Since it is a negative correlation, as age decreases, deck A is more likely to be 1. The opposite can also be said: moving from deck A = 0 to deck A = 1 slightly correlates with a decrease in age.

Correlations with Python

Correlations are commonly calculated during the data exploration phase of a data science project, usually right after the data has been cleaned. Since we have a nice clean dataset from the work in chapter 4, we can use it to calculate correlations.

```
import pandas as pd
```

```
import matplotlib.pyplot as plt
import numpy as np
import scipy.stats as stats

train = pd.read_csv("train_clean.csv", sep=",")
test = pd.read_csv("test_clean.csv", sep=",")
train['Set'] = 'train'
test['Set'] = 'test'
alldata = pd.concat([train.drop(['Survived'], axis=1), test], ignore_index=True)
```

Let us start by looking at the covariance and Pearson correlation between fare per person and age. This is not a very useful comparison, but these are the only numeric variables in this dataset. Note that age has missing values, but Python operates on pairwise complete observations by default.

```
print(alldata.cov()['Fare_Per_Person'])
print(alldata.corr(method='pearson')['Fare_Per_Person'])
```

There is a very small positive linear relationship between the two variables. An increase in age correlates with an increase in fare per person. If we scatter plot these variables, we can see that a very small relationship exists.

```
plt.scatter(alldata['Fare_Per_Person'], alldata['Age'])
plt.show()
```

Spearman's rank and Kendall's Tau correlations can be computed using the same function in Python. The default is Pearson's coefficient, but the method argument of the cov and corr functions can also be set to "kendall" or "spearman".

As an example of point biserial correlation, we can look at the correlation between age and deck A. It might be tempting to try to look at the correlation between age and survival, or fare and deck A, but remember that point biserial correlation assumes that the variables are independent. The result of a correlation between dependent variables would be spurious.

```
print(stats.pointbiserialr(np.asarray(alldata[~np.isnan(alldata['Age'])]['DeckA']),
np.asarray(alldata[~np.isnan(alldata['Age'])]['Age'])))
```

Looking at this relationship in the training set, we can see that there is a small negative correlation between age and deck A. Since it is a negative correlation, as age decreases, deck A is more likely to be 1. The opposite can also be said: moving from deck A = 0 to deck A = 1 slightly correlates with a decrease in age.

6. Linear Regression

Recall from chapter 5 how semi-partial correlation allows the variance of an outcome variable to be explained by a set of predictors. The concept is that all predictors except for one will be controlled for so that the effect of the one can be measured independently. In mathematical terms, regression can be thought of as fitting the line of best fit to data, in order to make predictions about future observations. Simple regression is when there is only 1 predictor variable. Multiple regression is when there are more than 1 predictor variables. The equation in figure 6.1a shows simple regression. Figure 6.1b shows a multiple regression equation with many predictor variables. Notice how these equations are in slope-intercept form, where alpha is the intercept, and beta is the slope:

$$y = \alpha + \beta x + \varepsilon$$

Figure 6.1a

$$y_i = \alpha + \beta_1 x_1 + \beta_2 x_2 + \cdots + \beta_n x_n + \varepsilon_i$$

Figure 6.1b

The epsilon in figure 6.1a and 6.1b is the error term. Epsilon represents any unobserved effects that are not explained by the model (regression equations are often called models). In mathematical terms, epsilon is the difference between what is predicted and what actually happens (the difference between the true value of an observation and a point on the regression line).

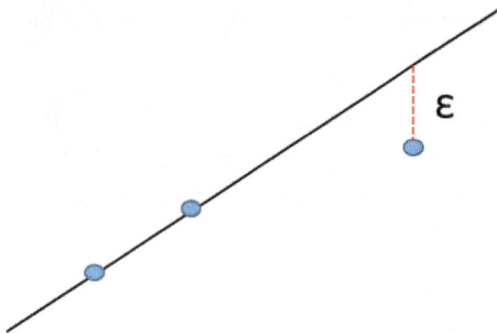

Figure 6.2

We could simply draw any line and call it a regression line, but a random line probably would not be accurate. An optimization method of fitting the best line to data is called **ordinary least squares (OLS)**. The least squares method fits a line to data by finding the line that results in the least amount of difference between the data points and the line. To do this, the vertical distance from each point to the line is squared, and then totaled. The result is called the **sum of squared residuals or sum of squared errors, SSR or SSE** (residuals are the differences between the point and the line; residual = error). The line with the lowest sum of squared residuals is the line of best fit. In figure 6.2, there is one point not on the regression line. If the line were fitted using OLS, the SSR would be the squared vertical distance of that point to the regression line. If there were multiple points that were not on the regression line, the squared vertical distances between each point and the line would be totaled. That would be the SSR. The SSR or SSE is the measure of variation that is not explained by the model.

Goodness of Fit: R2 and the F-test

The SSR is what is referred to as an **objective function or cost function** in data science. Every model seeks to optimize (often minimize) an objective function. The model that optimizes the objective function is the best fit to the data (James, Witten, Hastie, & Tibshirani, 2013). The method used for optimization is hugely important for the efficacy and interpretation of the model, so we will look at optimization in more detail in a later chapter.

In order to measure a model's fit, there needs to be some baseline measure to compare against. Recall from chapter 3 that the mean is the simplest model for describing data. Rather than using a line of best fit to describe data, we could just use the mean. Figure 6.3 shows how the mean would fit a dataset.

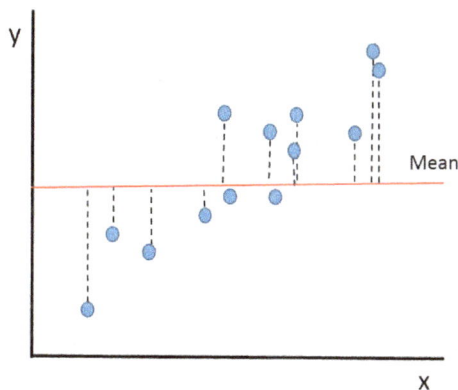

Figure 6.3

It is clear in figure 6.3 that the mean is not a very good model. Most of the time it will not be, but we can use it as a baseline to compare the line of best fit to. Just as we used the lowest SSR to determine the line of best fit, we can use the same technique to measure the fit of the mean. We cannot call the measure that we seek to optimize the SSR though, because we are already using that to measure the fit of the regression line. Instead, we can call it the **total sum of squares (SST)**. The sum the squared differences between each data point and the mean is therefore the total sum of squares (SST).

Given that the SSR and SST are the error measurements of different models, we could compare them directly. If the SSR < SST, then the regression line is a better fit than the mean. Dividing the SSR by the SST gives a measure of how well a model fits the data compared to the mean. If this ratio is subtracted from 1, it gives a measure called the **coefficient of determination**, often abbreviated as R^2. Some statistics texts refer to the R^2 as the multiple R^2. R^2 provides a good measure of the size of the relationship between the outcome and predictor variables (Field, Miles, & Field, 2012).

Another way of looking at a model's fit is to subtract the SSR from the SST. This produces the model sum of squares (SSM), otherwise known as the explained sum of squares (ESS). The SSM or ESS is a measure of explained variation (variation that is explained by the model). A large SSM implies that the model is a good fit. If the model sum of squares is divided by the total sum of squares, the result is the same as subtracting the ratio of the SSR to the SST from 1. Therefore, the SSM/SST also equals the coefficient of determination.

$$R^2 = \frac{(SST - SSR)}{SST} = \frac{SSM}{SST} = 1 - \frac{SSR}{SST}$$

Figure 6.4

As previously alluded to, the R^2 is the proportion of improvement that the model provides over the mean. R^2 is the amount of variance in the outcome explained by the model relative to how much variation there was to explain in the first place. $R^2 * 100$ is the percentage of variance in the outcome that is explained by the model. The square root of R^2 is Pearson's r, which as we saw in chapter 5 is the proportion of shared variance. Pearson's r provides a good estimate of the overall fit, and R^2 provides a good measure of the size of the relationship between the outcome and the predictors (Field, Miles, & Field, 2012). Since the square root of R2 is Pearson's r, for simple regressions with only 1 predictor variable, the square root of R2 is the same correlation coefficient that would result from calculating the Pearson correlation between the variables.

Chapter 3 explained how the goal of statistical modeling is to find the model that best describes the relationships between variables. Measures of accuracy, like confidence intervals and p-values, show the likelihood of a model being the true estimate of the population parameters, while measures like goodness of fit show how accurately the model reflects the sample data. Both accuracy and goodness of fit are required to adequately evaluate a statistical model. We have just seen how R2 can be used as a measure of goodness of fit. Now we need a test statistic to show the likelihood of there truly being a relationship between the outcome and predictor variables.

Recall that test statistics show the amount of variance explained by a model divided by the amount of variance not explained by it. The **F-test** is accomplishes this for linear regression (recall from chapter 3 that the F-test is also called the variance ratio test because it compares explained variance to unexplained variance). F is based on the ratio of the improvement due to the model (SSM) and the error in the model (SSR). To calculate F, the SSM is divided by the number of variables in the model minus one (the degrees of freedom). This gives the mean of the model sum of squares. The mean of the model sum of squares is then divided by the mean squared error (the SSR divided by the number of observations minus the number of variables).

$$F = \frac{\dfrac{SSM}{(k-1)}}{\dfrac{SSR}{(N-k)}}$$

Figure 6.5

In figure 6.5, k is the number of predictor variables and N is the number of observations. Recall that the SSM is equivalent to (SST − SSR). If a linear model is good, then F will be large (greater than 1). The F-test tables can assess whether F is statistically significant. If it is, then the null hypothesis, that there is no relationship between the outcome variable and the predictors, can be rejected. The strength of that relationship is summarized by R^2.

Assessing Individual Predictors: t-test and p-values
So far we have only looked at the overall fit of a linear regression model. Now we will examine how the predictor variables can be assessed so that their individual impact on the outcome variable can be determined.

If the value of a predictor variable changes, but the value of the outcome does not change, then that predictor variable will have a regression coefficient of 0. So if a variable is a good predictor of the outcome variable, then its coefficient should be large and significant. A **t-test** is used to determine variable significance: it tests the null hypothesis that the coefficient of the variable is 0.

The t-statistic is the ratio of the beta coefficient of a variable to the standard error of the beta estimate (recall from equation 6.1 that beta is the coefficient of a variable, or the slope in slope-intercept form). Standard error is used in the denominator of the t-statistic formula because it measures the variance across many samples in a sampling distribution. It therefore measures the similarity of beta values across samples. If the standard error in beta is small, then most samples of beta will be the same as the one that was estimated.

$$t = \frac{b_{observed} - b_{expected}}{SE_b} = \frac{b_{observed}}{SE_b}$$

Figure 6.6

Since the null hypothesis is that beta (b) is 0, the expected b is 0, so the equation can be simplified to the form on the far right of figure 6.6. The degrees of freedom are N-p-1, or the sample size minus the number of predictor variables minus 1. If the t-statistic is large, then the predictor is likely to make a significant contribution to the outcome variable, but it should be compared to the t-test tables to determine whether it is statistically significant.

In addition to the t-test, the **p-value** can be used to determine whether a predictor impacts the outcome variable in a statistically significant way. If the p-value is < 0.05, then the predictor is significant at the 95% confidence level.

Interaction Terms

Predictors may not always have singular effects on the outcome variable. It is possible for two or more predictors in a multiple regression to not affect the outcome individually, but still have a combined effect on the outcome. For example, even if caffeine and sunlight do not affect body temperature individually, their combined interaction could. When the interaction is thought to have an effect on the outcome, an interaction term should be included in the regression model. An **interaction term** is the product of the interacting predictors. It has its own coefficient.

Interaction terms should be included in a regression model when there is reason to think that the effect of a predictor on the outcome is different for different values of another predictor. For example, the effect that physical activity has on metabolic rate varies by gender. Therefore, an interaction term between physical activity and gender would test for a difference in the effect of physical activity on metabolic rate for different genders.

Interaction terms are commonly used to compare the effect of a continuous predictor on the outcome across different groups of a categorical variable (Hastie, Tibshirani, & Friedman, 2009). This type of analysis (ANOVA and ANCOVA) will be examined in a later chapter. It is important to note that when interaction terms are used, the interpretation of the coefficients changes. The unique effect of a predictor that is included in an interaction term is the sum of the coefficient for the predictor and the coefficient of the interaction term. The coefficient of the predictor, by itself, only represents the effect of the predictor in the absence of any interaction.

Preventing Overfitting: AIC, BIC, and Adjusted R2 Measures of Fit

In simple regression, the equation being calculated represents a 2-dimensional line. In multiple regression with two predictors, the equation being calculated represents a 3-dimensional plane. In multiple regression with three or more predictors, the equation being calculated represents a complex high-dimensional shape. None of this changes how regression is carried out, however it has implications for the R^2 measure of goodness of fit. Specifically, as the number of predictors increase, R^2 also increases (Hastie, Tibshirani, & Friedman, 2009). Refer back to figure 6.4 to see why. As the number of predictors increase, the SSM will inevitably increase as well. Therefore, models with many predictors appear to fit the data extremely well. This problem is called **overfitting**.

Overfitting occurs when a model is fit so tightly to a particular set of training data that it cannot generalize to new, unseen data (Hastie, Tibshirani, & Friedman, 2009). It can also result in a model that describes random noise rather than a true underlying relationship between variables. A model that cannot generalize and fits better to noise than actual signal is useless. So there need to be methods of combatting the overfitting problem. The methods described in this section are alternative measures of goodness of fit that reduce the likelihood of overfitting. In general, the process of penalizing models to reduce overfitting is called **regularization**. Figure 6.7 was created by Wikipedia user Ghiles (https://en.wikipedia.org/wiki/File:Overfitted_Data.png) and has a CC BY-SA 4.0 license (https://creativecommons.org/licenses/by-sa/4.0/deed.en). The figure, which has not been altered from its original form, shows the difference between a model that overfits (blue) and one that does not (black). The blue line fits the data perfectly but would not be able to generalize to new data, whereas the black line is not a perfect fit but would be better able to describe new data.

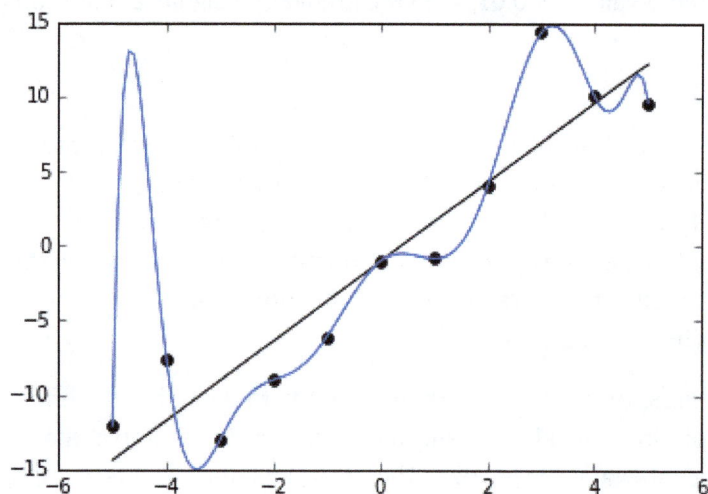

Figure 6.7

The **Akaike Information Criterion (AIC)** is a measure that rewards high goodness of fit and penalizes models that have more predictor variables (James, Witten, Hastie, & Tibshirani, 2013). The AIC is different than the measures of fit we have seen so far, as it does not indicate how well a model fits the data, but rather how much information is lost when a model is used to estimate (re-generate) the data. The concept is based in information theory, which is beyond the scope of this book, but we can still make use of it without understanding the inner workings.

$$AIC - N\ln\left(\frac{SSR}{N}\right) + 2k$$

Figure 6.8

In the equation in figure 6.8, N is the number of observations or sample size, SSR is the sum of squared residuals (the variance that is not explained by the model), and k is the number of predictor variables. Sometimes SSR/N is replaced by **mean squared error (MSE)**, because the sum of squared residuals (a.k.a. sum of squared errors) divided by the number of observations is the mean squared error. The MSE can be used by itself as a measure of fit for a model, as can its square root, the **root mean squared error (RMSE)**. The RMSE is often preferable to the MSE because the MSE is a squared measure, and the RMSE measures the error in the same units as the dependent variable, making it easy to interpret (James, Witten, Hastie, & Tibshirani, 2013). Notice that as the number of predictors, k, increases, so does the AIC. A large AIC means the model is a poor fit. There are no guidelines for how small the AIC should be, because it is different for every model. Tuning the AIC requires trial and error of fitting the regression with different combinations of predictors to see how the AIC changes. The best model would be the one with the best fit and lowest AIC.

The **Bayesian Information Criterion (BIC)** is an alternative to the AIC that also rewards high goodness of fit and penalizes models that have more predictor variables (James, Witten, Hastie, & Tibshirani, 2013). The BIC penalizes additional predictors more heavily than the AIC.

$$BIC = N\ln\left(\frac{SSR}{N}\right) + k\ln(N)$$

Figure 6.9

In the equation in figure 6.9, N is the number of observations or sample size, SSR is the sum of squared residuals (the variance that is not explained by the model), and k is the number of predictor variables. Sometimes SSR/N is replaced by mean squared error (MSE), because the sum of squared residuals (a.k.a. sum of squared errors) divided by the number of observations is the mean squared error. Notice that as the number of predictors, k, increases, so does the BIC. A large BIC means the model is a poor fit. There are no guidelines for how small the BIC should be, because it is different for every model. Tuning the BIC requires trial and error of fitting the regression with different combinations of predictors to see how the BIC changes. The best model would be the one with the best fit and lowest BIC.

It has been suggested in studies comparing the AIC and BIC that, for the purposes of regression, the AIC may be preferable to the BIC, because it is asymptotically optimal (the algorithm cannot be out-performed by more than a constant factor for large enough inputs).

The adjusted R^2 is another way to select one model out of many choices with different numbers of predictors. The larger the adjusted R^2, the better the model (it has smaller test error). The intuition is that once all "correct" variables have been included in the model, adding additional "noise" variables will slightly decrease the SSR and increase the number of predictors, k. This will decrease the adjusted R^2.

$$Adjusted\ R^2 = 1 - \frac{\frac{SSR}{(N-k-1)}}{\frac{SST}{(N-1)}}$$

Figure 6.10

In the equation in figure 6.10, N is the number of observations or samples, k is the number of predictors, the SSR is the sum of squared residuals, and the SST is the total sum of squares. Like the equation for the standard R^2, this equation compares the amount of variance in the outcome explained by the model to how much variation there was to explain in the first place. It measures the size of the relationship between the outcome and its predictors, but in this case adjusted for the number of predictors.

Clearing Up Confusion with the SSM, SSR, and SST, and Other Measures of Fit

The mislabeling of the sum of squared residuals, total sum of squares, and model sum of squares by statisticians is perhaps the greatest source of confusion for students studying regression (James, Witten, Hastie, & Tibshirani, 2013). To clear up any confusion, this entire section is dedicated to defining them again, and explaining how they lead to two other useful measures of fit: the standard error of the regression and the mean squared error.

Measurement Name	Model Sum of Squares	Sum of Squared Residuals	Total Sum of Squares
Alternate Measurement Name	Explained Sum of Squares	Sum of Squared Errors	
Abbreviation	SSM	SSR	SST
Alternate Abbreviation	ESS	SSE	
Definition	The variation in the data that IS explained by the model.	The variation in the data that is NOT explained by the model.	The variation in the data that is NOT explained by the mean (the baseline model).
Concept Graph			

Figure 6.11

$$SSM = ESS = SST - SSR = SST - SSE$$

Figure 6.12

SSR and SST are sometimes reversed so that they read RSS and TSS, which is ok. However, sometimes the SSE or ESS are reversed, which causes great confusion because they are two different measurements. The sum of squared errors (a.k.a. SSE or SSR) should never be labeled ESS, and the explained sum of squares (a.k.a. ESS or SSM) should never be labeled SSE. To make matters worse, sometimes the model sum of squares (a.k.a. SSM or ESS) is also labeled SSR, which makes it appear to be referring to the sum of squared residuals. This is bad practice and should be avoided.

The unexplained variation (SSR) can be used to find the **standard error of the regression, S**, as shown in figure 6.13. The standard error of the regression measures the regression's variability. It is the average standardized distance from the regression line that each observation lies.

$$S = \sqrt{\frac{SSR}{N - k}}$$

Figure 6.13

In the equation in figure 6.13, N is the number of observations and k is the number of predictors. S can be used as an alternative measure of goodness of fit for the regression model. If S is small, the model fits well.

Another measure of fit is the **mean squared error (MSE)**. The mean squared error is defined as the sum of squared residuals (unexplained variance) divided by the number of observations:

$$MSE = \frac{SSR}{N}$$

Figure 6.14

In the equation in figure 6.14, N is the number of observations. The smaller the MSE, the better the model. The square root of the MSE is called the **root mean squared error (RMSE)**. The RMSE is useful because it is in the same units as the outcome (dependent) variable. The MSE and RMSE are two more possible cost functions to use to determine the model of best fit. An important point to remember when using the MSE or RMSE as a measure of fit is that they are not standardized, so models trained on different datasets cannot be compared using these measurements. It is also hard to determine what a "good" measurement should be. One way of gauging how good the RMSE is, is to look at the unit of measure and range of the outcome variable. If the outcome variable has a range of 0-100 and the RMSE is 40, then the model probably is not very good. However, if the outcome variable has a range of 0-10,000, then a model with a RMSE of 40 might be pretty good. It is harder to determine what a good MSE should look like, so the RMSE is used more often (Field, Miles, & Field, 2012).

Another measure of fit that is similar to MSE and RMSE is the **mean absolute error (MAE)**. The MAE is defined as the mean of the absolute values of the errors.

$$MAE = \frac{\sum_{i=1}^{N}|y_i - \hat{y}_i|}{N}$$

Figure 6.15

In the equation in figure 6.15, N is the number of observations in the sample, y sub i is the actual value of the dependent variable for the ith observation, and y hat sub i is the predicted value of the dependent variable for the ith observation. Since the MAE does not square the errors like the MSE or RMSE, it gives less weight to predictions that are far off the mark. For this reason, the MAE may be advantageous to use if outliers are expected in a new dataset. Linear models never make predictions that are outliers, so penalizing a model because a new dataset has outliers that were not predicted may not always be prudent. So a rule of thumb is to use MAE if outliers are unimportant and MSE or RMSE if outliers are important (Hastie, Tibshirani, & Friedman, 2009). The MAE can be used to form boundaries for the RMSE. For example, the RMSE will always be greater than or equal to the MAE and always less than or equal to the MAE*sqrt(number of samples).

$$MAE \leq RMSE \leq MAE * \sqrt{N}$$

Figure 6.16

The equation in figure 6.16 shows that as the samples size increases, the RMSE tends to be increasingly larger than the MAE, which can be problematic when a model is evaluated on samples of different sizes. So another circumstance when the MAE is preferable to the MSE or RMSE is when the error of a model needs to be compared to the error for different sized samples.

Bias-Variance Tradeoff

So far we have seen that linear regression outputs several statistics and measurements. For example, the F-statistic tests the hypothesis that there is a statistically significant relationship between the dependent and independent variables. The R^2 and other goodness of fit measurements show how well the model fits the data and the strength of the relationship between the dependent variable and the independent variables. Finally, we have considered OLS as an optimization method for minimizing the cost function and finding the model that fits the data the best, but does not overfit. Thing brings us to the bias-variance tradeoff.

- **Bias** refers to the error introduced by making erroneous assumptions during the approximation of a relationship between variables, due to simplification. Models inherently simplify complex relationships between variables, but if that simplification causes the model to learn false heuristics or generalizations about the relationship, then it is said to be biased. Bias is essentially the error between the model and the true relationship; it measures the model's accuracy. A low bias model will fit a dataset extremely well, whereas a high bias model does not fit well and is said to be a poor learner. If the model fits so well than it cannot generalize to new data, it is considered to be overfitted. Overfitted models have very low bias but high variance.

- **Variance** refers to a model's error due to fluctuations in the data; it measures the model's precision. When a model is trained on a dataset, the model attempts to explain the most variation in that data that is possible. A model that perfectly fits the data would have a very high variance, because any changes in the data would greatly impact the model's accuracy. Such a model would be imprecise and be said to be a weak learner. A good model fits well enough to adequately explain much of the variance in the data, but not so much that it cannot generalize to new data. A good model should have consistent accuracy across training datasets.

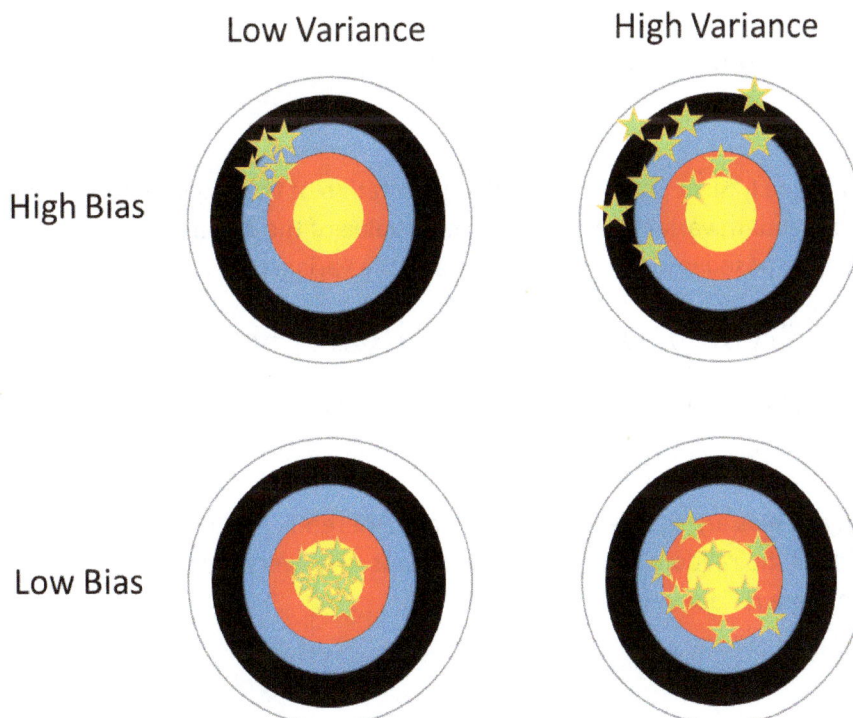

Figure 6.17

There is somewhat of a tradeoff between bias and variance, although it is not a complete inverse relationship. It is more of a slight backwards J shape (see figure 6.18). As a model is fit, the bias initially decreases faster than variance increases. Later on, variance catches up. The optimal model minimizes the error and balances the bias and variance as best possible (notice in figure 6.18 that minimizing the MSE does not quite match where the bias and variance intersect – that is intentional, as the smallest error might not be the perfect balance between bias and variance).

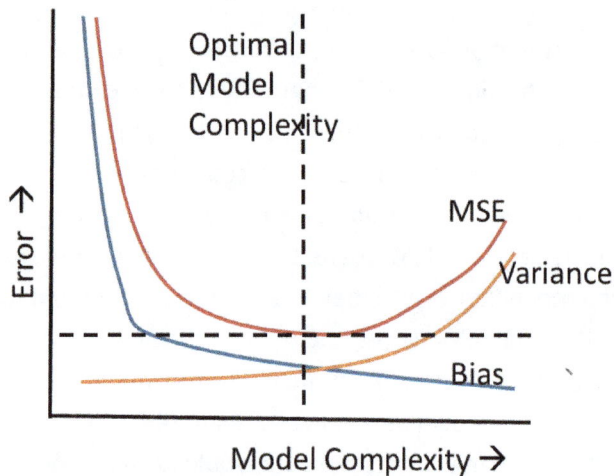

Figure 6.18

We have previously looked at the mean squared error (MSE) as a measure of fit for a regression model, but MSE can also be used to derive the bias variance tradeoff, if we think of the MSE as the expected squared error of a regression model:

$$MSE(x) = E\left[(Y - \hat{f}(x))^2\right]$$

Figure 6.19

The equation in figure 6.19 reads: the MSE at point x equals the expectation of the squared difference between the true value of the dependent variable and the predicted value of the dependent variable for point x. This equation can be decomposed into the squared bias and the variance, plus random error:

$$MSE(x) = \left(E[\hat{f}(x)] - f(x)\right)^2 + E\left[(\hat{f}(x) - E[\hat{f}(x)])^2\right] + \sigma_\epsilon^2$$

$$MSE(x) = bias^2 + variance + random\ error$$

Figure 6.20

Note that because the variance and bias are squared values, they are inherently non-negative, and can therefore never undo the effects of the random error term. This means that every model, regardless of how good it is, will always have some random error.

Linear Regression Assumptions

Good models must be able to generalize to new data. This means that they must not only fit the training data well, but when applied to a new dataset, they must also fit that dataset well. Models that cannot generalize are useless. In order for a linear regression model to generalize to a population, the following assumptions are made, in addition to the parametric assumptions listed in chapter 3:

1. All predictors must be either quantitative or categorical with no more than 2 categories, and the outcome variable must be quantitative (measured at the interval level), continuous, and unbounded. Unbounded means that there should be no restriction on the bounds. For example, if the dependent variable has a true range of 1-10, but the data collected in the sample only range from 3-7 due to some restriction of measurement, then the dependent variable is bounded and this assumption is not met.
2. The predictors should have a non-zero variance (they should not all be the same value).
3. The predictors should not be collinear (they should not correlate highly).
4. The predictors should not correlate with variables that are not included in the model.
5. The residuals of the predictors should have constant variance (recall from chapter 3 that this is called homoskedasticity). The parametric assumption in chapter 3 was that the predictors have constant variance though, whereas this assumption is that the errors have a constant variance.
6. The residuals in the model should be normally distributed.
7. The errors for any 2 observations should be independent or uncorrelated. The presence of correlation in the errors is referred to as **autocorrelation** (or serial correlation if the data is panel data or time series data). So there should be no autocorrelation.

If these assumptions and the parametric assumptions listed in chapter 3 are met, then the estimators (the coefficients and parameters) are unbiased. This doesn't always mean that the model will fit the population, but it increases the odds of it happening. The OLS method of finding the line of best fit is considered to be the **best linear unbiased estimator (BLUE)** if these assumptions are met.

The first two assumptions in the list above can be verified simply by looking at the data and understanding the way in which it was collected. The third assumption is that predictors should not be collinear (they should not strongly correlate with one another). Recall the 3 concentric circles from figure 5.10 in chapter 5 that had overlapping variances. If two of the circles overlap and account for the same variance in the dependent variable, there would be no way to know which one was statistically significant, because they would be interchangeable. Since they would be interchangeable, a simple solution would be to get rid of one. If collinearity or **multicollinearity** is present, (multicollinearity is when three or more predictors are correlated), then only one of the collinear variables should be kept in the model. The others can be discarded, because the variance that they explain is already captured by the variable that is left in the model.

If multicollinearity is present, it inflates the standard error for each beta coefficient of the predictors. The larger the standard error of the beta coefficients, the more they vary between samples and the less reliable the model will be.

One way to identify multicollinearity is to view a correlation matrix of all the predictor variables, and look for any that are correlated at 0.8 or higher. Another way is to look at the **variance inflation factor**

(VIF). The VIF measures how much larger the standard errors of the beta coefficients are than they would be if there were no multicollinearity. For example, a VIF of 8 indicates that the standard errors of the beta coefficients are 8 times larger than they would be if there were no multicollinearity. If the VIF is low, then multicollinearity is probably not present. The recommended maximum VIFs vary between statistics texts, but 5 and 10 have been commonly suggested as the thresholds. If the VIF is > 1 but under the maximum threshold, then there is some correlation between the predictors, but not enough to pose a problem.

The fourth assumption that the predictors should not correlate with variables that are not included in the model, is not easily tested, but can be validated by thinking about how the model is structured. For example, if the model regresses crime rate on air conditioner purchases, there might appear to be a relationship between two variables. One might be tempted to conclude that the more air conditioners people buy, the higher crime rate will become. But this model is not considering an outside variable: temperature. When the temperature rises, both the crime rate and the number of purchases of air conditioners will increase. In this example, the exclusion of temperature led to an inaccurate model.

The fifth assumption, that the residuals of the predictors should have a constant variance, is called homoskedasticity. In chapter 3, we tested the assumption that the predictors had constant variance around their means, using Levene's test and the variance ratio. Now we want to look at the errors. This assumption can be tested by plotting the residuals in a scatterplot. The scatterplot should appear to be random. If the plot looks like a funnel, then there is evidence of heteroskedasticity: the variance in the errors changes. If the plot looks like a curve, then the residuals are non-linear (biased) and the assumption is violated. Figure 6.21 shows different residual plots and what they mean.

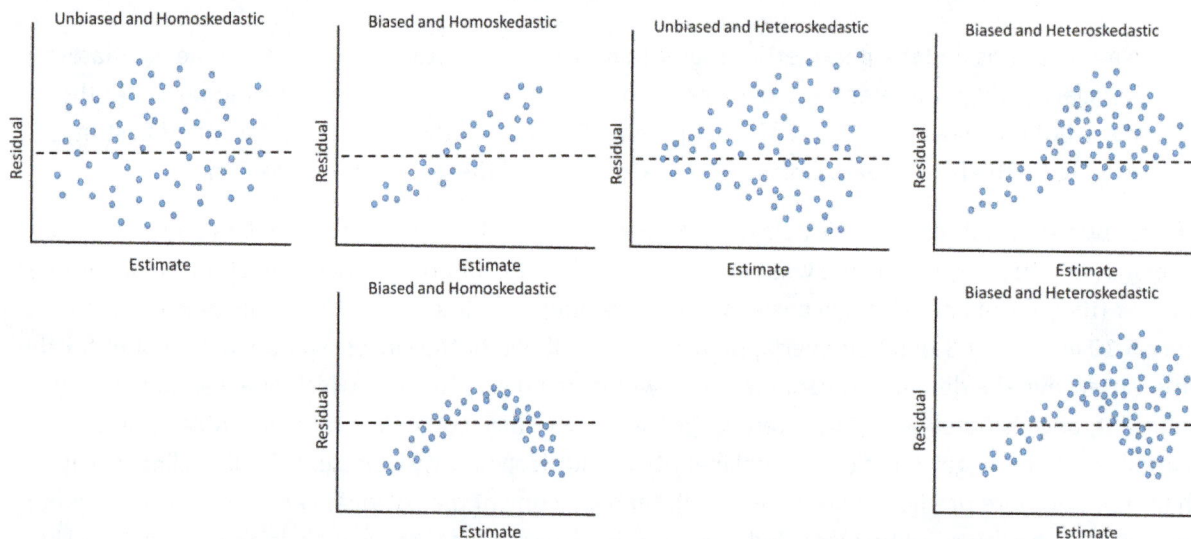

Figure 6.21

If there is heteroskedasticity or the residuals show evidence of non-linearity, a linear model may not be the best choice; a non-linear or non-parametric approach might be better. Aside from resorting to non-linear or non-parametric models, an approach to get rid of the heteroskedasticity in the errors might be to try transforming the raw data (see chapter 4), but there is no guarantee than transformation of the raw data would have any impact on the residuals. Another option is to use a more robust estimation method, such as WLS, GLS, or MLE, as we will see in later chapters. In general, if heteroskedasticity is the only problem, it may still be possible to use OLS estimation. A rule of thumb is that OLS produces estimates of the parameters that are in the ballpark as long as the maximum residual variance is less than 4 times the minimum residual variance (Field, Miles, & Field, 2012).

The sixth assumption, that the residuals be normally distributed can also be tested by plotting as in figure 6.21. Alternatively, the normality of the residuals could be tested by creating a QQ-Plot (see chapter 3 testing the normality assumption), or by plotting a frequency distribution or histogram of the residuals. The frequency distribution should resemble a bell curve is the residuals are normally distributed. This assumption is violated when outliers distort the skewness or kurtosis of the distribution of residuals. By transforming all the variables to make their distributions more normal, it is possible to correct for violations in this assumption. However, normality of the residuals does not really matter if the goal of the regression model is to simply produce good estimates of the coefficients. Therefore, violations of this assumption are not as detrimental to linear regression as violations of the other assumptions (Field, Miles, & Field, 2012).

The seventh assumption, that the errors for any 2 observations should be independent, is the most important of all of these assumptions. If this assumption cannot be met, then the results of the regression could be spurious. This assumption can be tested by looking at the **Durbin-Watson statistic (DW statistic)**. The DW statistic tests the null hypothesis that an observation is not correlated with its preceding observation. It is therefore dependent on the order of the observations. If we re-order the data, the DW statistic will change. The DW statistic is limited to a 1 observation lookback, unless it is generalized to higher order autocorrelations like 2, 3, or 4 lookbacks. The equation used to calculate the test statistic is different for higher order autocorrelations. The formulas are not included here, but the DW statistic is easily calculated in R or Python. The range of possible values for the DW statistic is 0-4, where a value close to 2 suggests that there is no autocorrelation. If the value of the DW statistic is closer to 0, then there is evidence for positive autocorrelation. If the value is closer to 4, then there is evidence of negative autocorrelation.

The DW statistic is not valid when the lagged dependent variable is used as a predictor, but a modified DW statistic called the **Durbin h-test** can be used instead, when the lagged dependent variable is a predictor. A **lag** is another word for the previous observation, so if a variable is lagged by 1, then it is as if the entire column has shifted down 1 row. The table in figure 6.22 shows an example of a lag. Lags are common occurrences when dealing with time series data.

Observation Number	Temperature	First Lag of Temperature	Second Lag of Temperature
1	35.6	NULL	NULL
2	35.9	35.6	NULL
3	36.1	35.9	35.6
4	36.2	36.1	35.9

Figure 6.22

Looking at the table in figure 6.22, it might seem like a simpler approach than using the DW statistic to test for autocorrelation would be to calculate the correlation coefficient between a variable and its lag. That is exactly what a correlogram does, but autocorrelation between observations is a phenomenon known as an autoregressive process. We will look at correlograms and other ways to detect autoregression in a later chapter. Remember that right now, the assumption being made is not for autoregression, but that the errors in a linear regression are not autocorrelated. A predictor could be correlated with its lags like in an autoregressive process, but that may not have any impact on the correlation of the errors between the predicted and actual values of 2 observations. The DW statistic specifically tests for autocorrelation of the errors.

If autocorrelation is detected in the residuals, then there are a few options to adapt a model to deal with it.

1. Expand the confidence intervals around the coefficients by using a **Newey-West estimator and heteroskedasticity and autocorrelation (HAC) robust standard errors**. This is the simplest approach, but the drawback is that it weakens the statistical power of the model by expanding the confidence intervals in exchange for being robust to autocorrelation in the residuals (Field, Miles, & Field, 2012).
2. Include lags of the dependent variable as predictors. This essentially transforms the regression into an autoregressive process, as we will learn about in a later chapter.
3. Build a regression model with an autoregressive component for the errors. We will learn about autoregressive models in a later chapter. So for now, just remember that they can be used to model autocorrelation in residuals.

If any of the parametric assumptions or the assumptions for linear regression cannot be met, and no method of transformation can make the data fit the assumptions, then a non-linear or non-parametric model may be more appropriate.

Dealing with Categorical Predictors

So far we have been assuming that all of the predictors are numeric variables. Categorical variables must be transformed before they can be included in a linear regression. Categorical variables can only be included in a linear regression if they have 2 categories (0 and 1). If the categorical variable has more than 2 categories, it must be dummy encoded. **Dummy variables, also known as binary explanatory variables**, are categorical variables with 2 categories (0 and 1). Dummy encoding involves splitting a categorical variable with k categories into k-1 new predictors. The one category that is left out becomes the baseline against which the other categories are compared (when all other categories have a value of 0, then the category left out must have a value of 1). Since the k-1 predictors are all binary, it should be evident why one category is always left out in dummy encoding: if all k variables were included, then they would be perfectly collinear, which would violate the assumption that the predictors are not collinear. For example, consider fiscal quarters as a categorical variable. To avoid multicollinearity, fiscal quarter should be split into 3 new dummy variables. In the example in figure 6.23, Q4 is left out, so Q4 becomes the baseline when the Q1, Q2, and Q3 dummies are all 0.

Fiscal Quarter	Q1	Q2	Q3
Q1	1	0	0
Q2	0	1	0
Q3	0	0	1
Q4	0	0	0

Figure 6.23

The beta coefficients for the dummies show the impact of changing the dummy from the baseline to the value of the dummy. So a Q1 coefficient of 0.5 means that moving from Q4 to Q1, the outcome variable will change by 0.5 units.

A similar technique to dummy encoding is called one-hot encoding. One-hot encoding does the same transformation as dummy encoding, except instead of k-1 new predictors, it creates k new predictors. This will cause multicollinearity, so a common practice is to get rid of the intercept when using one-hot encoding in a linear regression. One-hot encoding should not be done at all for linear regression, because the intercept is useful for determining what the value of the dependent variable is without any other effects. There are uses for one-hot encoding, but for linear regression it is recommended to dummy encode to avoid multicollinearity (James, Witten, Hastie, & Tibshirani, 2013).

Sample Size, Training/Testing Split, and Outliers
For small effects, a linear regression may miss the effect altogether if the observation sample size is not large enough. The smaller the effect being tested with a linear model, the larger the sample size needs to be. In general, large effects usually need minimum sample sizes of 80. Medium effects usually require at least 200 observations, and small effects require 600 or more observations. These are rules of thumb though (Field, Miles, & Field, 2012).

Common practice in data science is to randomly split a dataset into a training dataset on which models are trained, and a testing dataset on which models are tested. It is up to the researcher how to split the data, but common splits are 80% training, 20% test, or 75:25. The training set is always the larger of the two, and observations should be assigned to the datasets at random. Linear models should be able to generalize to new samples, and splitting the dataset into a training and test sets will help indicate whether a particular model can generalize.

If a linear model performs well on a training dataset and then performs poorly on a testing dataset, one reason may be the presence of outliers in the training set. Recall from chapter 4 that Cook's Distance is a good method for detecting outliers that affect a linear regression. Refer back to chapter 4 for more about detecting and dealing with outliers.

Linear Regression with R
Since the Titanic dataset does not have any good numeric variables to build models of, we are going to take a break from it this chapter and look at data collected from a Fitbit tracker. Fitbit is producer of wearable fitness trackers that track information like the number of steps taken, heart rate, distance traveled, hours slept, and estimated calories burned. More information can be found at Fibit.com (Fitbit Inc, 2018). The Fitbit dataset that we will be using can be downloaded here: https://github.com/nlinc1905/Fitbit-Regression/blob/master/Fitbit%20Data.csv. This dataset contains data collected by my own Fitbit in 2013 and 2014. Anybody can download their personal Fitbit data by following the instructions on Fitbit's website: https://help.fitbit.com/articles/en_US/Help_article/1133. Our goal will be to predict the number of calories burned, given the other variables in the dataset: steps, distance walked, active minutes (minutes where heart rate is in the target calorie burning zone), floors climbed, hours slept, and times awake (the number of times woken up or restless during the night). To start off, we need to examine the data with the same scrutiny that we gave the Titanic dataset.

```
alldata <- read.csv("fitbit.csv", header=T)
```

```
colnames(alldata) <- gsub("[.]", "_", colnames(alldata))  #Replace "." with "_" in
column name - this is just personal preference
```

```
str(alldata)
head(alldata)
```

The date field is read into R as a factor. It should be converted to a date. The integers should be converted to numeric fields, because they are all numbers rather than numbered categories.

```
alldata$Date <- as.Date(alldata$Date, format="%m/%d/%Y")
```

```
alldata$Steps <- as.numeric(alldata$Steps)
alldata$Calories <- as.numeric(alldata$Calories)
alldata$Active_Minutes <- as.numeric(alldata$Active_Minutes)
alldata$Floors_Climbed <- as.numeric(alldata$Floors_Climbed)
alldata$Times_Awake <- as.numeric(alldata$Times_Awake)
```

Check for null or NA values using the function defined in chapter 4. Note that running it as it is will throw an error because when the date field is compared to the strings, R attempts to convert the strings to dates. So the function must be modified in order to handle date fields.

```
#Define a function to count the nulls in every field
naCol <- function(x){
  y <- sapply(x, function(y) {
    if (any(class(y) %in% c('Date', 'POSIXct', 'POSIXt'))) {
      sum(length(which(is.na(y) | is.nan(y) | is.infinite(y))))
    } else {
      sum(length(which(is.na(y) | is.nan(y) | is.infinite(y) | y=='NA' | y=='NaN' |
y=='Inf' | y=='')))
    }
  })
  y <- data.frame('feature'=names(y), 'count.nas'=y)
  row.names(y) <- c()
  y
}
```

```
naCol(alldata)
```

The hours slept and times awake features are the only ones with nulls. Since there are < 500 rows, the number of null values in these columns is greater than 50% of the total rows. With so many missing values, it is probably worth dropping those two columns. But for curiosity's sake, and for the purpose of example, we will retain them. In this case, the missing values are missing at random. There were quite a few nights when I did not sleep with my Fitbit on, but I never systematically removed it on certain nights. Since the data is missing at random, multiple random imputation can be used to impute the missing values. There is an R package called MICE that makes multiple random imputation very simple. The first test we can run is to look at the pattern of missing values.

```
library(mice)
md.pattern(alldata)
```

The first row of the output of this line shows that 214 observations have no missing values. The second row shows that 18 observations are missing only times awake, and the third row shows that 241 observations are missing both hours slept and times awake. To perform multiple imputation, pass the entire data frame into the mice function, except for the date field.

```
mimp <- mice(alldata[,-which(colnames(alldata) %in% c('Date'))], seed=14)
imputed_vars <- complete(mimp, 1)
```

The frequency distributions of hours slept and times awake can be plotted in histograms and compared to the histograms of the date with the imputed values.

```
hist(alldata$Hours_Slept, main="Histogram of Hours Slept before Imputation")
hist(imputed_vars$Hours_Slept, main="Histogram of Hours Slept after Imputation")
hist(alldata$Times_Awake, main="Histogram of Times Awake before Imputation")
hist(imputed_vars$Times_Awake, main="Histogram of Times Awake after Imputation")
```

Notice that the imputations never include new values that are outside the range of pre-existing values. The distributions are very similar. When a dataset contains fields with so many missing values, it is often better to just discard the fields. In this case however, the reasoning for retaining the hours slept and times awake fields is that I have knowledge about what the missing values could be, and I am confident that none of the missing values are outside the range of existing values of these fields. I have this knowledge because I created the dataset. In most situations, knowledge about what the missing values could be is unobtainable, so it is better to just get rid of the fields that are so sparse. The drawback to imputing such a large percentage of the values of these fields is that any model fitted to the data will likely be overfitted.

Now the imputed data can be joined back with the date, and we can proceed to look for outliers. Since there are no categorical fields in this dataset, we only need to look at box plots, which apply the IQR test for outliers.

```
for (col in 1:ncol(alldata[,which(sapply(alldata, class) == 'numeric')])) {
  boxplot(alldata[,which(sapply(alldata, class) == 'numeric')][col],
          main=paste0("Box Plot for ", colnames(alldata[,which(sapply(alldata, class)
== 'numeric')])[col]))
}

#Inspect rows with potential outliers
head(alldata[alldata$Times_Awake>30,])
head(alldata[alldata$Hours_Slept<6,])
head(alldata[alldata$Floors_Climbed>100,])
head(alldata[alldata$Active_Minutes>100,])
```

There are no values that seem completely absurd, but there are a few outliers in every field. Upon inspection, they are all explainable. The observation where I was awake > 30 times in one night was also a night when I slept for over 10 hours. The observations where hours slept were < 6 were unexceptional days with little activity and probably busy at work or something. The observations with > 100 floors climbed were days that I either moved out of an apartment or hiked up volcanos on a trip to Easter Island. These days also correspond to the observations with > 100 active minutes and unusually high steps and calories burned.

The data does not need to be transformed since all of the fields are in nice numeric format. So the last step to take before fitting a regression model is to verify the assumptions that are made for linear regression. Before checking the assumptions, let us drop the date field since it will not be helpful in producing a linear model to predict calories burned. We can also plot a correlation matrix to see which features are related, if any.

```
nodate <- alldata
nodate$Date <- NULL
allCor <- cor(nodate, use="pairwise.complete.obs")
library('corrplot')
corrplot(allCor, method="circle")

#Drop distance
nodate$Distance <- NULL
```

From what we know about correlation and regression, we can make some predictions based on the correlation matrix. First, the hours slept and times awake do not have strong correlations with anything, so if we use them in a linear model for calories, we should not expect them to have very large coefficients. Second, the steps, calories, distance, and active minutes variables are all highly correlated. Steps and distance are nearly perfectly correlated, so it would probably be best to drop either steps or distance from the data so that there is no collinearity. Distance traveled is directly dependent on how many steps are taken, so to remove this dependent relationship, we will discard distance. We could just as well discard steps, but since the model will predict calories, it seems that a model that shows how many steps need to be taken to burn 1 additional calorie would be more useful. A step just seems like a more tangible action that can be taken than traveling some amount of distance. Now for the assumption testing...

```
#Are the predictors independent?
colnames(nodate[,-2])  #Yes - there's no reason to think any of these depend on the
others

#Are all predictors quantitative or categorical with no more than 2 categories?
str(nodate[,-2])  #Yes - assumption met

#Do the predictors have non-zero variance?
summary(nodate[,-2])  #Yes - assumption met

#Is there multicollinearity among the predictors?
cor(nodate[,-2], use="pairwise.complete.obs")  #No - assumption met

#Might the predictors correlate with variables that are not in dataset?
#Possibly - date was dropped, but steps and activity might be correlated with season,
since summer months are more conducive to activity

#Are the residuals homoskedastic?
#Cannot tell yet - need to build the model to get the residuals

#Are the residuals normally distributed?
#Cannot tell yet - need to build the model to get the residuals

#Are the residuals autocorrelated?
#Cannot tell yet - need to build the model to get the residuals
```

Notice that there may be a seasonal effect on activity. Summer and weekends are more conducive to physical activity. Since the timespan of the data only covers 1 full summer and part of another, there is not enough information to reliably draw conclusions about the effect of season on activity. However, we can add dummy variables for the days of the week.

```
library(caret)
alldata$Day <- as.factor(weekdays(alldata$Date))
dummies <- dummyVars( " ~ Day", data=alldata, fullRank=T)
dummyenc <- data.frame(predict(dummies, newdata=alldata))
colnames(dummyenc) <- gsub("[.]", "_", colnames(dummyenc))
alldata <- cbind(alldata, dummyenc)
nodate <- cbind(nodate, dummyenc)
head(nodate)
```

Note that we used the dummyVars function from the caret package to do the encoding. We used it in an earlier chapter to one-hot encode variables in the Titanic dataset. The difference between one-hot encoding, where no categories are omitted, and dummy encoding, where one category is chosen to be the baseline and is omitted, is that dummy encoding requires the fullRank argument to be true. The function sorts the categories alphabetically and uses the first one as the baseline, so Friday is chosen as the baseline. Therefore, when the coefficients of the weekday dummies are interpreted, they will be with respect to Friday.

The last few assumptions require the model to be built before they can be tested. So let us now build the linear model. Fitting and testing a model requires fitting on a training dataset and testing on a testing dataset, so the data is split 70:30 before the model is built.

```
set.seed(14)
train_indices <- createDataPartition(y=nodate$Calories, p=0.7, list=F)
train <- nodate[train_indices,]
test <- nodate[-train_indices,]

linear_reg <- lm(Calories ~ ., data=train)  #Note that "." is short in R for
"everything"
summary(linear_reg)
plot(linear_reg)
```

The summary function shows the results of the regression. The F-statistic (66.1) is very large and significant (p < 0.05), meaning there is a relationship between calories and the predictors. The R squared (0.69) shows that the model is a fairly good fit, as it is > 0.5. Six of the predictors are statistically significant: the intercept, steps, floors climbed, times awake, Saturday, and Sunday. The intercept should be correlated with calories because there is always a baseline number of calories burned, even without any activity. Since the value of the intercept is 1,980, this suggests that I burned 1,980 calories without activity. This number should be near my basal metabolic rate, which was about 1,776 calories according to myfitnesspal.com. So the model is a bit off, but being within 200 calories is definitely in the ballpark. The coefficient of steps is about 0.05, which suggests that all other factors being equal, taking one additional step burns 0.05 additional calories. What is most surprising is the large coefficient (6.76) on the times awake variable. Apparently waking up at night 1 more time burns roughly 6-7 more calories. Floors climbed increased calories burned by 1.8 for every additional floor. If the day is Saturday, 67 more calories are burned on average than on the baseline day (Friday). If the day is Sunday, about 51 more calories are burned on average than on the baseline day. Strangely, active minutes had no statistically significant effect on calories.

The residuals vs fitted plot shows a fairly flat fitted line, suggesting that the residuals are homoscedastic. The scale location plot is the standardized version of the residuals vs fitted plot, and its fitted line is a little less flat. The fitted line in this plot appears to be swayed by the outliers. There is no definite trend however, so it appears that the homoskedasticity assumption holds.

```
mean(linear_reg$residuals)  #This should be near zero if the assumption of residual
normality holds
library(car)
qqPlot(linear_reg, main="QQ Plot")
library(MASS)
sresid <- studres(linear_reg)
hist(sresid, freq=F, main="Distribution of Studentized Residuals Compared to Normal
Dist")
normfitx <- seq(min(sresid), max(sresid), length=40)
normfity <- dnorm(normfitx)
lines(normfitx, normfity)
```

The mean of the residuals is so small that it is essentially zero, which suggests that the assumption that the residuals are normally distributed holds. However, the QQ Plot is not entirely straight. When a histogram of the residuals is plotted against the normal distribution, it becomes apparent that the residuals are close to a normal distribution, but have a longer right tail. These signs point to a violation in the assumption of normality of the residuals, and the most likely cause is the outliers. The gvlma package has tests for skewness, kurtosis, and homoskedasticity that can confirm whether this assumption has been violated.

```
library(gvlma)
gvmodel <- gvlma(linear_reg)
summary(gvmodel)
```

It appears that the assumption is indeed violated. We could transform the variables to make them be more normally distributed, but the normality of residuals assumption is less important than the other assumptions. A robust estimation method, other than OLS, may be able to get around violations in this assumption. Alternatively, since the distribution is only slightly non-normal, we could ignore it altogether. This last course of action is what we will take here – we will ignore the violation of the normality of residuals assumption. Let us check the multicollinearity assumption using the VIF and the autocorrelation assumption using the Durbin Watson statistic. We previously found no evidence of multicollinearity, but now we can confirm that the assumption has not been violated by looking at the variance inflation factor for the predictors.

```
#VIF test for multicollinearity
vif(linear_reg)
```

```
#DW test for autocorrelation - note this uses the car package
durbinWatsonTest(linear_reg)
```

None of the VIF's are above 5, so multicollinearity is not an issue. The DW statistic tests the null hypothesis that there is no autocorrelation among the residuals. Since the p-value is 0, the null can be rejected: there is autocorrelation in the residuals. The test in the code above only checks the first lag. We can quickly look for other significant lags using a correlogram.

```
acf(linear_reg$residuals)
```

The correlogram shows several lags that are above the confidence interval. Since the data was originally ordered by date, it is likely that there is serial correlation. That is, the calories burned on one day is correlated with the calories burned on previous days. This makes sense, as I was more active on weekends than during weekdays, as shown by the plot below:

```
library(ggplot2)
ggplot(alldata, aes(Day, Steps)) + stat_summary(fun.y=mean, geom="bar",
position="dodge", fill="#56B4E9") + labs(x="Day of Week", y="Average Steps") +
ggtitle("Average Steps by Day of Week")
ggplot(alldata, aes(Day, Active_Minutes)) + stat_summary(fun.y=mean, geom="bar",
position="dodge", fill="#56B4E9") + labs(x="Day of Week", y="Average Active Minutes")
+ ggtitle("Average Active Minutes by Day of Week")
```

These bar plots make it very apparent that I was more active on weekends. The effect of autocorrelation is likely diluted because the data was divided up between the training and test sets. It is likely that the effect would be even more pronounced if the data were kept in its original form. This means that new datasets that the model could be applied to, and that would not be split into different sets, would have autocorrelation in the residuals too. So we need to either use the Newey-West HAC approach or add lags of the dependent variable as predictors, in order to deal with the autocorrelation in the residuals. Let's use Newey-West HAC estimates of the coefficients.

```
library(sandwich)
library(lmtest)
coeftest(linear_reg, vcov.=NeweyWest)
```

The changes in the coefficients and p-values are imperceptible. That is a good thing, because it suggests that even though there may be autocorrelation in the residuals, the model is robust enough to still be accurate and reliable. Using this model, it is possible to determine that, all else being equal, 20 steps need to be taken in order to burn 1 calorie, because 1/0.05 = 20.

Since active minutes, hours slept, and the days other than Saturdays and Sundays were not statistically significant, we can build a final model that excludes them. Doing this could help prevent overfitting, and we can test this idea by looking at the AIC and BIC for each model. The better model should have a good fit, as measured by R squared, and also a lower AIC and lower BIC.

```
reduced_linear_reg <- lm(Calories ~ Steps + Floors_Climbed + Times_Awake +
Day_Saturday + Day_Sunday, data=train)
summary(reduced_linear_reg)
AIC(linear_reg)
AIC(reduced_linear_reg)
BIC(linear_reg)
BIC(reduced_linear_reg)
```

After trying this idea out, it seems that removing the other variables did not affect the R squared at all, and it improved the AIC and BIC. Therefore, this slimmer model is better than the original. Now we can apply it to the test set to see how well it does with new data.

```
test$Pred <- round(predict(reduced_linear_reg, test[,-2]), 0)  #Rounding since target
var is rounded to whole number
mae <- mean(abs(test$Calories-test$Pred))
rmse <- sqrt(mean((test$Calories-test$Pred)^2))
baseline_model <- mean(train$Calories)
```

```
mae_baseline <- mean(abs(test$Calories-baseline_model))
rmse_baseline <- sqrt(mean((test$Calories-baseline_model)^2))
```

The mean absolute error is 114.45 calories and the root mean squared error is 155.05 calories. These values imply that the linear model we built is only off by an average of 114-155 calories, which is pretty good! That means we can use the model to predict how many calories would be burned on any given day, and the prediction would be within roughly 155 or fewer calories of the correct number. If we compare those numbers to the baseline model (the baseline is just the average number of calories), we can see that the model offers a substantial improvement over the baseline. If we used the baseline or mean to predict how many calories would be burned on any given day, the prediction would be within roughly 231.47 or fewer calories of the correct number.

Linear Regression with Python

Since the Titanic dataset does not have any good numeric variables to build models of, we are going to take a break from it this chapter and look at data collected from a Fitbit tracker. Fitbit is producer of wearable fitness trackers that track information like the number of steps taken, heart rate, distance traveled, hours slept, and estimated calories burned. More information can be found at Fibit.com (Fitbit Inc, 2018). The Fitbit dataset that we will be using can be downloaded here: https://github.com/nlinc1905/Fitbit-Regression/blob/master/Fitbit%20Data.csv. This dataset contains data collected by my own Fitbit in 2013 and 2014. Anybody can download their personal Fitbit data by following the instructions on Fitbit's website: https://help.fitbit.com/articles/en_US/Help_article/1133. Our goal will be to predict the number of calories burned, given the other variables in the dataset: steps, distance walked, active minutes (minutes where heart rate is in the target calorie burning zone), floors climbed, hours slept, and times awake (the number of times woken up or restless during the night). To start off, we need to examine the data with the same scrutiny that we gave the Titanic dataset.

```
import pandas as pd
```

```
alldata = pd.read_csv("fitbit.csv", sep=",")
alldata.columns = alldata.columns.to_series().str.replace('\s+', '_')
print(alldata.info())
print(alldata.head())
```

The date field is read into Pandas as an object. It should be converted to a date. The integers should be converted to numeric fields, because they are all numbers rather than numbered categories.

```
#Convert Date to a date
alldata['Date'] = pd.to_datetime(alldata['Date'])
```

```
#Convert integers to numeric
alldata['Steps'] = alldata['Steps'].astype(float)
alldata['Calories'] = alldata['Calories'].astype(float)
alldata['Active_Minutes'] = alldata['Active_Minutes'].astype(float)
alldata['Floors_Climbed'] = alldata['Floors_Climbed'].astype(float)
alldata['Times_Awake'] = alldata['Times_Awake'].astype(float)
```

Check for null or NA values using the functions defined in chapter 4.

```
#Define a function to count the nulls in every field
def naCol(df):
    y = dict.fromkeys(df.columns)
```

```
    for idx, key in enumerate(y.keys()):
        if df.dtypes[list(y.keys())[idx]] == 'object':
            y[key] = pd.isnull(df[list(y.keys())[idx]]).sum() +
(df[list(y.keys())[idx]]=='').sum()
        else:
            y[key] = pd.isnull(df[list(y.keys())[idx]]).sum()
    print("Number of nulls by column")
    print(y)
    return y
```

```
#Define a function to count the nulls by row to see if any rows have too many missing
values to be useful
def naRow(df, threshold=0.5):
    y = dict.fromkeys(df.index)
    for idx, key in enumerate(y.keys()):
        y[key] = sum(df.iloc[[idx]].isnull().sum())
    print("Rows with more than 50% null columns")
    print([r for r in y if y[r]/df.shape[1] > threshold])
    return y
```

```
naCol(alldata)
naRow(alldata)
```

The hours slept and times awake features are the only ones with nulls. Since there are < 500 rows, the number of null values in these columns is greater than 50% of the total rows. With so many missing values, it is probably worth dropping those two columns. But for curiosity's sake, and for the purpose of example, we will retain them. In this case, the missing values are missing at random. There were quite a few nights when I did not sleep with my Fitbit on, but I never systematically removed it on certain nights. Since the data is missing at random, multiple random imputation can be used to impute the missing values. There is a Python library called fancyimpute that can be used for multiple imputation. Note that fancyimpute needs to be compiled, and Windows machines are bad at this. I recommend either using Linux for this chapter, or cheating and reading in the train/test files for this dataset that were created with R. If the train/test files are imported, no data cleaning or imputation is necessary.

```
#Use multiple imputation to fill in the missing values for hours slept and times
awake
from fancyimpute import MICE
import random

def estimate_by_mice(df):
    df_estimated_variables = df.copy()
    random.seed(14)
    mice = MICE()  # model=RandomForestClassifier(n_estimators=100))
    result = mice.complete(np.asarray(df.values, dtype=float))
    df_estimated_variables.loc[:, df.columns] = result[:][:]
    return df_estimated_variables

impdata = estimate_by_mice(alldata.drop(['Date'], axis=1))
alldata = pd.concat([alldata[['Date']], impdata], axis=1)
```

It is fine to pass the imputation function the entire dataset, because only the missing values will be filled in – the rest of the data will not change. There is a comment in the code indicating that the MICE function from fancyimpute uses a random forest to impute the missing values. See the documentation for more details. The result of running our imputation function is a complete dataset with no missing values. We can plot the distributions of the original data and the imputed data to ensure they match closely.

```
#Plot distributions of original data and imputed to see if imputation is on track
fig = plt.figure()
plta = fig.add_subplot(211)
alldata.hist(column='Hours_Slept', ax=plta)
plta.set_title('Histogram of Hours Slept Before Imputation')
pltb = fig.add_subplot(212)
impdata.hist(column='Hours_Slept', ax=pltb)
pltb.set_title('Histogram of Hours Slept After Imputation')
fig.tight_layout()
plt.show()
plt.close(fig)
fig = plt.figure()
plta = fig.add_subplot(211)
alldata.hist(column='Times_Awake', ax=plta)
plta.set_title('Histogram of Times Awake Before Imputation')
pltb = fig.add_subplot(212)
impdata.hist(column='Times_Awake', ax=pltb)
pltb.set_title('Histogram of Times Awake After Imputation')
fig.tight_layout()
plt.show()
plt.close(fig)
```

Now we can proceed to look for outliers. Since there are no categorical fields in this dataset, we only need to look at box plots, which apply the IQR test for outliers.

```
#Boxplots for numeric variables to check for outliers
numeric_cols = [col for col in alldata.columns if alldata[col].dtype == 'float64']
for col in numeric_cols:
    sns.boxplot(y=alldata[~np.isnan(alldata[col])][col])
    plt.title("Box Plot for " + col)
    plt.show()

#Inspect rows with potential outliers
print(alldata[alldata['Times_Awake']>30].head())
print(alldata[alldata['Hours_Slept']<6].head())
print(alldata[alldata['Floors_Climbed']>100].head())
print(alldata[alldata['Active_Minutes']>100].head())
```

There are no values that seem completely absurd, but there are a few outliers in every field. Upon inspection, they are all explainable. The observation where I was awake > 30 times in one night was also a night when I slept for over 10 hours. The observations where hours slept were < 6 were unexceptional days with little activity and probably busy at work or something. The observations with > 100 floors climbed were days that I either moved out of an apartment or hiked up volcanos on a trip to Easter Island. These days also correspond to the observations with > 100 active minutes and unusually high steps and calories burned.

The data does not need to be transformed since all of the fields are in nice numeric format. So the last step to take before fitting a regression model is to verify the assumptions that are made for linear regression. Before checking the assumptions, let us drop the date field since it will not be helpful in producing a linear model to predict calories burned. We can also plot a correlation matrix to see which features are related, if any.

```
#Drop date and look at correlation matrix
nodate = alldata.drop(['Date'], axis=1)
allcor = nodate.corr()
cmap = sns.diverging_palette(220, 10, as_cmap=True)  #Sets up diverging palette for
seaborn
sns.heatmap(allcor, cmap=cmap, vmin=-1, vmax=1, center=0,
            square=True, linewidths=.5, cbar_kws={"shrink": .5})
plt.title('Correlation Matrix of Numeric Variables')
plt.show()

#Drop distance
nodate.drop(['Distance'], axis=1, inplace=True)
```

From what we know about correlation and regression, we can make some predictions based on the correlation matrix. First, the hours slept and times awake do not have strong correlations with anything, so if we use them in a linear model for calories, we should not expect them to have very large coefficients. Second, the steps, calories, distance, and active minutes variables are all highly correlated. Steps and distance are nearly perfectly correlated, so it would probably be best to drop either steps or distance from the data so that there is no collinearity. Distance traveled is directly dependent on how many steps are taken, so to remove this dependent relationship, we will discard distance. We could just as well discard steps, but since the model will predict calories, it seems that a model that shows how many steps need to be taken to burn 1 additional calorie would be more useful. A step just seems like a more tangible action that can be taken than traveling some amount of distance. Now for the assumption testing...

```
#Are the predictors independent?
print(nodate.drop(['Calories'], axis=1).columns)  #Yes - there's no reason to think
any of these depend on the others

#Are all predictors quantitative or categorical with no more than 2 categories?
print(nodate.drop(['Calories'], axis=1).info())  #Yes - assumption met

#Do the predictors have non-zero variance?
for col in list(nodate.drop(['Calories'], axis=1).columns):
    print(stats.variance(nodate[col]))  #Yes - assumption met

#Is there multicollinearity among the predictors?
print(allcor)  #No - assumption met

#Might the predictors correlate with variables that are not in dataset?
#Possibly - date was dropped, but steps and activity might be correlated with season,
since summer months are more conducive to activity

#Are the residuals homoskedastic?
#Cannot tell yet - need to build the model to get the residuals

#Are the residuals normally distributed?
#Cannot tell yet - need to build the model to get the residuals

#Are the residuals autocorrelated?
```

```
#Cannot tell yet - need to build the model to get the residuals
```

Notice that there may be a seasonal effect on activity. Summer and weekends are more conducive to physical activity. Since the timespan of the data only covers 1 full summer and part of another, there is not enough information to reliably draw conclusions about the effect of season on activity. However, we can add dummy variables for the days of the week.

```
alldata['Day'] = alldata['Date'].dt.dayofweek
dummies = pd.get_dummies(alldata['Day'], drop_first=True).add_prefix('Day_')
alldata = pd.concat([alldata, dummies], axis=1)
nodate = pd.concat([nodate, dummies], axis=1)
print(nodate.head())
```

Note that we used the getdummies function from Pandas to do the encoding. We used it in an earlier chapter to one-hot encode variables in the Titanic dataset. The difference between one-hot encoding, where no categories are omitted, and dummy encoding, where one category is chosen to be the baseline and is omitted, is that dummy encoding requires the fullRank argument to be true. Unlike R, which sorts the categories alphabetically and uses the first one as the baseline, Pandas is smart enough to recognize that the data is a weekday coming from a date, so Monday is ordered first and chosen as the baseline. Therefore, when the coefficients of the weekday dummies are interpreted, they will be with respect to Monday.

The last few assumptions require the model to be built before they can be tested. So let us now build the linear model. Fitting and testing a model requires fitting on a training dataset and testing on a testing dataset, so the data is split 70:30 before the model is built.

```
x = nodate.drop(['Calories'], axis=1).values
y = nodate['Calories'].values
x_train, x_test, y_train, y_test = train_test_split(x, y, test_size=0.3,
random_state=14)
train = np.concatenate((x_train, y_train.reshape(y_train.shape[0],1)), axis=1)
train = pd.DataFrame(data=train, columns=list(nodate.drop(['Calories'],
axis=1).columns)+['Calories'])
train.to_csv('train.csv', index=False)
test = np.concatenate((x_test, y_test.reshape(y_test.shape[0],1)), axis=1)
test = pd.DataFrame(data=test, columns=list(nodate.drop(['Calories'],
axis=1).columns)+['Calories'])
test.to_csv('test.csv', index=False)

x_train = train.drop(['Calories'], axis=1).values
y_train = train['Calories'].values
x_test = test.drop(['Calories'], axis=1).values
y_test = test['Calories'].values
train.columns = ['Steps', 'Calories', 'Active_Minutes', 'Floors_Climbed',
'Hours_Slept',
        'Times_Awake', 'Day_0', 'Day_5', 'Day_6',
        'Day_3', 'Day_1', 'Day_2']
test.columns = ['Steps', 'Calories', 'Active_Minutes', 'Floors_Climbed',
'Hours_Slept',
        'Times_Awake', 'Day_0', 'Day_5', 'Day_6',
        'Day_3', 'Day_1', 'Day_2']

#Build linear model
```

```
linear_reg = LinearRegression().fit(x_train, y_train)
print("Regression Parameters",
      "\nIntercept:", linear_reg.intercept_)
for idx, col in enumerate(train.drop(['Calories'], axis=1).columns):
    print(col, "Coefficient:", linear_reg.coef_[idx])
print("R^2:", linear_reg.score(x_train, y_train))
```

Unlike R, which prints out a nice summary of all the important statistics, sklearn produces barebones output. That is because sklearn was developed with a machine learning philosophy. In the past, before computers were strong enough for resampling techniques like boosting and cross validation (to be described in a later chapter), the only ways to assess regression models were to use the F-stat, R^2, and other statistics. Sklearn omits these, but we can write our own function to at least produce the R^2 and p-values for the predictors.

```
def get_lr_output(lr, x, y):
    #Takes a fitted sklearn linear regression as input and outputs results
    #Thanks to stackoverflow: https://stackoverflow.com/questions/27928275/find-p-
value-significance-in-scikit-learn-linearregression
    import numpy as np
    import scipy.stats as scistats
    parameters = np.append(lr.intercept_, lr.coef_)
    predictions = lr.predict(x)
    X = pd.DataFrame({"Constant":np.ones(len(x))}).join(pd.DataFrame(x))
    mse = (sum((y-predictions)**2))/(len(X)-len(X.columns))
    sd = np.sqrt(mse*(np.linalg.inv(np.dot(X.T, X)).diagonal()))
    ts = parameters/sd
    p_values = [2*(1-scistats.t.cdf(np.abs(i), (len(X)-1))) for i in ts]
    sd = np.round(sd, 4)
    ts = np.round(ts, 4)
    p_values = np.round(p_values, 4)
    parameters = np.round(parameters, 4)
    predictors = pd.Series('intercept')
    predictors = predictors.append(pd.Series(x.columns)).reset_index()[0]
    output_df = pd.DataFrame()
    output_df['predictor'] = predictors
    output_df['coefficient'] = parameters
    output_df['standard_error'] = sd
    output_df['t-statistic'] = ts
    output_df['p-value'] = p_values
    r_squared = round(lr.score(x, y), 4)
    adj_r_squared = round(1-(1-lr.score(x, y))*(len(y)-1)/(len(y)-x.shape[1]-1), 4)
    print(output_df, '\n R-squared:', r_squared, '\n Adjusted R-squared:',
adj_r_squared)
    return [output_df, r_squared, adj_r_squared]
```

```
linear_reg_output = get_lr_output(linear_reg, train.drop(['Calories'], axis=1),
train['Calories'])
```

The R squared (0.68) shows that the model is a fairly good fit, as it is > 0.5. Five of the predictors are statistically significant: the intercept, steps, floors climbed, times awake, and Saturday. The intercept should be correlated with calories because there is always a baseline number of calories burned, even without any activity. Since the value of the intercept is 1,980, this suggests that I burned 1,980 calories without activity. This number should be near my basal metabolic rate, which was about 1,776 calories according to myfitnesspal.com. So the model is a bit off, but being within 200 calories is definitely in the ballpark. The coefficient of steps is about 0.05, which suggests that all other factors being equal, taking one additional step burns 0.05 additional calories. What is most surprising is the large coefficient (6.76) on the times awake variable. Apparently waking up at night 1 more time burns roughly 6-7 more calories. Floors climbed increased calories burned by 1.8 for every additional floor. If the day is Saturday, 67 more calories are burned on average than on the baseline day (Monday). Strangely, active minutes had no statistically significant effect on calories.

Let's finish testing the assumptions by checking residual normality. We can get the residuals using the predict method for the linear regression. The mean of the residuals should be near 0 and the distribution should appear normal.

```
#Testing normally distributed residuals assumption
train_preds = linear_reg.predict(x_train)
residuals = y_train-train_preds
print(np.mean(residuals))  #This should be near zero if the assumption of residual
normality holds
sns.distplot(residuals)
plt.show()
```

The mean of the residuals is so small that it is essentially zero, which suggests that the assumption that the residuals are normally distributed holds. When a histogram of the residuals is plotted against the normal distribution, it becomes apparent that the residuals are close to a normal distribution, but have a longer right tail. This points to a violation in the assumption of normality of the residuals, and the most likely cause is the outliers. We could transform the variables to reduce the impact of outliers and make them be more normally distributed, but the normality of residuals assumption is less important than the other assumptions. A robust estimation method, other than OLS, may be able to get around violations in this assumption. Alternatively, since the distribution is only slightly non-normal, we could ignore it altogether. This last course of action is what we will take here – we will ignore the violation of the normality of residuals assumption. Let us check the autocorrelation assumption using an autocorrelation plot (correlogram) of the first 30 lags of the residuals.

```
#Correlogram to look for significant lags in the residuals
plot_acf(residuals, lags=30)
plt.show()
```

The correlogram shows several lags that are above the confidence interval. Since the data was originally ordered by date, it is likely that there is serial correlation. That is, the calories burned on one day is correlated with the calories burned on previous days. This makes sense, as I was more active on weekends than during weekdays, as shown by the plots below:

```
alldata.groupby(['Day'])['Steps'].mean().plot(kind='bar', title="Average Steps by
Day")
plt.xlabel('Day of Week (Monday=0)')
plt.ylabel('Average Number of Steps')
plt.show()
alldata.groupby(['Day'])['Active_Minutes'].mean().plot(kind='bar', title="Average
Active Minutes by Day")
plt.xlabel('Day of Week (Monday=0)')
plt.ylabel('Average Active Minutes')
plt.show()
```

These bar plots make it very apparent that I was more active on weekends. The effect of autocorrelation is likely diluted because the data was divided up between the training and test sets. It is likely that the effect would be even more pronounced if the data were kept in its original form. This means that new datasets that the model could be applied to, and that would not be split into different sets, would have autocorrelation in the residuals too. So it would be wise to use Newey-West HAC estimates of the coefficients. Unfortunately, there are no Newey West HAC estimates for sklearn. An alternative is to bootstrap the coefficient estimates. We will explore bootstrapping regression coefficients in a later chapter. Recall from the R section that the Newey West HAC estimates were imperceptibly different from the originals. That is a good thing, because it suggests that even though there may be autocorrelation in the residuals, the model is robust enough to still be accurate and reliable. Using this model, it is possible to determine that, all else being equal, 20 steps need to be taken in order to burn 1 calorie, because 1/0.05 = 20.

Since active minutes, hours slept, and the days other than Saturdays were not statistically significant, we can build a final model that excludes them. Doing this could help prevent overfitting, and we can test this idea by looking at the AIC for each model. The better model should have a good fit, as measured by R squared, and also a lower AIC.

```
#Remove unnecessary variables from the model to see the result
x_train_red = train[['Steps', 'Floors_Climbed', 'Times_Awake', 'Day_5']].values
x_test_red = test[['Steps', 'Floors_Climbed', 'Times_Awake', 'Day_5']].values
reduced_linear_reg = LinearRegression().fit(x_train_red, y_train)
reduced_linear_reg_output = get_lr_output(reduced_linear_reg, train[['Steps',
'Floors_Climbed', 'Times_Awake', 'Day_5']], train['Calories'])

def AIC(y, y_pred, k):
    '''
    Takes residuals of a model and number of predictors k, and outputs AIC
    '''
    resid = y - y_pred.ravel()
    sse = sum(resid ** 2)
    return 2*k - 2*np.log(sse+1e-32)

print(AIC(y_train, linear_reg.predict(x_train), x_train.shape[1]))
print(AIC(y_train, reduced_linear_reg.predict(x_train_red), x_train_red.shape[1]))
```

After trying this idea out, it seems that removing the other variables did not affect the R squared at all, and it improved the AIC. Therefore, this slimmer model is better than the original. Before applying the reduced model to the test set to see how well it does with new data, let's pause to consider the AIC values: they are negative. This is not a problem. The smaller, more negative, value produced by the reduced regression is still better.

```
test_preds = np.round(reduced_linear_reg.predict(x_test_red), 0)  #Rounding since
target var is rounded to whole number
mae = np.mean(abs(y_test-test_preds))
rmse = np.sqrt(np.mean((y_test-test_preds)**2))
baseline_model = np.round(np.mean(y_train), 0)
mae_baseline = np.mean(abs(y_test-baseline_model))
rmse_baseline = np.sqrt(np.mean((y_test-baseline_model)**2))
```

The mean absolute error is 114.63 calories and the root mean squared error is 157.02 calories. These values imply that the linear model we built is only off by an average of 115-157 calories, which is pretty good! That means we can use the model to predict how many calories would be burned on any given day, and the prediction would be within roughly 157 or fewer calories of the correct number. If we compare those numbers to the baseline model (the baseline is just the average number of calories), we can see that the model offers a substantial improvement over the baseline. If we used the baseline or mean to predict how many calories would be burned on any given day, the prediction would be within roughly 231.48 or fewer calories of the correct number.

7. Optimization

Before moving on to other types of regression, it would be helpful to acquire a deeper understanding of optimization. So far, only the ordinary least squares (OLS) method of optimization has been explained. Optimization algorithms for regression are also called estimators, so it is common to see OLS referred to as an **estimator** for example. Recall from chapter 6 that OLS optimizes the sum of squared residuals (SSR). The SSR is what is referred to as an **objective function or cost function** in data science. When we seek to minimize the objective function, as in the case of OLS, the objective function can also be called a cost function. There is no way to find a model that perfectly describes the relationships between variables under every possible set of conditions. Therefore, the best model must be estimated. Optimization is a broad area of mathematics, but for data science it is used to estimate the model of best fit. Some data scientists go so far as to say that every model fitted to data can be reduced to an optimization problem.

Optimization can be split into two types: convex and non-convex (Boyd & Vandenberghe, 2004). Most models in statistics and machine learning use convex optimization to minimize a cost function and find the optimal model parameters. **Convex optimization** deals with convex cost functions, which have the unique property of the local minimum also being the global minimum. Think about a quadratic equation, such as $y = x^2$, represented in figure 7.1. As we will soon see, there is only 1 minimum value of the function $y = x^2$. This property is extremely useful in data science because if there is only 1 local minimum, then when it is found, it can be considered the global minimum, and the parameters of a model that minimizes the cost function can be considered to be optimal.

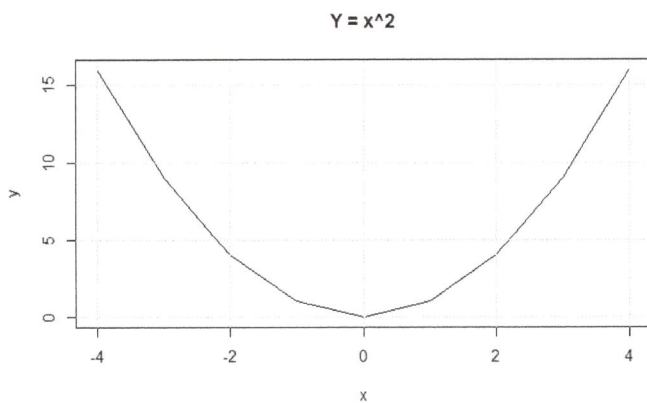

Figure 7.1

Non-convex optimization deals with non-convex objective functions where there could be multiple local minimums. The challenge to using non-convex optimization is that a solution can never be guaranteed to be the best possible. There are several tricks to improve the odds of finding the best possible solution to non-convex objective functions though. Sometimes non-convex functions can be transformed to convex functions. For example, consider the non-convex equation $y = \sin(x)$. If the sine function is the objective function, it could be transformed into a convex function by restricting its domain (the minimum and maximum values on the x-axis) to (pi, 2*pi). See the blue lines in figure 7.2 for a visual explanation. Transformations like this can be done when it makes sense to do so.

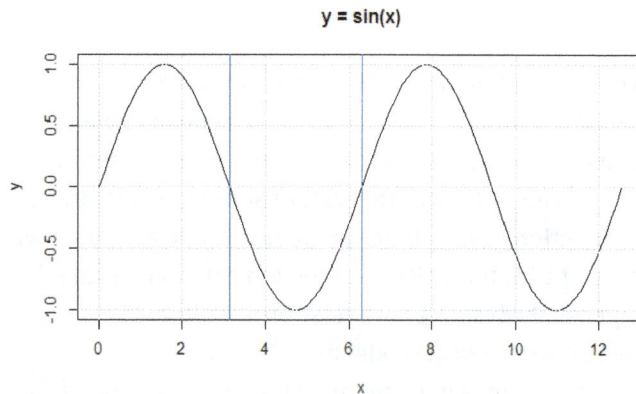

Figure 7.2

The optimization methods explained in this chapter are far from exhaustive, but they are some of the most common and will be useful in understanding the models introduced in this book's coming chapters, as well as many machine learning algorithms.

First and Second Derivative Tests

Let's begin with using standard calculus to find the minimum of the function $y = x^2$. Finding the first derivative of the function and setting it equal to zero shows critical points of the function (critical points are either minima or maxima). This is called the **first derivative test**. The reason we are interested in points where the first derivative is 0 is that these are points along the curve where the curve's slope is 0. That means that when the first derivative is 0, the curve has either reached a maximum or minimum.

$$y' = 2x$$

$$0 = 2x$$

$$x = 0$$

Now we know that $x = 0$ is a critical point, but is it a maximum or minimum? If the value of a critical point is passed to the second derivative, then it can be determined whether the critical point is a local maximum or local minimum. If the result is > 0, then the point is a local minimum, and if it is < 0, then the point is a local maximum. This is called the **second derivative test**.

$$y'' = 2$$

If there were an "x" in the equation above, then the value of the critical point could have been plugged into the equation, but since the second derivative is just 2, and 2 > 0, we know that the only critical point ($x = 0$) is a local minimum. Since there is only 1 critical point, and it is a minimum, it is confirmed that $x = 0$ is a global minimum for the function $y = x^2$.

Often there are constraints to optimization problems. **Constraints** are conditions that must be true. For example, suppose we need to build a storage box with a volume of 20 cubic inches. The length of the base is twice the width of the box. If material for the base costs $10 per square inch, and material for the sides cost $5 per square inch, what is the width of the box that minimizes the cost of materials?

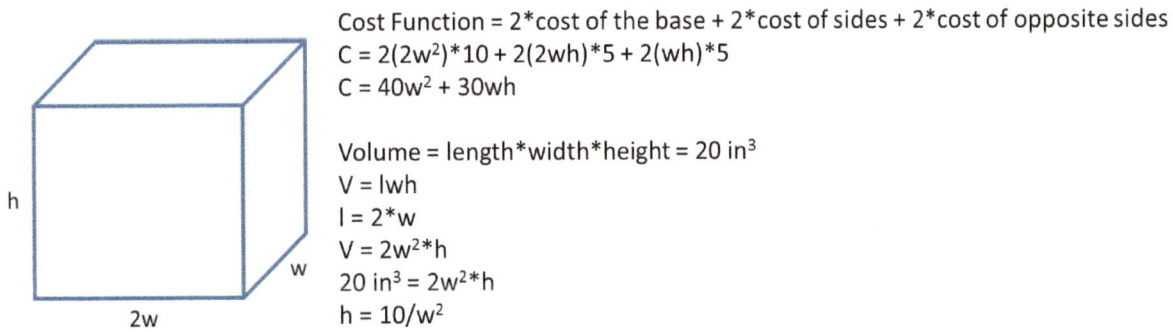

Cost Function = 2*cost of the base + 2*cost of sides + 2*cost of opposite sides

$C = 2(2w^2)*10 + 2(2wh)*5 + 2(wh)*5$

$C = 40w^2 + 30wh$

Volume = length*width*height = 20 in^3

$V = lwh$

$l = 2*w$

$V = 2w^2*h$

20 in^3 = 2w^2*h

$h = 10/w^2$

Figure 7.3

Using the information given, and the knowledge that the volume of a box is length * width * height, we can set up the cost function and solve for the height of the box as in figure 7.3. The constraint in this problem is the volume of the box. By solving for the height, the constraint can be substituted into the objective function to express it in terms of 1 variable. This process is referred to as optimizing the objective function subject to a constraint.

$C = 40w^2 + 30w*(10/w^2)$

$C = 40w^2 + 300/w$

$C = 40w^2 + 300w^{-1}$

The first derivative test will find critical points:

$C' = 80w - 300w^{-2}$, which can be re-written as:

$C' = 80w - 300/w^2$

$0 = 80w - 300/w^2$

$80w = 300/w^2$

$80w^3 = 300$

$w^3 = 3.75$

$w = 3.75^{1/3}$

Now using the second derivative test, we can determine whether that critical point is a minimum or maximum.

$C'' = 80 + 600w^{-3}$

$80 + 600(3.75^{1/3})^{-3}$

≈ 466.2

So the critical value is a global minimum for the cost function. Therefore, the width that minimizes the cost is 3.75^1/3. If we were building an optimum model for the cost of the box, 3.75^1/3 would be the parameter that optimizes the objective function. Similarly, we can think of the beta coefficients and intercept of a linear regression as the parameters to be optimized.

The first and second derivatives have been able to solve the optimization problems we have seen so far. The trouble with fitting a model to data is that the true equation that needs to be optimized is never known. Since it is impossible to take the derivative of an unknown function, we must estimate the equation of the curve and differentiate the difference between the estimate and the true values. This brings us back to ordinary leas squares, which seeks to minimize the sum of squared residuals.

Ordinary Least Squares (OLS)

The sum of squared differences between the true values and the estimated values (sum of squared residuals or SSR) can be expressed by the equation in figure 7.4.

$$SSR = \sum (y - \hat{y})^2$$

Figure 7.4

In the equation in figure 7.4, y is the true value and y-hat is the estimated value given by a linear model, such as y = alpha + beta*x. Thus, the equation can be re-written as:

$$SSR = \sum (y - \alpha - \beta x)^2$$

Figure 7.5

For simplicity, we are assuming that there is only 1 predictor variable (x), but the same idea applies to regression models with more predictors. The parameters to be optimized by OLS are the intercept (alpha) and the beta coefficient. Recall from chapters 5 that semi-partial correlation allows the variance of an outcome variable to be explained by a set of predictors. The concept is that all predictors except for one will be controlled for, so that the effect of the one can be measured independently. This is how and OLS linear regression estimates the beta coefficient for each predictor.

To measure the effect of 1 variable at a time, we assume all other variables are constant. So the partial derivative of the equation in figure 7.5 with respect to each variable will need to be computed and set equal to 0. By the nature of the model and standard assumptions for linear regression (see chapter 6), we know that OLS estimation is a convex optimization problem, so there is only 1 minimum. Therefore, only the first derivative needs to be calculated, and we can be sure that the critical point is the global minimum.

$$\frac{\partial SSR}{\partial \beta} = \frac{\partial (\sum (y - \beta x - \alpha)^2)}{\partial \beta} = 2 \sum ((y - \beta x - \alpha) * (0 - x - 0))$$

The equation above gives the first partial derivative of the SSR (figure 7.5) with respect to the beta coefficient. Now it can be set equal to zero:

$$2 \sum ((y - \beta x - \alpha) * (0 - x - 0)) = 0$$

We know that alpha and beta are constants, so they can be moved outside of the summation. The equation above simplifies to:

$$\sum(-xy) + \beta \sum x^2 + \alpha \sum x = 0$$

Further simplification results in the solution for alpha, in terms of beta. This is useful because we can later express the parameters of the regression model in terms of 1 variable. Note that the denominator simplifies from the sum of x (a constant) to N (the number of observations) times x, because the sum of any constant is N times the constant.

$$\alpha = \frac{\sum xy - \beta \sum x^2}{Nx}$$

Now we can find the first partial derivative of the SSR with respect to alpha and set it equal to zero:

$$\frac{\partial SSR}{\partial \alpha} = \frac{\partial(\sum(y - \beta x - \alpha)^2)}{\partial \alpha} = 2 \sum((y - \beta x - \alpha) * (0 - 0 - 1))$$

$$-2 \sum(y - \beta x - \alpha) = 0$$

$$-\sum y + \beta \sum x + \alpha \sum 1 = 0$$

Using summation rules to simplify the above expression, we arrive at the equation that defines y in terms of alpha and beta:

$$y = \beta x + \alpha$$

This is no surprise. But now we can substitute alpha with the expression that defines it in terms of beta.

$$y = \beta x + \frac{\sum xy - \beta \sum x^2}{Nx}$$

$$y = \beta\left(x - \frac{\sum x^2}{Nx}\right) + \frac{\sum xy}{Nx}$$

$$yNX = \beta\left(Nx^2 - \sum x^2\right) + \sum xy$$

Since we are searching for the parameters of the regression model, alpha and beta, and we know that alpha = y – beta*x, then the final parameter that we need is beta. So the expression above can be solved for beta:

$$\beta = \frac{Nxy - \sum xy}{(Nx^2 - \sum x^2)}$$

These equations produce the optimal parameters, alpha and beta, of a regression model estimated by OLS. Now we know what goes on behind the scenes of an OLS regression to produce the estimates we see in the output from linear regression libraries for Python and R.

Weighted Least Squares (WLS)

Estimation methods are said to be robust when they can stand up to violations in the linear regression assumptions listed in chapter 6. One of those assumptions is that the residuals of the predictors should have constant variance. This is called homoskedasticity. If there is evidence of heteroskedasticity, OLS may produce biased estimates of the parameters. One way to even out the variance of the residuals if they are non-constant is to weight them. Weighted least squares accomplishes this by adding a weight to the OLS cost function:

$$WSSR = \sum_{i=1}^{N} w_i(y_i - \beta x_i - \alpha)^2$$

Figure 7.6

In the equation in figure 7.6, WSSR stands for the weighted sum of squared residuals. The weight in the equation is inversely proportional to the error variance, meaning smaller error variances get larger weights. This evens out the variance in the residuals. To perform a WLS regression, the weights must be applied to the variables as a transformation. The weights (w_i in figure 7.6) are determined by estimating the covariance between each pair of residuals. There are many possible options for estimating the weights, but some common examples include:

- Divide the dependent variable and the independent variables by x.
- Divide the dependent variable and the independent variables by N, where N is the sample size.
- Divide the dependent variable and the independent variables by square root of N, where N is the sample size.
- Divide the dependent variable and the independent variables by some value that varies for different subsets of the sample.

Let us take a moment to consider why these weights must be estimated. If we look build a covariance matrix (a.k.a. variance-covariance matrix) of the residuals, in a standard OLS regression the values of the covariances would be 1 down the diagonal (every residual perfectly correlates with itself) and 0 everywhere else. This would reflect the assumption that the residuals are not correlated. If there is heteroskedasticity in the residuals, then the diagonal values would vary. So if we replace the 1's down the diagonal with sigma^2_{ei}, meaning the variance between residuals e and i, then we can visualize what we are estimating in WLS. In WLS, the estimates of the covariances between every pair of residuals are the weights. So $1/\text{sigma}^2_{ei} = w_i$ from figure 7.6. That is why all of the weighting options listed above involve dividing by the weight, because it is the same as multiplying by 1/weight.

$$OLS\ covariance\ matrix = \begin{bmatrix} 1 & 0 & 0 \\ 0 & 1 & 0 \\ 0 & 0 & 1 \end{bmatrix}$$

$$WLS\ covariance\ matrix = \begin{bmatrix} \sigma^2_{e1} & 0 & 0 \\ 0 & \sigma^2_{e2} & 0 \\ 0 & 0 & \sigma^2_{e3} \end{bmatrix}$$

$$\frac{1}{\sigma^2_{ei}} = w_i$$

Figure 7.7

Due to the challenge of determining the proper weights for WLS regression, WLS is often only used when the weights are known or can be assumed; that is, when it is known which weighting scheme will target the observations causing a large residual variance. WLS also cannot handle outliers or autocorrelation of the residuals. In general, if heteroskedasticity is the only problem, it may still be possible to use OLS. A rule of thumb is that OLS produces estimates of the parameters that are in the ballpark as long as the maximum residual variance is less than 4 times the minimum residual variance.

Generalized Least Squares (GLS)

Generalized least squares can treat both heteroskedasticity and autocorrelation of the residuals, making it a robust estimation method. GLS is a generalized version of WLS (WLS is actually a special case of GLS where the weights are known or assumed). Refer back to the WLS covariance matrix of the residuals in figure 7.7. In WLS, there is only heteroskedasticity (the diagonal values vary) but no autocorrelation. If there is autocorrelation, then the off-diagonal values will also vary (they will not be 0 as in WLS or OLS).

$$GLS\ covariance\ matrix = \begin{bmatrix} \sigma_{e1}^2 & Cov(e1,e2) & Cov(e1,e3) \\ Cov(e1,e2) & \sigma_{e2}^2 & Cov(e2,e3) \\ Cov(e1,e3) & Cov(e2,e3) & \sigma_{e3}^2 \end{bmatrix}$$

Figure 7.8

Like WLS, the idea behind GLS is to weight the dependent and independent variables so that the residuals become homogenous. In GLS, the weighting applies a transformation to get rid of both heteroskedasticity and autocorrelation in the residuals. It is tricky to determine what the weights should be, so if they are not known or they cannot be assumed like in WLS, then they can be estimated using **feasible generalized least squares (FGLS)** (Boyd & Vandenberghe, 2004).

Partial Least Squares (PLS)

Partial least squares is an estimation method that handles multicollinearity and situations when there are more predictor variables than observations in the dataset. PLS works by reducing the number of predictors to a smaller subset of uncorrelated latent features, and then estimates a regression using least squares and the latent features instead of the original predictors. As we will see later, the process of reducing the number of predictors is called **dimension reduction**. **Latent features** are hidden abstractions that account for the most variance in the predictors. We will look at latent features when we explore dimension reduction in a later chapter.

Maximum Likelihood Estimation (MLE)

Maximum likelihood estimation is a robust estimation method that estimates the parameters of a model such that when the values of the predictors are placed into the model, the resulting predicted values of the dependent variable are closest to the observed values. In essence, MLE finds the parameters that maximize the likelihood of observing the true values of the dependent variable. As indicated by its name, MLE is different from the other optimization methods we have seen in that it seeks to maximize, rather than minimize, an objective function. That objective function is called the likelihood function.

$$likelihood(\theta) = \prod_{i=1}^{N} f(x_i|\theta)$$

Figure 7.9

The equation in figure 7.9 reads: the likelihood of parameter theta equals the product of a function of x, for every observation i, given parameter theta. Note that the likelihood is equivalent to probability, so the equation may also be written with P(theta). Rather than maximizing the product in the equation in figure 7.9, it can be assumed that since the logarithm is an increasing function, it would be equivalent to maximize the log likelihood of the sum. Solving for the maximum likelihood is the same as solving for the maximum log likelihood, but solving for the log likelihood is mathematically easier because it involves a sum rather than a product, so it is more commonly used as the objective function for MLE. The reason it works so well is that the log of a function reaches its maximum value at the same points as the function itself.

$$\log likelihood(\theta) = \sum_{i=1}^{N} \log(f(x_i|\theta))$$

Figure 7.10

By taking the first derivative of the equation in figure 7.10 and setting it equal to 0, it is possible to determine the optimal parameter theta, in terms of x. After algebraic manipulation to put theta in terms of x, the result is shown in figure 7.11.

$$\hat{\theta} = \frac{\sum_{i=1}^{N} x_i}{N}$$

Figure 7.11

In the equation in figure 7.11, theta hat is the estimate of parameter theta, and N is the number of observations.

MLE should be used over OLS when the sample size is very large, because it produces more precise estimates of the parameters and lower standard error. Some data scientists prefer using MLE by default instead of OLS, since it is also more robust to non-normal residuals (Hastie, Tibshirani, & Friedman, 2009).

Gradient Descent

The optimization methods we have seen so far use differentiation of a cost function, and substitution of every observation into the result, to arrive at the optimal parameters for a model. As we have seen, using partial derivatives for models with more than 1 parameter is tedious, and this step is done before substituting the observation values into the equation. For models with hundreds or more features/predictors, and thousands or even millions of observations, it becomes too time consuming to use least squares methods, even for computers. Fortunately, there are several computational optimization techniques that take iterative approaches to converge on a solution more quickly. Gradient descent is one of the most well-known and intuitive of these approaches (Ruder, 2016).

Gradient descent starts with an initial set of model parameters and iteratively moves toward the set of parameters that minimize the cost function. As its name implies, gradient descent works by calculating the gradient (slope) of the cost function for each iteration, and continues as long as the gradient is negative (this would indicate the function is continuing to decrease). Figure 7.12 shows an example of how gradient descent iteratively finds the minimum of a function. Note that the w stands for weight, which is another word for a model parameter, just like how theta is used to represent a model parameter for MLE.

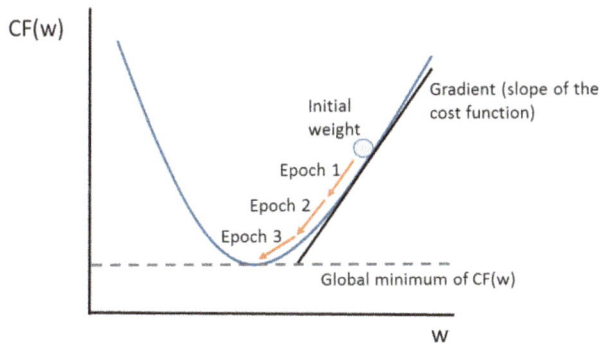

Figure 7.12

For simplicity, let us continue to use a linear model with two parameters, alpha and beta, as our example. The slope of the cost function can be calculated by taking the first partial derivative for each parameter, just like in least squares. To visualize this, we can think of every possible value for our two parameters existing on a 2D plane. As the slope of the cost function for each parameter decreases, the cost function approaches a minimum. If the parameters of the current iteration result in a smaller value for the cost function than the parameters of the previous iteration, then the parameters of the current iteration become the new best parameters. Each iteration or pass over the training dataset is called an **epoch**.

Gradient descent is like walking down a hill and checking the slope at each step. If the slope is negative, then we continue taking steps. But how do we know how big of a step to take? If we take very small steps, it would take us forever to walk down the hill, but if we took a giant leap, we might miss the bottom of the hill and end up on the other side (see figure 7.12 – too large a step could miss the global minimum). Step size is called the learning rate of the gradient descent algorithm. The learning rate must be optimized for gradient descent to work properly, so not only do the model parameters have to be optimized, but now we also need to optimize the optimization algorithm's higher level learning rate parameter. Higher level parameters for optimization algorithms, like the learning rate, are called **hyperparameters**.

Fortunately, the search space (the range of possible values) for hyperparameters is usually much smaller than the search space for model parameters, so optimizing them is a bit more trivial than optimizing the model parameters. We will look at methods for optimizing hyperparameters in the next chapter.

Gradient descent requires substituting every sample in the training dataset, per epoch/iteration, into the partial derivatives in order to update a single model parameter, just like least squares. This is very time consuming. A shortcut is to randomly select one sample or a subset of training samples to plug into the partial derivatives to determine the optimal parameters. Randomly sampling the training samples that are used to compute the optimal parameters is called stochastic gradient descent.

Stochastic Gradient Descent

Stochastic gradient descent (SGD) only uses a random sample of the training observations to update the optimal model parameters. Unlike gradient descent and least squares, which calculate the true values of the optimal parameters, SGD calculates estimates of the true optimal parameters. SGD does not minimize the cost function as well as gradient descent, but it takes much less time to converge than gradient descent, therefore it is the preferred method for large datasets (Ruder, 2016). This does not mean SGD is inaccurate though: the estimates are usually good enough so that once the optimal values are approached, the algorithm bounces around them a very slight amount, essentially indicating that the algorithm has converged on the optimal solution. Since the optimized parameters found by SGD are estimates of the true optimal parameters, they are said to be stochastic approximations, hence the name stochastic gradient descent.

Pure SGD only uses 1 observation to update the model parameters per epoch, but a more common approach is to use random subsets of the observations called mini-batches (the entire set of observations would be a batch). SGD with mini-batches is a compromise between pure SGD and gradient descent, because the larger samples allow for better approximations of the optimal parameters. The batch size is a hyperparameter that must be optimized. It has a range of possible values from 2 to N-1, where N is the number of observations in the training dataset (Ruder, 2016).

SGD is very common in machine learning, and often the goal is to find the optimal "weights" of a model. Weights are another word for the model parameters. In linear regression, it is more common to refer to alpha (the intercept) and the beta coefficients as parameters, but they could also be called weights. Although it is beyond the scope of this book to dive into, the width nomenclature comes from the training of neural networks, which gradient descent algorithms are well known for.

In pseudocode, the gradient descent and stochastic gradient descent algorithms work as in figure 7.13.

Gradient Descent
(all training observations used to update model parameters)

```
#Randomly initialize model parameters
params = (1, 42)

#Iterate over the training set
#Calculate slope of cost function and update parameters
for epoch in range(number_epochs):
    cf_slope = calculate_slope(cost_function, training_observations, params)
    params = params - (learning_rate * cf_slope)
```

Pure Stochastic Gradient Descent
(one random training sample used to update model parameters)

```
#Randomly initialize model parameters
params = (1, 42)

#Iterate over the training set
#Calculate slope of cost function and update parameters per sample
for epoch in range(number_epochs):
    cf_slope = calculate_slope(cost_function, training_observations, params)
    params = params - (learning_rate * cf_slope)
```

Mini-Batch Stochastic Gradient Descent
(random batch of training samples used to update model parameters)

```
#Randomly initialize model parameters
params = (1, 42)

#Iterate over the training set
```

```
#Calculate slope of cost function and update parameters per mini-batch
for epoch in range(number_epochs):
    for mbatch in create_batches(training_observations, batch_size=10):
        cf_slope = calculate_slope(cost_function, mbatch, params)
        params = params - (learning_rate * cf_slope)
```
Figure 7.13

Objective functions may or may not be linear. Often they are not. Suppose for example that SGD is being used to minimize the non-linear cost function shown in figure 7.14. The algorithm could converge on a local minimum that is not the global minimum.

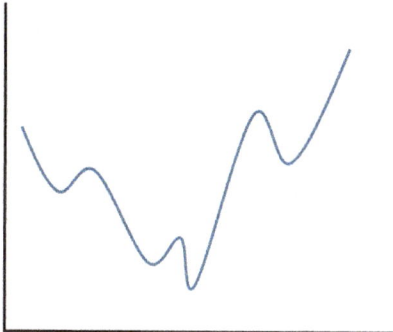

Figure 7.14

Figurer 7.14 shows a simplification of a cost function. When there are multiple parameters, the cost function is a multi-dimensional abstraction with many hills and troughs. For example, if there are 2 parameters, then the cost function is a plane. If one of the parameters slopes down while the other slopes up, it creates a saddle in the plane, which could "trap" the SGD algorithm in a sub-optimal local minimum. One way to get around these local minima is to give the algorithm momentum. **Momentum** is a technique that accelerates SGD toward a particular direction while diminishing the ability of local minima to trap the algorithm. If we think of a ball rolling down the cost function in figure 7.14, it could get stuck in one of the local minima. By giving it momentum to roll through the local minima, it would be more able to find the global minimum. Momentum is added by storing the parameter updates at each epoch for later us. In the next epoch, a fraction of the stored parameter update is added to the new update.

$$new\ w = w - (\eta \Delta CF(w) + \alpha \Delta w)$$

Figure 7.15

The equation in figure 7.15 reads: new weight = old weight − ((learning rate * cost function slope given weight w) + fraction of the previous change in the weight). Momentum is a hyperparameter that should be < 1. A good value to start with is 0.9. The equation in figure 7.15 means that momentum increases as the algorithm moves down a gradient, and then slows when the algorithm reaches a trough or starts moving up a gradient. This results in faster convergence of the algorithm.

A smarter version of momentum is called **Nesterov Accelerated Gradient (NAG).** With NAG, the cost function becomes a function of both the current weight and the momentum term. This gives an approximation of the next change in the weight, which allows the momentum to look ahead and cheat in a sense, to slow down in anticipation of a more gradual slope, and speed up in anticipation of a steeper slope. NAG prevents momentum from becoming too large and increases the algorithm's responsiveness to changes in the cost function's slope. It has been asserted that NAG outperforms SGD with momentum for certain kinds of neural networks, but it is up to the research to decide which method to use. In general, momentum and NAG are useful for non-linear cost functions with many features (Ruder, 2016). For the purposes of linear regression, momentum and NAG may only rarely be necessary.

$$new\ w = w - (\eta\Delta CF(w - \alpha\Delta w) + \alpha\Delta w)$$

Figure 7.16

Let us now return to the learning rate or step size. The learning rate is a hyperparameter that must be tuned so that the optimization algorithm does not overshoot the cost function's minimum, but also does not take forever to reach it. One way to configure the learning rate would be to take larger steps during the first few epochs and gradually take smaller steps as the gradient starts to level out. This method is called an **adaptive learning rate**. We could adjust the learning rate for all parameters at once, through a **learning rate schedule** where the learning rate decays over time as the algorithm converges. For example, rather than using a constant learning rate, we could set the learning rate to 0.1 − (0.001*epoch), which would make learning slow down over time. When a learning rate schedule is used, it is advisable to use a larger momentum hyperparameter to avoid getting stuck at a local minim when the learning rate gets small. Rather than adjusting the learning rate for all parameters at once, we could adjust the learning rate differently for different parameters, depending on their importance. This approach is called AdaGrad.

AdaGrad, AdaDelta, RMSprop, and Adam

AdaGrad is a flavor of gradient descent that adapts the learning rate so that parameters for features that appear more frequently are given lower learning rates, while parameters for features that appear less frequently are given higher learning rates. The idea is that when a rare feature occurs, the algorithm should take notice. Put another way, AdaGrad gives large learning rates to parameters that update less frequently or whose updates are very big changes, and smaller learning rates to parameters that update more frequently or whose updates are very small changes (Ruder, 2016).

Typically only categorical features are said to "appear more frequently". For example, a dummy variable that appears frequently will have a value of 1 for most of the observations in dataset. If a variable only appears rarely, it generally has more predictive power, so the AdGrad algorithm ensures that weights or parameters for these rare features are given large learning rates. Although AdaGrad favors infrequent categorical features, it can be used for mixed datasets with numeric features as well.

The nature of AdaGrad favoring rare features makes it useful for working with sparse data, or text or image data. The learning rate updates for each parameter, and the parameters are updated as in figure 7.17.

$$new\ w = w - \frac{\eta}{\sqrt{G_{e,jj}}} * \Delta CF(w)$$

Figure 7.17

The equation in figure 7.17 must be carried out for every individual parameter. The variable G is a matrix containing the sum of squares of all of the parameter's historical cost function gradients, and each diagonal element of the matrix j,j is the sum of squares of the past gradients with respect to the parameter w, up to epoch e. Delta CF(w) is the gradient of the cost function, given parameter w.

The greatest benefit to AdaGrad is that it eliminates the need to tune the learning rate. Most implementations of the AdaGrad algorithm set the learning rate to 0.01 and let it go, as it will adapt as necessary. The drawback to using AdaGrad is that the accumulation of the sum of squared gradients in the denominator (variable G in figure 7.17) causes the learning rate to continuously shrink and eventually become infinitesimal. When this happens, no more learning can occur. To prevent this from happening, a cap can be placed on the accumulated past gradients. This is what the AdaDelta algorithm does.

AdaDelta is nearly identical to AdaGrad, except it limits the number of past gradients used to update learning rates for individual parameters to some fixed size. Whereas AdaGrad stores the sum of squares of all past gradients, AdaDelta adjusts the learning rate by adding a decaying moving average of past gradients. AdaDelta entirely eliminates the need to tune a learning rate.

RMSprop and **Adaptive Moment Estimation (Adam)** are two other variants of AdaGrad that, like AdaDelta, solve the problem of the adaptive learning rate shrinking to near zero. Adam is considered to be a generalization of AdaGrad. Like AdaDelta, both Adam and RMSprop divide the learning rate by an exponentially decaying moving average of squared gradients. The differences between AdaDelta, RMSprop, and Adam are minimal – they all use different mathematical tricks to accomplish the same thing.

Simulated Annealing

So far we have looked at algorithmic and iterative computational approaches to optimization. Another group of computational optimization methods is called heuristics. Unlike the other approaches we have seen, heuristics are not mathematically guaranteed to converge on an optimal solution; they are used to approximate optimal solutions. In general, heuristics trade precision for speed, as they are faster than iterative computational methods (Segaran, 2007). The first heuristic we will explore is simulated annealing. Annealing is the process of heating and cooling a material, especially metal, to make it more pliable o that it can be more easily formed. Similarly, simulated annealing involves moving between optimal and suboptimal solutions in an attempt to find the global optimum of a function. Like momentum in SGD, simulated annealing is a way to avoid getting trapped in local minima.

Simulated annealing starts with a random solution to the problem. For curve fitting tasks, like linear regression, this would mean choosing random initial model parameters. Due to the way simulated annealing works, the starting point does not matter, as the algorithm will find a good solution no matter what the starting points are. After solving the cost function using the initial parameters, the algorithm generates random new parameters for a new solution to the cost function. If the new solution is better, then the new parameters replace the old as the optimal parameter set. If the new solution is worse, then the algorithm considers an acceptance probability, and it either stays with the old parameters, or moves to the new ones even if they are worse. It is this probability of sometimes choosing a worse solution that allows simulated annealing to escape local minima. Looking back at figure 7.14, choosing a worse solution would be like moving up the slope of one side of a local minimum in hopes of finding another minimum somewhere else that might be the global minim.

The key to this algorithm is the acceptance probability, which is calculated as in figure 7.18.

$$a = e^{\frac{CF_{old}-CF_{new}}{T}}$$

Figure 7.18

In the equation in figure 7.18, a is the acceptance probability, CF is the cost function (either old or new for the new set of parameters), e is Euler's number (2.71828), and T is temperature. Temperature is a hyperparameter inspired by the annealing process in metallurgy. It typically starts at 1.0 and decreases with every iteration by a factor of alpha, another hyperparameter. Alpha is typically a value between 0.8 and 0.99 (Segaran, 2007). Temperature multiplied by alpha gives the new temperature for the next iteration. Looking at figure 7.18, it should be clear that the acceptance probability decreases when temperature decreases (i.e. when the new solution is worse than the old one), and is always > 1 when the new solution is better than the old one. This means that simulated annealing is more likely to accept somewhat bad solutions more than really bad solutions, as the somewhat bad solutions occur earlier on in the training process, when the temperature is high (Segaran, 2007).

Genetic Algorithm

Simulated annealing generates a new solution to compare the old solution to, but the genetic algorithm generates many new solutions to compare to the old solution. The genetic algorithm is inspired by the process of natural selection and includes features like mutation and crossover (breeding). The genetic algorithm has a much higher computational cost than simulated annealing, so it takes more time to find a solution. Its advantage over simulated annealing is that it is easily parallelizable. This means that he algorithm's computation can be easily split between several computer processors, each acting as a search agent over the parameter space. The parallelized independent search agents reach an optimal solution more quickly than if there were only 1 search agent. The genetic algorithm has a wide range of use and there is even a branch of artificial intelligence called genetic programming, in which computer programs are "evolved" and optimized by a genetic algorithm to perform well at specific tasks. It is conceivable that a program could be written to build linear regression models, and optimize them through genetic programming (McCall, 2004).

The genetic algorithm starts with a population of randomly generated individuals (parameter sets that solve the cost function). This population is called the first generation. The next iteration of the algorithm produces the next generation of solution sets. Each solution set or individual is evaluated by solving the cost function for those parameters. The evaluation is called "fitness" as inspired by natural selection. More fit individuals from the current population are selected to be passed on to the next generation. There are generally 3 ways the individuals can be passed on, although other ways exist:

1. **Elitism** – choose the best fit individuals and pass them to the next generation without modification
2. **Mutation** – choose a sample of the best fit individuals and change something about them, such as one of the parameters, before passing them on to the next generation

3. **Crossover** – combine a sample of the individuals in some way, such as one parameter from a poorly fit individual with the parameters from a well fit individual, and pass the combined results to the next generation

One hyperparameter for this algorithm is the sampling methodology for individuals to be passed on. Usually the more fit individuals are favored, but they can be combined with the less fit individuals. On average, each generation should be more fit than the last. Another hyperparameter is the mutation methodology, as usually there is only a small probability of mutation.

The genetic algorithm terminates when solution that minimizes the cost function is found, when a pre-defined number of generations have been created, or when the fitness level of individuals has plateaued and no better fit individuals are being produced.

The genetic algorithm has several potential drawbacks:

- Unless computation is distribute din parallel, the algorithm will take a long time to find a solution.
- The algorithm does not scale well, so when there are more individuals or more parameter sets to test, the search space for optimal parameters blows up exponentially.
- There is no failsafe for the problem of running into a local minimum.
- Before the genetic algorithm can be used, the problem must be put into a representation that can be sued by the algorithm, such as representing the individuals as an array of bits (binary 1's and 0's). Then the results must be decoded back after the algorithm finds a solution, which can be tricky because some encoded results might not have a decoded counterpart. These "orphaned" results are simply ignored.

The genetic algorithm, like all optimization algorithms, can be used to find an optimized solution to a problem, but it cannot redefine the problem. For example, given a candle, an optimization algorithm can produce a better candle, but it cannot develop the lightbulb.

Particle Swarm

Like the genetic algorithm, particle swarm optimization (PSO) is a heuristic that generates many new solutions to compare to the old solution. Particle swarm is even a good choice to solve **multi-objective optimization** problems, in addition to **single objective** problems. Single objective problems have only 1 objective or cost function, whereas multi-objective problems have 2 or more. A situation in which there might be multiple objectives would be buying a refrigerator: the goal would be to minimize cost while maximizing quality, and typically there is a tradeoff between each objective. Solutions to multi-objective problems cannot minimize both cost functions at the same time, so a balance must be struck. For our purposes, we will lonely look at particle swarm for single objective optimization.

Particle swarm is inspired by the swarming f bees or flocking of birds, where several individuals of lesser intelligence seemingly move in unison as part of a more intelligent whole (Kennedy & Eberhart, 2011). To simulate a swarm, a number of particles is chosen and their positions are randomized in the search space. The particles are then given a random velocity and move toward more optimal positions. Collectively, their positions are evaluated to come up with the swarm's solution. In pseudocode, the algorithm can be implemented as in figure 7.19.

```
#Initialize parameters
swarm_size = 20
best_swarm_position = 1

#Randomly initialize the positions and velocities of the particles
for particle in range(swarm_size):
    particle_velocity = random_nbr()
    particle_position = random_nbr()
    particle_best_position = particle_position
    if (cost_function(particle_best_position) <= cost_function(best_swarm_position):
        best_swarm_position = particle_best_position

#Iterate over the particles in the swarm until the stop condition is met
#In each iteration, update each particle's position and velocity and compare it to the best
while current_condition != stop_condition:
    for particle in range(swarm_size):
        particle_velocity = update_velocity(particle_velocity, particle_position, best_swarm_position)
        particle_position = update_positiion(particle_position, particle_velocity)
        if cost_function(particle_position) <= cost_function(particle_best_position):
            particle_best_position = particle_position
            if cost_function(paticle_best_position) <= cost_function(best_swarm_position):
                best_swarm_position = particle_best_position
```

Figure 7.19

Notice that like the genetic algorithm, PSO terminates when some condition is met. The stop condition is determined by the researcher. There are two functions listed in figure 7.19 to update the particle's velocity and position. The equation to update the particle's velocity is given in figure 7.20.

$$v_i(t+1) = v_i(t) + \left(w_1 * random\_nbr * \left(p_i^{best} - p_i(t)\right)\right) + \left(w_2 * random\_nbr * \left(p_{s\_best} - p_i(t)\right)\right)$$

Figure 7.20

The equation in figure 7.20 reads: the velocity for particle i at the next iteration (the current iteration is iteration t) equals the particle's current velocity plus the adjustment to the weight of the particle's best position, plus the adjustment to weight of the swarm's best position. P sub i is the particle's position, and P sub s_best is the swarm's best position. The random number is a number between 0 and 1 that gives the probability of the weight adjustment. The purpose of the weights is to balance the particle's movement so that it moves on its own to some extent, but is also influenced by the rest of the swarm (Kenned & Eberhart, 2011). Looking back at the equation, the difference between the particle's best position and its current position is multiplied by a probability and weight 1, and the difference between the swarm's best position and the particle's current position is multiplied by a probability and weight 2. These parts of the equation accomplish the balancing between the particle's own movement and the swarm's influence. The equation to update a particle's position is as follows:

$$p_i(t+1) = p_i(t) + v_i(t)$$

Figure 7.21

The equation in figure 7.21 reads: the position of particle i at the next iteration equals the particle's current position plus the particle's current velocity.

The hyperparameters for PSO include the number of particles (typically 20-40), the speed that a particle can move (the maximum change in a particle's position during 1 iteration should be bounded so that it is not more than some percentage of the domain of possible positions in the search space, for example), and the weights for the particle and swarm influence (usually between 0 and 4) (Kennedy & Eberhart,

2011). It is possible that a particle could jet off towards empty search space, never to return. Although it is not shown in these equations, a wrapping function could be added to ensure that the particles do not leave the search space entirely; it would confine the particles within a known search space.

Figure 7.22 shows a representation of how particle swarm finds an optimal solution. The plot on the left shows the initial random configuration of the particles. The plot on the right shows how the particles collectively find the global minimum, even though some particles might get stuck in local minima.

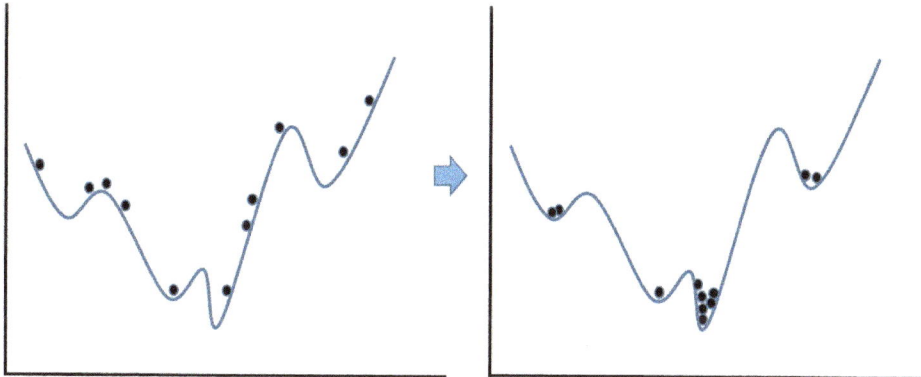

Figure 7.22

Optimization with Python

Most of the optimization algorithms in this chapter can be passed as an argument into a modeling function in R and Python. For this reason, I have omitted examples using data. But I would like to show alternative uses for a couple of the algorithms we looked at, using Python. This will be a digression from the rest of the book, but I hope to show here that these optimization algorithms have many use cases outside the area of model fitting.

Let's consider the **traveling salesman problem**. If we employ a salesman to visit a list of cities, what is the shortest possible path between all of the cities that would only require the salesman to visit each city once? The path distance is traditionally determined by miles or kilometers between cities, but it could also be determined by travel time or travel cost. We are going to stick to distance in miles. Let's start by getting the latitude and longitude coordinates for a list of 16 cities.

```
from math import sin, cos, sqrt, atan2, radians
from random import random
import numpy as np

#Create a dictionary of cities with lat/lon coordinates
cities = {
    'New York': (40.71, 74.00),
    'Washington D.C.': (38.91, 77.04),
    'Chicago': (41.88, 87.63),
    'Los Angeles': (34.05, 118.24),
    'San Francisco': (37.77, 122.42),
    'Dallas': (32.78, 96.80),
    'Miami': (25.76, 80.19),
    'Houston': (29.76, 95.37),
    'Atlanta': (33.75, 84.39),
    'Philadelphia': (39.95, 75.17),
```

```
    'Detroit': (42.33, 83.05),
    'Salt Lake City': (40.76, 111.89),
    'Seattle': (47.61, 122.33),
    'Austin': (30.27, 97.74),
    'Columbus': (39.96, 83.00),
    'Indianapolis': (39.77, 86.16)
    }
```

Now we can define a function to calculate the distance between two coordinate pairs.

```
#Create a function to calculate distance between coordinates
def distance(c1, c2):
    r = 3959.0  #Radius of the Earth in miles
    lat1 = radians(c1[0])
    lat2 = radians(c2[0])
    lon1 = radians(c1[1])
    lon2 = radians(c2[1])
    dist_lat = lat2 - lat1
    dist_lon = lon2 - lon1
    #Use Haversine formula to get shortest distance over earth's surface
    a = sin(dist_lat/2)**2 + cos(lat1) * cos(lat2) * sin(dist_lon/2)**2
    c = 2*atan2(sqrt(a), sqrt(1-a))
    return round(r*c, 2)
```

Next we can calculate the distance in miles between every pair of cities. These distances will help us calculate the cost function. Since we aim to minimize the total distance traveled, we will seek to find the sequence of cities resulting in the lowest total distance.

```
#Calculate distances between every pair of cities
distances = {}
for key_a, val_a in cities.items():
    distances[key_a] = {}
    for key_b, val_b in cities.items():
        if key_b == key_a:
            distances[key_a][key_b] = 0.0
        else:
            distances[key_a][key_b] = distance(val_a, val_b)
```

Now we create the cost function, which calculates the total distance for a sequence of cities:

```
#Define an objective function (the total distance for a sequence of cities)
def objective_fxn(city_list):
    total_dist = 0
    for i, c in enumerate(city_list):
        if i == len(city_list)-1:
            next
        else:
            total_dist = total_dist + distances[city_list[i]][city_list[i+1]]
    return round(total_dist, 2)
```

The next block of code initializes the sequence and creates neighboring solutions by swapping the positions of 2 cities in the sequence at a time.

```
#Define the starting sequence and build a function to swap 2 cities at a time
initial_seq = list(distances.keys())
```

```
def swap(city_list, city_a, city_b):
    city_list[city_a], city_list[city_b] = city_list[city_b], city_list[city_a]
    return city_list
```

Let's build our simulated annealing optimization algorithm now.

```
#Simulated annealing optimization function
def simulated_anneal(city_list, T=1.0, T_min=0.000001, alpha=0.9):
    old_cost = objective_fxn(city_list)
    while T > T_min:
        for i in range(len(city_list)):
            for j in range(len(city_list)):
                new_seq = swap(city_list, i, j)
                new_cost = objective_fxn(new_seq)
                acceptance_prob = np.exp((old_cost-new_cost)/T)
                if acceptance_prob > random():
                    best_seq = new_seq
                    old_cost = new_cost
        T = T*alpha
    return best_seq, old_cost
```

We can now compare the initial sequence of cities and its total distance traveled with the optimized sequence of cities and its total distance:

```
print(initial_seq, objective_fxn(initial_seq))
print(simulated_anneal(initial_seq))
```

After running this a few times, it becomes clear that simulated annealing produces good solutions but not necessarily the most optimal. This makes sense based on what we learned about heuristics. This code is far from optimal, as it re-calculates the distance for the entire sequence of cities during each iteration, rather than only the 2 that are swapped, but it illustrates the concept.

Now let's use the particle swarm algorithm to optimize flight schedules. Suppose we need to fly from Tampa, FL (TPA) to London Gatwick (LGW). A selection of possible flights is shown below in CSV format, in the order of origin, destination, departure time, arrival time, price in USD. Note that the times extend beyond 24 hours, but that is to keep things simpler.

```
TPA,ORD,08:55,11:50,145
TPA,JFK,09:15,11:55,115
TPA,ATL,12:10,13:40,83
TPA,ATL,14:45,16:15,90
TPA,ATL,08:41,10:18,85
TPA,CLT,11:25,01:03,200
TPA,IAH,07:00,09:10,110
ORD,YYZ,16:31,18:05,150
JFK,LGW,20:33,27:33,850
JFK,LGW,23:33,30:40,875
ATL,JFK,14:55,17:09,150
ATL,CDG,10:09,18:38,730
CDG,LGW,23:59,25:05,70
CLT,JFK,15:38,17:38,100
IAH,YYZ,13:05,16:15,250
YYZ,LGW,21:47,29:00,350
YYZ,LGW,22:47,29:59,325
```

```
YYZ,LGW,23:26,30:33,340
```

Our goal will be to get to LGW using the cheapest route possible with as few layovers as possible. We can start by creating a list of flights.

```python
from random import randint, random

class flight(object):
    def __init__(self, origin, dest, depart, arrive, price):
        self.origin = origin
        self.dest = dest
        self.depart = depart
        self.arrive = arrive
        self.price = int(price)

flights = []
with open('flight_schedule.txt') as file:
    for line in file:
        origin, dest, depart, arrive, price = line.strip().split(',')
        flights.append(flight(origin, dest, depart, arrive, price))
```

This problem has specific constrains: it is not possible to interchange any two flights in a sequence, because some flights must follow others. If we fly from Tampa to Orlando, then every next flight in the sequence must leave from Orlando and at a departure time that is later than the arrival time in Orlando. This limits the number of possible flight sequences to 6. The next block of code creates the 6 possible flight sequences.

```python
#Replace : in times with . to make them easier to work with
for f in flights:
    f.depart = float(f.depart.replace(':', '.'))
    f.arrive = float(f.arrive.replace(':', '.'))

#Determine possible sequences of flights
flight_seqs = []
for f in flights:
    seq = []
    if f.origin != 'TPA':
        break
    seq.append(f)
    airport = f.dest
    time = float(f.arrive)
    #print(f.origin, airport, time)
    #Second flight
    for ff in flights:
        if ff.origin == airport and ff.depart > time and len(seq) < 2:
            #print(ff.origin, ff.dest, ff.arrive)
            seq.append(ff)
            airport = ff.dest
            time = float(ff.arrive)
            if airport == 'LGW':
                flight_seqs.append(seq)
                #print(len(flight_seqs))
            #Third flight
            for fff in flights:
```

```
            if fff.origin == airport and fff.depart > time and len(seq) < 3:
                #print(fff.origin, fff.dest, fff.arrive)
                seq.append(fff)
                airport = fff.dest
                time = float(fff.arrive)
                flight_seqs.append(seq)
                #print(len(flight_seqs))
```

Now the objective function can be defined. Since we want something that minimizes both cost and number of layovers, multiplying them together is appropriate. We will just have to remember to divide if we want the answer in terms of USD.

```
#Define the objective function to be minimized, given a flight schedule
def objective_fxn(flight_sch):
    total_price = 0
    layovers = 0
    lgw_arrival_time = 0.0
    for f in range(len(flight_sch)):
        total_price = total_price + flight_sch[f].price
        layovers = layovers + 1
    return total_price * layovers
```

Now we can build the particle swarm algorithm, following the pseudo code and equations in figures 7.19, 7.20, and 7.21.

```
class particle(object):
    def __init__(self, max_step, max_pos):
        self.vel = randint(0, max_step)  #Move at most x steps in any direction
        self.pos = randint(0, max_pos) #Random start position in the sequence
        self.best_pos = 0

#Particle Swarm Optimization
#Note that the position is a index number for the flight_seqs
#flight_seqs lists all possible flight sequences
def particle_swarm_opt(d, swarm_size=min(len(flight_seqs), 20), best_swarm_pos=0):
    particles = []
    for p in range(swarm_size):
        particles.append(particle(max_step=2, max_pos=len(flight_seqs)-1) )
        particles[p].best_pos = particles[p].pos
        #print(particles[p].best_pos, best_swarm_pos)
        if objective_fxn(flight_seqs[particles[p].best_pos]) <=
objective_fxn(flight_seqs[best_swarm_pos]):
            best_swarm_pos = particles[p].best_pos
    #Custom stop condition will be at 50 iterations
    i = 1
    while (i < 51):
        for p in range(swarm_size):
            particles[p].vel = particles[p].vel +
int(round((0.5*random()*(particles[p].best_pos-particles[p].pos)) +
(0.5*random()*(best_swarm_pos-particles[p].pos)), 0))
            particles[p].pos = particles[p].pos + particles[p].vel
            #The next if statement handles positions that stray outside the range of
values
            #5 is hardcoded because there are only 6 possible sequences
            if particles[p].pos > 5:
```

```
                    particles[p].pos = 5
            if objective_fxn(flight_seqs[particles[p].pos]) <=
objective_fxn(flight_seqs[particles[p].best_pos]):
                    particles[p].best_pos = particles[p].pos
                if objective_fxn(flight_seqs[particles[p].best_pos]) <=
objective_fxn(flight_seqs[best_swarm_pos]):
                    best_swarm_pos = particles[p].best_pos
            i = i + 1
    return best_swarm_pos, objective_fxn(flight_seqs[best_swarm_pos])
```

Note that since there are only 6 possible sequences, there can be a maximum of 6 particles, and the algorithm converges on the exact optimal solution. These 6 sequences are like the parameters sets that optimize statistical models. For larger problems with more parameter sets, PSO does not always converge on the best possible solution, as it is still a heuristic. Run the following code to see the best schedule:

```
opt_flight_sch, lowest_cost = particle_swarm_opt(flight_seqs)
print('Best flight sequence: ', opt_flight_sch)
for f in flight_seqs[opt_flight_sch]:
    print(f.origin, f.dest)
print('The cost of this flight schedule is: ',
lowest_cost/len(flight_seqs[opt_flight_sch]))
```

8. Hyperparameter Optimization

In chapter 7, we saw that computational optimization methods have hyperparameters that must be tuned. So far we have not looked at methods to optimize or tune hyperparameters. One might be tempted to use any of the optimization algorithms described in chapter 7 to tune another optimization algorithm's hyperparameters, but hyperparameter tuning is a non-convex optimization problem. This means that the local minimum is not guaranteed to be the global minimum. Heuristics like the genetic algorithm and PSO could be used to tune hyperparameters, but it is much easier and thus more common to use one of the techniques described in this chapter.

The challenge with hyperparameter tuning is that hyperparameters cannot be put into an equation that can be optimized; therefore, the hyperparameter function and its derivative are unknown. When the function to be optimized is unknown, it is referred to as a **black box function**. Black box functions cannot be optimized by techniques like gradient descent. Since there is no known function to optimize, a workaround is to choose hyperparameters that optimize a metric like MSE, RMSE, classification accuracy, or log-likelihood. Grid search and random search take this approach. Another option is to use a surrogate function to approximate the black box function that optimizes the metric of choice, as we will soon see.

Before going further in this chapter, it should be noted that there is often a great deal of emphasis put on tuning models in data science literature. It is of the utmost importance however to remember that no amount of model tuning can make up for bad data, a lack of data, or a lack of feature engineering. The best models are always the ones that are built with due diligence; they have novel features and a good quantity of clean data that comes from extensive exploratory data analysis. Hyperparameter tuning should be used to nudge a model that is already a good fit to an even higher level.

Grid Search and Random Search

Grid search is a brute force optimization method, meaning it involves trying every possible combination of hyperparameters to determine the best ones. It works by sequentially checking every combination of hyperparameters in the search space and returning the combination that results in the best evaluation metric. Grid search is never used as an optimization algorithm for model parameters, because the number of parameter combinations could be infinite, but it is an option for tuning hyperparameters. Data scientists use grid search when the search space is small, such as when it is possible to list a handful of good choices for the hyperparameter values (Gonzalez, 2017).

Random search is similar to grid search, except the hyperparameters are chosen from a random distribution. Random search starts from a random position and continues choosing random nearby solutions with each iteration, until a stopping criterion like the number of iterations or a desired fitness of the evaluation metric is met, or until no better solutions can be found. It has been shown that if the approximately optimal region of hyperparameters occupies 5% of the search space, then random search will likely find them in under 60 iterations (Gonzalez, 2017). Random search is considered to be better than grid search in terms of accuracy and scalability. For these reasons, random search is considered the de-facto hyperparameter tuning method by many data scientists. Like grid search, random search is not ideal for estimating the model parameters, but it is useful for tuning a model's hyperparameters.

Bayesian Optimization

Bayesian optimization uses a **surrogate** function to approximate the black box objective function (Brochu, Cora, & de Freitas, 2010). Techniques like Bayesian optimization that use a surrogate to model a black box function are part of a class of algorithms known as **sequential model-based global optimization (SMBO)**. SMBO's do not require calculating derivatives (although some may require integration). The surrogate function for Bayesian optimization is a Gaussian (random) process, which will be defined momentarily. First it is necessary to define Bayes' theorem.

Bayesian optimization is based on **Bayes' theorem**. We will explore Bayes' theorem and Bayesian statistics at the end of this book, but for now it is only necessary to know that Bayes' theorem states that the posterior probability of a model M given evidence E is proportional to the probability of E given M times the prior probability of M. This is represented by figure 8.1.

$$P(M|E) \propto P(E|M)P(M)$$

Figure 8.1

The prior probability represents the researcher's belief about the search space for objective functions. If the objective function of the hyperparameters (the black box function) is considered to be from a distribution of functions, then the prior probability distribution, P(M), is a distribution that embodies the researcher's beliefs about the function, before any evidence is considered. For example, this would be the researcher's guess about the initial probability of drawing the black box function from a Gaussian (random) distribution of black box functions. A distribution of black box functions, or any functions for that matter, is called a **Gaussian Process**. The initial prior probability (a.k.a. the first prior) is a hyperparameter, because its components are chosen by the researcher. The posterior probability, P(M|E) is the distribution of functions that results from combining the prior with the likelihood of distribution E, given model M. It can be thought of as the updated belief after new evidence has been combined with prior belief.

A Gaussian process is a distribution of functions with mean m and covariance k (recall from chapter 3 that covariance shows how well the variance of random variables, or in this case random functions, move together). It can be represented as in figure 8.2.

$$f(x) \sim GP(m(x), k(x, x'))$$

Figure 8.2

The mean of a Gaussian process, for the purpose of defining the prior, is assumed to be a zero function: m(x) = 0. The covariance can be defined any number of ways, so it is a hyperparameter of sorts (it is actually the choice of a function rather than the choice of a single value). A common choice is the squared exponential function in figure 8.3.

$$k(x_i, x_j) = e^{-\frac{1}{2}\|x_i - x_j\|^2}$$

Figure 8.3

In the equation in figure 8.3, x sub i and j are the variances of functions i and j. Note that the covariance approaches 1 as the x values get closer together and 0 as the x values get further apart (points that are closer together influence each other more). The covariance function in figure 8.3 measures the similarity between the variances of functions i and j. Functions that form a distribution of similarity measures like this are called kernel functions. A **kernel function** is a function that shows similarity between two distributions. Aside from the function in figure 8.3, there are other choices for the kernel for Bayesian optimization, such as the Matern kernel.

The Gaussian process in figure 8.1 is highly abstract, but by substituting the mean and covariance and using algebraic manipulation, we can arrive at the equation below:

$$P(f_{t+1}|D_{1:t}, x_{t+1}) = \mathcal{N}(\mu_t(x_{t+1}), \sigma^2(x_{t+1}))$$

Figure 8.4

The equation in figure 8.4 can be read as the posterior probability of model f (the next probability), given the distribution of evidence from iteration 1 to t (the evidence of all previous iterations) equals the multivariate normal distribution ($\mathcal{N}$) of mean mu and variance sigma.

Similar to how MLE estimates the model that would most likely generate a dataset, fitting a Gaussian process involves fitting a distribution of functions to a dataset such that he functions generate predicted data the is close to the observed data. After the prior probability is assigned, every point in the search space is considered, and the one that gives the highest expected improvement will be evaluated next. As this process continues, the information from all previous evaluations will be used to determine what point in the search space to try next. Figure 8.4 shows how this is done mathematically and figure 8.5 illustrates this visually. Each iteration pinches the uncertainty together, as more points from the distribution are calculated.

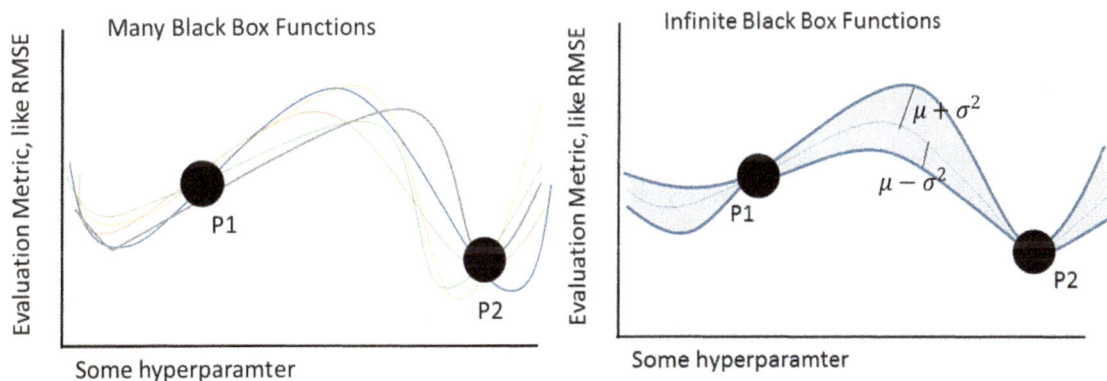

Figure 8.5

In order to determine which point in the search space of black box functions to evaluate next, Bayesian optimization uses a utility function of the evidence so far (e.g. in figure 8.5, the utility function would consider all of the points in order to find the next one). A **utility function** is any function that represents the fit of a selection from a set of options. Whereas a cost function is an objective function to be minimized, a utility function is an objective function to be maximized. In Bayesian optimization, the utility function is also called the **acquisition function**. One acquisition function that could be used in Bayesian optimization is the probability of improvement. The **probability of improvement** gives a unit reward for points in the search space that improve the current minimum, however it does not account for the size of the improvement. **Expected improvement** is an acquisition function that accounts for both the direction and size of an improvement, so it is the most common acquisition function used in Bayesian optimization. Figure 8.6 shows the expected improvement acquisition function.

$$E(I) = \sigma(x)\left[\frac{\mu(x) - f(x^+)}{\sigma(x)}\Phi\left(\frac{\mu(x) - f(x^+)}{\sigma(x)}\right) + \phi\left(\frac{\mu(x) - f(x^+)}{\sigma(x)}\right)\right]$$

Figure 8.6

In the equation in figure 8.6, capital phi is the cumulative density of the normal distribution, lowercase phi is the probability density function of the normal distribution, and x superscript + is a positive value of x. This equation is maximized in order to find which point in the black box function search space to evaluate next, and it gives the point of maximum expected improvement over the current evaluation.

When the improvement is maximized, the next point should balance the mean of the Gaussian process with the variance. The balancing of the mean and variance of a Gaussian process is called the **exploration-exploitation tradeoff**. The tradeoff exists because Bayesian optimization seeks to maximize improvement over the prior (exploitation) while maintaining enough uncertainty to prevent finding local minima (exploration).

In summary, Bayesian optimization uses a distribution of black box functions called a Gaussian process, as a surrogate to approximate the objective function. The Gaussian process is represented by figure 8.4, which consists of a mean (figure 8.2) and covariance (figure 8.3). In order to determine which point in the search space (in other words, which black box function in the distribution) to evaluate next, an acquisition function is used so that the next chosen point gives the best possible improvement over the prior. When the next point is chosen, the Gaussian process is evaluated at that point, and the point becomes the prior for the next iteration.

There are limitations to Bayesian optimization:

- In order to optimize a hyperparameter, it is generally required to provide a range of possible values for the hyperparameter, which is subjective. This limitation applies to grid search, random search, and Tree-Structured Parzen estimation too though.
- Due to the mathematics involved, Bayesian optimization does not scale to high dimensional problems where there are many hyperparameters to estimate.
- Bayesian optimization is vulnerable to poor choices for its own hyperparameters, like the prior probability and covariance function. The wrong hyperparameters for Bayesian optimization can result in poorly optimized solutions (Brochu, Cora, & de Freitas, 2010).

There are only a few implementations of Bayesian optimization, but Spearmint is a popular one for Python and rBayesianOptimization exists for R. The pseudocode below shows how a Bayesian optimizer can be constructed with the help of the Gaussian process module from Python's scikit-learn :

```
import sklearn.gaussian_process as gp

def bayesian_optimization(nbr_iterations, evaluation_func, prior):
    #Define the Gaussian process using scikit's built in function
    model = gp.GaussianProcessRegressor(kernel=gp.kernels.Matern())

    hyperparameters_list = []
```

```
fitness_list = []

#Initialize the algorithm with a prior and evaluate it
#Evaluation function could calculate the model's log loss for example
hyperparameters_list = hyperparameters_list.append(prior)
fitness_list = fitness_list.append(evaluation_func(prior))

for i in range(nbr_iterations):
    #Fit the Gaussian process on the priors, using fitness to evaluate
    #hyperparameters_list contains the black box functions of hyperparameters
    model.fit(hyperparameters_list, fitness_list)
    #Use the acquisition function to determine which point to try next
    next_point = acquisition(model, fitness_list)
    #Evaluate the fitness of the next function of hyperparameters
    next_fit = evaluate_fitness(next_point)

    hyperparameters_list = hyperparameters_list.append(next_point)
    fitness_list = fitness_list.append(next_fit)

return hyperparameters_list, fitness_list
```

Figure 8.7

Tree-Structured Parzen Estimation

Tree-Structured Parzen estimation (TPE) is another SMBO optimization algorithm, like Bayesian optimization. Whereas Bayesian optimization uses a Gaussian process and models P(M|E), TPE uses a **Gaussian Mixture Model** and models P(E|M) and P(M). The difference between a Gaussian process and a Gaussian mixture model (GMM) is that a Gaussian process is a distribution of functions and a GMM is a mixture distribution of several subpopulations within the overall population (Bergstra et. al., 2011). A mixture distribution is the probability distribution of a population that contains subpopulations with fundamentally different probability distributions. For example, consider the normal distribution of human height. Mean height is fundamentally different for men and women, so the distribution of all human height is a Gaussian mixture of two subpopulations. Mean height for all humans is some combination of the mean heights of men and women. It is possible to make statistical inferences about the parameters of subpopulation, given only the observations of the whole population. That is, mixture models can infer properties about the subpopulations, even when the models are not told which observations belong to which subpopulation. This is a strong assertion and it is the basis for unsupervised learning. Although unsupervised learning is beyond the scope of this book, it is important to understand the GMM as it is used by TPE.

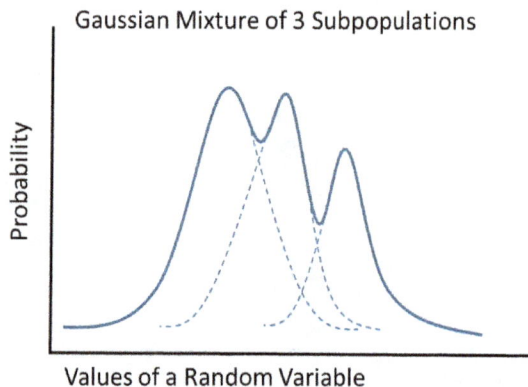

Figure 8.8

Figure 8.8 shows a multi-modal distribution, which is a sign that the distribution might be a mixture of many unimodal distributions for different subpopulations. Gaussian mixtures may not always be so easily spotted as in figure 8.8 though.

Since inference can be made about the subpopulations, given only the population distribution, it is possible to estimate the probability densities of the subpopulation distributions. Recall from chapter 2 that a probability density function is the function that gives the likelihood that a sample drawn from the distribution of a continuous random variable equals some value. In this case, the samples are functions. Density estimation is therefore the estimation of the probabilities that sample functions drawn from a Gaussian mixture equal a certain function (the optimal black box function).

A nonparametric way to estimate the density of a distribution is the Parzen window, otherwise known as kernel density estimation (KDE). KDE involves approximating the density function f hat using a mixture of continuous distributions K centered at x sub i, with a bandwidth equal to h. This is represented in figure 8.9

$$\hat{f}(x) = \frac{1}{Nh} \sum_{i=1}^{N} K\left(\frac{x - x_i}{h}\right)$$

Figure 8.9

The hyperparameters in KDE are the choice of the kernel function and the bandwidth. The bandwidth scales the kernel and determines the degree of mixture between the functions. If h < 1, the kernels are narrower, meaning there is a lower measure of similarity between the functions. If h > 1, the kernels are wider, meaning there is a higher measure of similarity between the functions (Bergstra et. al., 2011).

Now we know where the "Parzen Estimation" comes from in TPE. What about the "Tree-Structure"? A random forest of regression trees is used to fit the Parzen estimation of a black box function to the evaluation metric of choice, such as log loss or RMSE. The trees in the forest have the evaluation metric as the continuous variable to be fitted against. Random forests are beyond the scope of this book, but we will look at regression trees in greater detail in a later chapter.

One of the advantages TPE has over Bayesian optimization is that it can retain dependencies between hyperparameters (Bergstra et. al., 2011). So if any of the hyperparameters depend on any other hyperparameters, as is usually the case, then TPE is a good choice. The limitation is that TPE is inefficient at retaining these dependencies, so one of the hyperparameters should be fixed while the others are allowed to vary and be optimized. Bayesian optimization has an advantage over TPE in that it generally requires fewer searches to reach an optimal solution than TPE.

TPE is more likely than random search to find the best solution in a high dimensional problem. So the more hyperparameters there are to choose from, the better TPE performs than random search (Bergstra et. al., 2011).

There is no implementation of TPE for R, but Hyperopt is an implementation for Python (Bergstra, Yamens, & Cox, 2013).

Hyperparameter Optimization with R

To show how hyperparameter optimization can be used to tune models, we can reuse the Fitbit data from chapter 6, and fit a linear regression using gradient descent instead of OLS. Gradient descent is an iterative algorithm with hyperparameters, and we should expect that the results of a tuned linear regression fitted using gradient descent should produce similar coefficients to a model fitted using OLS.

```
train <- read.csv('train.csv', header=T)
test <- read.csv('test.csv', header=T)
```

There are no libraries for gradient descent for R, so we will build the algorithm from scratch. We can do the same for stochastic gradient descent for comparison. These algorithms are defined as:

```
gd <- function(df, target_var, cost, learning_rate, num_iters, test_set=NULL) {
  #Split data and target variable
  x <- df[,which(colnames(df) != target_var)]
  y <- df[,target_var]
  #These will keep a running history of cost and the coefficient estimates
  cost_history <- double(num_iters)
  coef_matrix_history <- list(num_iters)
  #Initialize coefficient estimates (a 1 column matrix of k predictors + 1 rows)
  coef_matrix <- matrix(c(0,0), nrow=ncol(x)+1)
  #Convert the input data to matrix form and bind it to a column of 1's for the
intercept coefficient
  x_matrix <- cbind(1, as.matrix(x))
  #Implement gradient descent
  for (i in 1:num_iters) {
    error <- (x_matrix %*% coef_matrix) - y
    gradient <- (t(x_matrix) %*% error) %*% (1/length(y))
    coef_matrix <- coef_matrix - (learning_rate * gradient)
    cost_history[i] <- cost(x_matrix, y, coef_matrix)
    coef_matrix_history[[i]] <- coef_matrix
  }
  #Make predictions if test set was provided
  if (!is.null(test_set)) {
    preds <- coef_matrix[1]
    for (i in seq_along(1:ncol(x))) {
      preds <- preds + coef_matrix[i+1]*x[,i]
    }
  } else {
    preds <- "Cannot make predictions because no test set was provided to the
function."
```

```
  }
    return (list(cost_hist=cost_history, final_cost=tail(cost_history, n=1),
  coef_hist=coef_matrix_history, coef=coef_matrix, preds=preds))
  }

  sgd <- function(df, target_var, cost, learning_rate, num_iters, batch_size, seed=14,
  test_set=NULL) {
    #Split data and target variable
    x <- df[,which(colnames(df) != target_var)]
    y <- df[,target_var]
    #These will keep a running history of cost and the coefficient estimates
    cost_history <- double(num_iters)
    coef_matrix_history <- list(num_iters)
    #Initialize coefficient estimates (a 1 column matrix of k predictors + 1 rows)
    coef_matrix <- matrix(c(0,0), nrow=ncol(x)+1)
    #Convert the input data to matrix form and bind it to a column of 1's for the
  intercept coefficient
    x_matrix <- cbind(1, as.matrix(x))
    #Implement stochastic gradient descent
    for (i in 1:num_iters) {
      sample_indices <- sample(nrow(x), batch_size)
      x_matrix_sample <- x_matrix[sample_indices, , drop=F]
      y_sample <- y[sample_indices, drop=F]
      error <- (x_matrix_sample %*% coef_matrix) - y_sample
      gradient <- (t(x_matrix_sample) %*% error) / length(y_sample)
      coef_matrix <- coef_matrix - (learning_rate * gradient)
      cost_history[i] <- cost(x_matrix_sample, y_sample, coef_matrix)
      coef_matrix_history[[i]] <- coef_matrix
    }
    #Make predictions if test set was provided
    if (!is.null(test_set)) {
      preds <- coef_matrix[1]
      for (i in seq_along(1:ncol(x))) {
        preds <- preds + coef_matrix[i+1]*x[,i]
      }
    } else {
      preds <- "Cannot make predictions because no test set was provided to the
  function."
    }
    return (list(cost_hist=cost_history, final_cost=tail(cost_history, n=1),
  coef_hist=coef_matrix_history, coef=coef_matrix, preds=preds))
  }
```

These algorithms are designed to minimize a cost function. We will use the mean squared error as the cost function.

```
#Define the cost function to be MSE
mse_cost <- function(x_matrix, y, coef_matrix) {
  sum((y - (x_matrix %*% coef_matrix))^2)/length(y)
}
```

Now we would run the algorithms as they are to build linear models, however if this is attempted, the resulting coefficients will be returned as NaN. Why might this be? If we look at the coefficient matrix history that is returned by the functions we just defined, we can see that they quickly become very large and then go to NaN. This is because the first few iterations are producing large errors, and since the values of our data are large, the numbers become infinitely large. This is called the **exploding gradient problem**, and it is common with any variant of gradient descent. To overcome this problem, we need to scale the data so that it is much smaller and the numbers do not blow up to infinity before have a chance to come back down and stabilize in later iterations. Standardization was described in chapter 4, and we will apply it here to transform the numeric variables in the dataset to have a mean of 0.

```
train_std <- cbind(train[,-which(colnames(train) %in% c('Steps', 'Calories',
'Active_Minutes', 'Floors_Climbed', 'Hours_Slept', 'Times_Awake'))],
scale(train[,which(colnames(train) %in% c('Steps', 'Calories', 'Active_Minutes',
'Floors_Climbed', 'Hours_Slept', 'Times_Awake'))]))
test_std <- cbind(test[,-which(colnames(test) %in% c('Steps', 'Calories',
'Active_Minutes', 'Floors_Climbed', 'Hours_Slept', 'Times_Awake'))],
test[,which(colnames(test) %in% c('Steps', 'Calories', 'Active_Minutes',
'Floors_Climbed', 'Hours_Slept', 'Times_Awake'))])
```

Now we can perform gradient descent and stochastic gradient descent. For this first round, we will use hardcoded hyperparameters. A good starting point for the learning rate is 0.01, and we will stop after 500 iterations. Looking at the final MSE for gradient descent, we can see it is about 0.31. For stochastic gradient descent, it is about 0.45. Both of these are very good scores for MSE. By looking at linear_reg_gd$cost_hist[1], we can see the MSE after the first iteration as a basis for comparison.

```
#Perform gradient descent and stochastic gradient descent with hardcoded
hyperparameters
#Be sure to remove the target from the test set when passing it as the test_set
argument
set.seed(14)
linear_reg_gd <- gd(train_std, 'Calories',
                cost=mse_cost, learning_rate=0.01, num_iters=500,
                test_set=test_std[,which(colnames(test) != 'Calories')])
linear_reg_sgd <- sgd(train_std, 'Calories',
                cost=mse_cost, learning_rate=0.01, num_iters=500,
batch_size=50,
                test_set=test_std[,which(colnames(test) != 'Calories')])

linear_reg_gd$final_cost
linear_reg_sgd$final_cost
linear_reg_gd$coef
linear_reg_sgd$coef
```

If the cost function is plotted over time, it shows how the cost stabilizes near the later iterations. Notice how the cost function is variable for stochastic gradient descent. This is due to the fact that SGD uses mini-batches to train the data, rather than fitting the gradient over the entire dataset. For large datasets, this makes it much more efficient, as explained in chapter 7.

```
plot(linear_reg_gd$cost_hist, type='line', col='blue', lwd=2,
    main='Cost function for gradient descent', ylab='cost', xlab='Iterations')
plot(linear_reg_sgd$cost_hist, type='line', col='blue', lwd=2,
    main='Cost function for stochastic gradient descent', ylab='cost',
xlab='Iterations')
```

Before continuing, look back at the coefficients of the final model: linear_reg_gd$coef and linear_reg_sgd$coef. Since the data is standardized, they are much different than the coefficients for the linear model built in chapter 6. Our interpretation of them must also change. For example, the coefficient for Steps in the SGD fitted model is 0.61. That means that a change of 1 standardized unit in steps results in a 0.61 standardized unit change in calories burned. This is much more difficult to understand than the interpretation of the coefficients produced by the OLS fitted linear model in chapter 6. Therefore, if it is more important to the researcher to understand how individual predictors effect the target variable, then OLS regression is preferred to gradient descent. However, if all that matters is that a good model is fitted, then either approach would work.

Now we can tune the hyperparameters. To do this, we will use a technique called k-fold cross validation, which will be explained in a later chapter. Until then, just remember that k-fold cross validation is useful for hyperparameter tuning, because the hyperparameters are trained on "folds" or validation sets that are carved out of the training set. For the purpose of example, we will tune the hyperparameters for gradient descent, using grid search and random search. R does not have any packages available to do automated hyperparameter tuning, such as Bayesian optimization or Tree-Structured Parzen estimation.

```
library(caret)
cv_folds <- createFolds(train_std$Calories, k=5)
linear_reg_gd_cv <- function(lrate, niters=50) {
  #Manually perform cross validation
  cv_scores <- list()
  cv_preds <- list()
  for (fold in cv_folds) {
    trainingdata <- unlist(cv_folds[-fold])  #Use the other folds for training
    #print(nrow(trainingdata))
    fold_res <- gd(train_std, 'Calories',
                   cost=mse_cost, learning_rate=lrate, num_iters=niters,
                   test_set=train_std)
    cv_scores <- c(cv_scores, list(fold_res$final_cost))
    cv_preds <- c(cv_preds, list(fold_res$preds))
  }
  #Return a list of the average score for the k folds and the predictions of the fold
with the best score
  return (list(Score=mean(unlist(cv_scores)),
Pred=unlist(cv_preds[which(cv_scores==max(unlist(cv_scores)))])))
}


#Grid search best hyperparameters
bestScore <- 100
bestParams <- list()
for (i in c(0.001, 0.01, 0.1)) {
  for (j in c(10, 100, 500)) {
    res <- linear_reg_gd_cv(lrate=i, niter=j)
    if (res$Score < bestScore) {
      bestScore <- res$Score
      bestParams <- list(lrate=i, niter=j)
    }
  }
}
```

```
bestScore
bestParams

#Random search best hyperparameters
bestScore <- 100
bestParams <- list()
hyperparams <- list(lrate=seq(1:100)/1000, niters=seq(from=10, to=500, by=10))
for (i in seq(1:30)) {
  set.seed(14)
  lr <- sample(hyperparams$lrate, 1)
  n <- sample(hyperparams$niters, 1)
  res <- linear_reg_gd_cv(lrate=lr, niters=n)
  if (res$Score < bestScore) {
    bestScore <- res$Score
    bestParams <- list(lrate=lr, niter=n)
  }
}

bestScore
bestParams
```

Notice that the best score is similar between the two methods.

Hyperparameter Optimization with Python

To show how hyperparameter optimization can be used to tune models, we can reuse the Fitbit data from chapter 6, and fit a linear regression using gradient descent instead of OLS. Gradient descent is an iterative algorithm with hyperparameters, and we should expect that the results of a tuned linear regression fitted using gradient descent should produce similar coefficients to a model fitted using OLS. We can use the SGDRegressor function from sklearn and the hyperopt library for hyperparameter optimization. But first, we should train a model with the default hyperparameters to see how the optimized model compares.

```
import numpy as np
import pandas as pd
from sklearn.preprocessing import StandardScaler
from sklearn.model_selection import cross_val_score, KFold
from sklearn.linear_model import SGDRegressor
from hyperopt import hp, fmin, tpe, rand, Trials, STATUS_OK

#Ignore all deprecation warnings from hyperopt
import warnings
warnings.filterwarnings("ignore")

train = pd.read_csv('train.csv')
test = pd.read_csv('test.csv')

train.columns = ['Steps', 'Calories', 'Active_Minutes', 'Floors_Climbed',
'Hours_Slept',
        'Times_Awake', 'Day_0', 'Day_5', 'Day_6',
        'Day_3', 'Day_1', 'Day_2']
test.columns = ['Steps', 'Calories', 'Active_Minutes', 'Floors_Climbed',
'Hours_Slept',
        'Times_Awake', 'Day_0', 'Day_5', 'Day_6',
```

```
         'Day_3', 'Day_1', 'Day_2']

#Convert to numpy arrays and standardize all values, otherwise SGD error will be
large
x_train = StandardScaler().fit_transform(train.drop(['Calories'], axis=1).values)
y_train = StandardScaler().fit_transform(train['Calories'].values.reshape(-1,1))
x_test = StandardScaler().fit_transform(test.drop(['Calories'], axis=1).values)
y_test = StandardScaler().fit_transform(test['Calories'].values.reshape(-1,1))

#Set up k-fold cross validation
kfold = KFold(n_splits=10, shuffle=True, random_state=14)

#Train SGD regression with default hyperparameters
sgd_reg = SGDRegressor()
sgd_results = cross_val_score(sgd_reg, x_train, y_train, cv=kfold,
scoring='neg_mean_squared_error')
```

Now that we have a baseline model, we can set up our function to optimize the hyperparameters using TPE optimization. This requires specifying an objective function for hyperopt to minimize using the fmin function. We will use negative MSE for the objective function. We also need to set up the search space for the hyperparameters. So we define the penalty as L1, L2, elasticnet, or none. This penalty does not make sense now, but it will be explained in a later chapter. The penalty enables regularized regression models to be tested. The L1 ratio specifies the degree of regularization and alpha is a constant multiplied by the regularization ratio. Different learning rates should also be tested. Our function tunes the hyperparameters over 10-fold cross validation and returns the best model with the mean of the objective function over the 10-folds for that best model. We can add an option to the function to accept either "bayesian" or "random" for the types of hyperparameter optimization. So if we ever wanted to use random search in the future, we could. After running the function, the results are printed so they can be compared to the baseline without optimization.

```
#Train model with optimized hyperparameters

def sgd_model_opt(opt_type='bayesian'):


    #Define objective function to optimize

    def obj_func(params):

        clf = SGDRegressor(**params)

        return {'loss': -cross_val_score(clf, x_train, y_train, cv=kfold,
scoring='neg_mean_squared_error').mean(), 'status': STATUS_OK}


    #Define search space of hyperparameters

    search_space = {

        'loss': hp.choice('loss', ['squared_loss', 'huber', 'epsilon_insensitive',
'squared_epsilon_insensitive']),

        'penalty': hp.choice('penalty', [None, 'l2', 'l1', 'elasticnet']),
```

```python
        'alpha': hp.uniform('alpha', 0.0001, 10.0),

        'l1_ratio': hp.uniform('l1_ratio', 0.0, 1.0),

        'epsilon': hp.uniform('epsilon', 1.00001, 20.0),

        'learning_rate': hp.choice('learning_rate', ['constant', 'optimal',
'invscaling'])

    }


    #Minimize the objective function over the search space using hyperopt

    #Then train the model with the optimized hyperparameters

    trials = Trials()

    if (opt_type=='bayesian'):

        best_params = fmin(obj_func, search_space, algo=tpe.suggest, max_evals=100,
trials=trials)


        #If condition to account for each item in the list passed to loss

        if (best_params['loss'] == 0):

            best_params['loss'] = 'squared_loss'

        elif (best_params['loss'] == 1):

            best_params['loss'] = 'huber'

        elif (best_params['loss'] == 2):

            best_params['loss'] = 'epsilon_insensitive'

        else:

            best_params['loss'] = 'squared_epsilon_insensitive'

        #If condition to account for each item in the list passed to penalty

        if (best_params['penalty'] == 0):

            best_params['penalty'] = None

        elif (best_params['penalty'] == 1):

            best_params['penalty'] = 'l2'

        elif (best_params['penalty'] == 2):

            best_params['penalty'] = 'l1'

        else:

            best_params['penalty'] = 'elasticnet'
```

```
    #If condition to account for each item in the list passed to learning_rate
    if (best_params['learning_rate'] == 0):

        best_params['learning_rate'] = 'constant'
    elif (best_params['learning_rate'] == 1):

        best_params['learning_rate'] = 'optimal'
    else:

        best_params['learning_rate'] = 'invscaling'


    best_model = SGDRegressor(loss=best_params['loss'],
                            penalty=best_params['penalty'],
                            alpha=best_params['alpha'],
                            l1_ratio=best_params['l1_ratio'],
                            epsilon=best_params['epsilon'],
                            learning_rate=best_params['learning_rate'])
    best_results = cross_val_score(best_model, x_train, y_train, cv=kfold,
scoring='neg_mean_squared_error')
    return best_model, best_results
elif (opt_type=='random'):
    best_params = fmin(obj_func, search_space, algo=rand.suggest, max_evals=100,
trials=trials)


    #If condition to account for each item in the list passed to loss
    if (best_params['loss'] == 0):

        best_params['loss'] = 'squared_loss'
    elif (best_params['loss'] == 1):

        best_params['loss'] = 'huber'
    elif (best_params['loss'] == 2):

        best_params['loss'] = 'epsilon_insensitive'
    else:

        best_params['loss'] = 'squared_epsilon_insensitive'
    #If condition to account for each item in the list passed to penalty
    if (best_params['penalty'] == 0):
```

```python
            best_params['penalty'] = None
        elif (best_params['penalty'] == 1):
            best_params['penalty'] = 'l2'
        elif (best_params['penalty'] == 2):
            best_params['penalty'] = 'l1'
        else:
            best_params['penalty'] = 'elasticnet'
        #If condition to account for each item in the list passed to learning_rate
        if (best_params['learning_rate'] == 0):
            best_params['learning_rate'] = 'constant'
        elif (best_params['learning_rate'] == 1):
            best_params['learning_rate'] = 'optimal'
        else:
            best_params['learning_rate'] = 'invscaling'


        best_model = SGDRegressor(loss=best_params['loss'],
                                  penalty=best_params['penalty'],
                                  alpha=best_params['alpha'],
                                  l1_ratio=best_params['l1_ratio'],
                                  epsilon=best_params['epsilon'],
                                  learning_rate=best_params['learning_rate'])
        best_results = cross_val_score(best_model, x_train, y_train, cv=kfold,
scoring='neg_mean_squared_error')
        return best_model, best_results
    else:
        print('Invalid opt_type')
        return


sgd_best_model, sgd_best_results = sgd_model_opt(opt_type='bayesian')

print('Stochastic Gradient Descent Cross Validated MSE:', abs(sgd_results.mean()))
print('Optimized Stochastic Gradient Descent Cross Validated MSE:',
abs(sgd_best_results.mean()))
```

In this case, hyperparameter optimization did nothing to improve the results of the model. In fact, the default model had a slightly lower MSE than the optimized model. This illustrates how hyperparameter optimization is not a silver bullet for data modeling.

9. Logistic Regression

Standard linear regression models can predict continuous outcome variables, but they cannot predict categorical outcomes. Logistic regression is a type of regression that can predict categorical dependent variables. The simplest form of logistic regression involves a binary dependent variable. This type of model is called a binary classifier. A **classifier** is any model that predicts a categorical outcome or class. There are also multinomial classifiers that predict dependent variables with multiple categories. Logistic regression can be used to predict multiple categorical outcomes through a form called **multinomial logistic regression**. Note that multinomial logistic regression is not the same as multiple logistic regression: multinomial logistic regression deals with dependent variables with many categories, whereas multiple logistic regression is logistic regression with multiple predictors. To begin this chapter, let us look at logistic regression for binary classification.

The plot on the left side of figure 9.1 shows a binary variable. It is either 1 or 0 and nothing in between. The equation for the plot on the left is nonlinear because it is not possible to fit a line to the data. The plot in the middle of figure 9.1 shows what would happen if a line were fitted. By thinking of the outcome in terms of probability, it becomes more linear. The probability of an outcome could take any value between 0 and 1, so the plot would look like an S shape. An S shape is still nonlinear, but it is certainly closer to looking like a line than a piecewise step function. Taking the logarithm of the probability results in a linear looking plot as shown on the right of figure 9.1.

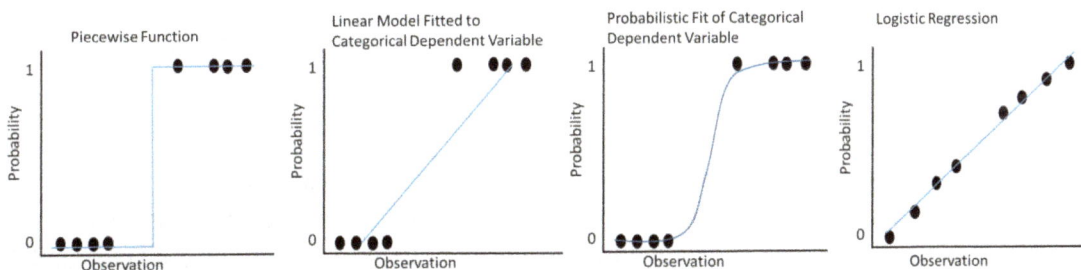

Figure 9.1

The log of the probability is called the **logit** (a.k.a. log odds). The logit allows nonlinear functions with binary categorical dependent variables to be transformed so that the logit can be estimated by a linear equation. In other words, the logit fits input to a binary output (0 or 1). Note that it is the logit that is being estimated by solving the **logistic function** in logistic regression, so any inferences made about the resulting model and its parameter coefficients must be interpreted differently than a standard linear regression. The logit is the inverse of the logistic function that is solved for in a logistic regression. Figure 9.2 shows the difference between the logit and the logistic function.

Rather than predicting the value of the outcome directly, like in standard regression, logistic regression predicts the log probability of the outcome. The formula for a multiple logistic regression is shown in the logistic function in figure 9.2:

Logit	Logistic Function (Logistic Regression Equation)	Relationship Between Logit and Logistic Function
$logit(y) = \ln\left(\dfrac{y}{1-y}\right)$	$P(y) = \dfrac{1}{1 + e^{-(\beta_0 + \beta_1 x_1 + \beta_2 x_2 + \cdots \beta_n x_n)}}$	$logit^{-1}(y) = logistic(y)$

Figure 9.2

In the equation in figure 9.2, y is some number, e is Euler's number, P(y) is a value between 0 and 1 that represents the probability of y occurring.

When fitting a logistic regression, maximum likelihood estimation (MLE) is used instead of OLS (see chapter 7). MLE produces coefficients of the predictors such that when the values of the predictors are placed into the model, the resulting values of y are closest to the observed values. In logistic regression, MLE optimizes the logistic function that results in the highest log-likelihood ratio (Hastie, Tibshirani, & Friedman, 2009).

Goodness of Fit: Log-Likelihood Ratio

The **likelihood ratio** shows how many times more likely the data are to appear if they are generated by one model compared to another (baseline) model. It is another way of looking at the ratio of explained variance to unexplained variance, and can be computed as in figure 9.3.

$$\Lambda(x) = \frac{L(\theta_0 | x)}{L(\theta_1 | x)}$$

Figure 9.3

In the equation in figure 9.3, lambda stands for the likelihood ratio, and L(theta | x) represents the likelihood of parameter theta, for model 0 or 1, given data x. The numerator is the null or baseline model.

The log of the likelihood ratio is called the **log-likelihood ratio**, or simply log likelihood (LL). Log-likelihood is a way of assessing the fit of a logistic regression, much like R^2 is a measure of fit for OLS linear regressions. The log-likelihood is the sum of the probabilities associated with the predicted and actual outcomes. Just like the sum of squared residuals in multiple linear regression, the log-likelihood is a measure of how much unexplained information there is after the model has been fitted. Since log-likelihood is taking the log of probabilities that range in size from 0 to 1, and the log of numbers in that range is negative, the log-likelihood is often negative. Therefore, the higher the value, the better the fit (-1 would be a better fit than -10 for example, and 0 would be a perfect fit). Sometimes, because the log-likelihood is a sum and some observations will have a positive log-likelihood, the log-likelihood can be positive, and when this happens, smaller values indicate better fit (so if LL is positive, 1 would be a better fit than 10, and 0 would be a perfect fit) (Field, Miles, & Field, 2012).

$$log\ likelihood = \sum_{i=1}^{N} [y_i \ln(P(y_i)) + (1 - y_i) \ln(1 - P(y_i))]$$

Figure 9.4

Recall from chapter 7 that MLE is the estimator that maximizes the log likelihood. MLE therefore maximizes the equation in figure 9.4 across all observations, and if we look back to figure 7.10 in chapter 7, it should be obvious that figure 9.4 is an expansion of figure 7.10. Log-likelihood is a function of sample size however, so it must be standardized in order to be useful as a measure of fit.

Standardization for sample size is done by averaging the log-likelihood. For this reason, most textbooks will show a 1/N in front of the summation in figure 9.4. The average log-likelihood allows different models for datasets of different sizes to be compared.

When the log-likelihood is averaged and multiplied by -1, the result is called the **log loss** or **cross entropy loss**. The phrase cross entropy comes from information theory, but cross entropy loss is the same as the negative log likelihood, which is also called log loss. When the dependent variable only has 2 categories, the cross entropy loss may be called binary cross entropy loss. The lower the log loss, the better the model (log loss of 0 would be a perfect model). If the log-likelihood is positive, then the log loss will be negative and larger log losses (less negative or closer to 0) would be better. Figure 8.5 shows the log loss equation for a binary classification problem on top and the generalization to multiple classes on the bottom.

$$binary\ classifiction\ log\ loss = -\frac{1}{N}\sum_{i=1}^{N}[y_i\ln(P(y_i)) + (1 - y_i)\ln(1 - P(y_i))]$$

$$multiclass\ classification\ log\ loss = -\frac{1}{N}\sum_{i=1}^{N}\sum_{j=1}^{M}y_{ij}\ln P(y_{ij})$$

Figure 9.5

In the equations in figure 9.5, y sub i equals 1 when the true class of the dependent variable for observation i is 1, N is the total number of observations, P(y sub i) is the probability of the model assigning observation i to class 1, M is the number of classes, y sub ij equals 1 when the true class of the dependent variable for observation i is class j, and P(y sub ij) is the probability of the model assigning observation I to class i.

When the log likelihood is multiplied by -2, it gives the **deviance statistic (-2LL or D)**, which allows logistic regression models to be compared to one another because the deviance follows a Chi-square distribution.

$$D = -2 * (log\ likelihood)$$

Figure 9.6

The Chi-square statistic for the model can be calculated by subtracting the deviance of the fitted model from the deviance of the baseline model. In OLS linear regression, the baseline model is the mean. In MLE logistic regression, the baseline model is the category with the greatest frequency of occurrences. So for a binary outcome, if the outcome is positive 51/100 times, then the baseline model is the one with a positive outcome. No matter which category has the greatest frequency, the equation for the baseline model is always the same: it is the model with only the intercept included. So the Chi-square statistics represents the change in deviance from the model with only an intercept to the model with 1 or more predictors included.

$$\chi^2 = \left(-2LL(baseline)\right) - \left(-2LL(model)\right)$$

$$= 2LL(model) - 2LL(baseline)$$

Figure 9.7

The degrees of freedom for the Chi-square statistic are the number of predictors in the model minus the number of predictors in the baseline. The number of predictors in the baseline will always be 1, since the baseline model only includes a constant.

$$df = k_{model} - 1$$

Figure 9.8

Recall from chapter 6 that the F-test tested for any relationship between the outcome and the predictors, and the R^2 measured the strength of that relationship. Similarly, the Chi-square test tests for any relationship between a categorical outcome and the predictors in a logistic regression, and the log likelihood or deviance statistic measures the strength of that relationship.

Other Measures of Fit: R, AIC, BIC, Confusion Matrix, ROC Curve

In logistic regression, R is the partial correlation between a predictor and the outcome. It can vary between -1 and 1. When it's positive, then as the predictor increases, the probability of the outcome occurring does as well. When it's negative, then as the predictor increases, the probability of the outcome occurring decreases. The closer R is to 0, the less impact the predictor has. Unlike OLS linear regression, R cannot be squared to give a measure of goodness of fit. There are several approaches to calculating R^2 for logistic regression, such as Hosmer and Lemeshow's R^2, McFadden's R^2, Cox and Snell's R^2, and Nagelkerke's R^2, but there is no consensus on a single method (Field, Mils, & Field, 2012).

Recall from chapter 6 that as the number of predictors increases, the fit increases too. The AIC and BIC were introduced in chapter 6 as measures that penalize models with more predictors. They can be calculated for logistic regression as in figure 9.9.

$$AIC = -2LL + 2k$$

$$BIC = -2LL + 2k * \log n$$

Figure 9.9

The **confusion matrix** is another way to evaluate a classification model. Recall from chapter 3 that the confusion matrix is used for hypothesis testing. The confusion matrix in figure 3.5 is the same as the confusion matrix for a binary classifier. For a classifier with K classes, the confusion matrix dimensions will be K*K. Figure 9.10 shows a confusion matrix layout for binary and trinary classifiers. In a confusion matrix for a multiclass classifier, the positives and negatives must be assessed in a one vs all approach, for each class. For example, the matrix in figure 9.10 is using 0 as the class for comparison. This means that observations that are not predicted to be 0 when they are truly not 0 would be true negatives, even if the model misclassified some 1's as 2's or vice versa.

Confusion Matrix for		True Classification	
Binary Classifier		0	1
Model Predicted Classification	0	True Negatives – this is what the model got right	False Negatives – this is what the model got wrong
	1	False Positives – this is what the model got wrong	True Positives – this is what the model got right

Confusion Matrix for		True Classification		
Trinary Classifier		0	1	2
Model Predicted Classification	0	True Positives – this is what the model got right, using 0 as the baseline class	False Negatives – this is what the model got wrong, using 0 as the baseline class	False Negatives – this is what the model got wrong, using 0 as the baseline class
	1	False Positives – this is what the model got wrong, using 0 as the baseline class	True Negatives – this is what the model got right, using 0 as the baseline class	True Negatives – this is what the model got right, using 0 as the baseline class
	2	False Positives – this is what the model got wrong, using 0 as the baseline class	True Negatives – this is what the model got right, using 0 as the baseline class	True Negatives – this is what the model got right, using 0 as the baseline class

Figure 9.10

From the confusion matrix, the following measures can be calculated:

$$Accuracy = \frac{True\ Positives + True\ Negatives}{Number\ Observations}$$

$$True\ Positive\ Rate\ (sensitivity\ or\ recall) = \frac{True\ Positives}{Number\ Observations\ Positive}$$

$$False\ Positive\ Rate\ (type\ I\ error) = \frac{False\ Positives}{Number\ Observations\ Negative}$$

$$True\ Negative\ Rate\ (specificity) = \frac{True\ Negatives}{Number\ Observations\ Negative}$$

$$False\ Negative\ Rate\ (type\ II\ error) = \frac{False\ Negatives}{Number\ Observations\ Positive}$$

$$Positive\ Predictive\ Value\ (precision) = \frac{True\ Positives}{True\ Positives + False\ Positives}$$

$$Negative\ Predictive\ Value = \frac{True\ Negatives}{True\ Negatives + False\ Negatives}$$

$$False\ Discovery\ Rate = 1 - Positive\ Predictive\ Value$$

$$False\ Omission\ Rate = 1 - Negative\ Predictive\ Value$$

Figure 9.11

Due to the difficulty of interpreting a confusion matrix for a multiclass problem, it is better to use other methods to evaluate the model. Even for a binary classifier, the **classification accuracy** (the top equation in figure 9.11) is not a good metric to use to evaluate the model, because the classes could be imbalanced. For example, if 99% of the observations were truly in the 0 class, then a model that predated that every observation should be classified as 0 would be 99% accurate. In this scenario, the model would not have really learned anything about the data. For this reason, log loss is typically a better choice of measure to evaluate classification models than accuracy (Hastie, Tibshirani, & Friedman, 2009).

Another way of evaluating a classification model is by looking at the **receiver operator characteristic (ROC) curve**. The ROC curve plots the true positive rate on the y-axis and the false positive rate on the x-axis. Figure 9.12 shows a sample ROC curve.

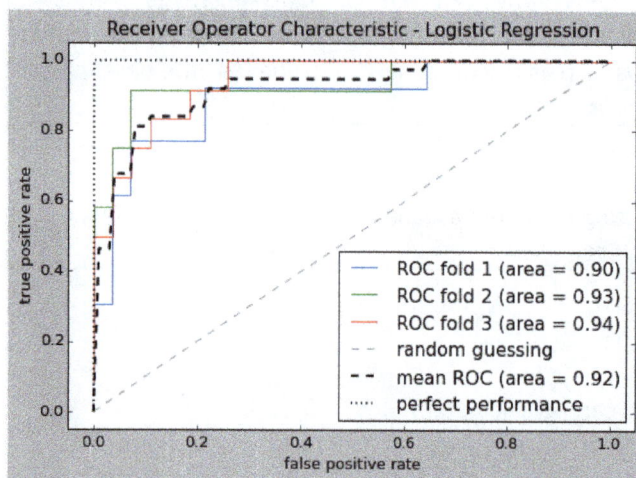

Figure 9.12

The diagonal line represents the baseline model that is equivalent to randomly guessing the class for each observation. A flawless model would form a perfect right angle in the top left of the graph, getting 100% true positives and 0% false positives. So the closer a model looks to forming a right angle, the better it is. The slope of the tangent line (derivative) of a cut point on the ROC curve is the log likelihood ratio for that cut point. The area under the ROC curve is equivalent to the accuracy of the model, so a perfect model would have an **area under the curve (AUC)** equal to 1, and the baseline model (the diagonal line) would have an AUC of 0.5.

For multiclass models, one ROC curve would need to be plotted per class, using the one vs all approach described previously. So a 5 class model would have 5 ROC curves (Raschka, 2015).

Residual Diagnostics
Logistic regression models, like standard linear regression models, can also be assessed by looking at their residuals. **Diagnostics** are measures of fitness for a model based on the model's residuals. Residuals can show whether or not there are outliers that are adversely influencing the model. For example, if the absolute values of the standardized residuals are examined, only 5% should be > 1.96, and only 1% should be > 2.58. Observations lying close to 3 standard units away from the linear model

should be inspected, and any observations greater than 3 standard units away are cause for concern (Field, Miles, & Field, 2012).

Recall that the deviance statistic is -2LL (see figure 9.6). The deviance statistic can also be expressed as the sum of the deviance residuals. The **deviance residuals** measure the contribution of each point to the log-likelihood. By comparing the deviance of the model with all observations to a model with observations with large deviance residuals removed, it is possible to see whether the observations with a high deviance are adversely influencing the model.

$$\Delta D_i = D - D_{-i}$$

Figure 9.13

The equation in figure 9.13 shows that the change in deviance for the ith observation is the deviance minus the deviance when the ith observation is removed.

Another way the effect of single observations on the model can be assessed is by looking at their leverage. An observation with an extreme value for a predictor variable is said to have high leverage. More generally, **leverage** is a measure that can identify extreme observations in a multidimensional feature space. The range of values for leverage is between 0 (the observation has no influence) and 1 (the observation has complete influence). The expected average leverage should be approximately $(k+1)/N$, where k is the number of predictors and N is the number of observations. Observations whose leverages are 2-3 times larger than the expected average leverage should be inspected as possible outliers. Observations with this much influence could be throwing off the logistic regression model.

Assessing Individual Predictors: Odds Ratio, Wald Statistic, Likelihood Ratio Test
The **odds ratio** is an effect size for an individual predictor while all others are held constant. Recall from chapter 3 that the effect size shows the strength and direction of the relationship between variables. The odds ratio shows the ratio of the odds after a 1 unit change in a predictor. For example, if the binary variable y is regressed on the continuous variable x, then the odds ratio would show the change in the ratio of y being 1 to y being 0 for a 1 unit change in x. The odds ratio is not the same as probability. Probability measures the likelihood of some value out of all possible values, whereas the odds ratio measures the ratio of some value to another value, such as the frequency of the dependent variable being 1 divided by the frequency of the dependent variable being 0. Unlike probability, odds are not limited to a range of (0, 1). Figure 9.14 illustrates the difference.

$$Odds\ Ratio = \frac{P(x)}{1 - P(x)}$$

$$P(x) = \frac{Odds\ Ratio(x)}{1 + Odds\ Ratio(x)}$$

Favorite Colors	Probability of Red	Odds Ratio of Red to Blue
Red = 9 students Blue = 11 students	9/20 = 45% chance of favorite color being red	9/11 = 0.82 students whose favorite color is red for every student whose favorite color is blue

Figure 9.14

The odds ratio shows the strength and direction of the relationship between a predictor and the outcome variable, but it is in terms of the units of the predictor, so it is non-standardized. This means that odds ratios for different predictors cannot be directly compared, unless the predictors have been standardized by replacing their values with z-scores.

Notice that the odds ratio equation is very similar to the logit equation in figure 9.2. Referring back to equation 9.2, if y were a probability instead of a number, then the equation in figure 9.14 would give odds of that probability. Therefore, the logarithm of the odds (log-odds) equals the logit, which is exactly how the logit was previously defined.

The **Wald statistic** can be used to determine the statistical significance of a predictor. For large sample sizes, it is roughly equivalent to the t-test. So a t-test can also be used to determine the statistical significance of a predictor in a logistic regression.

The **likelihood ratio test (LRT)** is another option for testing the significance of an individual predictor. This test works by building two models; one with all of the predictors, and a second model without the predictor whose significance is to be tested. The LRT divides the log-likelihood of the model with the smaller predictor set, s, by the model with all of the predictors, g, as in figure 9.15.

$$LRT = -2\ln\left(\frac{L_s(\theta)}{L_g(\theta)}\right)$$

Figure 9.15

The LRT approximates the Chi-square distribution with k-1 degrees of freedom, where k is the number of predictors. The result can be compared to the Chi-square statistical table, and if it is larger than the Chi-square critical value, then the model with the predictor included is a significant improvement over the model without it.

Logistic Regression Assumptions

In order for a logistic regression model to generalize to a population, the following assumptions are made, in addition to the parametric assumptions listed in chapter 3:

1. There must be a linear relationship between the predictors and the logit of the outcome variable. Note that this does not mean that the outcome and the predictors need to be linearly related, only that the predictors and the logit be linearly related. This assumption is tested by determining whether or not the interaction between the predictor and its log transformation is significant.
2. The predictors should not be collinear (they should not correlate highly).
3. The errors for any 2 observations should be independent or uncorrelated. The presence of correlation in the errors is referred to as **autocorrelation** (or serial correlation if the data is panel data or time series data). So there should be no autocorrelation.

The first assumption can be tested by determining whether or not the interaction between the predictor and its log transformation is significant. This would require adding an interaction term to the regression equation for the predictor and its log transformation. An alternative visual way would be to scatterplot the outcome on each predictor, one at a time. If a locally weighted scatterplot smoothing (LOWESS or

LOESS) curve is fitted to the scatterplot, it should look like an S, suggesting that the logistic regression accurately describes the relationship between the outcome and predictor.

Another visual way to check the first assumption is to plot the observed vs predicted probabilities after the logistic regression is run. The plot should look linear.

The second and third assumptions listed above can be checked using the same methods that were described in chapter 6 for linear regression. The same methods for testing for collinearity (correlation matrix, VIF) and autocorrelation (DW statistic) can be applied before the logistic regression is carried out.

Complete Separation

When the dependent variable can be perfectly predicted by 1 independent variable or some combination of them, **complete separation** can occur. If complete separation occurs, the standard error of the logistic regression and the predictor coefficients will be very large. Complete separation is a risk chiefly in small datasets. To avoid it, more data can be collected or the predictors causing complete separation can be dropped if they can be identified.

Probit Regression

Probit regression is rarely used, but it is worth exploring briefly. Probit regression is logistic regression's twin, except probit regression is meant only for binary classification. Probit regression is carried out and evaluated the same way binary class logistic regression is, and it requires making the same assumptions. The difference is the function that is optimized by MLE. In logistic regression, the logit or likelihood function is maximized. In probit regression, the probit or cumulative normal probability density function is maximized. The probit, like the logit, transforms a regression equation for a categorical variable into a continuous representation, but the probit curve has slightly steeper tails than the logit.

Probit would be used in the case when the categorical outcome is actually normally distributed in the population, even though it appears categorical in the sample data. This is referred to as a hidden Gaussian variable. For example, suppose we are trying to predict blood pressure, and blood pressure is specified as a categorical variable with two categories: high and normal. In the whole population of humans, blood pressure is normally distributed, but in the sample data, there are only 2 possible values.

Logistic Regression with R

In the last few chapters, we have been preparing the Titanic dataset for classification. Now we will use logistic regression to finally predict whether passengers lived or died. Before we build a model though, there are a few additional features that could be built. Since feature engineering almost always produces better models, it pays to be diligent. The first new feature we can make is family size. We previously calculated group size using passengers who shared the same ticket, but the dataset has a variable to show the number of parents or children on board, Parch, and a variable to show the number of siblings or spouses on board, SibSp. Summing these and adding 1 to count the passenger himself/herself produces the family size.

```
train <- read.csv("train_clean.csv", header=T)
test <- read.csv("test_clean.csv", header=T)
train$Set <- 'train'
test$Set <- 'test'
alldata <- rbind(train[,-which(colnames(train) %in% 'Survived')], test)
```

```
#Add calculated family size (siblings + spouse + parents + children + 1 for self)
alldata$Family_Size <- alldata$SibSp + alldata$Parch + 1
```

A binary flag for whether or not the passenger has a family might be useful as well.

```
alldata$Has_Family <- as.factor(ifelse(alldata$Family_Size > 1, 1, 0))
```

So far we have ignored the text data (the Name variable). The name contains people's titles, which may be useful. Titles convey respect and social status, so it is reasonable to think they may be useful in predicting who survived. Some titles are only used for children, so if we want to fill in the missing age values, the person's title may provide a clue for how old they might be. We can split off titles from the name using the code below.

```
alldata$Name <- as.character(alldata$Name)
alldata$Title <- sapply(alldata$Name, FUN=function(x) {strsplit(x,
split='[,.]')[[1]][2]})
#Remove the first space in the title
alldata$Title <- sub(' ', '', alldata$Title)

table(alldata$Title)
```

Looking at the frequency of each title, it seems that many only appear once or twice. These could be considered outliers. An argument could be made to leave them as they are, because some titles might only be given to very high status people who could have been given priority in the Titanic's life boats. However, features that only appear once in a dataset usually are not useful for prediction because they only apply to one observation. If more data could be collected, features that appear once may appear more often and become useful for prediction, but it is not possible to collect more data for the Titanic dataset, because the dataset we have already contains the entire population of Titanic passengers. So the titles that appear once are combined with more frequent titles.

```
alldata$Title[alldata$Title %in% c('Dona', 'Jonkheer', 'Lady', 'Mlle', 'Mme', 'the
Countess')] <- 'Mme'
alldata$Title[alldata$Title %in% c('Capt', 'Col', 'Don', 'Sir', 'Major')] <- 'Sir'
alldata$Title[alldata$Title %in% c('Miss', 'Ms')] <- 'Miss'
alldata$Title <- factor(alldata$Title)
```

Now we can fill in the missing ages, using title as a guide. Anyone whose title is master or miss is more likely a child. In chapter 4, we surmised that age might not be missing at random, because people who survived may be more likely to have an age value. For this reason, we did not use single random imputation to fill in the values. Now that we have title though, we can use a piecewise imputation model to fill in the age. A new binary column can be added to indicate whether age was originally missing. That way, if missing age has anything to do with the passenger's survival, the binary column will capture it. We will use age 16 as the cutoff between children and adults.

```
alldata$Missing_Age <- ifelse(is.na(alldata$Age), 1, 0)

#Create distributions to sample from, and then perform piecewise single random
imputation
age_dist_child <- alldata[!is.na(alldata$Age) & alldata$Age < 16, 'Age']
age_dist_adult <- alldata[!is.na(alldata$Age) & alldata$Age >= 16, 'Age']
```

```
hist(alldata$Age, main="Histogram of Age Before Piecewise Single Random Imputation")
alldata[is.na(alldata$Age) & alldata$Title %in% c('Master', 'Miss'), 'Age'] <-
sapply(alldata[is.na(alldata$Age) & alldata$Title %in% c('Master', 'Miss'), 'Age'],
FUN=function(x) {sample(age_dist_child, 1)})
alldata[is.na(alldata$Age), 'Age'] <- sapply(alldata[is.na(alldata$Age), 'Age'],
FUN=function(x) {sample(age_dist_adult, 1)})
hist(alldata$Age, main="Histogram of Age After Piecewise Single Random Imputation")
```

The last feature engineering step is to dummy encode title, create a binary feature to indicate whether the passenger is a child, and split the data back into training and test sets.

```
#Child indicator
alldata$Is_Child <- ifelse(alldata$Age < 16, 1, 0)

#Dummy encode title - be sure to set fullRank=T to dummy encode instead of one-hot
library(caret)
dummies <- dummyVars( " ~ Title", data=alldata, fullRank=T)
dummyenc <- data.frame(predict(dummies, newdata=alldata))
alldata <- cbind(alldata, dummyenc)
colnames(alldata) <- gsub("[.]", "_", colnames(alldata))
rm('dummies', 'dummyenc')

#Remove the title column and one column from each group of the one-hot encoded column
groups
alldata$Title <- NULL
alldata$Sex_female <- NULL
alldata$Embarked_C <- NULL
alldata$DeckA <- NULL

#Split back into training and test sets
train_clean_feats <- alldata[alldata$Set=='train',]
train_clean_feats$Set <- NULL
train_clean_feats$Survived <- train$Survived
test_clean_feats <- alldata[alldata$Set=='test',]
test_clean_feats$Set <- NULL
write.csv(train_clean_feats, file="train_clean_feats.csv", row.names=F)
write.csv(test_clean_feats, file="test_clean_feats.csv", row.names=F)
```

To perform logistic regression, we must first check the assumptions that need to be made for the results of a logistic regression to be able to be generalized.

```
#Are all predictors quantitative or categorical with no more than 2 categories?
str(train)  #Oops!  Forgot to get rid of name
train$Name <- NULL
test$Name <- NULL
str(train)  #Yes - assumption met
str(test)   #Yes - assumption met

#Is there a linear relationship between the predictors and the logit?
#Cannot tell yet - need to build the model to get the observed vs predicted
probabilities

#Is there multicollinearity among the predictors?
corm <- cor(train, use="pairwise.complete.obs")
library('corrplot')
```

```
corrplot(corm, method="circle")     #Possible problems here

#Are the residuals homoskedastic?
#Cannot tell yet - need to build the model to get the residuals

#Are the residuals autocorrelated?
#Cannot tell yet - need to build the model to get the residuals
```

Some variables are highly correlated. This could point to multicollinearity. Let us keep them for now to see what happens when we try to run a logistic regression.

```
logistic_reg <- glm(Survived ~ ., family=binomial(link='logit'), data=train)
summary(logistic_reg)
```

R threw a warning when the glm function was run, because there is multicollinearity. The results show that family size has NAs for every parameter – this is bad. If we take another look at the correlation matrix, we can see that family size is highly correlated with SibSp, Parch, and Group Size. This is no surprise, but since the other variables capture the same variance, we can safely remove family size. Fare is also highly correlated with fare per person. Since we calculated fare per person to more accurately capture the fare of groups, we can get rid of fare. DeckZ is causing problems because there were only first class passengers on Deck Z. Since DeckZ is part of the dummies we created for deck, we should go back and recode them to use DeckZ as the baseline instead of DeckA. The problem is that DeckA also only had first class passengers. So instead of recoding, we can simply get rid of DeckZ, because both DeckA and DeckZ had first class passengers only. Getting rid of DeckZ will cause the combination of DeckA and DeckZ to be the baseline for the other dummies, which is perfectly fine in this case, because it essentially compares the decks to the first class only decks. The titles are causing issues with the sex variable, because titles are gender specific. That is no surprise. We could get rid of sex, but it would make more sense to keep it, because when we interpret the results it seems more useful to say that being male altered the log odds of survival by x, rather than to say that having a title of Mr. altered the log odds of survival, for instance. Let us get rid of all of the title dummies. The titles capture combinations of age, class, and gender anyway, and since there are other variables to capture this information, it will not be a problem to dispose of the titles.

```
train <- train[,-which(colnames(train) %in% c('Fare', 'Family_Size', 'DeckZ',
                                  'Title_Master', 'Title_Miss',
                                  'Title_Mme', 'Title_Mr',
                                  'Title_Mrs', 'Title_Rev',
                                  'Title_Sir'))]
test <- test[,-which(colnames(test) %in% c('Fare', 'Family_Size', 'DeckZ',
                                  'Title_Master', 'Title_Miss',
                                  'Title_Mme', 'Title_Mr',
                                  'Title_Mrs', 'Title_Rev',
                                  'Title_Sir'))]
corm <- cor(train, use="pairwise.complete.obs")
corrplot(corm, method="circle")
logistic_reg <- glm(Survived ~ ., family=binomial(link='logit'), data=train)
summary(logistic_reg)
```

That is much better. It is known that the Titanic instituted a "women and children first" policy with regards to the lifeboats. Glancing at the results of this regression, it seems we have confirmed exactly that. Being a woman or a child, or particularly, being a first class woman or child, increased the log odds of survival. Passenger class, age, SibSp, Parch, group size, sex, DeckE, ticket, the family flag, the missing age flag, and the child flag were all found to be statistically significant. Their effect on the log odds of survival are as follows:

- A 1 unit increase in passenger class reduces the log odds of survival by 0.68, meaning first class passengers were most likely to survive and third class were least likely.
- A 1 unit increase in age decreases the log odds of survival by 0.02, meaning older people were slightly less likely to survive.
- A 1 unit increase in SibSp (number of siblings and spouses on board) reduces the log odds of survival by 0.07, meaning people with families were less likely to survive.
- A 1 unit increase in Parch (number of parents and children on board) reduces the log odds of survival by 0.04, which adds support to the idea that people with families were less likely to survive.
- A 1 unit increase in group size increases the log odds of survival by 0.01, which seems to contradict the idea that people with families were less likely to survive. However, group size measures people who shared tickets, not necessarily families.
- Being male decreases the log odds of survival by 2.57. This is the biggest statistically significant impact so far.
- Being on deck E increases the log odds of survival by 1.22, compared to decks A and Z. This is a surprise.
- A 1 unit increase in ticket reduces the log odds of survival by 0.0007, a hardly perceptible impact. A unit change in ticket has no real meaning because tickets are simply label encoded, but it does seem that tickets influence survival. This could back up the finding with group size, since it is based on people who share tickets.
- Having a family increases the log odds of survival by 0.08. This directly contradicts the findings of SibSp and Parch. Some more investigation into families should be done.
- Missing a value for age reduces the log odds of survival by 0.04. This confirms our idea that missing data was related to survival.
- Being a child increases the log odds of survival by 0.9.

By doing analysis of variance, we can see the impact each predictor has on the deviance statistic and the AIC for the model compared to the baseline (the baseline is a model with only the intercept). We will learn more about ANOVA in a later chapter. Recall from figure 9.6 that the deviance statistic is -2*log likelihood and since it follows the Chi-square distribution, it can be directly compared across different models. The column titled Deviance shows the drop in the deviance attributable to each variable. The larger the decrease in deviance, the better. The column titled Resid.Dev shows the AIC as each variable is added to the model. The lower, the better.

```
anova(logistic_reg, test="Chisq")
```

The predictors with large p-values do not explain any additional variance in the log odds of the outcome, so they are less valuable in the model. A model without the insignificant predictors might be better. We can test this by specifying a new model without the insignificant predictors, and running an ANOVA that compares it to the original model.

```
#New model without insignificant predictors
new_logistic_reg <- glm(Survived ~ Pclass + Age + SibSp + Parch + Group_Size +
                        Fare_Per_Person + Sex_male + DeckE + Ticket_Enc +
                        Has_Family + Missing_Age + Is_Child,
                        family=binomial(link='logit'), data=train)
summary(new_logistic_reg)
anova(new_logistic_reg, logistic_reg, test="Chisq")
```

The new model fails to outperform the original model, because the p-value for the Chi-square statistic is > 0.05 and the deviance is higher for the new model. So sticking with the original model, we can now check it for multicollinearity using the variance inflation factor.

```
library(car)
vif(logistic_reg)
```

None of the values are > 5 so we can be sure that multicollinearity is not a problem. The Durbin-Watson statistic and the correlogram can be used to check for autocorrelation in the residuals.

```
#DW test for autocorrelation - note this uses the car package
durbinWatsonTest(logistic_reg)
```

```
#Correlogram to look for significant lags in the residuals
acf(logistic_reg$residuals)
```

Since the p-value for the DW test is > 0.05 and the correlogram has no lags that stray far beyond the confidence interval, we can be sure that there is no autocorrelation in the residuals. We can also verify the assumption that the residuals are homoscedastic by looking at the QQ plot provided when we called plot(logistic_reg). Since the QQ plot was fairly linear, we can confirm the assumption holds.

At this point, all we have left to do is determine how well the model fits the data as a whole. One of the best ways to do this for classification models is to calculate the log loss or cross entropy loss. Since this is a binary classification problem, we can also look at the area under the ROC curve. The closer the AUC is to 1, the better the model. There is a problem however. We do not have the survival data for the test set. To make up for this, we can carve a validation set out of the training set and apply these methods. In theory, the performance on the validation set should match the performance on the test set.

```
#Split training data into training and validation sets
set.seed(14)
train_indices <- createDataPartition(y=train$Survived, p=0.7, list=F)
train_new <- train[train_indices,]
validation <- train[-train_indices,]
logistic_reg <- glm(Survived ~ ., family=binomial(link='logit'), data=train_new)
```

```
#Make survival predictions for validation set
validation$Pred <- predict(logistic_reg, validation[,-21], type="response")
validation$Survived_Pred <- ifelse (validation$Pred >= 0.5, 1, 0)
```

```
#Evaluate the model using the log loss and AUC
library(MLmetrics)
LogLoss(validation$Survived_Pred, validation$Survived)
AUC(validation$Survived_Pred, validation$Survived)
Accuracy(validation$Survived_Pred, validation$Survived)
#View the confusion matrix with predicted values on the left
```

```
prop.table(table(validation$Survived_Pred, validation$Survived))
```

In the example above, we have used 0.5 as the cutoff to make a survival prediction of 1. This can be tuned to any value, depending on whether false positives or false negatives are less desirable. The choice of cutoff will alter the log loss and AUC. The model has an AUC of 0.76 and an accuracy of 78%, which is a decent model for this dataset.

Logistic Regression with Python

In the last few chapters, we have been preparing the Titanic dataset for classification. Now we will use logistic regression to finally predict whether passengers lived or died. Before we build a model though, there are a few additional features that could be built. Since feature engineering almost always produces better models, it pays to be diligent. The first new feature we can make is family size. We previously calculated group size using passengers who shared the same ticket, but the dataset has a variable to show the number of parents or children on board, Parch, and a variable to show the number of siblings or spouses on board, SibSp. Summing these and adding 1 to count the passenger himself/herself produces the family size.

```
import pandas as pd
import re
import numpy as np
import matplotlib.pyplot as plt
import random
from sklearn.linear_model import LogisticRegression
from sklearn import metrics
import seaborn as sns
from statsmodels.stats.outliers_influence import variance_inflation_factor
from statsmodels.stats.stattools import durbin_watson

train = pd.read_csv("train_clean.csv", sep=",")
test = pd.read_csv("test_clean.csv", sep=",")
train['Set'] = 'train'
test['Set'] = 'test'
alldata = pd.concat([train.drop(['Survived'], axis=1), test], ignore_index=True)
print(alldata.info())
print(alldata.head())

#Add calculated family size (siblings + spouse + parents + children + 1 for self)
alldata['Family_Size'] = alldata.SibSp + alldata.Parch + 1
```

A binary flag for whether or not the passenger has a family might be useful as well.

```
alldata['Has_Family'] = alldata['Family_Size'].apply(lambda x: 1 if (x > 1) else 0)
```

So far we have ignored the text data (the Name variable). The name contains people's titles, which may be useful. Titles convey respect and social status, so it is reasonable to think they may be useful in predicting who survived. Some titles are only used for children, so if we want to fill in the missing age values, the person's title may provide a clue for how old they might be. We can split off titles from the name using the code below.

```
alldata['Title'] = alldata['Name'].apply(lambda x: re.split(',|\.', x)[1].strip())

print(alldata['Title'].value_counts())
```

Looking at the frequency of each title, it seems that many only appear once or twice. These could be considered outliers. An argument could be made to leave them as they are, because some titles might only be given to very high status people who could have been given priority in the Titanic's life boats. However, features that only appear once in a dataset usually are not useful for prediction because they only apply to one observation. If more data could be collected, features that appear once may appear more often and become useful for prediction, but it is not possible to collect more data for the Titanic dataset, because the dataset we have already contains the entire population of Titanic passengers. So the titles that appear once are combined with more frequent titles.

```
myfilter = alldata['Title'][alldata['Title'].apply(lambda x: x in ['Dona',
'Jonkheer', 'Lady', 'Mlle', 'Mme', 'the Countess'])]
alldata.loc[myfilter.index, 'Title'] = 'Mme'
myfilter = alldata['Title'][alldata['Title'].apply(lambda x: x in ['Capt', 'Col',
'Don', 'Sir', 'Major'])]
alldata.loc[myfilter.index, 'Title'] = 'Sir'
myfilter = alldata['Title'][alldata['Title'].apply(lambda x: x in ['Miss', 'Ms'])]
alldata.loc[myfilter.index, 'Title'] = 'Miss'
```

Now we can fill in the missing ages, using title as a guide. Anyone whose title is master or miss is more likely a child. In chapter 4, we surmised that age might not be missing at random, because people who survived may be more likely to have an age value. For this reason, we did not use single random imputation to fill in the values. Now that we have title though, we can use a piecewise imputation model to fill in the age. A new binary column can be added to indicate whether age was originally missing. That way, if missing age has anything to do with the passenger's survival, the binary column will capture it. We will use age 16 as the cutoff between children and adults.

```
alldata['Missing_Age'] = alldata['Age'].apply(lambda x: 1 if np.isnan(x) else 0)

#Create distributions to sample from, and then perform piecewise single random
imputation
age_dist_child = alldata['Age'][(alldata.Age < 16) & (~np.isnan(alldata.Age))]
age_dist_adult = alldata['Age'][(alldata.Age >= 16) & (~np.isnan(alldata.Age))]

alldata.hist(column='Age')
plt.suptitle('Histogram of Age Before Piecewise Single Random Imputation')
#plt.show()
myfilter = alldata['Age'][(alldata['Title'].apply(lambda x: x in ['Master', 'Miss']))
& (np.isnan(alldata['Age']))]
alldata.loc[myfilter.index, 'Age'] = alldata.loc[myfilter.index, 'Age'].apply(lambda
x: random.sample(list(age_dist_child), 1)[0])
alldata.loc[pd.isnull(alldata.Age), 'Age'] =
alldata['Age'][pd.isnull(alldata.Age)].apply(lambda x:
random.sample(list(age_dist_adult), 1)[0])
alldata.hist(column='Age')
plt.suptitle('Histogram of Age After Piecewise Single Random Imputation')
plt.show()
```

The last feature engineering step is to dummy encode title, create a binary feature to indicate whether the passenger is a child, and split the data back into training and test sets.

```
#Child indicator
alldata['Is_Child'] = alldata['Age'].apply(lambda x: 1 if x < 16 else 0)
```

```
#Dummy encode title - be sure to set dummy encode instead of one-hot
title_dummy =
pd.get_dummies(alldata['Title']).drop(alldata.sort_values('Title')['Title'].unique()[
0], axis=1).add_prefix('Title_')
alldata = pd.concat([alldata, title_dummy], axis=1)
del title_dummy

#Remove the title column and one column from each group of the one-hot encoded column
groups
del alldata['Title'], alldata['Sex_female'], alldata['Embarked_C'], alldata['DeckA']

#Split back into training and test sets
train_clean_feats = alldata.copy().loc[alldata['Set']=='train']
del train_clean_feats['Set']
train_clean_feats['Survived'] = train['Survived']
test_clean_feats = alldata.copy().loc[alldata['Set']=='test']
del test_clean_feats['Set']
train_clean_feats.to_csv("train_clean_feats.csv", index=False)
test_clean_feats.to_csv("test_clean_feats.csv", index=False)
```

To perform logistic regression, we must first check the assumptions that need to be made for the results of a logistic regression to be able to be generalized.

```
#Are all predictors quantitative or categorical with no more than 2 categories?
print(train.info())  #Oops!  Forgot to get rid of name
del train['Name'], test['Name']
print(train.info())  #Yes - assumption met
print(train.info())  #Yes - assumption met

#Is there a linear relationship between the predictors and the logit?
#Cannot tell yet - need to build the model to get the observed vs predicted
probabilities

#Is there multicollinearity among the predictors?
corm = train.corr()
plt.matshow(corm)
#plt.show()  #Possible problems with collinearity here

#Are the residuals homoskedastic?
#Cannot tell yet - need to build the model to get the residuals

#Are the residuals autocorrelated?
#Cannot tell yet - need to build the model to get the residuals
```

Some variables are highly correlated. This could point to multicollinearity. Let us keep them for now to see what happens when we try to run a logistic regression.

```
#Split training dataset into x and y numpy arrays
array = train.values
x = array[:, :train.shape[1]-1]
y = array[:, train.shape[1]-1]

#Build logit model
logistic_reg = LogisticRegression(fit_intercept=True, random_state=14).fit(x, y)
```

```
logistic_reg_preds = logistic_reg.predict(x)
logistic_reg_prob_preds = logistic_reg.predict_proba(x)
print(logistic_reg.get_params, '\n')
print('Intercept:', logistic_reg.intercept_)
for idx, c in enumerate(logistic_reg.coef_[0]):
    print('Coefficient for', train.columns[idx], c)
print('Classification Accuracy:', logistic_reg.score(x, y))
```

Notice that unlike with R, the logistic regression in scikit-learn produces no warning about multicollinearity. The accuracy on the training dataset is roughly 83%. This is good enough to get into the top 2% (200 out of 10,000 teams) of Kaggle's leaderboard for the Titanic competition. But remember that this is the training set. We have no way of knowing what our model would score on the test set, unless we create a validation set. Let's proceed with inspecting the results of the model in more detail, using the confusion matrix, log loss, and area under the ROC curve. Then we can check the assumption that the predictors are independent using the variance inflation factor.

```
#Confusion matrix
cm = metrics.confusion_matrix(y, logistic_reg_preds)
sns.heatmap(cm, annot=True, fmt=".2f", square=True)
plt.xlabel('Predicted Class')
plt.ylabel('Actual Class')
plt.title('Confusion Matrix')
plt.show()

#Log loss (a.k.a. negative log likelihood)
print('Log loss:', metrics.log_loss(y, logistic_reg_preds))

#Plot ROC curve
fpr, tpr, threshold = metrics.roc_curve(y, logistic_reg_preds)
roc_auc = metrics.auc(fpr, tpr)
plt.plot(fpr, tpr, lw=2, label='ROC Curve (area = %0.2f)' % roc_auc)
plt.legend(loc='lower right')
plt.plot([0, 1], [0, 1], 'b--')  #Diagonal line
plt.xlabel('False Positive Rate')
plt.ylabel('True Positive Rate')
plt.title('Receiver Operating Characteristic')
plt.show()

#VIF test for multicollinearity
vif = pd.DataFrame()
vif['VIF_Factor'] = [variance_inflation_factor(x, i) for i in range(x.shape[1])]
vif['Feature'] = train.columns[:train.shape[1]-1]
print(vif)
```

As we saw with R, there are a few variables that are collinear. If we take another look at the correlation matrix, we can see that family size is highly correlated with SibSp, Parch, and Group Size. This is no surprise, but since the other variables capture the same variance, we can safely remove family size. Fare is also highly correlated with fare per person. Since we calculated fare per person to more accurately capture the fare of groups, we can get rid of fare. DeckZ is causing problems because there were only first class passengers on Deck Z. Since DeckZ is part of the dummies we created for deck, we should go back and recode them to use DeckZ as the baseline instead of DeckA. The problem is that DeckA also only had first class passengers. So instead of recoding, we can simply get rid of DeckZ, because both

DeckA and DeckZ had first class passengers only. Getting rid of DeckZ will cause the combination of DeckA and DeckZ to be the baseline for the other dummies, which is perfectly fine in this case, because it essentially compares the decks to the first class only decks. The titles are causing issues with the sex variable, because titles are gender specific. That is no surprise. We could get rid of sex, but it would make more sense to keep it, because when we interpret the results it seems more useful to say that being male altered the log odds of survival by x, rather than to say that having a title of Mr. altered the log odds of survival, for instance. Let us get rid of all of the title dummies. The titles capture combinations of age, class, and gender anyway, and since there are other variables to capture this information, it will not be a problem to dispose of the titles. We can get rid of these variables, calculate the VIF again, and re-run the logistic regression.

```
#Get rid of the highly correlated variables and try again
train.drop(['Fare', 'Family_Size', 'DeckZ',
            'Title_Master', 'Title_Miss',
            'Title_Mme', 'Title_Mr', 'Title_Mrs',
            'Title_Rev', 'Title_Sir'], axis=1, inplace=True)
test.drop(['Fare', 'Family_Size', 'DeckZ',
            'Title_Master', 'Title_Miss',
            'Title_Mme', 'Title_Mr', 'Title_Mrs',
            'Title_Rev', 'Title_Sir'], axis=1, inplace=True)

array = train.values
x = array[:, :train.shape[1]-1]
y = array[:, train.shape[1]-1]

vif = pd.DataFrame()
vif['VIF_Factor'] = [variance_inflation_factor(x, i) for i in range(x.shape[1])]
vif['Feature'] = train.columns[:train.shape[1]-1]
print(vif)   #Looks better

logistic_reg = LogisticRegression(fit_intercept=True, random_state=14).fit(x, y)
logistic_reg_preds = logistic_reg.predict(x)
logistic_reg_prob_preds = logistic_reg.predict_proba(x)
print(logistic_reg.get_params, '\n')
print('Intercept:', logistic_reg.intercept_)
for idx, c in enumerate(logistic_reg.coef_[0]):
    print('Coefficient for', train.columns[idx], c)
print('Classification Accuracy:', logistic_reg.score(x, y))
```

Next we check the assumption that residuals are not serially correlated using the Durbin Watson statistic.

```
residuals = y - logistic_reg_preds
print('DW Statistic:', durbin_watson(residuals))
```

Everything looks good. Finally, we can apply the logistic regression to the test dataset.

```
x_test = test.values

logistic_reg_preds = logistic_reg.predict(x_test)
logistic_reg_prob_preds = logistic_reg.predict_proba(x_test)
```

To submit the results to Kaggle, write the logistic_reg_preds to a CSV file.

10. Regression for Count or Integer Variables

Standard linear regression predicts a continuous numerical outcome. Logistic regression predicts a categorical outcome. Sometimes the outcome to be predicted is a discrete count or integer, such as the number of goals scored in a hockey season or the number of votes for a presidential candidate. Standard linear regression assumes that the outcome is continuous, and when the outcome is a count variable, this assumption does not hold. Discrete count or integer variables are bounded at zero, meaning they always have positive values. Continuous variables are unbounded. It may be that the outcome for some regression experiments is continuous and just happens to always be positive. This would not mean that the methods described in this chapter would need to be used. The methods described in this chapter only apply when the outcome is not continuous.

To clear things up, **continuous variables** are numeric variables that can take any value. Discrete variables can only take certain values. **Discrete variables** could be categorical or numeric. The methods in this chapter are targeting discrete numeric variables. If a continuous variable is grouped into bins, it is said to be **discretized**. For example, if we were predicting the number of pennies held by a bank at the end of any given day, would it make any practical difference if the result were 2,000,000 versus 2,000,005? If not, it may be worth discretizing pennies into groups containing ranges of values, such as 1-50,000, 50,000-100,000, and so on.

If a discrete count outcome variable were treated as a continuous numeric variable, and a standard linear regression were run, the results would probably be very close to the results of running the flavors of regression in this chapter. The opposite is also true – the regression methods in this chapter could be run on continuous outcomes and result in regression models that are very close to the models produced by linear regression. The difference is in the predicted outcomes. For example, if predicting votes, a standard linear regression could return a value like -25.2 votes as the predicted outcome for some value of inputs. The methods in this chapter will always predict positive values however, making vote prediction a good example of when to use these methods.

Poisson Regression

Poisson regression uses a log-linear model and MLE to predict the likelihood of a count, much like logistic regression (we will explore log-linear analysis more in depth in a later chapter). The key assumption for Poisson regression is that the mean of the outcome variable is equal to its variance. This is not always true, and when it is not, negative binomial regression (see the next section) may be used instead (Pennsylvania State University, 2018).

Poisson regression is based on the **Poisson distribution**, which is a discrete probability distribution that shows the probability of a certain number of events occurring in a fixed interval of time or space, if these events occur with a known average rate, and independently of the time since the last event. An example of a process that follows this distribution is the number of phone calls received by a call center in 1 hour.

The equation for a Poisson distribution is as follows:

$$P(X = k) = \frac{\lambda^k e^{-\lambda}}{k!}$$

Figure 10.1

This equation means that the probability of a random variable X being equal to some value k, is the product of the average number of times X = k raised to the kth and Euler's constant raised to the negative mean number of times X=k, divided by k factorial. So lambda is the average number of successes, or the average number of times X takes on some value k, that the model computes the odds of. The assumptions are that the mean of the outcome variable's distribution equals lambda, and the variance of that variable is also equal to lambda.

Recall from chapter 2 that the equation for a probability distribution can provide the probability of a random variable taking any given value along the distribution. To illustrate how this can be done with the Poisson distribution, consider the following example. Brigg's Construction builds an average of 35 bridges per day. If we want to know the probability that exactly 40 bridges will be built tomorrow, then our lambda equals 35 and k = 40. Substituting these values into the equation in figure 10.1 gives:

P(X = 40) = (35^40 * e^-35) / 35! = 0.0447

So the probability that exactly 40 bridges will be built tomorrow is 4.47%. Just like any other distribution, the Poisson distribution has a cumulative distribution function that can be used to calculate the probability of the value of the random variable being at least some number, or no more than some number. If we wanted to know, for example, the probability of Brigg's Construction building no more than 30 bridges tomorrow, we could use the cumulative distribution function, or simply sum the probabilities of building 1 bridge, 2 bridges and so on, up to 30 bridges.

The regression equation for a Poisson regression is shown in figure 10.2 below. Notice that the equation computes the natural log of the dependent variable, so in order to interpret the parameter estimates of a Poisson regression, they must be transformed from a log scale back into a normal scale.

$$\ln(E(Y \mid x)) = \alpha + \beta'x$$

Figure 10.2

We have seen many equations like the one in figure 10.2 and so far we have referred to them as regression equations or models. They can also be called **link functions**, because they link the predictors to the outcome.

Instead of OLS, MLE should be used to calculate the estimates for a Poisson regression.

Negative Binomial Regression
The Poisson distribution is actually a special case of the negative binomial distribution in which the mean equals the variance. When the mean does not equal the variance, or specifically, when the variance is larger than the mean, the data is said to have over-dispersion. When data is over-dispersed there is more variance than would be expected, for any given model. The negative binomial regression can be used to deal with over-dispersion. The negative binomial regression adds flexibility by allowing the mean to be a random variable drawn from a gamma distribution (Introduction to SAS, 2018a).

An example of when a negative binomial model would be preferable to a Poisson model is predicting the number of times people smoke cigarettes. The variance of such a model would be high due to over-dispersion, because most people never smoke, smoke once or twice to try it, or only smoke occasionally. There is a small percentage of the population who smoke a lot and give the distribution a large variance.

Like the Poisson regression, MLE should be used to calculate the estimates of a negative binomial regression. The equation for a negative binomial regression is shown below in figure 10.3.

$$P(Y = y_i | \mu_i, \alpha) = \frac{\Gamma(y_i + \alpha^{-1})}{\Gamma(\alpha^{-1})\Gamma(y_i + 1)} \left(\frac{1}{1 + \alpha\mu_i}\right)^{\alpha^{-1}} \left(\frac{\alpha\mu_i}{1 + \alpha\mu_i}\right)^{y_i}$$

Figure 10.3

In the equation in figure 10.3, gamma represents the gamma distribution from which the mean of the negative binomial distribution is drawn. Alpha is the constant, mu is the mean, and y is some value that Y can take.

On occasion, over-dispersion of data can be caused by excessive zeroes in the outcome variable. When excessive zeroes are the cause of over-dispersion, a zero-inflated regression model would provide better parameter estimates (better means the distribution of values that the parameter estimates can take will be narrower if they are estimated with a zero-inflated model).

Zero Inflation

Both the Poisson regression and negative binomial regression can be fitted to work with data with a lot of zeroes. This process is called zero inflation (Introduction to SAS, 2018b). Referring back to the example of the number of cigarettes that people smoke, many people do not smoke. It is possible that when counting cigarettes, non-smokers inflate the number of zeros in the data. In this case, a zero-inflated negative binomial regression would give the best estimates of the model parameters.

Zero inflated is performed by combining the logit function with the probability distribution of the model. Recall from chapter 9 that the logit function produces a binary output (0 or 1). The addition of this function allows a sampling of zeros to be factored into the model, on top of the zeros that the distribution includes by itself. In practice, zero inflating a Poisson model is often not as good as zero inflating a negative binomial model, but the differences can be trivial. The equation in figure 10.4 below shows how zero inflation applies to negative binomial regression.

$$P(y_i = j) = \begin{cases} \pi_i + (1 - \pi_i)g(y_i = 0) & \text{if } j = 0 \\ (1 - \pi_i)g(y_i) & \text{if } j > 0 \end{cases}$$

Figure 10.4

In the equation in figure 10.4, g(y sub i) is the equation for the negative binomial regression given in figure 10.3, and pi sub I is the logistic function shown in figure 9.2.

Count and Integer Regression with R

Given what was explained in this chapter, a good question to ask would be if these methods could be used for the Fitbit regression we did in chapter 6. Calories are a count/integer variable, so yes! We already did the assumption checks in chapter 6, so all we need to do is fit the model. First we will check to see if the mean of the outcome variable equals the variance.

```
train <- read.csv("train.csv", header=T)
test <- read.csv("test.csv", header=T)

mean(train$Calories)
var(train$Calories)
```

No it does not. So the negative binomial regression is a better choice than Poisson regression. In the model specification below, we are re-using the reduced model that was found to be a better fit in chapter 6. We are also using the identity as the link function, which means the resulting coefficients will be comparable to the coefficients from standard linear regression in chapter 6. If we wanted, we could also use the log link function, which would return coefficients in terms of the log count of the outcome variable. In this case, we would interpret a 1 unit increase in each predictor to affect the log count of the outcome by the amount of the coefficient.

```
library(MASS)
negb_reg <- glm.nb(Calories ~ Steps + Floors_Climbed + Times_Awake + Day_Saturday +
Day_Sunday, data=train, link=identity)
summary(negb_reg)
deviance(negb_reg)
plot(negb_reg)

#Get the Newey-West HAC estimates of the model
library(sandwich)
library(lmtest)
coeftest(negb_reg, vcov.=NeweyWest)
```

The results show that all of the predictors were statistically significant, and the coefficients are very similar to what were found in chapter 6. The difference between the interpretation of the results here is that each 1 unit increase in the predictors produce a change in the count of the outcome equivalent to the coefficients. So for example, one additional step increases the count of calories burned by about 0.05. From here we would make our predictions on the test data the same way we did in chapter 6.

Count and Integer Regression with Python

Given what was explained in this chapter, a good question to ask would be if these methods could be used for the Fitbit regression we did in chapter 6. Calories are a count/integer variable, so yes! We already did the assumption checks in chapter 6, so all we need to do is fit the model. First we will check to see if the mean of the outcome variable equals the variance.

```
import pandas as pd
import numpy as np
from sklearn.preprocessing import StandardScaler, MinMaxScaler
import statsmodels.formula.api as smf
import statsmodels.api as sm

train = pd.read_csv("train.csv")
test = pd.read_csv("test.csv")

print("Mean:", np.mean(train.Calories))
print("Variance:", np.var(train.Calories))
```

No it does not. So the negative binomial regression is a better choice than Poisson regression. In the model specification below, we are re-using the reduced model that was found to be a better fit in chapter 6. We are also using the identity as the link function, which means the resulting coefficients will be comparable to the coefficients from standard linear regression in chapter 6. If we wanted, we could also use the log link function, which would return coefficients in terms of the log count of the outcome variable. In this case, we would interpret a 1 unit increase in each predictor to affect the log count of the outcome by the amount of the coefficient.

```
negb_reg = smf.glm("Calories ~ Steps + Floors_Climbed + Times_Awake + Day_Saturday +
Day_Sunday", data=train,
family=sm.families.NegativeBinomial(link=sm.genmod.families.links.identity)).fit()
print(negb_reg.summary())
```

There were a few warnings generated by the code above. These warnings are caused by the optimization algorithm struggling to converge. To make them go away, we can scale the data so that the numbers do not get too large. We can also get the Newey-West Hac estimates of the standard errors by specifying "HAC" as the covariance argument in the fit method.

```
train_std_feats = MinMaxScaler().fit_transform(train)
test_std_feats = MinMaxScaler().fit_transform(test)
#Convert arrays back to dataframes
train_std = pd.DataFrame(train_std_feats, index=train.index, columns=train.columns)
test_std = pd.DataFrame(test_std_feats, index=test.index, columns=test.columns)
#Re-run on scaled data and get the Newey-West Hac estimates of the standard errors
negb_reg = smf.glm("Calories ~ Steps + Floors_Climbed + Times_Awake + Day_Saturday +
Day_Sunday", data=train_std,
family=sm.families.NegativeBinomial(link=sm.genmod.families.links.identity)).fit(cov_
type='HAC', cov_kwds={'maxlags':1})
print(negb_reg.summary())
```

The results show that all of the predictors were statistically significant, and the coefficients are very similar to what were found in chapter 6. The difference between the interpretation of the results here is that each 1 unit increase in the predictors produce a change in the count of the outcome equivalent to the coefficients. So for example, one additional normalized step increases the normalized count of calories burned by about 0.08. From here we would make our predictions on the test data the same way we did in chapter 6.

11. Regression with Censored Data

Censored data is data with one or more fields of observations that are only partially known. This can occur in experimental settings where there is a limit to measurement. For example, if an experiment involving super obese people were carried out using a weight scale with a maximum measurement of 250 pounds, then anyone weighing more than 250 pounds will be recorded as 250. The data will show several people weighing 250 pounds but nobody with a weight more than 250. This is an example of **truncation**, but censoring can also occur when observations have unknown values. For example, recall the example of a repeated measures design experiment from chapter 1, where 1 group of individuals showered after the gym and the other group did not, and the following day the groups were reversed. If some people did not show up for the experiment on day 2, there would be observations with unknown values included in the data.

Censoring is only a problem when it occurs with the dependent variable (the outcome). If observations are missing values of the independent variables (the predictors), then they should be treated using the methods described in chapter 4.

Recall from chapter 6 that one of the assumptions that is required to be met in order to generalize standard linear regression results to a population is that the outcome be continuous, and unbounded. Censored data prevents that assumption from being met. There are generally four ways to handle censored data. The first is to simply run a standard linear regression using OLS and accept the fact that the parameter estimates could be inconsistent between samples and may not approach the true population values (in other words they will be biased). The other three options are described in the following sections.

Tobit Model

The Tobit model, named after economist James Tobin, is useful when there is left or right (a.k.a. below or above) censoring in the data (Tobin, 1958). Censoring from above occurs when the data is capped at some value or threshold, like in the aforementioned obesity example. Values above the threshold are automatically relabeled to take the value of the threshold; the highest possible value. Censoring from below occurs when the data has a minimum value threshold. Values below the threshold are automatically relabeled to take the value of the threshold; the lowest possible value. If these observations are kept in the data, then the Tobit model is appropriate. For situations when these observations are excluded from the data, see the next section on truncated regression.

The Tobit model works by supposing there is a latent variable y* that is linearly dependent on x through the vector of parameters beta. In mathematical terms, the model is the same as a standard regression model except that y is replaced with y*. When the latent variable is > 0, then the observed variable y equals the latent variable y*. Otherwise, the observed variable y equals 0. A **latent variable** is a variable that is not directly observed but whose existence is inferred from a model's independent variables. Latent variables can represent physical phenomena or abstract constructs.

Truncated Regression

When data is truncated as described in the obesity example, and the truncated observations are excluded from the data entirely, then truncated regression may be appropriate. When data is truncated from below, then the mean of the truncated variable will be greater than the mean of untruncated data. When data is truncated from above, then the mean of the truncated variable will be less than the mean of untruncated data. Therefore, when a model is applied to new data that is untruncated, the mean estimate will need to be adjusted to compensate for truncation in the training data.

Note that untruncated and nontruncated are interchangeable in statistics texts, although neither term is officially recognized as an antonym of truncated.

Interval Regression

Interval regression is used to model outcomes that have interval censoring. Interval censoring occurs when the ordered category into which each observation falls is known, but the exact value of the observation is unknown. Sometimes left and right censored data is also interval censored. For example, if a dataset contains income buckets but not the true income values, and the goal is to predict annual income, then there is left, right, and interval censoring, because the lowest and highest buckets are grouping everything below and above thresholds. Another example is modeling grade point average on a set of variables, and the outcome consists of the GPA ranges: 2.0-3.0 and 3.0-4.0.

Note that ordered probit regression is an acceptable alternative to interval regression. Modeling the GPA buckets for example, could be done using ordered pairs of categories: 0 (0.0-2.0), 1 (2.0-2.5), 2 (2.5-3.0), 3 (3.0-3.4), 4 (3.4-3.8), and 5 (3.8-4.0).

Censored Data Regression with R

There is no section for Python in this chapter because there are no Python libraries for regression with censored data. For the examples in this chapter, we will reuse the Fitbit data, but manually alter the values of the dependent variable to simulate censoring from above and interval censoring. So first, let us clip the values of calories to 2600 to simulate censoring from above (a.k.a. right censoring). Before we modify the data, we will view the frequency distribution of calories and construct a negative binomial model to use for comparison.

```
#This gives us a baseline to compare against
hist(train$Calories, breaks=50, main="Freq Dist of Calories Before Censoring")
library(MASS)
negb_reg <- glm.nb(Calories ~ Steps + Floors_Climbed + Times_Awake + Day_Saturday +
Day_Sunday, data=train, link=identity)
summary(negb_reg)


#Apply censoring
train[train$Calories>2600, 'Calories'] <- 2600
test[test$Calories>2600, 'Calories'] <- 2600


#See how censoring impacts the freq dist and the regression model
hist(train$Calories, breaks=50, main="Freq Dist of Calories After Censoring")
negb_reg_cens <- glm.nb(Calories ~ Steps + Floors_Climbed + Times_Awake +
Day_Saturday + Day_Sunday, data=train, link=identity)
summary(negb_reg_cens)
```

Notice how after the data was censored, the model's standard error increased, predictors became statistically insignificant, and the coefficients were altered quite a bit. Now let us see how the Tobit model handles censoring. Note that the Tobit function from the AER package has arguments for left and right censoring. Since there is no left censoring (censoring from below), we set left = -Inf. Since the right censoring occurred at 2600, we set right=2600. In situations where it is not known what value was used for censoring, the frequency distribution of the outcome can be used to make a good guess.

```
library(AER)
tobit_reg <- tobit(Calories ~ Steps + Floors_Climbed + Times_Awake + Day_Saturday +
Day_Sunday, data=train, left=-Inf, right=2600)
summary(tobit_reg)
```

Notice how the output of the Tobit regression is much more in line with what we saw in the negative binomial regression. Note that while we are using Tobit regression on a count variable, Tobit regression can be used for continuous (unbounded) variables too. Next we can bin the values of the outcome variable to simulate interval and right censoring.

```
train$Calories <- cut(train$Calories, c(0, 2100, 2200, 2300, 2400, 2500, Inf))
test$Calories <- cut(test$Calories, c(0, 2100, 2200, 2300, 2400, 2500, Inf))
plot(train$Calories, main="Calories After Binning")
```

Now we can use interval regression. Be sure to use Inf as the boundary.

```
library(intReg)
interval_reg <- intReg(Calories ~ Steps + Floors_Climbed + Times_Awake + Day_Saturday
+ Day_Sunday, data=train, boundaries=c(0, 2100, 2200, 2300, 2400, 2500, Inf),
method="probit")
summary(interval_reg)
```

Notice how the output of the interval regression is a bit worse than what we have seen in other models, because binning the outcome eliminates some information, but the model is not horrendous. From here we would make the predictions for the test set, ensuring that the outcome is binned the same way for the test data.

12. Generalized Linear Models: t-Test, ANOVA, ANCOVA, Chi-Square Test, and Log-Linear Analysis

The regression methods introduced in chapters 10 and 11 fall under the large umbrella of generalized linear models. Generalized linear models extend linear regression to be able to model dependent variables with non-normal error distributions. For example, count or integer variables are not normally distributed. Likewise, the models in this chapter are generalized linear models.

t-Test: Comparing 2 Means

A t-test can be used to compare the means of the distributions of two categories of observations. When a dependent variable is modeled only on 1 independent variable, and that independent variable is categorical with only two categories (meaning it is binary), then a t-test is an appropriate linear model (Field, Miles, & Field, 2012). Here are a few examples of when a situation like this could arise, and a t-test would be required to model the relationship between the variables:

- We want to see if the Kingda Ka roller coaster is more frightening than Top Thrill Dragster, so we measure the heart rates of riders of the coasters. In this scenario, heart rate is the dependent variable (it is used as a proxy for fear) and roller coaster is the binary independent variable.
- We want to see if music affects running speed, so we carry out an experiment with one group of people running with music and another group running without music. In this scenario, speed is the dependent variable and music/no music is the binary independent variable.

From these examples, it should be clear that the t-test is a popular model for experimental research. Recall from chapter 1 that experiments can be carried out with one group of participants taking part in each condition (repeated measures design), or with two groups of participants where each group only takes part in one condition (independent measures design). There are two types of t-tests used for each type of experimental design: independent means for independent measures design, and dependent means for repeated measures design.

An **independent means t-test** is used for independent measures design experiments, where there are 2 groups of participants; 1 group for each condition. Recall the equation for a t-test from figure 3.4 in chapter 3:

$$t = \frac{\bar{x} - \mu_0}{\frac{s}{\sqrt{n}}}$$

Figure 3.4 from chapter 3

Figure 3.4 shows that the t-statistic equals the difference between the sample mean, x bar, and the population mean, mu sub 0, divided by the quotient of the sample standard deviation and the square root of the number of observations in the sample. When we used this equation for hypothesis tests in chapter 3, we were comparing the sample mean to some value. For example, we explored whether or not the mean Titanic passenger age differed from 34 (34 was the average age of the UK population in 1974, which was as far back as the data went). Now however, we want to compare the mean of one group to the mean of another group composed of different samples or participants.

The same equation from figure 3.4 is used for paired t-tests, as shown by figure 12.1:

$$t = \frac{\overline{X_1} - \overline{X_2}}{\sqrt{\frac{s_1^2}{N_1} + \frac{s_2^2}{N_2}}}$$

Figure 12.1

At their cores, the equations in figures 3.4 and 12.1 give the difference between the observed and the expected mean, divided by standard error. In figure 3.4, the observed mean is the sample mean. In figure 12.1, the observed mean is the mean of the first variable. It does not matter which variable is first, as the result is interpreted the same regardless of which tail of the t-distribution it falls into. The expected mean in figure 12.1 is replaced by the mean of the second variable. In the denominator of 12.1, the standard error is calculated across both variables, assuming each group of the binary independent variable contains the same number of observations. If one group has more observations than the other, then the equation in figure 12.1 is modified to weight the variance of each observation, as shown by figure 12.2.

$$t = \frac{\overline{X_1} - \overline{X_2}}{\sqrt{\frac{s_p^2}{n_1} + \frac{s_p^2}{n_2}}}$$

Figure 12.2

The weighted average variance (s_p^2) in the equation in figure 12.2 is found by taking the weighted average of each variance multiplied by its degrees of freedom, and divided by the sum of the weights.

A **dependent means t-test** is used for repeated measures design experiments, where one group of participants is tested on each condition. In a dependent means t-test, observations are paired together because they contain information about the same sample or participant, but in two categories. For this reason, it is common to see dependent t-tests comparing the means of 2 groups referred to as **paired t-tests**. Figure 12.3 illustrates the differences between a single sample t-test, independent samples t-test, and a dependent paired t-test.

Single Sample t-test: Compare the mean of a random variable to some value	Independent Samples t-test: Compare the means of 2 groups of different samples or participants across categories	Dependent (Paired) t-test: Compare the means of 2 groups of the same samples or participants, across categories

y			y	x		y	x	
18			18	Group A		11	Before	Paired observations (same sample across 2 groups)
25	34		25	Group A		20	After	
51			51	Group B		54	Before	Paired observations (same sample across 2 groups)
						60	After	

Figure 12.3

The equation for dependent means t-tests follows the same idea as the equations for single sample and independent means t-tests:

$$t = \frac{\bar{D} - \mu_D}{\frac{s_D}{\sqrt{N}}}$$

Figure 12.4

In the equation in figure 12.4, the difference between one group mean, D bar, and the group's expected mean, mu (we are expecting one group's mean to be the same as the other group's mean), is divided by the standard error of the differences. If the null hypothesis is true, then there is no significant difference between the group means.

The equation in figure 12.4 shows that the differences between each participant's score under each condition in an experiment are added up and divided by the total number of participants to get the average difference, D bar. D bar is a measure of systematic variation in the data (the experimental effect). This variation is compared to the measure of systematic variation that would occur naturally. The standard error is a measure of naturally occurring systematic variation because it is the standard deviation of a sampling distribution of differences between sample means.

The t-statistic for dependent paired t-tests should be > 1 if there is any kind of effect from the condition in an experiment.

When a t-test is carried out to compare means, the following assumptions are made:

- The sampling distribution is normal. For dependent t-tests, this means that the sampling distribution of the differences between scores should be normal, not the scores themselves.
- Data is measured at the interval level or lower.
- For independent t-tests, it is also assumed that scores in different effects or conditions are independent because they come from different participants, and that there is homogeneity in variance. If the sample size is the same between groups, then the homogeneity assumption does not matter. If the sample size is different, Welch's t-test can be used to correct for heteroskedasticity (Field, Miles, & Field, 2012).

ANOVA: Comparing Several Means

ANOVA stands for analysis of variance. Just like t-tests show whether or not 2 groups of observations have the same mean, ANOVA shows whether or not 3 or more groups of observations have the same mean. ANOVA produces an F-statistic, which is the ratio of systematic variance to unsystematic variance, just like the t-statistic. ANOVA is an omnibus test, meaning it tests for an overall effect but does not specify which groups are affected. For example, if the F-ratio shows that the means across groups are not the same, it is unknown whether only 1 group is different, 2 groups, or all of them have different means. ANOVA also does not indicate which groups have different means, if a difference is detected. In order to determine which groups have different means, post hoc analysis is required. We will look at post hoc analysis later in this chapter. Note that when we refer to ANOVA in this chapter, it is one-way ANOVA: comparing group means of a single categorical predictor.

A reasonable question might be: Why not use multiple t-tests for every combination of groups, instead of using ANOVA? The answer is that doing this would produce several independent probabilities of type I error, and the overall probability would be the product of all of them. So if n t-tests were carried out for every combination of groups at the 95% confidence level, then the error probability would be $1-(0.95^n)$. The more groups there are, the more t-tests would be required, and the greater the overall probability of type I error.

In order to understand how analysis of variance compares group means, it is necessary to revisit the F-ratio introduced in chapter 3 to compare two variances. Recall from chapter 6 that the F-ratio is also used as a measure of goodness of fit for a regression model; it shows how well the model can predict an outcome compared to the error within the model. If we think of the task of comparing means as fitting generalized linear regression model consisting of only categorical predictors, then the F-ratio shows how well the model fits the data. ANOVA is really just a special case of regression in which all of the predictors are categorical variables.

The F-ratio is defined as the ratio of explained variance to unexplained variance. If we want to test whether group means are similar, then the group means should have low variance. If the variance between the group means and the overall mean is large and statistically significant, according to the F-test, then then the groups can be said to have different means. Thus, the F-statistic in ANOVA shows the ratio of the variation between group means to the variation within the groups.

As an example for how ANOVA is carried out, suppose that there are 3 groups (3 levels of 1 categorical predictor), as shown in figure 12.5. One of the groups can be used as the base group. We will use X1 as the base group. In a regression model, the beta coefficient for the base group is just the y-intercept. The beta coefficient for group X2 is the difference between the group's mean and the base group's mean:

Beta coefficient = mean(X2)-mean(X1) = 38.5 − 18 = 20.5

The beta coefficient for group X3 is the difference between the group's mean and the base group's mean:

Beta coefficient = mean(X3)-mean(X1) = 25.5 − 18 = 7.5

The null hypothesis is that the three groups share the same mean. If the groups do not have the same means, then the predictors affect the dependent variable, y, in some way. If we were conducting an experiment with 3 conditions, then a rejection of the null would mean that the conditions we tested did have an effect on the dependent variable. The table in figure 12.5 shows how a single categorical predictor with 3 conditions can be represented as 3 binary columns. Note that these are not 3 different predictors, only categories of one predictor. Recall from chapter 4 that representing a categorical variable with K levels as K binary predictors is called one-hot encoding.

y	X1	X2	X3
18	1	0	1
26	0	1	0
51	0	1	0
33	0	0	1

Figure 12.5

To calculate the F-statistic for ANOVA, we need the degrees of freedom for each group and for the combination of all three groups. Since all 3 groups have the same number of observations, N, the degrees of freedom is N-1 or 3. The degrees of freedom for all three groups is the sum of the degrees of freedom for the individual groups: 9. The between groups DoF, 2, and the within groups DoF, 6 can then be looked up in an F statistical table to determine the critical value for the F-test. For this example, the critical value is 5.078. Now to calculate the F-statistic, we can use the formula in figure 12.6.

$$F = \frac{between\ group\ variance\ a.k.a.\ expalined\ variance}{within\ group\ variance\ a.k.a.\ unexplained\ variance}$$

$$F = \frac{\frac{\sum_{i=1}^{K} n_i (\bar{x}_i - \bar{x}_{grand})^2}{K - 1}}{\frac{\sum_{i=1}^{K} \sum_{j=1}^{n_i} (x_{ij} - \bar{x}_i)^2}{N - K}}$$

Figure 12.6

In the equation in figure 12.6, x bar sub i is the sample mean of the ith group, n sub i is the number of observations in the ith group, x bar sub grand is the overall mean of all samples in all groups, K is the number of groups, x sub ij is the jth observation of the ith group, and N is the total number of observations in the dataset. Inserting the values and calculating everything out, we end up with an F-statistic of approximately 0.19, which is less than the critical value of 5.14, meaning the null hypothesis that the groups share the same mean cannot be rejected. Since the F-statistic is < 1, it means that the variation within groups is larger than the variation between the groups.

After an ANOVA model has been fit, it can be assessed by calculating deviation as in figure 12.6:

$$deviation = \sum (observed - model)^2$$

Figure 12.6

Deviation can be calculated by the sum of squares. Recall the different variations of the sum of squares from chapter 6:

1. Total Sum of Squares – expresses deviation as the difference between each observation and the mean
2. Sum of Squared Residuals – expresses deviation as the difference between each observation and the model (the group means)
3. Model Sum of Squares – expresses deviation as the difference between the mean and the model (the group means)

The total sum of squares (SST) expresses deviation as the difference between each observation and the mean. It is the total variation in the data itself. The total variation in the data can be found by summing the squared differences between each observation and the overall mean.

$$SST = \sum_{i=1}^{N} (x_i - \bar{x}_{grand})^2$$

Figure 12.7

The SST can also be found using the variance, because the variance equals the sum of squares divided by N-1, where N is the number of observations.

$$s^2 = \frac{SS}{(N-1)}$$

Figure 12.8

Therefore, the variance equation can be rearranged to find the total sum of squares.

$$SST = s^2_{grand}(n-1)$$

Figure 12.9

The model sum of squares (SSM) measures the total variation that can be explained by the model. It is a measure of model deviation based on comparing the means of the groups with the overall mean. In the equation in figure 12.10 below, n sub k is the number of participants in group k, and x bar sub k is the mean of group k.

$$SSM = \sum_{n=1}^{k} n_k \left(\bar{x}_k - \bar{x}_{grand} \right)^2$$

Figure 12.10

The degree of freedom is 1 less than the number of parameters estimated (the number of groups – 1).

Once the total variation in the data, SST, and the variation explained by the model, SSM have been calculated, the next step in ANOVA is to find out how much variation cannot be explained by the model. The sum of squared residuals (SSR) can be calculated either by taking the difference between the SST and SSM, or by summing the product of each group's variance and 1 less than the number of observations within the group.

$$SSR = (SST - SSM) = \sum s_k^2(n_k - 1)$$

Figure 12.11

The SSM and SSR are summed values, so they are influenced by the number of values summed. By averaging these values (by dividing them by the degrees of freedom), they can be directly compared. Recall from chapter 6 (figure 6.14) that the result of dividing the SSR by the degrees of freedom is the average variance, or mean squared error.

The F-ratio is the ratio of the explained variance to unexplained variance, which in terms of sum of squares, is the model sum of squares divided by the sum of squared residuals:

$$F = \frac{SSM}{SSR}$$

Figure 12.12

If the F-ratio is < 1, then the effect is non-significant (because that would mean that the SSR is greater than SSM, like in our previous example). If the F-ratio is > 1, then the F-statistical table can show the maximum values that would be expected to occur by chance, if the group means were equal in an F-distribution with the same degrees of freedom. If the F-ratio is larger than the critical value from the F-table, then the result of ANOVA is most likely not due to chance, and the independent variables do have an effect on the dependent variable; the group means are different.

Since ANOVA is a linear model, is requires the standard parametric assumptions. However, ANOVA is somewhat robust to violation in the assumptions, as long as the group sizes are the same.

As an example use case for ANOVA, suppose a production manager at a warehouse that fills customer orders needs to cut back on the number of shifts worked by warehouse employees. To decide which shift to cut, the manager looks at the number of daily packing mistakes made over a three month period for each shift. The shift that makes the most mistakes will be cut. Intuitively, the simplest way for the manager to determine which shift makes the most mistakes would be to simply look at the percentage of packing mistakes made over the three month period. The shift with the highest percentage of orders that were packed incorrectly would be the one to be cut. So why use ANOVA at all? The answer lies in the homogeneity of variance assumption that ANOVA makes. Suppose that the shift with the highest percentage of packing mistakes had a really bad day, and nearly all of the mistakes they made during the three month period occurred on that day. This would give that shift a high variance in the number of packing mistakes. Would it be fair to cut that shift when another shift made mistakes more consistently? ANOVA would compare the mean number of mistakes made by each shift while accounting for variance. Therefore, it would give a more accurate measure of differences between mistakes made by each shift.

Post Hoc Analysis

Recall that ANOVA indicates whether there is a difference between group means, but it does not indicate which groups are different. Post hoc analysis must be carried out to examine specific groups. There are two types of post hoc analysis that can be carried out: planned contrasts a.k.a. planned comparison, and post hoc tests. **Planned contrasts** are like one-tailed tests in the sense that they are applied when there is a known, specific comparison that is desirable. Planned contrasts are useful when there are specific groups that are planned to be compared. **Post hoc tests** are like two-tailed tests in the sense that they are good at finding any effects between group comparisons, regardless of which groups are compared. While the analogy between planned contrasts, post hoc tests, and one and two tailed tests is useful for understanding how they work, there are no one or two tailed tests that can be applied to ANOVA because directional analysis in the comparison of several group means would be meaningless.

The logic behind post hoc tests is to carry out tests that are similar to the independent t-test for comparing 2 group means, but corrected for the inflated type I error rate that multiple t-tests creates (recall from the beginning of the section on ANOVA why ANOVA is preferred to multiple t-tests). There are too many variations of post hoc tests available to describe in this chapter, however two of the most common are **Tukey's test** a.k.a. **Tukey's honest significant difference (HSD) test**, and the **Newman-Keuls method**. These post hoc tests require similar assumptions to the t-test: observations are independent within and among the groups, the groups are normally distributed, and there is equal within-group variance across the groups (homogeneity of variance). Of the post hoc tests, the Newman-Keuls method is the most robust to violations in these assumptions. However, the tradeoff for being robust is a loss in statistical power. The Tukey test is better able to detect differences in group means.

The Newman-Keuls method makes stepwise comparisons between groups. The method starts by comparing the group with the largest mean to the group with the smallest, and checking the significance of the test. The process continues, comparing the group with the next largest mean to the group with the smallest and next smallest, and so on, until the test indicates that a comparison is insignificant. The method uses the same formula to calculate that test statistic as the Tukey test, which is shown in figure 12.13.

$$q = \frac{\bar{X}_A - \bar{X}_B}{\sqrt{\frac{MSE}{n}}}$$

Figure 12.13

In the equation in figure 12.13, q is the studentized range test statistic, X bar sub A is the mean of group A (the larger of the two groups being compared), X bar sub B is the mean of group B (the smaller of the two groups being compared), the MSE is the mean squared error of an ANOVA model, and n is the number of observations in the dataset. Note that the denominator, the square root of the MSE divided by n, is equivalent to the standard error of an ANOVA model. If the groups being compared do not have the same sample size, then the equation in figure 12.14 is used instead.

$$q = \frac{\bar{X}_A - \bar{X}_B}{\sqrt{\frac{MSE}{2} * \left(\frac{1}{n_A} + \frac{1}{n_B}\right)}}$$

Figure 12.14

The test statistic is compared to a q table of critical values using the same alpha level of significance and DoF as ANOVA.

Post hoc tests perform pairwise comparisons; they compare 2 groups at a time for every possible combination (or until there are no more statistically significant comparisons). They correct for type I error inflation by using a correction term to determine the p critical value. One such correction term, the **Bonferroni Correction**, is simply the alpha level of significance divided by the number of comparisons, k:

$$p_{critical} = \frac{\alpha}{k}$$

Figure 12.15

The tradeoff for Bonferroni's correction is that type II error is inflated. Other correction terms account for type II error as well. One example is the **Benjamini-Hochberg correction**, which involves computing the p-values for every predictor, ordering them largest to smallest, and assigning a sequential index value, j. The index, j, is divided by the number of comparisons, k, and multiplied by the alpha level of significance to create the p critical value. The p-value of the predictors are compared to the critical p-value, starting from the highest index and progressing to smaller indices until a statistically significant predictor is found. At that point, it can be assumed that all of the remaining predictors are significant too, and no more tests need to be carried out.

$$\rho_{critical} = \left(\frac{j}{k}\right)\alpha$$

Figure 12.16

In summary, if it is more important to conservatively control type I error, a Bonferroni correction is preferable. However, if it is more important to control the ratio of type I error to the total rejections of the null, then the Benjamini-Hochberg correction is better.

A reasonable question might be: why should ANOVA be used at all if the post hoc tests correct for type I error inflation and are able to show which specific groups are different? It is perfectly acceptable to skip ANOVA altogether, however the post hoc tests have less statistical power to detect differences between group means than ANOVA. Thus, it is customary to only perform these tests post hoc (after ANOVA) so that it is more certain than differences between the group means exist.

Planned contrasts are post hoc comparisons between specific groups. Typically, orthogonal contrasts are the best planned contrasts that can be made. Orthogonal contrasts must sum to 0 and the groups being compared must negate one another. As an example, suppose there are 4 groups that we wish to compare. We could make two of them 1 and two of them -1. If we have 5 groups to compare, we could make three of them 1/3 and two of them -1/2. Which three we make into 1/3 depends on the groupings that we choose, based on which groupings we think would give the most predictive power. Groups that we have already compared in previous contrasts, that we do not want to compare twice, are set to 0. To visualize how this would work, see figure 12.17.

Group	Contrast 1	Contrast 2	Contrast 3
X1	1	0	-1
X2	-1	-1	0
X3	-1	1	0
X4	1	0	1

Figure 12.17

In figure 12.17, the first contrast compares the grouping X1 and X4 with the grouping of X2 and X3. The second contrast compares X2 with X3. The third contrast compares X1 with X4. Unlike post hoc tests, planned comparisons are made manually, and the comparisons can be set up somewhat arbitrarily. We will see how this works in practice at the end of this chapter.

ANOVA is commonly used to test drug effectiveness. Pharmaceutical companies use ANOVA to compare the effectiveness of different drugs, doses, or other experimental conditions. One example is a study carried out by SAS that used ANOVA to determine the effectiveness of different chronic obstructive pulmonary disease (COPD) drugs (Chowdavarapu et. al, 2017). The data for the test consisted of patients who took 1 of 5 drug doses: Advair, Symbicort, Spiriva, Advair and Spririva, and Symbicort and Spiriva. The length of hospital stays were measured, so the goal was to compare the mean length of stay for each group. According to the results of the test, there was a statistically significant difference between the group mean lengths of stay. Post hoc tests were then carried out to determine which specific groups had longer or shorter mean lengths of stay. The study found that the average length of stay was least for the patients in the group who only took Advair, with a mean of 5.72 days. No significant difference was found between the mean lengths of stay of patients who took Symbicort and Spiriva, and the combination of Symbicort and Spiriva had the longest mean length of stay.

ANCOVA: Comparing Several Means with Covariates

ANCOVA stands for analysis of covariance. Standard linear regression involves modeling a continuous dependent variable on several other continuous variables. ANOVA models a continuous dependent variable one or more categorical variables to test for differences in the mean of the dependent variable between groups. ANCOVA is a combination of the two: it compares the means of several groups while accounting for 1 or more continuous variables, thereby modeling a continuous dependent variable on one or more categorical and continuous variables. The continuous variables that ANCOVA controls for can be included in the features of observational research or one of the variable in an experiment. However, they can also be extraneous variables that influence the dependent variable in some way. These variables are called **covariates**. So covariate is a synonym for a predictor or independent variable. As an example of an extraneous covariate, suppose we want to compare the resting heart rates of men and women. Our dataset consists of observations of heart rates and a dummy variable indicating whether the person is a man (if the dummy is 0 then it can be assumed the person is a woman). It is likely that a person's age also affects their resting heart rate, even though age is not included in the dataset. Therefore, age is an extraneous covariate.

Covariates can be continuous or categorical, but for ANCOVA, they are always continuous. Covariates are included in the model like any other continuous predictor, regardless of whether they were part of the dataset or extraneous. Since extraneous variables are often confused with confounding variables or latent variables, figure 12.18 compares the definitions of these types of variables.

Extraneous Covariate	Confounding Variable	Lurking Variable	Latent Variable
A variable that is not manipulated in an experiment or included in a dataset, but has an effect on the dependent variable. Covariates are added to a model.	A variable whose effect on the dependent variable cannot be distinguished from other predictors. Confounding variables are added to a model.	A variable whose effect on the dependent variable cannot be distinguished from other predictors. Lurking variables are **not** added to a model.	A variable that is not directly observed but whose existence is inferred from other predictors. Latent variables can represent physical phenomena or abstract constructs. One example is political philosophy.

Figure 12.18

To understand how covariates are useful in regression, consider starting a model with only a continuous variable; the covariate. When a binary/dummy independent variable is added to the model, its effect can be determined after the effect of the covariate, which partials it out. Recall from chapter 5 that partial correlation shows how the effects of several variables can be quantified, and that correlation is just standardized covariance. The inclusion of a covariate in ANCOVA has two main benefits:

- It reduces error variance within the groups: it does this by explaining some of the variance that was previously unexplained by the model (the SSR), because some of the SSR is explained by covariates.
- It eliminates confounding effects caused by extraneous variables.

The assumptions made for ANCOVA are the same as for ANOVA, except for these two additional assumptions:

- The covariate and treatment effect that are measured by the categorical variables must be independent (the must not be no collinear), because in order to add any explanatory value to the model, the covariate must be independent of the dummy variables that are already in the model.
- The relationship between the dependent variable and the covariates must be the same between groups, meaning there must be homogeneity in the regression slopes.

An important restriction on the choice of covariates is that they must never be differentiators between the groups of the categorical variables in the model. For example, if the groups are separated based on age, then age cannot be a covariate or it will introduce spurious effects into the model. This holds true even if age is found to be an unintentional differentiator. To ensure that covariates meet this criterion, the participants or observations in each group should be randomized (Field, Miles, & Field, 2012).

Chi-Square Test: Comparing 2 Groups of Categories

So far we have looked at GLMs with continuous numeric dependent variables, but what if the dependent variable is categorical? How could groups of categories be compared? This is where the Chi-square test and loglinear analysis come into play. The Chi-square test is like the t-test for categorical dependent variables: it compares 2 groups of categories. Loglinear analysis, as we will soon see, is like ANOVA for categorical dependent variables: it compares several groups of categories.

An example of when the Chi-square test for independence would be appropriate is if we wanted to compare favorite colors between male and female students. We would start by calculating the expected frequencies of each level of one categorical variable for each level of the other categorical variable, as in figure 12.19.

$$E_{a,b} = \frac{n_a * n_b}{n}$$

Figure 12.19

In the equation in figure 12.19, E sub a, b is the expected frequency count for level a of the first variable and level b of the second variable, and n is the number of sample observations.

The chi-square statistic is based on the idea of comparing the observed frequencies to the expected frequencies of each category occurring by chance. The Chi-square test statistic standardizes the sum of squared deviations by dividing by the sum of the expected frequencies for the model, as seen in figure 12.20.

$$\chi^2 = \sum \frac{(observed_{ab} - expected_{ab})^2}{expected_{ab}}$$

$$\chi^2 = \sum_{a=1}^{c} \sum_{b=1}^{d} \frac{(O_{a,b} - E_{a,b})^2}{E_{a,b}}$$

Figure 12.20

The equation at the top of figure 12.20 shows the general concept of the Chi-square test. In the equation at the bottom of figure 12.20, O sub a, b is the observed frequency count for level a of the first variable and level b of the second variable. The sum is computed over every combination of levels for each variable up until the final pair (levels c and d). Like the t-test, the resulting test statistic can be compared to a table of critical values. In this case, it would be the Chi-square table. Alternatively, the p-value could be examined and if it were below the threshold for a given significance level, then the null hypothesis that the groups are independent can be rejected. The alternative hypothesis for the Chi-square test for independence is that the groups are not independent. The alternative hypothesis implies that the variables are related, although it does not imply causation.

The expected frequencies for the Chi-square test can be found by dividing the total number of observations by the number of categories, meaning the assumption is that there should be the same number of observations in each category. An alternative method is to calculate each expected frequency for each combination of levels separately, as in figure 12.21.

$$expected_{ab} = \frac{row\ total_a * column\ total_b}{n}$$

Figure 12.21

The row and column variables in figure 12.21 refer to the rows and columns of a contingency (frequency) table, as defined in chapter 2. A cross-tabulation or contingency table shows the frequency counts for each combination of levels for two categorical variables.

The degrees of freedom for the Chi-square test are found by $(r-1)(c-1)$ where r = rows and c = columns of a contingency table.

There are only two assumptions that are made for the Chi-square test:

1. The observations must be independent of one another. This means that the Chi-square test cannot be used with repeated measures experiments, because the participants would not be independent.
2. The expected frequencies of each level should be greater than 5 (there should be 5 or more observations per category). For larger samples, it is acceptable to have up to 20% of the categories with fewer than 5 observations, but there is a loss of statistical power if there are. There should never be a time when the expected frequencies are below 1 (Field, Miles, & Field, 2012).

Notice that it is not necessary for either variable to have a normal distribution, because that would not make sense for categorical variables.

Loglinear Analysis: Comparing Several Groups of Categories

Loglinear analysis is like ANOVA for categorical dependent variables: it compares several groups of categories at the same time, thereby modeling a categorical dependent variable on several categorical independent variables. Since the independent variables are categorical, their log must be taken in order to make the regression equation linear. To see how this would be expressed as an equation, consider a linear model with 2 categorical variables and an interaction term. Recall from chapter 6 that interaction terms are included in a regression model to test for effects of one predictor on the outcome that vary based on different values of another predictor.

$$Ln(y) = ln(b_0 + b_1x_1 + b_2x_2 + b_3x_1x_2) + ln(error)$$

Figure 12.22

In the equation in figure 12.22, x1 and x2 represent different categories or groups of the same variable. Their values are either 0 or 1. Since the data is categorical, group means cannot be used for comparison like in t-tests and ANOVA. Instead, group frequencies are used like in the Chi-square test. The observed frequency is the outcome. For example, if we observe 18 observations that fall into 2 categories, then ln(y) in the equation in figure 12.22 would become ln(18) and the right side of the equation would represent the 2 observed categories x1 and x2.

The intercept in the model in figure 12.22 represents the log of the observed value when all categories are 0 (when x1 and x2 are both 0 in 12.22). If we have a contingency table of outcomes, then each beta coefficient can be calculated independently by using variable substitution with the frequencies. Thus, loglinear analysis is a way for Chi-square test to be conceptualized as a linear model.

When loglinear analysis is performed, the standard error will be 0, because there is no error. This means that all of the combinations of coded variables completely explain the observed values (which they truly do since there are no observations outside of the combinations in the contingency table). Since the standard error is 0, loglinear analysis results in a saturated model. A **saturated model** is a model in which all combinations of coded variables completely explain the observed values.

There is a slight problem with using the equation in figure 12.22 as a linear representation of the Chi-square test however. Recall that the Chi-square test checks for the independence of categorical variables. The Chi-square test does not make use of interaction terms between the variables. So if we remove the interaction term from the equation in figure 12.22, it becomes:

$$Ln(y) = ln(b_0 + b_1x_1 + b_2x_2) + ln(error)$$

Figure 12.23

Without the interaction term, it is no longer possible to predict the observed values. Instead, the expected values are predicted. Therefore, the intercept in the model in figure 12.23 represents the log of the expected value when all other categories are 0. The other coefficients of 12.23 are the log values of the expected frequencies of each category, rather than the observed frequencies as they were in figure 12.22.

At this point, it should be clear that the two categorical variable form of the Chi-square test can be expressed linearly, as in figure 12.23. When there are more than 2 categorical variables, the equation in figure 12.23 expands to accommodate them all. It is worth describing how the equation in figure 12.22 morphs into the equation in 12.23 when the interaction term is removed, because loglinear analysis starts with a model that includes each variable and every combination of their interactions. In other words, loglinear analysis starts with a saturated model. The goal of loglinear analysis is to simplify the model while minimizing the increase in error. To accomplish this, hierarchical backward elimination is used. **Hierarchical backward elimination** is a **dimension reduction** technique that starts by removing the highest order interaction term first (we will explore dimension reduction in a later chapter). After the highest order interaction term is removed, the error is compared to the error of the model without the removal of the term, using the likelihood ratio test. If the error has not increased such that the result of the likelihood ratio test is significant, then the next highest order interaction term is removed. The process continues until the removal of an interaction term significantly changes the log likelihood ratio. At that point, there is no need to remove other terms. For each term that is eliminated, the comparison is always made between the original saturated model and the model with the eliminated terms.

The order of an interaction term is determined by how many variables make up the interaction. For example, an interaction between 3 predictors is higher order than an interaction between 2.

Recall from chapter 9 that the likelihood ratio test (LRT) can be calculated as follows:

$$LRT = -2\ln\left(\frac{\mathcal{L}_s(\theta)}{\mathcal{L}_g(\theta)}\right)$$

Figure 9.15

The LRT approximates the Chi-square distribution with k-1 degrees of freedom, where k is the number of predictors. The result can be compared to the Chi-square statistical table, and if it is larger than the critical value, then the model with the predictor included (L sub g) is a significant improvement over the model without it (L sub s). So if the LRT > Chi-square critical value, then the current interaction term should be left in the model and no more interaction terms should be removed.

The assumptions required for loglinear analysis are the same as the assumptions for the Chi-square test:

1. The observations must be independent of one another. This means that loglinear analysis cannot be used with repeated measures experiments, because the participants would not be independent.
2. The expected frequencies of each level should be greater than 5 (there should be 5 or more observations per category). For larger samples, it is acceptable to have up to 20% of the categories with fewer than 5 observations, but there is a loss of statistical power if there are. There should never be a time when the expected frequencies are below 1.

GLMs with R
Since we are very familiar with the Titanic dataset by this point, and we have already tested the standard assumptions, it will be easy to use it for example yet again. To start off, let us perform a t-test to compare group means between males and females. We will use fare price per person as the quantitative variable (suppose fare per person is the outcome that we want to predict). So the t-test will test for a difference between the mean fare price per person between males and females. Recall from earlier in the chapter that this type of t-test is an independent means t-test.

```
train <- read.csv("train_clean_feats.csv", header=T)
test <- read.csv("test_clean_feats.csv", header=T)
train$Set <- 'train'
test$Set <- 'test'
alldata <- rbind(train[,-which(colnames(train) %in% 'Survived')], test)

#Compare the mean fare per person between males and females
```

```
t.test(alldata$Fare_Per_Person~alldata$Sex_male)
```

Looking at the results, the p-value is < 0.05. The group means shown are different, and the fare per person is higher for Sex_male = 0, suggesting that women had a higher average fare price per person. The t-test says nothing about the cause of this difference. It could be that female passengers were unfairly charged more, on average, than their male counterparts. But in this case, the more likely reason is that there were many more men in third class than women, and the low prices of third class tickets probably dragged the average down for men. Without further analysis, we can only speculate as to the reasons the means between groups differ.

We can perform an independent t-test to compare the means of two numeric variables as well. For example, if we test for a difference between the means of fare per person and fare, we can see that fare per person is significantly lower than the mean of fare (the first variable entered is labeled x and the second is labeled y). Note the syntax difference between comparing means of a numeric variables across categories of a binary variable versus comparing the means of two numeric variables.

```
#Compare the mean fare per person with the mean fare
t.test(alldata$Fare_Per_Person, alldata$Fare)
```

Moving on to ANOVA, let us compare the fare per person across several groups. We will look at comparing group means for a categorical feature, and comparing group means for a categorical feature that has been one-hot encoded. For the categorical feature that is still a single column, we will use group size and simply convert it to a factor to make R treat it as a categorical variable. Before ANOVA can be carried out, the homogeneity of variance assumption much be checked for the groups. If there is heteroskedasticity of the numeric variable across the groups, then the assumption is violated.

```
alldata$Group_Size <- as.factor(alldata$Group_Size)
```

```
#Test for heteroskedasticity across groups
library(car)
leveneTest(alldata$Fare_Per_Person, alldata$Group_Size, center=median)
bartlett.test(alldata$Fare_Per_Person, alldata$Group_Size)
```

It is clear that the homogeneity of variance assumption is violated in this case, as both Levene's test and Barlett's test had p-values < 0.05. The variance of fare per person varies widely between the different group sizes. To cope with the violation of this assumption, we can use Welch's F-test, as it is a robust method. The oneway.test function uses Welch's F-test by default (the function has an argument var.equal with defaults to false).

```
oneway.test(Fare_Per_Person ~ Group_Size, data=alldata)
```

Oh no! The test produced NAs. This points to a problem with the math behind the function. Since a one-way ANOVA computes the F-statistic by dividing the between group variance by the within group variance, the problem is most likely that the within group variance was 0. Dividing by 0 is not possible, so NAs are returned. Why might the within group variance be 0? If the values of any group are all the same, then the group has a variance of 0. So we need to check to see if any of the groups have only one distinct fare per person.

```
aggregate(data=alldata, Fare_Per_Person ~ Group_Size, function(x) length(unique(x)))
```

Yes, fare per person was exactly the same for all 11 people for group size = 11. Recall from chapter 4 that this was because the group of 11 was a large family who paid one price for all of their tickets. We can redo the ANOVA by omitting that group from the analysis.

```
oneway.test(Fare_Per_Person ~ Group_Size, data=alldata[alldata$Group_Size != 11,])
```

The results show that the mean fare per person was significantly different between group sizes. Note that the group sizes varied widely in the number of observations contained in each group, so this is actually a bad example of using ANOVA. The group sizes should really be more even. But often times, real world data is not as perfect as we wish it to be. Given the imbalance in group size, it is no wonder the means are significantly different.

Now we will use embarkation port as the categorical variable to show how ANOVA can be performed on a variable that has been one-hot encoded into multiple columns. There are 3 groups of embarkation ports and they are split into two binary features. Recall that we removed port C from the data when we dummy encoded it back in chapter 4. We need to add it back to the dataset for use here, and then recombine all three binary variables into the original variable. If we had one-hot encoded embarkation port, then all that would be required would be to recombine the binary variables.

```
#Add embarkation port c back into the data
alldata$Embarked_C <- ifelse(alldata$Embarked_Q + alldata$Embarked_S == 0, 1, 0)
alldata$Embarked <- 0
alldata[alldata$Embarked_Q==1, 'Embarked'] <- 'Q'
alldata[alldata$Embarked_S==1, 'Embarked'] <- 'S'
alldata[alldata$Embarked_C==1, 'Embarked'] <- 'C'
alldata$Embarked <- as.factor(alldata$Embarked)

#Perform ANOVA to compare means by embarkation port
anova_reg <- aov(Fare_Per_Person ~ Embarked, data=alldata)
summary(anova_reg)
```

ANOVA shows that the mean fare per person is significantly different between ports. We could not perform a test for homogeneity of variance beforehand, because the categories were split into multiple columns. Instead, we can perform this check by looking at the QQ plot after running ANOVA. The aov function that was used above is the standard ANOVA function in R. We only used the oneway.test for the last example because we found evidence of heteroskedasticity. If there is no heteroskedasticity, then aov can be used instead.

```
qqPlot(anova_reg, main="QQ Plot")   #From car package
```

Once again, there is evidence of heteroskedasticity, so we can use the oneway.test function to use Welch's F.

```
oneway.test(Fare_Per_Person ~ Embarked, data=alldata)
```

The results show that there is a significant difference between the mean fare per person for different embarkation ports. So far, ANOVA has told us that there are differences in the mean fare per person for different groups of two categorical variables, but we do not know which groups are different. Therefore, we need to perform post hoc tests to find out. The first post hoc test we can try is Tukey's test. Tukey's test relies on the assumption that the variances between groups are homoscedastic, so it will be useless to interpret because heteroskedasticity was present in both of the ANOVA models. Nevertheless, it is useful to see how the test would be performed if this were not the case.

```
#Tukey test - good when groups are the same size and have and homogeneous variance
library(multcomp)
postHocs <- glht(anova_reg, linfct=mcp(Embarked="Tukey"))
summary(postHocs)
confint(postHocs)
```

Notice that the summary compares each combination of categories and provides the p-value to show which levels of the categorical variable were statistically significant. If a comparison is found to be statistically significant, then it impacted the mean of the dependent variable by the amount of the coefficient estimate shown in the summary. The confint function shows the 95% confidence intervals around the coefficient estimates.

Pairwise comparisons of the two variables using the Bonferroni and Benjamini-Hochberg corections can be carried out as follows:

```
#Pairwise comparison using Bonferroni correction of p-values
pairwise.t.test(alldata$Fare_Per_Person, alldata$Embarked,
p.adjust.method="bonferroni")
```

```
#Pairwise comparison using Benjamini-Hochberg correction of p-values
pairwise.t.test(alldata$Fare_Per_Person, alldata$Embarked, p.adjust.method="BH")
```

The resulting table shows the p-values of the pairwise comparisons between each of the levels of the categorical variable. If any are significant, then the comparison is statistically significant. In this case, they are all < 0.05 so they are all significant, but remember that the heteroskedasticity assumption is violated, so the results here cannot be trusted.

Now let us run ANCOVA to compare several means with a covariate. But we need to deal with the heterskedastic variance between groups. We tested for differences between the mean fare per person between group sizes and had to rely on robust estimates of the model parameters because of heteroskedasticity. But let us think about why the fare per person might have different variances between group sizes. One reason is probably because we are not separating passenger classes, and we saw from our exploratory analysis in chapter 4 that fare is distributed differently for each class. We also saw from our t-test that fare per person varies between men and women. So there are several groups of people lumped together in the group size categories. By separating these groups, or by only focusing on one of them, we can get rid of the heteroskedasticity! I have cherry picked the example below to make sure the categories of group size have homoscedastic variances. I chose this group by simply plotting the fare per person for different cross sections until I found one that graphically appeared to have constant variance between group sizes.

```
#Look for heteroskedasticity
plot(alldata[alldata$Pclass==2 & alldata$Sex_male==1, 'Fare_Per_Person'],
alldata[alldata$Pclass==2 & alldata$Sex_male==1, 'Group_Size'])
#Second class male passengers with a fare price > 0 seem OK
#There are a couple group sizes with only 1 observation with these criteria though,
so make sure to filter them out too
```

```
#Test for heteroskedasticity
```

```
leveneTest(alldata[alldata$Pclass==2 & alldata$Sex_male==1 & alldata$Fare>0 &
!(alldata$Group_Size %in% c(5, 6, 7)), 'Fare_Per_Person'], alldata[alldata$Pclass==2
& alldata$Sex_male==1 & alldata$Fare>0 & !(alldata$Group_Size %in% c(5, 6, 7)),
'Group_Size'], center=median)
bartlett.test(alldata[alldata$Pclass==2 & alldata$Sex_male==1 & alldata$Fare>0 &
!(alldata$Group_Size %in% c(5, 6, 7)), 'Fare_Per_Person'], alldata[alldata$Pclass==2
& alldata$Sex_male==1 & alldata$Fare>0 & !(alldata$Group_Size %in% c(5, 6, 7)),
'Group_Size'])

sub <- alldata[alldata$Pclass==2 & alldata$Sex_male==1 & alldata$Fare>0 &
!(alldata$Group_Size %in% c(5, 6, 7)), ]
head(sub)

#Show ANOVA to see how ANCOVA is different
anova_reg <- aov(Fare_Per_Person ~ Group_Size, data=sub)
summary(anova_reg)
postHocs <- glht(anova_reg, linfct=mcp(Group_Size="Tukey"))
summary(postHocs)
confint(postHocs)
#PostHocs show that fare per person for groups sizes of 1 and 2 are different from
the rest

library(ggplot2)
boxplot <- ggplot(sub, aes(Group_Size, Fare_Per_Person)) + geom_boxplot() +
xlab('Group Size') + ylab('Fare Per Person')
boxplot
#Box plot confirms what the post hoc tests reported
```

We will use age as the covariate for ANCOVA.

```
#Create the ANCOVA regression
ancova_reg <- aov(Fare_Per_Person ~ Group_Size + Age, data=sub)
#Print the model summary with type III sums of squares
Anova(ancova_reg, type="III")
```

The results of ANCOVA show that the covariate, age has a statistically significant effect on fare per person. This means that the coefficient of the categorical variable, group size, cannot be interpreted until the effect of the covariate is controlled for. The effects library has a function that adjusts the means of the groups while controlling for the effect of the covariate in an ANCOVA regression.

```
#Cannot yet interpret group means because the effect of the covariate hasn't been
adjusted for, so look at adjusted means using the effect function
library(effects)
adjustedMeans <- effect("Group_Size", ancova_reg, se=TRUE)
summary(adjustedMeans)
adjustedMeans$se
```

The group means, adjusted for the covariate's effect, show that as group size increases, mean fare per person decreases from 12.37 in group size 1, to 11.76 in group size 2, to 10 in group size 3, and to 9.45 in group size 4. Note that the standard error of the mean increases with group size. That is because there are fewer observations in the larger group size, and the error of small samples is inherently higher than for larger samples. ANCOVA indicates that there is a statistically significant different between the mean fare per person across the group sizes, but it does not tell us which group sizes are different. So it is time to do post hoc tests to see which specific groups have different mean fares per person.

```
#Testing differences between adjusted means requires using glht() instead of
pairwise.t.test()
#Using glht() limits post hoc tests to Tukey or Dunnett's tests
library(multcomp)
postHocs <- glht(ancova_reg, linfct=mcp(Group_Size="Tukey"))
summary(postHocs)
confint(postHocs)

plot(ancova_reg)
```

The post hoc tests for ANCOVA show that the mean fare per person is similar for group sizes 1 and 2, and that those two categories are different from group sizes 3 and 4. Also, mean fare per person varies between group sizes 3 and 4. The post hoc results control for the effect of the covariate, age, and parrot the results we got from the earlier ANOVA test that we did for comparison purposes.

We are not out of the woods yet though. It is possible that the interaction between group size and its covariate, age, exhibits heteroskedasticity. To test this, we can look for homogeneity in the regression slopes with the interaction term added to the ANCOVA model.

```
#Add interaction term to ANCOVA model to look for homogeneity in the regression
slopes
ancova_reg_homo_rs <- aov(Fare_Per_Person ~ Group_Size + Age + Group_Size:Age,
data=sub)

#Get the type III sums of squares
Anova(ancova_reg_homo_rs, type="III")
```

If interaction term of this model is statistically significant, then the assumption of homogeneity is broken and robust methods must be used to estimate the model parameters. In this case, the interaction term is not statistically significant, so all is well: the original analysis without the interaction term can be trusted. If the assumption were broken, and a robust test were needed, the simplest robust test would be bootstrapped ANCOVA. The ancboot function from the WRS2 package can accomplish this.

Before moving on to the Chi-square test, remember that the results we just saw for ANCOVA only apply to a subset of the data: second class males whose fare price was > 0 and whose group sizes were 1-4. It is important to remember that a subset was used so that we do not mistakenly generalize the results to the entire dataset.

For the Chi-square test, we will compare passenger class and embarkation port. It will be interesting to see whether wealthier passengers boarded from a particular port. The Chi-square test could not be easier in R:

```
#Perform Chi-Square test to compare passenger class by embarkation port
library(gmodels)
CrossTable(alldata$Pclass, alldata$Embarked, expected=TRUE, prop.c=FALSE,
          prop.t=FALSE, prop.chisq=FALSE, chisq=TRUE, sresid=TRUE,
          format="SPSS")
```

The results show a statistically significant (< 0.05) p-value for the Chi-square test. This means we can reject the null that passenger class and embarkation port are independent. Specifically, since we tested passenger class on embarkation port, we can say that passenger class has a statistically significant effect on embarkation port.

For loglinear analysis, we will compare passenger class, embarkation port, and gender. We start off with a saturated loglinear model, which will have a log-likelihood ratio of 0 because it perfectly describes the data. Note that the loglinear function loglm requires a contingency table as input, rather than a data frame.

```
#Create contingency table of the categorical variables to be analyzed
#Note that A*B is a shortcut for specifying all possible interactions between A and B
contTable <- xtabs(~ Pclass + Sex_male + Embarked, data=alldata)

#Start off with saturated loglinear model with all interactions
loglinear_reg <- loglm(~ Pclass*Sex_male*Embarked, data=contTable)
summary(loglinear_reg)
```

Since loglinear analysis uses hierarchical backwards elimination, the 3-way interaction term is removed first. The likelihood ratio test is used to compare the models. If the p-value for the likelihood ratio test is significant, then removing the interaction term had a statistically significant negative impact on the model. If the p-value is significant (< 0.05) then the model with the interaction term should be used. The likelihood ratio test can be performed by using the anova function.

```
#Remove the highest order interaction effect and compare models
loglinear_reg_red <- update(loglinear_reg, .~. -Pclass:Sex_male:Embarked)
summary(loglinear_reg_red)
#Compare models by subtracting the likelihood ratios and DOF, or just use the anova
function to do the LR test automatically
anova(loglinear_reg, loglinear_reg_red)
```

The results show that the removal of the 3-way interaction term did not affect the model's error significantly. Now we repeatedly use hierarchical backwards elimination to simplify the model further, while using the likelihood ratio test to see if the error has increased too much after each variable is removed.

```
#Continue removing other interaction terms until the p-value of the reduced model is
significant (< 0.05)
loglinear_reg_red_further <- update(loglinear_reg_red, .~. -Sex_male:Embarked)
anova(loglinear_reg, loglinear_reg_red_further)
```

After removing the very next interaction term, the interaction of Sex_male and Embarked, the model's error significantly changed. Therefore, the model should keep the interaction between Sex_male and Embarked, and the final model is the one that only excludes the 3-way interaction term.

```
#Final model analysis
summary(loglinear_reg_red)
mosaicplot(loglinear_reg_red$fit, shad=TRUE, main="Loglinear Model of Pclass, Sex,
and Embarked, without 3-Way Interaction")
```

The summary does not tell us much, but the mosaic plot does. The mosaic plot of the fitted model shows the expected versus actual frequencies for each combination of variables. Larger boxes correspond to higher expected frequencies. The color of the box indicates the standardized residual between the expected versus the actual frequency. The darker the color, the greater the standardized error in that direction (i.e. dark blue means that there were far more than expected, dark red means there were far fewer than expected). If the line around a box is solid, then the actual frequency > expected frequency. If the line around a box is dashed, then the actual frequency < expected frequency. The box line and color essentially mean the same thing, but the color gives magnitude. For this plot, there were far more passengers of both sexes in first class who embarked from port C than were expected. There were far more third class women who embarked from port Q than were expected. The other categories were not as different from their expected frequencies.

GLMs with Python

Since we are very familiar with the Titanic dataset by this point, and we have already tested the standard assumptions, it will be easy to use it for example yet again. To start off, let us perform a t-test to compare group means between males and females. We will use fare price per person as the quantitative variable (suppose fare per person is the outcome that we want to predict). So the t-test will test for a difference between the mean fare price per person between males and females. Recall from earlier in the chapter that this type of t-test is an independent means t-test.

```
import pandas as pd
from scipy import stats

train = pd.read_csv('train_clean_feats.csv')
test = pd.read_csv('test_clean_feats.csv')
train['Set'] = 'train'
test['Set'] = 'test'
alldata = train.drop('Survived', axis=1).append(test, ignore_index=True)
print(alldata.info())

#Two sample independent t-test (compare group means for 2 groups)
t, p = stats.ttest_ind(alldata[alldata['Sex_male']==0]['Fare_Per_Person'],
alldata[alldata['Sex_male']==1]['Fare_Per_Person'], equal_var=True)

print("ttest_ind: t = %g  p = %g" % (t, p))
```

Looking at the results, the p-value is < 0.05. The group means shown are different, and the fare per person is higher for Sex_male = 0, suggesting that women had a higher average fare price per person. The t-test says nothing about the cause of this difference. It could be that female passengers were unfairly charged more, on average, than their male counterparts. But in this case, the more likely reason is that there were many more men in third class than women, and the low prices of third class tickets probably dragged the average down for men. Without further analysis, we can only speculate as to the reasons the means between groups differ.

We can perform an independent t-test to compare the means of two numeric variables as well. For example, if we test for a difference between the means of fare per person and fare, we can see that fare per person is significantly lower than the mean of fare (the first variable entered is labeled x and the second is labeled y). Note the syntax difference between comparing means of a numeric variables across categories of a binary variable versus comparing the means of two numeric variables.

```
#Compare the mean fare per person with the mean fare
t, p = stats.ttest_ind(alldata['Fare_Per_Person'], alldata['Fare'], equal_var=True)
```

```
print("ttest_ind: t = %g   p = %g" % (t, p))
```

Moving on to ANOVA, let us compare the fare per person across several groups. We will look at comparing group means for a categorical feature, and comparing group means for a categorical feature that has been one-hot encoded. For the categorical feature that is still a single column, we will use group size. Before ANOVA can be carried out, the homogeneity of variance assumption much be checked for the groups. If there is heteroskedasticity of the numeric variable across the groups, then the assumption is violated. To perform Levene's test, we can resurrect the function we defined in chapter 3. The same process can be used to define a function to carry out Bartlett's test. The reason we need to define these functions is because scipy's Levene and Bartlett functions take lists of arrays as input. That is fine as long as it is easy enough to manually type out every group, but when there are many groups to test, it becomes tedious. The functions below allow Levene's and Bartlett's tests to be run with one line each by automatically handling grouping variables of any number of groups.

```
alldata$Group_Size <- as.factor(alldata$Group_Size)

#Test for heteroskedasticity across groups
def levenes_test(num_variable, *group_variables, center='median'):
    temp = list(num_variable.groupby(group_variables))
    temp = [temp[i][1] for i,v in enumerate(temp)]
    return stats.levene(*temp, center=center)
print(levenes_test(alldata['Fare_Per_Person'], alldata['Group_Size']))
def bartlett_test(num_variable, *group_variables):
    temp = list(num_variable.groupby(group_variables))
    temp = [temp[i][1] for i,v in enumerate(temp)]
    return stats.bartlett(*temp)
print(bartlett_test(alldata['Fare_Per_Person'], alldata['Group_Size']))
```

It is clear that the homogeneity of variance assumption is violated in this case, as both Levene's test and Barlett's test had p-values < 0.05. The variance of fare per person varies widely between the different group sizes. To cope with the violation of this assumption, we used Welch's F-test in R. Welch's test is not available in Python, but the Kruskal-Wallis test is a suitable substitute. The Kruskal-Wallis test is a nonparametric test though, so rather than doing it now, we will hold off until a later chapter.

When we ran Welch's test with R, an error pointed to the possibility of one or more groups having one distinct value for fare per person. We do not have any helpful pointers with Python though, so we will just have to remember to always verify this in Python. We need to check to see if any of the groups have only one distinct fare per person.

```
print(alldata.groupby(['Group_Size'])['Fare_Per_Person'].nunique())
```

The output from the line above shows that fare per person was exactly the same for all 11 people for group size = 11. Recall from chapter 4 that this was because the group of 11 was a large family who paid one price for all of their tickets. We can run ANOVA by omitting that group from the analysis.

```
def oneway_test(num_variable, *group_variables):
    temp = list(num_variable.groupby(group_variables))
    temp = [temp[i][1] for i,v in enumerate(temp)]
    return stats.f_oneway(*temp)
F, p = oneway_test(alldata[alldata['Group_Size']!=11]['Fare_Per_Person'],
alldata[alldata['Group_Size']!=11]['Group_Size'])
```

```
print("one-way ANOVA: F = %g  p = %g" % (F, p))
```

The results show that the mean fare per person was significantly different between group sizes. Note that the group sizes varied widely in the number of observations contained in each group, so this is actually a bad example of using ANOVA. The group sizes should really be more even. But often times, real world data is not as perfect as we wish it to be. Given the imbalance in group size, it is no wonder the means are significantly different. Astute readers may notice that the results of the one-way test differ from R. That is because scipy's one-way test assumes equality of variance. Unlike with R, there is no argument for the one-way test to specify whether the variance is homogenous, so we are stuck with relying on the Kruskal-Wallis test if the assumption is violated and we are using Python.

Now we will use embarkation port as the categorical variable to show how ANOVA can be performed on a variable that has been one-hot encoded into multiple columns. There are 3 groups of embarkation ports and they are split into two binary features. Recall that we removed port C from the data when we dummy encoded it back in chapter 4. We need to add it back to the dataset for use here, and then recombine all three binary variables into the original variable. If we had one-hot encoded embarkation port, then all that would be required would be to recombine the binary variables.

```
alldata['Embarked_C'] = np.where((alldata['Embarked_Q']+alldata['Embarked_S'])==0, 1,
0)
alldata['Embarked'] = 0
alldata['Embarked'] = np.where(alldata['Embarked_Q']==1, 'Q', alldata['Embarked'])
alldata['Embarked'] = np.where(alldata['Embarked_S']==1, 'S', alldata['Embarked'])
alldata['Embarked'] = np.where(alldata['Embarked_C']==1, 'C', alldata['Embarked'])

#Perform ANOVA to compare means by embarkation port
anova_reg = ols("Fare_Per_Person ~ Embarked", alldata).fit()
anova_results = anova_lm(anova_reg)
print('\nANOVA results\n', anova_results)
```

ANOVA shows that the mean fare per person is significantly different between ports. We could not perform a test for homogeneity of variance beforehand, because the categories were split into multiple columns. Instead, we can perform this check by looking at the QQ plot after running ANOVA. Since ANOVA is a type of regression, we can fit an OLS model using the statsmodels library. By using the OLS function, we will be able to view the residuals.

```
sm.qqplot(anova_reg.resid, line='s')
plt.show()
```

Once again, there is evidence of heteroskedasticity. So it would be best to use a nonparametric test. We will return to this problem in a later chapter.

So far, ANOVA has told us that there are differences in the mean fare per person for different groups of two categorical variables, but we do not know which groups are different. Therefore, we need to perform post hoc tests to find out. The first post hoc test we can try is Tukey's test. Tukey's test relies on the assumption that the variances between groups are homoscedastic, so it will be useless to interpret because heteroskedasticity was present in both of the ANOVA models. Nevertheless, it is useful to see how the test would be performed if this were not the case.

```
#Tukey test - good when groups are the same size and have and homogeneous variance
```

```
postHoc = pairwise_tukeyhsd(alldata['Fare_Per_Person'], alldata['Embarked'],
alpha=0.05)
print(postHoc)
```

Notice that the summary compares each combination of categories and provides the p-value to show which levels of the categorical variable were statistically significant. If a comparison is found to be statistically significant, then it impacted the mean of the dependent variable by the amount of the coefficient estimate shown in the summary.

Pairwise comparisons of the two variables using the Bonferroni correction can be carried out as follows:

```
#Pairwise comparison using Bonferroni correction of p-values
mc = MultiComparison(alldata['Fare_Per_Person'], alldata['Embarked'])
#print(mc.allpairtest(stats.ttest_rel, method='Holm')[0])  #For paired t-test
print(mc.allpairtest(stats.ttest_ind, method='b')[0])      #For independent t-test
```

The resulting table shows the p-values of the pairwise comparisons between each of the levels of the categorical variable. If any are significant, then the comparison is statistically significant. In this case, they are all < 0.05 so they are all significant, but remember that the heteroskedasticity assumption is violated, so the results here cannot be trusted.

Now let us run ANCOVA to compare several means with a covariate. But we need to deal with the heterskedastic variance between groups. We tested for differences between the mean fare per person between group sizes and had to rely on robust estimates of the model parameters because of heteroskedasticity. But let us think about why the fare per person might have different variances between group sizes. One reason is probably because we are not separating passenger classes, and we saw from our exploratory analysis in chapter 4 that fare is distributed differently for each class. We also saw from our t-test that fare per person varies between men and women. So there are several groups of people lumped together in the group size categories. By separating these groups, or by only focusing on one of them, we can get rid of the heteroskedasticity! I have cherry picked the example below to make sure the categories of group size have homoscedastic variances. I chose this group by simply plotting the fare per person for different cross sections until I found one that graphically appeared to have constant variance between group sizes.

```
#Look for heteroskedasticity
plt.plot(alldata[(alldata['Pclass']==2) &
(alldata['Sex_male']==1)]['Fare_Per_Person'], alldata[(alldata['Pclass']==2)
&(alldata['Sex_male']==1)]['Group_Size'], 'bo')
plt.show()
#Second class male passengers with a fare price > 0 seem OK
#There are a couple group sizes with only 1 observation with these criteria though,
so make sure to filter them out too

#Test for heteroskedasticity
print(levenes_test(alldata[(alldata['Pclass']==2) & (alldata['Sex_male']==1) &
(alldata['Fare']>0) &
(alldata['Group_Size'].isin([1,2,3,4,8,9,10,11]))]['Fare_Per_Person'],
alldata[(alldata['Pclass']==2) & (alldata['Sex_male']==1) & (alldata['Fare']>0) &
(alldata['Group_Size'].isin([1,2,3,4,8,9,10,11]))]['Group_Size']))
print(bartlett_test(alldata[(alldata['Pclass']==2) & (alldata['Sex_male']==1) &
(alldata['Fare']>0) &
(alldata['Group_Size'].isin([1,2,3,4,8,9,10,11]))]['Fare_Per_Person'],
```

```
alldata[(alldata['Pclass']==2) & (alldata['Sex_male']==1) & (alldata['Fare']>0) &
(alldata['Group_Size'].isin([1,2,3,4,8,9,10,11]))]['Group_Size']))

sub = alldata[(alldata['Pclass']==2) & (alldata['Sex_male']==1) & (alldata['Fare']>0)
& (alldata['Group_Size'].isin([1,2,3,4,8,9,10,11]))]
print(sub.head())

#Show ANOVA to see how ANCOVA is different
anova_reg = ols("Fare_Per_Person ~ Group_Size", data=sub).fit()
anova_results = anova_lm(anova_reg)
print('\nANOVA results\n', anova_results)
mc = MultiComparison(alldata['Fare_Per_Person'], alldata['Embarked'])
print(mc.allpairtest(stats.ttest_ind, method='b')[0])     #For independent t-test
#PostHocs show that fare per person for groups sizes of 1 and 2 are different from
the rest

import seaborn as sns
sns.boxplot(sub['Group_Size'], sub['Fare_Per_Person'])
plt.show()
#Box plot confirms what the post hoc tests reported
```

We will use age as the covariate for ANCOVA.

```
#Create the ANCOVA regression
ancova_reg = smf.ols("Fare_Per_Person ~ Age + Group_Size", data=sub).fit()
#print(ancova_reg.summary())
#Print the model summary with type III sums of squares
ancova_results = anova_lm(ancova_reg, typ="III")
print('\nANCOVA results\n', ancova_results)
sm.qqplot(ancova_reg.resid, line='s')
plt.show()
```

The results of ANCOVA show that the covariate, age has a statistically significant effect on fare per person. The results do not indicate which group sizes are different. So it is time to do post hoc tests to see which specific groups have different mean fares per person.

```
postHoc = pairwise_tukeyhsd(sub['Fare_Per_Person'], sub['Embarked'], alpha=0.05)
print(postHoc)
```

The post hoc tests for ANCOVA show that the mean fare per person is similar for group sizes 1 and 2, and that those two categories are different from group sizes 3 and 4. Also, mean fare per person varies between group sizes 3 and 4. The post hoc results control for the effect of the covariate, age, and parrot the results we got from the earlier ANOVA test that we did for comparison purposes.

We are not out of the woods yet though. It is possible that the interaction between group size and its covariate, age, exhibits heteroskedasticity. To test this, we can look for homogeneity in the regression slopes with the interaction term added to the ANCOVA model.

```
#Add interaction term to ANCOVA model to look for homogeneity in the regression
slopes
ancova_reg = smf.ols("Fare_Per_Person ~ Age * Group_Size", data=sub).fit()
#print(ancova_reg.summary())
#Print the model summary with type III sums of squares
ancova_results = anova_lm(ancova_reg, typ="III")
```

If interaction term of this model is statistically significant, then the assumption of homogeneity is broken and robust methods must be used to estimate the model parameters. In this case, the interaction term is not statistically significant, so all is well: the original analysis without the interaction term can be trusted. If the assumption were broken, and a robust test were needed, the simplest robust test would be bootstrapped ANCOVA.

Before moving on to the Chi-square test, remember that the results we just saw for ANCOVA only apply to a subset of the data: second class males whose fare price was > 0 and whose group sizes were 1-4. It is important to remember that a subset was used so that we do not mistakenly generalize the results to the entire dataset.

For the Chi-square test, we will compare passenger class and embarkation port. It will be interesting to see whether wealthier passengers boarded from a particular port.

```
#Perform Chi-Square test to compare passenger class by embarkation port
freq_tbl = pd.crosstab(alldata['Pclass'], alldata['Embarked'])
print(freq_tbl)
freq_tbl = np.array(freq_tbl)
chi2, p, dof, expected = stats.chi2_contingency(freq_tbl)
print("Chi-square: chi2 = %g  p = %g  dof = %g" % (chi2, p, dof))
print("Expected frequencies:", expected)
print("Actual frequencies:", freq_tbl)
```

The results show a statistically significant (< 0.05) p-value for the Chi-square test. This means we can reject the null that passenger class and embarkation port are independent. Specifically, since we tested passenger class on embarkation port, we can say that passenger class has a statistically significant effect on embarkation port.

Unfortunately, there is no easy way to do loglinear analysis in Python, so that is all for this chapter.

13. Factorial ANOVA

In the last chapter, we looked at ANOVA to compare group means. When ANOVA is used to compare group means for a single independent variable it is called one-way ANOVA. In this chapter, we will discover how to use ANOVA to compare the group means of several independent variables. This is called factorial ANOVA. Factorial ANOVA gets its name because predictors are also sometimes referred to as factors.

A one-way ANOVA has one independent variable that splits a sample into several groups. Factorial ANOVA has two or more independent variables that split a sample into several groups (there must be at least 4 groups, since there are at least 2 independent variables). The independent variables in factorial ANOVA are usually systematically manipulated in an experiment in order to test for an effect.

To carry out factorial ANOVA, tests are performed to look at the effects of each predictor, as well as every combination of predictors for combined effects (the interactions between predictors). Recall from chapter 6 that interaction terms test for effects of predictors on the outcome that vary based on different groups of other predictors. Recall from chapter 1 that there are 3 types of experimental designs. There are three types of factorial ANOVA; one type for each kind of experiment that can be performed:

- **Independent factorial design** – This type of ANOVA is used for independent measures experiments, where there are several independent variables that have each been measured using different entities (the experiment compares conditions between groups).
- **Repeated-measures factorial design** – This type of ANOVA is used for repeated measures experiments, where there are several independent variables that have been measured using the same entity for all conditions (each entity has different measures for all of the conditions/independent variables).
- **Mixed design** – This type of ANOVA is used for mixed design experiments, where there are several independent variables that have been measured, some using the same entity, and others using different entities.

ANOVA is named for how many predictor variables it involves. In chapter 12, we looked at one-way ANOVA, which involved one predictor. Factorial ANOVA can involve 2 predictors and be called two-way independent ANOVA (for independent design experiments), or it can involve 3 predictors and be called three-way repeated measures ANOVA (for repeated measure experiments). In general, ANOVA for k predictors is named by using k-way, plus whatever type of experiment has been used, plus ANOVA. Factorial ANOVA is carried out the same way for all 3 types of experiments described above. The difference between how ANOVA is applied for the 3 types of experiments depends on the assumption of sphericity, as we will soon see.

Factorial ANOVA by hand is tedious, so rather than expound on it here, the application of factorial ANOVA will be left to the code at the end of the chapter. In general, it is carried out by performing one-way ANOVA between the effect of 1 independent variable and the individual levels of the others. Post hoc tests are used to look at the differences between groups in the main effect between the 1 independent variable and the dependent variable, but there could also be effects between the interactions of the predictors. It is good practice to include several orders of interaction terms in factorial ANOVA, even going so far as to include an interaction term for every combination of predictors. When an interaction term is significant, there is no reason to consider the individual effects making up the interaction, because they will be encompassed by the interaction.

The assumptions for factorial ANOVA are the same as for one-way ANOVA; they are the standard parametric assumptions from chapter 3 plus the linear regression assumptions from chapter 6. One of the standard assumptions is that the predictors are independent (not collinear). In the context of experimentation, this means that the measurements under different experimental conditions are independent. Repeated-measures experiments inherently violate this assumption, so F-tests would not be valid for repeated measures ANOVA unless sphericity is assumed (Field, Miles, & Field, 2012).

Sphericity

Sphericity is a condition in which the variances between all combinations of related groups are equal. Put simply, sphericity is the assumption that, for repeated measures factorial ANOVA, the variance of each condition applied to the individuals in the experiment is homogenous. If the variance for one of the conditions is extremely different from the other, then the dependent variable would be very hard to model with a linear model like ANOVA. Sphericity is chiefly a concern for repeated measures factorial design.

Sphericity is like the homogeneity of variance assumption that is made for standard linear regression, except that in the context of ANOVA it refers to groups of the dependent variable. When sphericity is assumed, then the relationship between pairs of experimental conditions are assumed to be similar (the level of dependence between experimental conditions is the same). The assumption is similar to the homogeneity of variance assumption for ANOVA (the variances between groups must be the same). If sphericity holds, then the F-test can be used to assess the result of ANOVA. Sphericity does not hold when the variances of the differences between all combinations of groups are not equal.

Sphericity is a case of compound symmetry. **Compound symmetry** is a condition in which the variances across conditions are equal, providing there are at least 3 conditions. Sphericity can be envisioned as an equation of the variance of each difference between pairs of observations, for every possible combination:

$$variance_{A-B} \approx variance_{A-C} \approx variance_{B-C}$$

Figure 13.1

Fingure 13.1 shows that sphericity is when the variance between each combination of groups (A-B, A-C, and B-C, for 3 conditions) is equal.

Sphericity can be tested using Mauchly's test. **Mauchly's test** looks at the differences between combinations of group variances to see if any are significantly different enough to break the sphericity assumption. As long as none of the p-values are lower than some alpha level of significance, such as 0.05, then the sphericity assumption holds (the null is that the assumption holds). When sphericity holds, then Tukey's test (refer to chapter 12) can be used to compare group means post hoc. Mauchly's test is similar to Levene's test, but applied to group variances (Field, Miles, & Field, 2012).

If sphericity cannot be assumed, then F-tests should only be carried out using a correction term, such as Bonferroni's correction (refer to chapter 12). Another option is to use multivariate analysis of variance

(MANOVA) as we will see in a later chapter. Like any significance test, large samples may have small but significant deviations that imply that sphericity cannot be assumed, while small samples may have large but non-significant deviations that imply that sphericity can be assumed. Therefore, it is a task for the researcher to determine whether sample size is influencing the test for sphericity.

There are a couple of options for correcting the F-test when sphericity does not hold. These corrections allow the F-test to still be used when sphericity fails, as they reduce the likelihood of type I error by estimating sphericity. One such correction is the **Greenhouse-Geisser** correction, often abbreviated GGe. The GGe equals 1 when there is homogeneity in the variances of the differences (when the data is spherical). The lower bound is 1/(number of conditions-1). The closer the GGe is to the lower bound, the greater the deviation from sphericity. The **Huynh-Feldt** estimate is another correction that is less conservative than the GGe when it comes to the odds of making a type I error. It too can be compared to the lower bound to determine deviation from sphericity. If the Huynh-Feldt estimate differs from the GGe, then the two measures can be averaged together, and the estimate that is closer to the average may be compared to the lower bound.

Factorial ANOVA with R

In this chapter, we will use the adult dataset from the UCI Machine Learning Repository: https://archive.ics.uci.edu/ml/datasets/adult (Kohavi, 1996). This dataset was curated by Ron Kohavi and contains attributes about a 1994 survey of adults in different countries. Typically, this dataset is used to classify adults in the sample as those having a salary > $50k and those having a salary <= $50k. We will use it to compare the mean working hours per week across several groups. This type of factorial ANOVA is independent factorial design, because the independent variables are measured using different entities (each observation or row is a different individual, and there are no repeats).

```
d <- read.table("adult.data", header=F, sep=",")
colnames(d) <- c('age', 'workclass', 'fnlwgt', 'education', 'education_nbr',
                 'marital_status', 'occupation', 'relationship', 'race',
                 'sex', 'capital_gain', 'capital_loss', 'hours_per_week',
                 'native_country', 'salary_bin')str(d)
head(d)

d$hours_per_week <- as.numeric(d$hours_per_week)
```

We will skip the data cleaning and exploratory analysis phases to save time. The first thing we need to do is determine if the homogeneity of variance assumption is met. Since the leveneTest and bartlett.test functions only take 2 arguments, they have to be run for each categorical variable to be used in ANOVA.

```
#Test for heteroskedasticity across groups
library(car)
leveneTest(d$hours_per_week, d$education, center=median)
leveneTest(d$hours_per_week, d$relationship, center=median)
bartlett.test(d$hours_per_week, d$education)
bartlett.test(d$hours_per_week, d$relationship)

library(ggplot2)
boxplot <- ggplot(d, aes(education, hours_per_week)) + geom_boxplot() +
xlab('Education') + ylab('Hours Per Week')
boxplot
```

There appears to be heteroskedasticity. Rather than finding a sub-section of the data with homoskedasticity, let us proceed with factorial ANOVA anyway. The results will not be trustworthy, but we will fix this in the next chapter. The next chapter shows how nonparametric tests can overcome heteroskedasticity. Below we perform two-way independent ANOVA and run the post hoc tests for one of the variables.

```
#Ignore the heteroskedasticity for now, and proceed
tw_ind_anova_reg <- aov(hours_per_week ~ education + relationship +
education:relationship, data=d)
Anova(tw_ind_anova_reg, type='III')

#Post Hoc tests for education only
library(multcomp)
pairwise.t.test(d$hours_per_week, d$education, p.adjust.method="bonferroni")
pairwise.t.test(d$hours_per_week, d$education, p.adjust.method="BH")
postHocs <- glht(tw_ind_anova_reg, linfct=mcp(education="Tukey"))
summary(postHocs)
confint(postHocs)
```

There is no sense in interpreting anything since we know the assumptions of ANOVA have not been met. We will return to this in the next chapter.

Factorial ANOVA with Python

In this chapter, we will use the adult dataset from the UCI Machine Learning Repository: https://archive.ics.uci.edu/ml/datasets/adult (Kohavi, 1996). This dataset was curated by Ron Kohavi and contains attributes about a 1994 survey of adults in different countries. Typically, this dataset is used to classify adults in the sample as those having a salary > $50k and those having a salary <= $50k. We will use it to compare the mean working hours per week across several groups. This type of factorial ANOVA is independent factorial design, because the independent variables are measured using different entities (each observation or row is a different individual, and there are no repeats).

```
import pandas as pd
from scipy import stats
import seaborn as sns
import matplotlib.pyplot as plt
from statsmodels.formula.api import ols
from statsmodels.stats.anova import anova_lm
import statsmodels.api as sm
from statsmodels.stats.multicomp import pairwise_tukeyhsd, MultiComparison

#Load adult dataset
d = pd.read_csv('C:/Users/Nick/Documents/Word Documents/Data Science Books/DSILT
Stats Code/12-16 Titanic and Adult Salaries/adult.data',
               names=['age', 'workclass', 'fnlwgt', 'education', 'education_nbr',
'marital_status', 'occupation', 'relationship', 'race',
                      'sex', 'capital_gain', 'capital_loss', 'hours_per_week',
'native_country', 'salary_bin'])
print(d.info())
print(d.head())
```

We will skip the data cleaning and exploratory analysis phases to save time. The first thing we need to do is determine if the homogeneity of variance assumption is met. Since the Levene and Bartlett test functions only take 2 arguments, they have to be run for each categorical variable to be used in ANOVA.

```python
#Test for heteroskedasticity across groups
def levenes_test(num_variable, *group_variables, center='median'):
    temp = list(num_variable.groupby(group_variables))
    temp = [temp[i][1] for i,v in enumerate(temp)]
    return stats.levene(*temp, center=center)
print(levenes_test(d['hours_per_week'], d['education']))
print(levenes_test(d['hours_per_week'], d['relationship']))
def bartlett_test(num_variable, *group_variables):
    temp = list(num_variable.groupby(group_variables))
    temp = [temp[i][1] for i,v in enumerate(temp)]
    return stats.bartlett(*temp)
print(bartlett_test(d['hours_per_week'], d['education']))
print(bartlett_test(d['hours_per_week'], d['relationship']))

sns.boxplot(d['education'], d['hours_per_week'])
plt.show()
```

There appears to be heteroskedasticity. Rather than finding a sub-section of the data with homoskedasticity, let us proceed with factorial ANOVA anyway. The results will not be trustworthy, but we will fix this in the next chapter. The next chapter shows how nonparametric tests can overcome heteroskedasticity. Below we perform two-way independent ANOVA and run the post hoc tests for one of the variables.

```python
#Ignore the heteroskedasticity for now, and proceed
#n-way Factorial ANOVA
#Note that C() forces a variable to be treated as categorical
anova_reg = ols("hours_per_week ~ C(education) + C(relationship) +
C(education):C(relationship)", data=d).fit()
#print(anova_reg.summary())
aov_table = anova_lm(anova_reg, typ="III")
print(aov_table)

#QQ Plot of residuals
sm.qqplot(anova_reg.resid, line='s')
plt.show()

#Post Hoc tests for education only
mc = MultiComparison(d['hours_per_week'], d['education'])
print(mc.allpairtest(stats.ttest_ind, method='b')[0])     #For independent t-test
```

There is no sense in interpreting anything since we know the assumptions of ANOVA have not been met. We will return to this in the next chapter.

14. Nonparametric Tests for Comparing Group Means

Recall from chapter 3 that parametric tests have four standard assumptions:

1. The sampling distribution is normally distributed. Notice that this does not mean that the observations in a single sample must be normally distributed, but that the sampling distribution is. Most large samples can meet this assumption by the Central Limit Theorem.
2. The variance in the dependent variable is homogenous throughout the dataset. This means that every sample comes from a population with the same variance, or dispersion around the mean.
3. The dependent variable is continuous. In other words, it is at least measured at the interval or ratio level.
4. Observations are independent of one another. This assumption also implies that the errors should not be correlated in regression analysis.

In all of the chapters so far, we have looked at parametric methods for describing the relationships between variables. Quite frequently, one or more of the parametric assumptions does not hold. When this happens, we can either use robust estimation methods (e.g. using MLE instead of OLS for a standard linear regression is more robust), or we can use non-parametric methods. The non-parametric tests described in this chapter are generalized linear models that make fewer assumptions about the underlying distribution of the data. Most of them work by sorting the data and assigning ranks to the values, as we will see.

Nonparametric tests sacrifice some predictive power in their abilities to detect genuine effects, in order to relax the parametric assumptions. Nonparametric tests have greater likelihood of type II error than parametric tests when the sampling distribution is normal (Field, Miles, & Field, 2012). Therefore, they should only be used when it is verified that the parametric assumptions cannot be met.

Wilcoxon Rank-Sum Test and Mann-Whitney Test

The first two tests we will look at are essentially the same, but the Wilcoxon rank-sum test is more commonly implemented in statistical software. These tests are the nonparametric equivalents of the independent t-test, as they are used to compare the means of the distributions of two categories of observations. When a dependent variable is modeled only on 1 independent variable, and that independent variable is categorical with only two categories (meaning it is binary), then a t-test is an appropriate linear model. When the parametric assumptions fail, then either the Wilcoxon rank-sum test or Mann-Whitney test is a suitable substitute for the t-test. Since the Wilcoxon rank-sum is more common, that is the one we will examine here.

The Wilcoxon rank-sum test works by ranking observations by the index of their sorted values of the dependent variable (Krzywinski & Altman, 2014). For example, suppose we are searching for differences in overall happiness between American and Canadian citizens, using the world happiness index as the dependent variable. If we sort happiness from highest to lowest, then we would expect American and Canadians to be evenly (randomly) disbursed between the ranks. If the dispersion were uneven, then one group would be clustered around one end of the ranking spectrum.

If there are ties in the happiness score between multiple observations, then the observations sharing the scores are assigned the same value: the average of the potential ranks for those observations. For example, if the 3rd, 4th, and 5th observations all have a score of 0.5, then their rank would be 4, because (3+4+5)/3 = 4.

The next step in carrying out the Wilcoxon rank-sum test is to sum the ranks per group. Since larger groups would have more observations and skew the ranks, they are weighted by subtracting the group mean from the total. The resulting values are called W, and the group with the lowest W is used as the test statistic.

$$W = sum\ of\ ranks - mean\ rank$$

$$mean\ rank = \frac{N(N+1)}{2}$$

Figure 14.1

In the equation in figure 14.1, N is the number of observations in a group, the sum of ranks refers to the sum of the ranks within a group, and W is the test statistic.

In statistical software, the test statistic and its associated p-value is found by using a Monte Carlo method. A **Monte Carlo method** is any method that uses simulated data to test the model. For the Wilcoxon rank-sum test, the method is to assign observations to groups at random and calculate the test statistic, W, assuming the null hypothesis (that there is no difference between the groups) is true. This process is carried out many times. If there are any ties in the ranks of the data, a Monte Carlo simulation cannot be used. Instead, the sampling distribution of W is assumed to be normal. This is called **normal approximation**. Normal approximation is preferable to the Monte Carlo method when the sample size is large, because there are more likely to be ties in the rank.

Wilcoxon Signed-Rank Test

The Wilcoxon signed-rank test is the nonparametric version of the dependent (paired) t-test: it is used to compare the means of 2 groups of the same samples or participants, across categories. If the data comes from a repeated measures design experiment where the same participants are measured for both conditions, and the parametric assumptions fail, then the Wilcoxon signed-rank test is appropriate (Field, Miles, & Field, 2012).

This test is carried out just like the Wilcoxon rank-sum test, except the groups are ranked separately. Rather than ranking observations based on the values of the dependent variable, the ranks are assigned based on the absolute value of the difference between groups. For example, if person A had a score of 0.5 for the first condition of an experiment and a score of 0.9 for the second condition, then the absolute value of the difference between their scores would be: |0.5-0.9| = 0.4. It is important to store the signage (positive or negative) though, because the positive and negative scores are summed separately. The absolute value is only used to rank the observations. If the scores between conditions are the same, they are omitted from the ranking.

When the positive and negative scores are totaled, separately, the results are the total positive differences and the total negative differences. The test statistic is the smaller absolute value of these two totals.

Kruskal-Wallis Test

The Kruskal-Wallis test is the nonparametric equivalent of the 1-way independent ANOVA: it is used to compare group means of a single categorical predictor (Glen, 2018).

This test is carried out by ordering the scores from highest to lowest, and ignoring which group they are in. The lowest score is assigned the rank of 1. The next lowest score is assigned the rank of 2, and so on. After the observations have been ranked, the scores are totaled by group. The test statistic, H, is calculated as follows:

$$H = \frac{12}{N(N+1)} \sum_{i=1}^{k} \frac{R_i^2}{n_i} - 3(N+1)$$

Figure 14.2

In the equation in figure 14.2, R_i is the sum of the ranks for group i, N is the total sample size, k is the number of groups, and n_i is the group sample size. The test statistic H has a Chi-square distribution with k-1 degrees of freedom.

Jonckheere-Terpstra Test

The Jonckheere-Terpstra test is an alternative to the Kruskal-Wallis test that assumes that there is an a priori ordering to the population. The Jonckheere-Terpstra test is another nonparametric equivalent of the 1-way independent ANOVA: it is used to compare group means of a single categorical predictor. The tests works by looking for trends between the groups. For example, it may be that the higher the dose of an anti-depressant drug a person takes, the less depressed they will feel. So if doses are put into groups, then there should be a trend that starts with less depression for the groups with the higher anti-depressant drug dosage, and goes to higher depression for the groups with lower anti-depressant drug dosage. In cases like this, the population may be ordered from lower depression to high, or from higher drug dosage to lower. For ordered populations, the Jonckheere-Terpstra test is preferred over the Kruskal-Wallis test.

The test is carried out by ordering the observations in the predicted order (as determined a priori). The number of observation scores to the right of each ordered score that are larger than the score are counted and assigned to the variable P. The number of observation scores to the left of each ordered score that are less than the score are counted and assigned to the variable Q. The test statistic, S, is calculated as P-Q.

Friedman's ANOVA

Friedman's ANOVA is the nonparametric equivalent to repeated-measures factorial ANOVA: it is used to compare group means for a categorical dependent variable for repeated measures experiments, where there are several independent variables that have been measured using the same entity for all conditions (each entity has different measures for all of the conditions/independent variables).

The test is carried out by putting each condition into a new column, and ranking the scores of the conditions for each subject. For example, for 3 conditions, there should be 3 new columns with rows ranked 1-3. The ranks for each condition are totaled across all observations. The test statistic, F, is calculated as follows:

$$F = \left[\frac{12}{Nk(k+1)}\sum_{i=1}^{k} R_i^2\right] - 3N(k+1)$$

Figure 14.3

In the equation in figure 14.3, R_i is the sum of the ranks for group i, N is the total sample size, and k is the number of conditions. The test statistic, F, has a Chi-square distribution with k-1 degrees of freedom.

Nonparametric Tests with R

In the last chapter, we saw that the homogeneity of variance assumption could not be met for a two-way independent ANOVA of hours per week on education and relationship. Now, with our knowledge of the nonparametric tests, we can run the ANOVA regression without worrying about the broken assumption. There is a slight problem however: there are no nonparametric tests for k-way factorial ANOVA! This problem can be overcome by either running each group/condition as its own 1-way ANOVA, or by redefining the problem in terms of another type of regression model. The latter option is by far the better of the two. For example, if our goal is to run ANOVA for hours per week on education and relationship, we could instead dummy encode the categorical variables and run a Poisson or negative binomial regression. This would work well because hours per week is a count/integer variable. For the purpose of example, we will do a 1-way ANOVA for hours per week on education.

```
d <- read.table("adult.data", header=F, sep=",")
colnames(d) <- c('age', 'workclass', 'fnlwgt', 'education', 'education_nbr',
                 'marital_status', 'occupation', 'relationship', 'race',
                 'sex', 'capital_gain', 'capital_loss', 'hours_per_week',
                 'native_country', 'salary_bin')
str(d)
head(d)

d$hours_per_week <- as.numeric(d$hours_per_week)

kruskal.test(hours_per_week ~ education, data=d)
#If p value for test stat (H or chi-squared) is < 0.05, then the dependent var does
significantly affect the outcome
#Post hoc tests are needed to see which groups were responsible for the diff

#Obtain the mean rank per group
d$ranks <- rank(d$hours_per_week)
by(d$ranks, d$education, mean)

#The function below shows differences between the mean ranks for the dataset
library(pgirmess)
kruskalmc(hours_per_week ~ education, data=d, cont='two-tailed')
```

The Kruskal-Wallis test showed a statistically significant p-value, meaning that education level has a statistically significant effect on the hours per week. Post hoc tests are needed to find out which groups of education affect hours per week. Notice that the kruskalmc function compares each group to a baseline. In this case, the baseline (or the control group) is 10th grade education. When the Jonckheere-Terpstra test is carried out, the results are not statistically significant at the 95% confidence level, but they are still significant at the 90% confidence level. Note that the data has to be ordered according to which direction the trend is expected to go. In this case, 10th grade is ordered first, and the education levels are ordered alphabetically. This may not be the best way, and the results of the Jonckheere-Terpstra test depend on ordering.

```
#Post Hoc tests
library(multcomp)
pairwise.t.test(d$hours_per_week, d$education, p.adjust.method="bonferroni")
pairwise.t.test(d$hours_per_week, d$education, p.adjust.method="BH")
postHocs <- glht(kw_test, linfct=mcp(education="Tukey"))
summary(postHocs)
confint(postHocs)

#Jonckheere test to look for trends across groups (must order the independent
variable in terms of an expected increasing or decreasing trend)
library(clinfun)
d <- d[with(d, order(education)),]
jonckheere.test(d$hours_per_week, as.numeric(d$education))
```

We will now try a Wilcoxon rank-sum test (the nonparametric independent t-test) to compare the mean hours per week by sex.

```
wilcox_ranksum <- wilcox.test(hours_per_week ~ sex, data=d)
wilcox_ranksum
```

The results show that there is a statistically significant difference between the mean hours per week worked by men and women.

Nonparametric Tests with Python

In the last chapter, we saw that the homogeneity of variance assumption could not be met for a two-way independent ANOVA of hours per week on education and relationship. Now, with our knowledge of the nonparametric tests, we can run the ANOVA regression without worrying about the broken assumption. There is a slight problem however: there are no nonparametric tests for k-way factorial ANOVA! This problem can be overcome by either running each group/condition as its own 1-way ANOVA, or by redefining the problem in terms of another type of regression model. The latter option is by far the better of the two. For example, if our goal is to run ANOVA for hours per week on education and relationship, we could instead dummy encode the categorical variables and run a Poisson or negative binomial regression. This would work well because hours per week is a count/integer variable. For the purpose of example, we will do a 1-way ANOVA for hours per week on education.

```
import pandas as pd
from scipy import stats
from statsmodels.stats.multicomp import pairwise_tukeyhsd, MultiComparison
from sklearn.preprocessing import LabelEncoder

#Load adult dataset
d = pd.read_csv('C:/Users/Nick/Documents/Word Documents/Data Science Books/DSILT
Stats Code/12-16 Titanic and Adult Salaries/adult.data',
                names=['age', 'workclass', 'fnlwgt', 'education', 'education_nbr',
'marital_status', 'occupation', 'relationship', 'race',
                        'sex', 'capital_gain', 'capital_loss', 'hours_per_week',
'native_country', 'salary_bin'])
print(d.info())
```

```
print(d.head())

######
#Kruskal-Wallis test (nonparametric ANOVA)

def kruskal_test(num_variable, *group_variables):
    temp = list(num_variable.groupby(group_variables))
    temp = [temp[i][1] for i,v in enumerate(temp)]
    return stats.kruskal(*temp)
print(kruskal_test(d['hours_per_week'], d['education']))
#If p value for test stat (H or chi-squared) is < 0.05, then the independent var does
significantly affect the outcome
#Post hoc tests are needed to see which groups were responsible for the diff
```

The Kruskal-Wallis test showed a statistically significant p-value, meaning that education level has a statistically significant effect on the hours per week. Post hoc tests are needed to find out which groups of education affect hours per week.

```
#Post Hoc tests
mc = MultiComparison(d['hours_per_week'], d['education'])
print(mc.allpairtest(stats.ttest_ind, method='b')[0])      #For independent t-test
```

We will now try a Wilcoxon rank-sum test (the nonparametric independent t-test) to compare the mean hours per week by sex.

```
le = LabelEncoder()
d['sex'] = le.fit_transform(d['sex'])

t, p = stats.wilcoxon(d['hours_per_week'], d['sex'])
print("Wilcoxon: t = %g  p = %g" % (t, p))
```

The results show that there is a statistically significant difference between the mean hours per week worked by men and women.

15. Discriminant Function Analysis for Classification

Discriminant function analysis (DFA) is a method for modeling a categorical dependent variable on one or more continuous independent variables. It is also possible to include a few binary independent variables to be in the model and perform DFA, although if there are more binary predictors than continuous, or if there are only binary predictors, then another method should be used. DFA is an alternative to logistic regression (LR), but there are key differences between the two (Poulsen & French). DFA works by finding the combinations of predictors that most accurately classify observations into their correct classes, whereas LR works by predicting the likelihood that an observation belongs to a certain class. The key differences between the two methods are summarized in figure 15.1.

Differences between DFA and LR	
Discriminant Function Analysis	**Logistic Regression**
Focuses on finding the most accurate prediction of the outcome variable, making it ideal for situations where correct classification is of high importance (e.g. if we want to know the classification of an observation and do not care as much about which predictors push an observation into a particular class).	Focuses on finding the predictors' influences on the outcome variable, making it ideal for situations where the effect of the predictors is of high importance (e.g. if we want to know which predictors are important in influencing an observation's class).
Can be used when the outcome variables has 2 or more classes. DFA is sometimes preferred over LR when there are more than 2 classes.	Can be used when the outcome variable has 2 classes, and can be extended to work with more classes (multinomial logistic regression). LR is sometimes preferred over DFA when there are only 2 classes.
Can only handle categorical predictors if they are binary, and binary predictors should only be included in the analysis when the classes of the outcome variable are balanced (see the multivariate normality assumption in the next section).	Can handle categorical predictors of any number of categories, and is robust to class imbalance in the outcome variable.
Can handle complete separation.	Recall from chapter 9 that complete separation is a problem for LR. If complete separation occurs, the standard error of the logistic regression and the predictor coefficients will be very large. Complete separation is not a problem for DFA.

Figure 15.1

In general, if the sample size is small, and the distributions of the predictors are normal (see the multivariate normality assumption in the next section), then DFA is preferred over LR (Field, Miles, & Field, 2012).

While DFA can be used for classification, it can also be used for dimension reduction. In fact, it is more common for DFA to be used for dimension reduction than classification. We will explore dimension reduction in a later chapter.

Discriminant function analysis is performed by creating combinations of predictors to maximize the differences between classes of the dependent variable, while minimizing the differences between observations within each class (Hastie, Tibshirani, & Friedman, 2009). A combination of predictors is simply a function, and since the functions generated are the ones that best discriminate between classes, the method is called discriminant function analysis. The functions are almost always linear, but they can also be quadratic (nonlinear). In DFA, the first discriminant function maximizes the differences between classes of the dependent variable. The second discriminant function does the same thing, but it is independent of the first function. We will get into the details of how this works in the following sections.

Each discriminant function is represented by a latent variable. Recall from figure 12.18 in the ANCOVA section of chapter 12 that a latent variable is a variable that is not directly observed but whose existence is inferred from a model's independent variables. Latent variables can represent physical phenomena or abstract constructs. Since these latent variables are used to distinguish between classes of the dependent variable, they are useful features to use in a new linear model with fewer features (or fewer dimensions) than the original. This is reasoning behind using DFA for dimension reduction, as we will see in a later chapter.

Discriminant Function Analysis Assumptions
DFA requires the following assumptions to be met:

- Multivariate normality: each combination of the predictors results in a random normal distribution. This assumption requires all of the predictors to be continuous, which should raise a flag since it was previously stated that DFA works with binary independent variables. It is not possible for binary predictors to be normally distributed. Nevertheless, it has been shown that DFA works even when this assumption is violated through the inclusion of binary predictors. The only times when this assumption must be strictly obeyed is when there are severe imbalances in the categories of the predictor groups (Hastie, Tibshirani, & Friedman, 2009).
- Homogeneity of covariance matrices: the variances among the classes of the outcome are the same across levels of the binary predictors.
- No multicollinearity: continuous predictors should not be correlated.
- Independence: observations are assumed to be independent of one another. This assumption means that DFA cannot be applied to data from repeated measures experiments.

The first assumption (multivariate normality) can be tested using **Henze-Zirkler's multivariate normality test**. The HZ test works by comparing the distances between two distribution functions. If the data is normally distributed, the test statistic will be lognormal. Recall from chapter 3 that the Jarque-Bera test, the Shapiro-Wilk test, and skewness/kurtosis tests are all methods for testing univariate normality. If all of the individual predictors are normally distributed, which can be determined by applying the univariate normality tests to the predictors, one at a time, then any combination of them will also be normally distributed. Therefore, another way to test for multivariate normality is to apply univariate normality tests to every predictor. The problem with doing this is that if one of the predictors is non-normally distributed, multivariate normality might still hold (the non-normality of one predictor may not be enough to cause statistically significant multivariate non-normality, according to the HZ test).

The second assumption (homogeneity of covariance matrices) can be tested with **Box's M statistic**. Recall from chapter 3 that homogeneity of variance can be tested with Levene's test. Levene's test cannot be used for the multivariate case however, as it does not take covariance into consideration. Like Levene's test, Box's M statistic is sensitive to sample size. If Box's M statistic detects heteroskedasticity (the null of homogeneity of the covariance matrix is rejected) for a large dataset, it is likely due to non-normality. Therefore, if the assumption is violated, it is often because the assumption of multivariate normality is also violated. The problem can be fixed by collecting more data, or by transforming the data as described in chapter 4 to try to obtain multivariate normality. Another option is to use linear discriminant analysis when the assumption holds, and quadratic discriminant analysis when the assumption does not hold, as we will soon see.

The assumptions for DFA are stricter than the assumptions required for LR. Recall from chapter 9, the assumptions for logistic regression are somewhat lenient, making LR more robust than DFA. Therefore, if any of the assumptions of DFA are violated, LR is a suitable substitute.

Linear Discriminant Analysis

Linear discriminant analysis (LDA) creates a linear function of the predictors that maximizes the differences between classes of the dependent variable. The differences between classes of the dependent variable can be defined by the discriminant rules listed below.

- **Bayes' Discriminant Rule**: an observation is assigned to the class that maximizes the product of the prior probability of that class and the class density. Bayes discriminant rule is based on Bayes' Theorem, which we will explore in a later chapter, but since this is the most commonly used rule, we will define it in this chapter.
- **Fisher's Linear Discriminant Rule**: an observation is assigned to the class that maximizes the ratio of the between-class sum of squares to the within-class sum of squares.

The discriminant rule to use for LDA is a matter of personal preference, but Bayes's discriminant rule is more general and common libraries like scikit-learn use it, so that is the one we will examine here. The equation for Bayes' Theorem is presented formally in a later chapter, so for now it must be taken at face value:

$$P(y_k|x) = \frac{P(x|y_k)P(y_k)}{P(x)}$$

Figure 15.2

In the equation in figure 15.2, P represents probability, so P(y sub k|x) can be read as the probability of y belonging to class k, given observation x. P(y sub k|x) is also called the posterior, as it is the posterior probability (the probability resulting from the equation to the right of the equals sign). P(y sub k) is the prior probability that the class for observation x is k. Prior probability is the probability without any analysis. For example, if we are trying to calculate the probability that a person has red hair, the prior probability would be the probability of red hair out of the entire population. For LDA, all classes are usually treated equally, so P(y sub k) = 1/number of classes. Since we are observing observation x, P(x) is essentially 1, so the denominator can be dropped completely for the purpose of LDA.

The probability of observation x, given class k, P(x|y sub k), cannot be found because LDA works with continuous variables, rather than discrete, and P(x|y sub k) is a discrete value. So instead of the discrete probability, it is possible to use a continuous probability density function, PDF(x|y sub k), which is proportional to the exact probability P(x|y sub k). Recall from chapter 2 that a PDF is the probability that a continuous variable equals some discrete value. Thus, the PDF allows P(x|y sub k) to be estimated. When there are only 2 classes, the PDF is:

$$PDF(x|y_k) = \frac{1}{\sqrt{2\pi} * \sigma_k} e^{-\frac{1}{2\sigma_k^2}(x-\mu_k)^2}$$

Figure 15.3

In the equation in figure 15.3, PDF(x|y sub k) is the probability density function of observation x in class k, sigma sub k is the standard deviation of class k, sigma squared sub k is the variance of class k, and mu sub k is the mean of class k. When there are 3 or more classes, the PDF is:

$$PDF(x|y_k) = \frac{1}{(2\pi)^{p/2} * |\Sigma|^{1/2}} e^{-\frac{1}{2}(x-\mu)^T \Sigma^{-1}(x-\mu)}$$

Figure 15.4

In the equation in figure 15.4, p is the number of predictors and the T that looks like an exponent is indicating a transposition of the matrix of values (x − mu). When this equation, which represents the PDF of x given class k, is substituted into Bayes' function, the result is the **Bayes Classifier**:

$$P(y_k|x) = \frac{\left(\frac{1}{\sqrt{2\pi} * \sigma_k} e^{-\frac{1}{2\sigma_k^2}(x-\mu_k)^2}\right) * P(y_k)}{\sum_{l=1}^{k}\left[\left(\frac{1}{\sqrt{2\pi} * \sigma_k} e^{-\frac{1}{2\sigma_k^2}(x-\mu_k)^2}\right) * P(y_l)\right]}$$

Figure 15.5

The Bayes classifier assigns observation x to the class that maximizes the equation in figure 15.5. Since LDA is dealing with PDF, rather than probability, it is helpful to step back and look at the equation from a higher level. The equation in figure 15.3 is the PDF for class k, and it must be calculated for every class. Since the PDF is proportional to the exact probability, it must be normalized by dividing it by the sum of the PDFs for all classes. For example, for 2 classes A and B, the result of would be:

$$P(y_k|x) = \frac{PDF(x|y_k) * P(y_k)}{[(PDF(x|y_A) * P(y_A)) + (PDF(x|y_B) * P(y_B))]}$$

Figure 15.6

The equation in figure 15.6 would be the same as the equation in figure 15.5, if 15.5 were being calculated for 2 classes. For more classes, 15.6 would have more terms in the denominator. When the natural log of the Bayes classifier in figure 15.5 is taken, the equation becomes linear. The log of the Bayes classifier is therefore a linear discriminant function.

If the covariance matrices between classes are substantially different, observations will tend to be assigned to the classes with greater variance. This is one of the reasons LDA requires homogeneity of the covariance matrices. If the assumption cannot be met, then an alternative form of DFA can be used: quadratic discriminant analysis.

Quadratic Discriminant Analysis

Quadratic Discriminant Analysis (QDA) is similar to LDA except that it assumes each class has its own covariance matrix. As a consequence of this assumption the equation for the estimated Bayes classifier becomes quadratic, rather than linear, when the natural log is taken. Unlike LDA, which has 2 choices for discriminant rules (Bayes' or Fisher's), QDA works only with Bayes.

QDA is preferred to LDA when the homogeneity of covariance matrices assumption cannot hold, or when the number of predictors is large (James, Witten, Hastie, & Tibshirani, 2013). This is because the estimation of the covariance matrix requires estimating $p(p+1)/2$ parameters, which becomes cumbersome when there are many predictors, p. In general, LDA is preferred over QDA when there are few observations or when the assumptions required for LDA can be met (QDA sacrifices predictive power for flexibility). LDA is less flexible because it makes stronger assumptions, so it produces smaller variance. QDA is better to use when there are a lot of observations, the variance is less important, or when there are a lot of predictors (James, Witten, Hastie, & Tibshirani, 2013).

LDA and QDA Classification with R

The adult dataset is a perfect example of a classification problem. The goal is to predict who earns more than $50k per year and who earns <= $50k per year. We have used this dataset in the last couple of chapters but have not done any data cleaning. We should do this now, as well as any feature engineering, before moving on to classification.

```
d <- read.table("adult.data", header=F, sep=",")
colnames(d) <- c('age', 'workclass', 'fnlwgt', 'education', 'education_nbr',
                 'marital_status', 'occupation', 'relationship', 'race',
                 'sex', 'capital_gain', 'capital_loss', 'hours_per_week',
                 'native_country', 'salary_bin')
str(d)
head(d)

#Convert variables to desired data types
d$age <- as.numeric(d$age)
d$fnlwgt <- as.numeric(d$fnlwgt)
d$education_nbr <- as.factor(d$education_nbr)
d$capital_gain <- as.numeric(d$capital_gain)
d$capital_loss <- as.numeric(d$capital_loss)
d$hours_per_week <- as.numeric(d$hours_per_week)

#Define a function to count the nulls in every field
naCol <- function(x){
  y <- sapply(x, function(y) {
    if (any(class(y) %in% c('Date', 'POSIXct', 'POSIXt'))) {
      sum(length(which(is.na(y) | is.nan(y) | is.infinite(y))))
    } else {
      sum(length(which(is.na(y) | is.nan(y) | is.infinite(y) | y=='NA' | y=='NaN' |
y=='Inf' | y=='' | y==' ?')))
    }
  })
  y <- data.frame('feature'=names(y), 'count.nas'=y)
```

```r
    row.names(y) <- c()
    y
}

naCol(d)

#Since there are 32k rows and <2k rows with nulls, it is safe to discard them
d <- d[(d$workclass!=' ?' & d$occupation!=' ?' & d$native_country!=' ?'),]

#Boxplots for numeric variables to check for outliers
for (col in 1:ncol(d[,which(sapply(d, class) == 'numeric')])) {
  boxplot(d[,which(sapply(d, class) == 'numeric')][col],
          main=paste0("Box Plot for ", colnames(d[,which(sapply(d, class) ==
'numeric')])[col]))
}

#Look at correlation matrix
allCor <- cor(d[,which(sapply(d, class) == 'numeric')], use="pairwise.complete.obs")
library('corrplot')
corrplot(allCor, method="circle")

#Dummy encode categorical variables
d[,which(sapply(d, class)=='factor')] <- sapply(d[,which(sapply(d,
class)=='factor')], substring, 2, 100)  #Gets rid of leading whitespaces that show up
in factor levels
d[,which(sapply(d, class)=='character')] <- lapply(d[,which(sapply(d,
class)=='character')], as.factor)
library(caret)
cols_to_encode <- c('workclass', 'education', 'marital_status', 'occupation',
'relationship', 'race', 'sex', 'native_country')
for (c in cols_to_encode) {
  dummies <- dummyVars(paste0(" ~ ", c), data=d, fullRank=T)
  dummyenc <- data.frame(predict(dummies, newdata=d))
  d <- cbind(d, dummyenc)
}
colnames(d) <- gsub("[.]", "_", colnames(d))
str(d)
#Get rid of redundant columns now that they have been dummy encoded
d$workclass <- NULL
d$education <- NULL
d$education_nbr <- as.numeric(d$education_nbr)
d$marital_status <- NULL
d$occupation <- NULL
d$relationship <- NULL
d$race <- NULL
d$sex <- NULL
d$native_country <- NULL

#Split data into training and test sets
set.seed(14)
train_indices <- createDataPartition(y=d$salary_bin, p=0.7, list=F)
train <- d[train_indices,]
test <- d[-train_indices,]
write.csv(train, 'adult_train.csv', row.names=F)
write.csv(test, 'adult_test.csv', row.names=F)
```

By now, the code should be self-explanatory, but there are a couple things to note. First, the naCol function was adjusted to search for ' ?' because there were several values that were question marks in this dataset. These should be treated as NAs. Since there are 32k records in this dataset, and fewer than 2k NAs, it is safe to discard the rows containing NA values. The dataset is large enough that any information lost by discarding these rows should be made up for by other observations. The other thing to make note of is that the categorical variables (factors in R) have a single whitespace at the beginning of every value. This is stripped away before the categorical variables are dummy encoded, which requires converting the variables to characters and then back to factors.

Performing LDA is quite simple. However, R displays a warning that some of the variables are collinear. Although there was no indication of multicollinearity in the correlation plot, it could be that some of the dummies are correlated to strongly. When the percentage of between class variance explained by the linear discriminants is examined, its value is 1, which is too good to be true. There must be some kind of multicollinearity issue that is overfitting the model. We will fix this in the next chapter by reducing the dimensions of the feature set.

```
library(MASS)
lda_model <- lda(formula=salary_bin ~ ., data=train)
lda_model

#Explore the percentage of between class variance explained by each linear
discriminant
lda_model$svd^2/sum(lda_model$svd^2)
```

Note that the lda function uses default prior probabilities for each of the classes that are equivalent to the classes' percentage of the observations. So if one class made up 7 out of 100 observations, then its default prior probability would be 0.07.

```
#Make survival predictions for validation set
post_lda <- predict(lda_model, test[,-7])
summary(post_lda)
test$pred <- post_lda$class
test$baseline <- '<=50K'  #Baseline assigns everything to majority class

#Evaluate the model using the log loss, AUC, and accuracy
library(MLmetrics)
LogLoss(as.integer(test$pred), as.integer(test$salary_bin))
AUC(as.integer(test$pred), as.integer(test$salary_bin))
Accuracy(test$pred, test$salary_bin)
Accuracy(test$baseline, test$salary_bin)
#View the confusion matrix with predicted values on the left
prop.table(table(test$pred, test$salary_bin))
```

When the model is evaluated, it has an 83% classification accuracy on the test set, compared to a 75% classification accuracy of the baseline model. The baseline model is simply assigning everything to the majority class. So the model is better than guessing, but not by much. Surprisingly though, 83% accuracy is actually a good score for this dataset (Deepajothi & Selvarajan, 2012).

Performing QDA is just as easy, but the collinearity problem arises again.

```
qda_model <- qda(formula=salary_bin ~ ., data=train)
```

```
#Look at complete correlation matrix
allCor <- cor(d[,which(sapply(d, class) == 'numeric')], use="pairwise.complete.obs")
allCor > 0.6
allCor < (-0.6)
```

The error message that comes up when the qda function is called says that the rank is deficient. This usually means that the observation to feature ratio is too low, however, we have more than 20k observations, which should be plenty for this problem. The issue is with the collinearity that appears. Looking at the correlation matrix, it is not obvious that there is any collinearity. So we will have to revisit this issue after performing dimension reduction in the next chapter, because it is probably the fault of the dummies.

LDA and QDA Classification with Python

The adult dataset is a perfect example of a classification problem. The goal is to predict who earns more than $50k per year and who earns <= $50k per year. We have used this dataset in the last couple of chapters but have not done any data cleaning. We should do this now, as well as any feature engineering, before moving on to classification.

```python
import pandas as pd
import numpy as np
from sklearn.preprocessing import LabelEncoder, OneHotEncoder, StandardScaler
from sklearn.model_selection import train_test_split
import matplotlib.pyplot as plt
import seaborn as sns
from sklearn.discriminant_analysis import LinearDiscriminantAnalysis,
QuadraticDiscriminantAnalysis
from sklearn import metrics

#Set seed for repeatability
seed = 14
np.random.seed(seed)

#Load adult dataset
d = pd.read_csv('C:/Users/Nick/Documents/Word Documents/Data Science Books/DSILT
Stats Code/12-16 Titanic and Adult Salaries/adult.data',
               names=['age', 'workclass', 'fnlwgt', 'education', 'education_nbr',
'marital_status', 'occupation', 'relationship', 'race',
                      'sex', 'capital_gain', 'capital_loss', 'hours_per_week',
'native_country', 'salary_bin'])
print(d.info())
print(d.head())

#Fixes a bug in printing output from IDLE, it may not be needed on all machines
import sys
sys.__stdout__ = sys.stdout

######
#Data Cleaning

#Define a function to count the nulls in every field
def naCol(df):
    y = dict.fromkeys(df.columns)
    for idx, key in enumerate(y.keys()):
```

```
        if df.dtypes[list(y.keys())[idx]] == 'object':
            y[key] = pd.isnull(df[list(y.keys())[idx]]).sum() +
(df[list(y.keys())[idx]]=='').sum() +(df[list(y.keys())[idx]]==' ?').sum()
        else:
            y[key] = pd.isnull(df[list(y.keys())[idx]]).sum()
    print("Number of nulls by column")
    print(y)
    return y

naCol(d)

#Since there are 32k rows and <2k rows with nulls, it is safe to discard them
d = d.dropna()
d = d[(d != '?').all(1)]
d = d[(d != ' ?').all(1)]

#Convert variables to desired data types
d[['age', 'fnlwgt', 'capital_gain', 'capital_loss',
'hours_per_week']].apply(pd.to_numeric)
encoder = LabelEncoder()
categorical_vars = ['workclass', 'marital_status', 'occupation', 'relationship',
                    'race', 'sex', 'native_country', 'salary_bin']
categorical_var_mapping = dict()
for cv in categorical_vars:
    d[cv] = d[cv].str[1:]                    #Gets rid of the leading white spaces in
the text
    d[cv] = encoder.fit_transform(d[cv])  #Encodes as integer
    categorical_var_mapping[cv] = list(encoder.classes_)  #Saves integer to category
mapping

#Boxplots for numeric variables to check for outliers
for col in ['age', 'fnlwgt', 'capital_gain', 'capital_loss', 'hours_per_week']:
    sns.boxplot(d[col])
    plt.title('Box Plot for ' + col)
    plt.show()

#Look at correlation matrix
corm = d[['age', 'fnlwgt', 'capital_gain', 'capital_loss', 'hours_per_week']].corr()
plt.matshow(corm)
plt.show()

#Get rid of education because it is already encoded as integer in 'education_nbr'
d.drop(['education'], axis=1, inplace=True)

#Save the version of the dataset without dummies
d_no_dummies = d.copy()

print(d.columns)
#Dummy encode categorical variables except for the target (salary_bin)
#Note that pd.get_dummies automatically removes the originals after dummy encoding
d = pd.get_dummies(d, columns=categorical_vars[:6], drop_first=True)
print(d.columns)
print(d.info())

#Perform z-score standardization on the numeric features
```

```
#Standardization is only needed if the matrix of the data will be decomposed instead
of the covariance matrix
d[['age', 'fnlwgt', 'education_nbr', 'capital_gain', 'capital_loss',
'hours_per_week']] = StandardScaler().fit_transform(d[['age', 'fnlwgt',
'education_nbr', 'capital_gain', 'capital_loss', 'hours_per_week']])
d_no_dummies[['age', 'fnlwgt', 'education_nbr', 'capital_gain', 'capital_loss',
'hours_per_week']] = StandardScaler().fit_transform(d_no_dummies[['age', 'fnlwgt',
'education_nbr', 'capital_gain', 'capital_loss', 'hours_per_week']])

#Convert the dataframe into numpy array and specify dependent variable
x = d.drop(['salary_bin'], axis=1).values.astype(float)
y = d[['salary_bin']].values
x_nd = d_no_dummies.drop(['salary_bin'], axis=1).values.astype(float)
x_numeric = d_no_dummies[['age', 'fnlwgt', 'education_nbr', 'capital_gain',
'capital_loss', 'hours_per_week']].values.astype(float)

#Split data into training and test sets - be sure to stratify since this is for
classification
x_train, x_test, y_train, y_test = train_test_split(x, y, test_size=0.3, stratify=y,
random_state=seed)
x_nd_train, x_nd_test, y_nd_train, y_nd_test = train_test_split(x_nd, y,
test_size=0.3, stratify=y, random_state=seed)
xn_train, xn_test, yn_train, yn_test = train_test_split(x_numeric, y, test_size=0.3,
stratify=y, random_state=seed)
```

By now, the code should be self-explanatory, but there are a couple things to note. First, the naCol function was adjusted to search for ' ?' because there were several values that were question marks in this dataset. These should be treated as NaNs. Since there are 32k records in this dataset, and fewer than 2k NaNs, it is safe to discard the rows containing NaN values. The dataset is large enough that any information lost by discarding these rows should be made up for by other observations. The other thing to make note of is that the categorical variables (the variables with object data types) have a single whitespace at the beginning of every value. This is stripped away before the categorical variables are dummy encoded.

Performing LDA is quite simple. When the percentage of between class variance explained by the linear discriminants is examined, its value is 1, which is too good to be true. There must be some kind of multicollinearity issue that is overfitting the model. We will fix this in the next chapter by reducing the dimensions of the feature set.

```
#Run LDA for classification
#Note if n_components=None, then all of them are kept
lda = LinearDiscriminantAnalysis(n_components=None, solver='svd')
lda.fit(x_train, y_train)
print(lda.get_params())
print('Priors:', lda.priors_)        #Class prior probabilities
print('Classification Accuracy:', lda.score(x_train, y_train))

#Explore the percentage of between class variance explained by each linear
discriminant
print('Explained variance:', lda.explained_variance_ratio_)

######
#Evaluating the model on new data
```

```
#Make income predictions for validation set
post_lda = lda.predict(x_test)
post_lda = post_lda.reshape(post_lda.shape[0], 1)
print('Classification Accuracy:', lda.score(x_test, y_test))

#Confusion matrix
cm = metrics.confusion_matrix(y_test, post_lda)
sns.heatmap(cm, annot=True, fmt=".2f", square=True)
plt.xlabel('Predicted Class')
plt.ylabel('Actual Class')
plt.title('Confusion Matrix')
plt.show()

#Log loss (a.k.a. negative log likelihood)
print('Log loss:', metrics.log_loss(y_test, post_lda))

#Plot ROC curve
fpr, tpr, threshold = metrics.roc_curve(y_test, post_lda)
roc_auc = metrics.auc(fpr, tpr)
plt.plot(fpr, tpr, lw=2, label='ROC Curve (area = %0.2f)' % roc_auc)
plt.legend(loc='lower right')
plt.plot([0, 1], [0, 1], 'b--')  #Diagonal line
plt.xlabel('False Positive Rate')
plt.ylabel('True Positive Rate')
plt.title('Receiver Operating Characteristic')
plt.show()
```

Note that the lda function uses default prior probabilities for each of the classes that are equivalent to the classes' percentage of the observations. So if one class made up 7 out of 100 observations, then its default prior probability would be 0.07.

When the model is evaluated, it has an 83% classification accuracy on the test set, compared to a 75% classification accuracy of the baseline model. The baseline model is simply assigning everything to the majority class. So the model is better than guessing, but not by much. Surprisingly though, 83% accuracy is actually a good score for this dataset (Deepajothi & Selvarajan, 2012).

Performing QDA is just as easy, but a problem with collinearity causes an error to be thrown.

```
#Run QDA for classification
qda = QuadraticDiscriminantAnalysis()
qda.fit(x_train, y_train)
print(qda.get_params())
print('Priors:', qda.priors_)        #Class prior probabilities
print('Classification Accuracy:', qda.score(x_train, y_train))

#Look at complete correlation matrix
print(corm)
```

Looking at the correlation matrix, it is not obvious that there is any collinearity. So we will have to revisit this issue after performing dimension reduction in the next chapter, because it is probably the fault of the dummies.

16. Dimension Reduction

LDA and QDA introduced the concept of latent variables that are useful for classification. Latent variable creation is also a dimension reduction technique. Dimension reduction is the practice of reducing the number of dimensions or features that are included in a model. Dimension reduction is useful because it results in simpler models, reduces the chances of overfitting a model to a particular dataset, reduces model training time, and eliminates redundancy that might occur if two or more variables are strongly correlated. All of these problems that are solved by dimension reduction are collectively referred to as the **curse of dimensionality**, because they are exacerbated by including more dimensions. Dimension reduction can be broken down into two tracts of procedures: feature selection and feature extraction. Feature selection and feature extraction are not mutually exclusive; it is actually common to see feature extraction performed on a subset of selected features.

Feature Selection Methods

One way to reduce the number of dimensions in a model is to simply hand pick the features that should be used to build the model, and leave out the rest. This is called feature selection. The most common feature selection methods are forward, backward, and best subsets. **Forward selection** involves starting the model building process with only one feature, and adding one additional feature to see how it affects the model. Features are continuously added until some evaluation metric either fails to improve any further or begins to deteriorate. **Backward selection** is the opposite of forward selection; the model building process starts with all features included, and one feature is removed until the evaluation metric fails to improve any further or begins to deteriorate. The **best subsets** method starts with one feature, and additional features are added based on how much they improve the evaluation metric. In the best subsets method, the best subset of features is the one that optimizes the evaluation metric. Any of the optimization algorithms described in chapter 7 can be used to find the best subset of features.

The three feature selection methods described in the last paragraph employ an evaluation metric to determine whether a feature improves the model or not. The two most common evaluation metrics are the AIC and BIC, as described in chapters 6 and 9. Other evaluation metrics have been proposed, but there is no clear best metric, so the AIC and BIC are sufficient. The decision to use forward, backward, or best subsets feature selection is left to the researcher. None of these methods consistently outperform the others in every situation.

When feature selection is applied to linear models, predictors can change their statistical significance. For example, predictor A may be insignificant, but when predictor B is added to the model, A suddenly becomes significant. This is caused by an interaction effect between A and B. Sometimes the opposite happens. The opposite is especially common when binary predictors are added. For example, if A is significant and the introduction of B greatly reduces or eliminates the significance of A, it is likely because adding B introduced collinearity into the model. This is true even if the addition of B improves the fit of the model. In this situation, B does not explain any additional variance in the dependent variable that is not already explained by A. It is up to the research to decide whether the additional fit, which may be overfit, is worth including B in the model.

Feature Extraction Methods

The remainder of this chapter will focus on feature extraction methods. Feature extraction is the practice of using derived features to build a model. Feature extraction is essentially unsupervised learning. As we have seen, a latent variable is an example of a derived or extracted feature. Derived features are produced by combining the initial features in some way. As we saw in chapter 15 for example, LDA derives latent features by creating a linear function of the initial features that maximizes the differences between classes of the dependent variable. These latent features can completely replace the original features in the model, so LDA will be explored as a dimension reduction technique. But before we examine LDA for dimension reduction, we must first look at matrix decompositions, because they are the foundation of nearly every linear feature extraction technique. Later in this chapter, we will also look at a couple nonlinear feature extraction techniques.

Eigendecomposition

In linear algebra, a single number is called a **scalar**. A **vector** is an array of scalars (numbers). Each number in a vector can be referenced by its index in the vector. The index is the ordered place of that number in the vector. For example, in the vector [56, 12, 18], 56 is the first number so its index is 1 (or 0 as in some programming languages, like Python). A **matrix** is a 2D array of numbers, so the location of each number requires 2 indices to identify it. The indices are ordered as: [row index, column index]. A **tensor** is a multidimensional array of numbers (an array of more than 2 dimensions), so the location of each number requires as many indices as there are dimensions to identify it. The number of dimensions of a tensor is its order. Matrices are actually tensors of the 2nd order because they have 2 dimensions. Likewise, vectors are 1st order tensors and scalars are 0th order tensors.

Scalar	Vector	Matrix
1	[1, 2, 3]	$\begin{bmatrix} 1 & 2 \\ 3 & 4 \end{bmatrix}$

Figure 16.1

A full treatment of linear algebra is beyond the scope of this book, so if the reader is not familiar with basic operations and transformations of the entities described in this section, the remainder of this chapter will have to be taken at face value.

Prime factorization is a method for decomposing an integer into its fundamental components. For example, the prime factorization of 12 is 2*2*3. Similar to how integers can be decomposed by prime factorization, matrices can be decomposed into component vectors. Eigendecomposition is one method for decomposing a matrix into its component vectors.

When a square matrix (a matrix with the same number of rows as columns), A, is multiplied by a vector v, the result is Av. Suppose A is the matrix shown in figure 16.2 and v is the vector. The product Av is the vector [3, 3]. It is possible to draw a line from the origin through vector v and the new vector Av, as shown in figure 16.4. This means that the new vector Av is simply a scaled version of the original vector v, because it was multiplied by a scalar.

$$A = \begin{bmatrix} 2 & 1 \\ 1 & 2 \end{bmatrix}$$

$$v = [1, 1]$$

$$Av = \begin{bmatrix} 2 & 1 \\ 1 & 2 \end{bmatrix} * [1,1] = [3,3]$$

$$[3,3] = \lambda * [1,1]$$

$$\lambda = 3$$

Figure 16.2

The definition of an **eigenvector** is a vector that when multiplied by a square matrix, is altered only in scale. The magnitude by which an eigenvector is scaled when multiplied by a square matrix is called an **eigenvalue**. Therefore, in the equations in figure 16.2, v is an eigenvector and lambda is an eigenvalue. Lambda is a scalar that shows how much the matrix A scaled the vector v. The equation in figure 16.3 formally defines the relationship between a square matrix, an eigenvector, and an eigenvalue. .

$$Av = \lambda v$$

Figure 16.3

The equation in figure 16.3 shows that when a square matrix A is multiplied by an eigenvector v, only the length or magnitude of the eigenvector changes, and the amount that it changes is represented by the eigenvalue lambda.

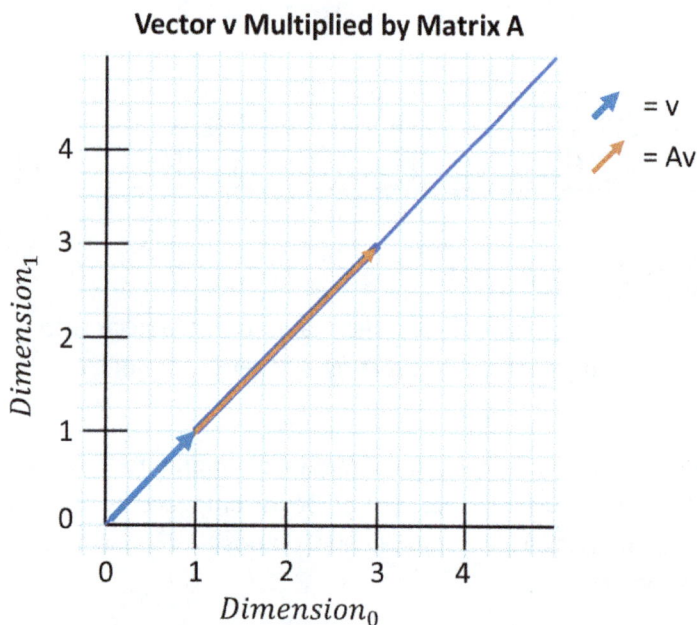

Figure 16.4

Square matrices can be decomposed into multiple eigenvectors, each with their own eigenvalues. For example, the matrix A in figure 16.2 can be decomposed into eigenvectors [1,1] with eigenvalue 3 and eigenvector [-1,1] with eigenvalue 1. The math behind the decomposition is beyond the scope of this book, as we only need to know what eigendecomposition has to do with dimension reduction, but curious readers can decompose matrix A and see the math behind it at https://www.symbolab.com.

Eigenvectors are especially useful in dimension reduction when they are orthogonal (orthogonal means they are at right angles to one another). Eigenvectors for square matrices are not always orthogonal, but eigenvectors for symmetrical matrices are always orthogonal. Therefore, it is common to perform eigendecomposition on either the covariance matrix or the correlation matrix of the features of a dataset, because both of these will always be square matrices that are symmetrical about their diagonals. Performing eigendecomposition on the covariance matrix (or on the correlation matrix) is the method used in principal component analysis, as we will soon see. Figure 16.5 shows what happens to the 2D space of dimensions 0 and 1 when a square matrix is multiplied by orthogonal eigenvectors.

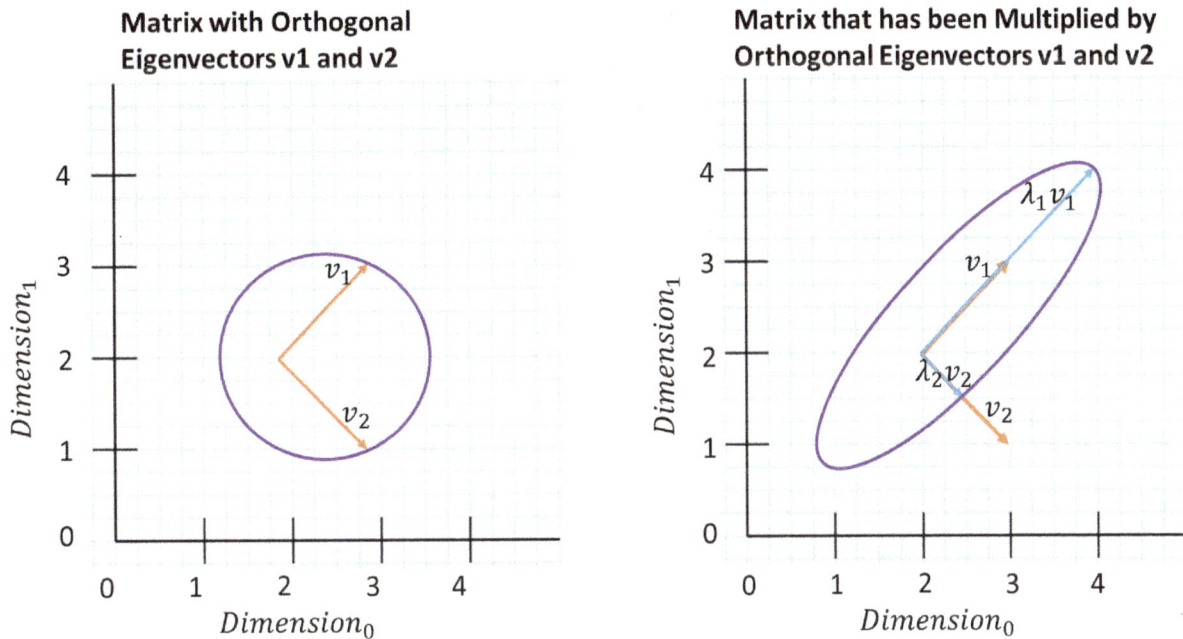

Figure 16.5

Notice how the product of the matrix and each eigenvector scales the space of the dimensions by the amounts represented by lambda sub 1 and lambda sub 2. As we will soon see, this space scaling is the driver behind principal component analysis (Powell & Lehe, 2018).

Singular Value Decomposition

Eigendecomposition is not always practical because matrices are not always square, or their eigenvectors are not orthogonal. Singular value decomposition (SVD) is a technique for decomposing a matrix that works for all matrices, regardless of shape. It is the generalization of eigendecomposition. SVD decomposes matrixes into singular vectors and singular values. The difference between eigendecomposition and SVD is that SVD decomposes a matrix into the product of three other matrices, rather than eigenvectors and eigenvalues (Barber, 2012).

$$A = UDV^{\mathsf{T}}$$

Figure 16.6

In the equation in figure 16.6, A is an m*n matrix, U is an m*m matrix, D is an m*n matrix, and V is an n*n matrix. The T superscript indicates that the matrix is transposed, meaning the rows and columns are swapped. The matrices U and V are orthogonal to one another. The matrix D is the diagonal matrix in which the elements along the diagonal of D are called **singular values** of the matrix A. The columns of matrix U are the left singular vectors because they correspond to the rows of matrix A. The columns of matrix V are the right singular vectors of A because they correspond to the columns of matrix A. The columns of matrices U and V are ordered so that the leftmost column of each matrix explains the most variance in the data (so the columns are the principal components, as we will soon learn).

The singular values are the square roots of the eigenvalues. So by squaring D and representing the results as a proportion of the total, it is possible to see the percentage of variance that is explained by the singular vector corresponding to each singular value.

SVD is an operation that reduces the rank of a matrix, meaning it produces a smaller matrix that approximates the original. When dimension reduction is performed with SVD, a selected number of singular values that explain the most variance are kept and the rest are discarded.

In addition to dimension reduction, SVD can be used to compress images, because images can be envisioned as matrices of pixels. By reducing the size of the matrix to a smaller approximation of the original, some information is lost, but the benefit is that models using the matrix take less time to train. It is up to the researcher to decide how much information to sacrifice in order to realize improvements in modeling time. The original matrix can be reconstructed from the UDV matrices by calculating the dot product of U, the diagonal of D, and the transpose of V.

Principal Component Analysis

Principal component analysis (PCA) is a dimension reduction technique that decomposes either the covariance matrix or the correlation matrix to find the principal components. PCA seeks to maximize the explained variance using a linear combination of principal components (Field, Miles, & Field, 2012). Since it involves maximizing explained variance, standardization should be performed to scale the data before the covariance or correlation matrix is formed. The method of matrix decomposition used in PCA can be either eigendecomposition or SVD. **Principal components** are the linear combinations of features that account for the most variance in the data. Principal components are eigenvectors. They can also be referred to as a type of latent variable, as the points that make up the line describe the underlying model. Consider the scatterplot of two variables, as shown in figure 16.7. If the variables are correlated, the scatterplot will have an ellipse shape like in the figure. If two orthogonal lines were drawn to measure the length and width of the ellipse, these lines would represent eigenvectors. The fact that they are orthogonal means they are independent of one another. The eigenvalues of these eigenvectors are simply the lengths of the lines. Notice how the length of the ellipse is longer than the width; this means that the eigenvector representing the length of the ellipse accounts for more variance than the eigenvector for the width, because variance measures the spread of data and there is greater spread along the length of the ellipse. An eigenvector is a principal component if it accounts for the most variance, so the length of the ellipse would be the first principal component.

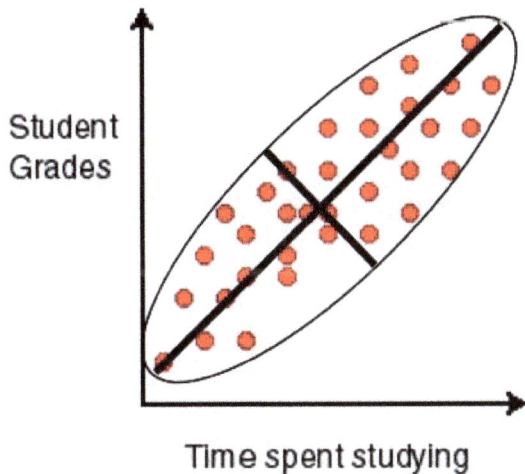

Figure 16.7

The relationship between the two variables shown in figure 16.7 can be represented by a linear model. We can image this model would look similar to the length eigenvector, which would make the length eigenvector be a great descriptor of the model. In fact, all eigenvectors are descriptors of the underlying relationship between variables. They describe the relationship in terms of different linear combinations of the variables that account for the most variance in the data.

Figure 16.7 shows eigenvectors for a dataset with only two features or dimensions. For every additional dimension, a new eigenvector is created. For example, a 3rd dimension would have an eigenvector for depth. Since the eigenvalues are the lengths of the eigenvectors, they show how evenly the variances are distributed. The eigenvector with the largest eigenvalue is the first principal component. Eigenvalues can therefore be thought of as being similar to the beta coefficients of a linear model if the principal components are the predictors, as they tell how influential particular principal components are.

- When there is no correlation between variables, the eigenvalues will all be the same. Consider the simple case of correlation between two variables, such as in figure 16.7. If the variables were not correlated, their scatterplot would be circular instead of elliptical, because variance would be evenly disbursed. If the largest eigenvalue were divided by the smallest, the result would be 1.
- When there is correlation between variables, the eigenvalues are different. Consider the simple case of correlation between two variables, such as in figure 16.7. Since the variables in figure 16.7 are correlated, their scatterplot is elliptical. The stronger the correlation, the more linear the ellipse will appear. If the largest eigenvalue were divided by the smallest, the result would be very large. This ratio will approach infinity as the correlation approaches 1.

PCA extracts (keeps) the eigenvectors with the highest eigenvalues, as these are the ones that explain the most variance in the data. Sometimes the researcher has a preconceived notion of how many principal components they would like to extract. For example, if the goal is to reduce the feature space to 2D for plotting purposes, then the first two principal components are extracted. At other times, the goal is to build a model on the principal components instead of the original features (the goal of dimension reduction is to produce a model with fewer parameters, so by replacing the features with the principal components, the resulting model will be a good approximation of the true relationship, using fewer parameters). For the situations when the goal is to use the principal components for modeling, a scree plot can be used as a visual way to judge how many components to extract. A **scree plot** is a graph of the eigenvalues on the y-axis and the corresponding eigenvectors on the x-axis. Figure 16.8 shows an example. Note that the x-axis of the scree plot is sorted in order of the descending eigenvalue, so that the plot has a negative exponential shape, as in figure 16.8.

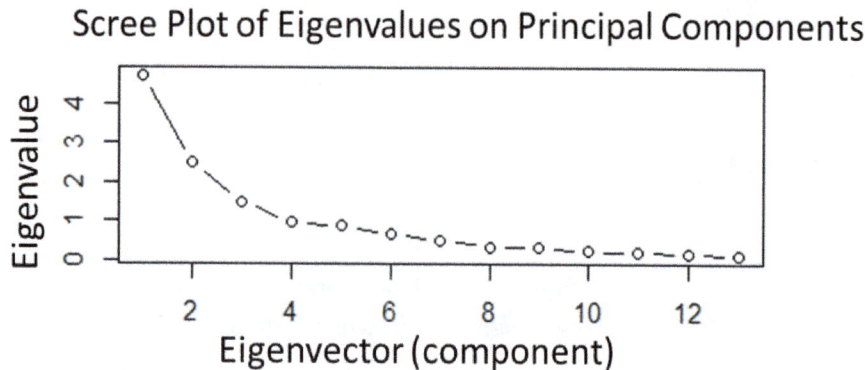

Figure 16.8

Every scree plot has an inflexion point or knee at which the slope of the curve changes dramatically. In figure 16.8, the inflection occurs at component 4. The components to the left of this point can be extracted. The component of the inflexion point itself should not be extracted, and the components to the right of the point should also not be extracted. So if the scree plot looks as it does in figure 16.8, then components 1, 2, and 3 should be extracted, and the rest can be discarded. As an alternative to the scree plot, **Kaiser's Criterion** (keep all components with eigenvalues > 1) or **Jolliffe's Criterion** (keep all components with eigenvalues > 0.7) can be used (Field, Miles, & Field, 2012). Figure 16.9 shows a decision tree to help determine when each criterion could be useful:

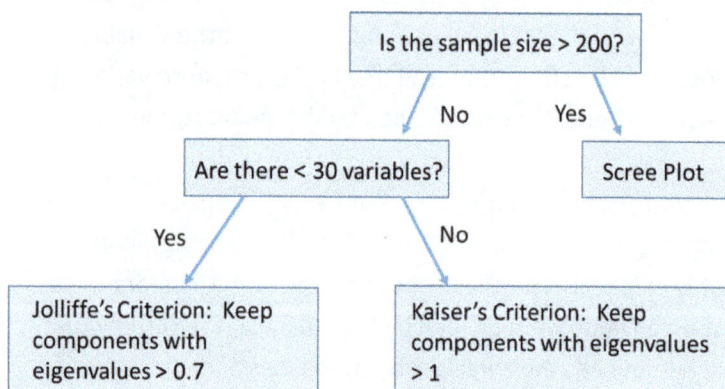

Figure 16.9

For fastidious researchers, **parallel analysis** is an alternative way to determine how many components to extract. Parallel analysis entails performing PCA on a randomized dataset. If the data is truly random, then there will be no underlying principal components and the eigenvalues will all be close to 1 (because none of the variables will be correlated). If the eigenvalues of the principal components of the non-random dataset are higher than the eigenvalues of their corresponding components of the randomized dataset, then they can be extracted.

After PCA is performed, it is often useful to look at the linear combinations of the original features that produced the principal components. To do this, some kind of rotation must be performed on the matrix of loadings. In PCA, the eigenvalues are the loadings for each eigenvector. The eigenvalues cannot be used to determine the linear combination of the original features however, because PCA rotates the coordinate axes behind the scenes. To understand how this works, look at figure 16.10.

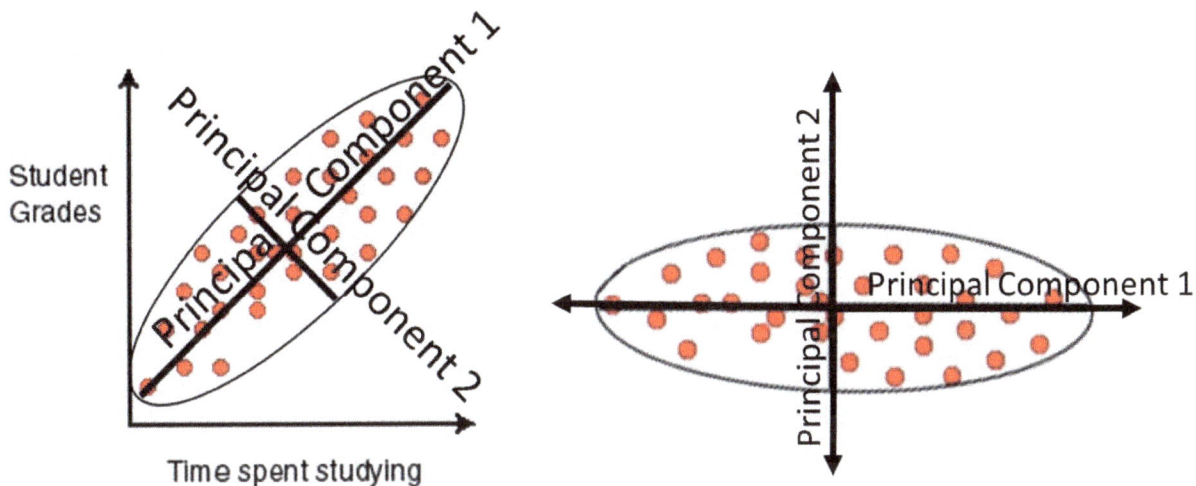

Figure 16.10

The figure on the left of 16.10 shows a 2D feature space with orthogonal eigenvectors. We know that PCA finds the eigenvectors that account for the most variance. Visually, the easiest way to see which vectors account for the most variance is to rotate the axes such that the eigenvectors are perpendicular to one another and parallel to each corresponding axis. The figure on the right of 16.10 shows the rotated coordinate plane with the same eigenvectors as on the left. Notice that the axes no longer represent the original features; they now represent the principal components. For this reason, the loadings of the principal components must be rotated back to the original dimensions in order to see how the original features contributed to the principal components.

There are two types of rotation transformations that can be applied. **Orthogonal rotation**, a.k.a. **varimax rotation**, keeps the components independent. It transforms only the variance of the components by maximizing the sum of the variances on the squared loadings (the squared correlations between variables and components). Varimax rotation tries to associate each variable to only one component. **Oblique rotation** allows components to correlate. The method of choice depends on whether there is any theoretical reason to suspect that the components could be related.

- If the underlying components are not thought to be related, then orthogonal/varimax rotation should be used.
- If the underlying components are suspected to be related, oblique rotation should be used.

Varimax rotation improves the interpretation of components by maximizing the loading of each variable on 1 of the extracted factors, while minimizing the loading on all others. This makes it clear which variables relate to which components. If varimax rotation is performed and the loadings on one or more principal components are nearly uniform, or there are no variables that are clearly loaded more on a particular principal component, then it is likely that the components are correlated and oblique rotation should be used instead.

Since varimax/orthogonal rotation maintains the orthogonality of the axes that are rotated, correlated components may be kept further apart than they should be (Hastie, Tibshirani, & Friedman, 2009). Correlated components should trend together so their axes should be closer together. This is how oblique rotation gets its name. It may be confusing to think of allowing axes to correlate, but remember that the axes of principal components are just vectors.

The principal components found through PCA are latent variables, because they typically represent abstractions, but they can sometimes represent concrete things. For example, suppose we are modeling student grades on number of practice problems solved, time spent studying, and several other variables. If the loadings show that the number of practice problems solved and time spent studying contribute to the first principal component far more than any of the other features, then we can assume that the first principle component likely represents time spent studying. In reality, it might be worth dropping the number of practice problems feature before performing PCA, but it serves the purpose of example here.

PCA is non-parametric, making it a flexible dimension reduction technique, but it does require making a few assumptions. The foremost assumption is that there are linear combination of the features that can closely approximate the underlying relationship. It is possible that a nonlinear combination of the features would better approximate the relationship. Kernel PCA is an extension of PCA that can account for nonlinearity, and we will explore this technique momentarily. Another assumption required for PCA is that the features are normally distributed. The final required assumption is that the features are sufficiently correlated so that the dimensions can actually be reduced through PCA.

The **Kaiser-Meyer-Olkin (KMO) statistic** is a number between 0 and 1 that measures how appropriate PCA is, based on the degree of correlation between the features. The closer the KMO statistic is to 1, the more suitable PCA is, because the variables are sufficiently correlated. However, it may be better to discard variables that are too strongly correlated (>0.8), rather than including them in PCA, because the principal components could end up being just a reflection of the correlated variables. Calculating the determinate can help determine whether variables are too strongly correlated. If the determinate is > 0.00001, then there is no need to get rid of any variables before performing PCA (Field, Miles, & Field, 2012).

Bartlett's test for sphericity is an alternative way to determine if the variables are sufficiently correlated for PCA to be useful. Recall from chapter 13 that sphericity is a condition in which the variances between variables are equal. In chapter 13, we learned that Mauchly's test was used to detect sphericity, but Mauchly's test is used specifically to look for homogeneity of variance across groups of the dependent variable for ANOVA. For PCA, there are no categories for which to compare variances amongst, but there are covariances that must be compared. The needs to compare covariances and check for sphericity also rule out Levene's test and the variance ratio test as candidates to check for sufficient correlation (Raschka, 2015).

If there is evidence to suggest that the relationship between the dependent and independent variables is nonlinear, then kernel PCA should be used (Hastie, Tibshirani, & Friedman, 2009). **Kernel PCA** is an extension of linear PCA that can account for nonlinearity. It works by using a technique called the kernel trick, where data is transformed implicitly through a kernel function. The kernel function can be any nonlinear function. Using this trick, multidimensional data can be separated into categories (or principal components) by the kernel. If we think about the dimensionality of the relationship between variables, the relationship between the dependent variable and one independent variable can be represented in 2D. As more independent variables are added, the representation requires more dimensions. Calculating a boundary that separates data that is related nonlinearly into principal components in a multidimensional space is extremely difficult. Instead, the kernel acts as a similarity function between the data in the multidimensional space and a lower dimensional space where it can be separated by a line. The separating line can be nonlinear, and that is where the power of the kernel is. The mathematics of kernels functions is beyond the scope of this book, but we will implement kernel PCA in the R and Python sections of this chapter, as it is easy to do.

Multiple Correspondence Analysis

Since PCA decomposes either the covariance matrix or the correlation matrix, it requires all of the independent variables to be numeric. Multiple correspondence analysis (MCA) is the categorical variable version of PCA, as it is only applied to categorical independent variables. MCA is computationally intensive, especially if the categorical variables have many levels or if there are a lot of them. MCA can be used on mixed datasets if the numeric variables are binned into groups so that they become categorical. However, if the number of numeric variables is more than a handful (say 5-10), then the computational expense of binning them and then performing MCA may not be worthwhile. Instead, a feature selection method may be preferable (Husson, Le, & Pages, 2017).

Since MCA is simply correspondence analysis applied to 3 or more variables, it is necessary to understand correspondence analysis (CA) in order to understand MCA.

Correspondence analysis decomposes the chi-squared statistic associated with a contingency table of two variables. Recall from chapter 2 that the chi-square distribution approximates the multinomial distribution, making it perfect for dealing with the frequencies of categorical variables. Contingency tables, or cross tabulation tables, were also introduced in chapter 2. Suppose we want to compare students' favorite colors by gender, to look for any patterns in the differences in favorite colors between male and female students. Figure 16.11 shows the contingency table for randomly chosen numbers.

		Favorite Color		
		Green	Blue	Red
Gender	Male	7	39	60
	Female	41	23	9

Figure 16.11

The first step in CA is to calculate the observed proportions of each cell relative to the total number of observations. To do this, each cell is divided by the total observations.

		Favorite Color		
		Green	Blue	Red
Gender	Male	0.039	0.218	0.335

	Female	0.229	0.128	0.05

Figure 16.12

The next step is to create vectors containing the row and column sums. These sums are referred to as masses in correspondence analysis.

[Male, Female] = [0.592, 0.407]

[Green, Blue, Red] = [0.268, 0.346, 0.385]

Next, the expected proportions of each cell can be calculated. The expected proportions are found by multiplying the row sum by the column sum, because we should expect the proportion of each cell to be the row and column percentages. This operation assumes that there is no relationship between the variables. In linear algebra, this operation is called an outer product of arrays.

		Favorite Color		
		Green	Blue	Red
Gender	Male	0.159	0.205	0.228
	Female	0.109	0.141	0.157

Figure 16.13

Now the residuals can be calculated as the differences between the observed and the expected proportions: residual = observed − expected.

		Favorite Color		
		Green	Blue	Red
Gender	Male	-0.12	0.013	0.107
	Female	0.12	-0.013	-0.107

Figure 16.14

At this point, we can see for example that male students tend to like green less frequently than female students, because the negative sign shows an inverse association between the categories. We can also see that female students tend to like the color red less frequently than male students. The residuals can be standardized by dividing them by the expected proportions and multiplying the result by the square root of the expected proportions: standardized_residuals = sqrt(expected_proportions) * (residuals / expected_proportions)

		Favorite Color		
		Green	Blue	Red
Gender	Male	-0.3	0.028	0.224
	Female	0.362	-0.034	-0.27

Figure 16.15

At this point, we could compare the standardized residuals to the critical values of the normal distribution for a given level of significance. For example, since none of the standardized residuals are < -1.96 or > 1.96, we can say none are significant at the 5% level. Therefore, if we were carrying out a hypothesis test to see if favorite color varied by gender, we could not reject the null that favorite colors do not differ between male and female students. But since our goal here is dimension reduction, we can perform SVD on the matrix of standardized residuals. SVD will produce the vector of eigenvalues, d, the matrix of left singular vectors, u, and the matrix of right singular vectors, v.

```
$d
[1] 5.881941e-01 4.874363e-17

$u
              [,1]      [,2]
Male    -0.6386088 0.7695315
Female   0.7695315 0.6386088

$v
              [,1]        [,2]
Green    0.79961917  0.07503535
Blue    -0.07503535  0.99687139
Red     -0.59580104 -0.02484200
```

As we can see from the output above, the matrix of left singular vectors, u, represents the rows of the original matrix, and the matrix of right singular vectors, v, represents the columns. Squaring vector d and expressing the values as proportions of the total gives the percentage of variance that the singular vectors explain. In this example, the first singular vector accounts for 100% of the variance. The second singular vector appears to account for some small percentage of the variance, but it is essentially 0.

```
[1] 1.000000e+00 6.867432e-33
```

The left and right singular vectors can be used to compute the factor scores, which can then be plotted to show how the categories relate to one another, similar to how the principal component scores can be used to plot the observations. Figure 16.16 shows how the categories of the original features are plotted on the new dimensions. As we saw in the table of standardized residuals, female students tend to prefer green more than male students so they are plotted closer together.

CA on Gender and Favorite Color

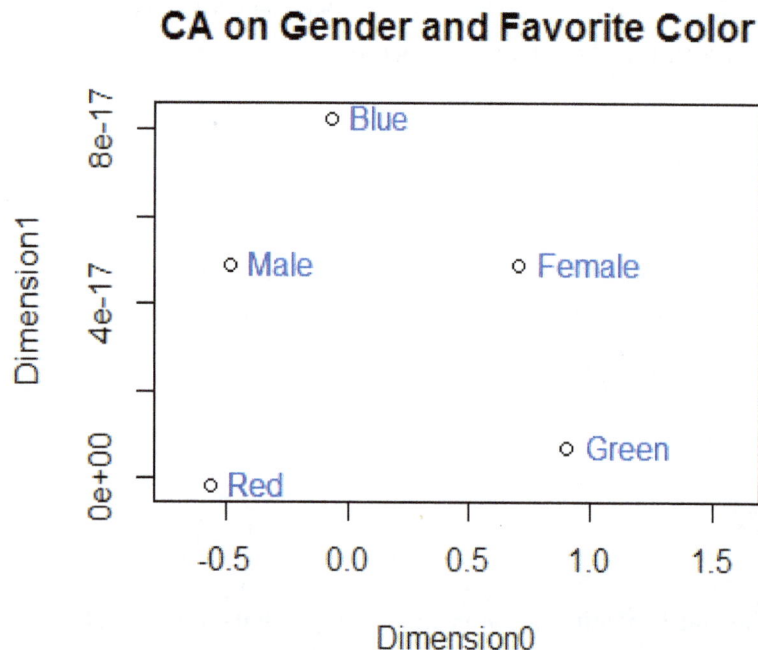

Figure 16.16

Now let us move on to MCA. MCA starts with a Burt Matrix instead of a contingency table. A **Burt Matrix** can be created by one-hot encoding the levels of every categorical variable, and forming a contingency table of the result with itself. This type of encoding is called **crisp encoding**. As an example, let us refer back to our example of students' favorite colors. If we were to represent the data in a Burt Matrix, it would look like figure 16.17.

	Male	Female	Green	Blue	Red
Male	106	0	7	39	60
Female	0	73	41	23	9
Green	7	41	48	0	0
Blue	39	23	0	62	0
Red	60	9	0	0	69

Figure 16.17

Refer back to figure 16.11 to see how the original contingency table compares to the Burt Matrix. The [Male, Male] cell contains the sum of observations who are males. The [Male, Female] cell contains the sum of the observations who are male and female. The [Male, Green] cell contains the sum of the observations who are male and whose favorite color is green. Notice that the Burt Matrix, like a covariance matrix, is square. Also notice that the number of columns in the Burt Matrix is greater than the number of categorical variables K. This presents a problem, because if MCA were carried out exactly the same way as CA, the variances explained by each singular vector would be underestimated (i.e. the singular values are too low). To correct for this underestimation, the singular values <= 1/K are zeroed out, while the ones > 1/K are calculated as in figure 16.18.

$$sv_{correction} = \begin{cases} \left(\dfrac{K}{K-1} * \left(sv - \dfrac{1}{K}\right)\right)^2, & sv > 1/K \\ 0, & sv \leq 1/K \end{cases}$$

Figure 16.18

For example, if we carried out MCA without correcting the singular values, we would find that the first singular vector explains 68% of the variance. Since there are only 2 categorical variables, we should expect the results of MCA to match the results of CA. After applying the correction equation in figure 16.18, the first singular vector is found to explain 98.7% of the variance, which is much closer to the 100% that we found using CA.

To summarize, MCA start with the creation of the Burt Matrix from the categorical independent variables. For the purpose of MCA, all of the independent variables must be categorical. We will get to mixed data shortly. The next step is to compute the standardized residuals, following the same steps as in CA. Then SVD can be applied to the matrix of standardized residuals. The resulting singular values are corrected using the formula in figure 16.18. The factor scores are calculated the same way as in CA.

Factor Analysis

Factor analysis is a technique for identifying groups of related variables. It has three uses:

1. Understand the structure of a set of variables (i.e. Are variables x, y, and z related?)
2. Determine how to measure an underlying variable (i.e. How can an abstract feature like intelligence be measured?)
3. Reduce a dataset to a manageable size while retaining as much of the original info as possible (i.e. How can the dimensions of the feature space be reduced, or how can collinear variables be combined to get rid of multicollinearity?)

In this chapter, we will focus on the third use, but we will see how the first two uses extend from the general procedure.

If a correlation matrix is constructed and there are a few highly correlated pairs, they could all be measuring aspects of the same underlying dimension or factor. A **factor** is a latent variable that represents some linear combination of the original features. This is identical to the definition we used for principal components, and indeed principal components can also be called factors, but factor analysis and PCA are slightly different methods, as we will soon see. The goal of factor analysis is to explain the maximum common variance with the smallest number of variables. Variables that make up a factor are highly correlated with one another, but not correlated strongly with any variables outside of the correlated group. Factors can be described by a linear equation as shown in figure 16.11. Factors are vectors, just like eigenvectors/principal components.

$$Factor_i = b_1 Variable_{1i} + b_2 Variable_{2i} + \cdots + b_n Variable_{ni} + \varepsilon_i$$

Figure 16.19

In the equation in figure 16.19, there is no intercept because factors always intersect at the origin. They are essentially axes, and they can be visualized just like the eigenvectors/principal components were shown in the figure to the right in 16.10.

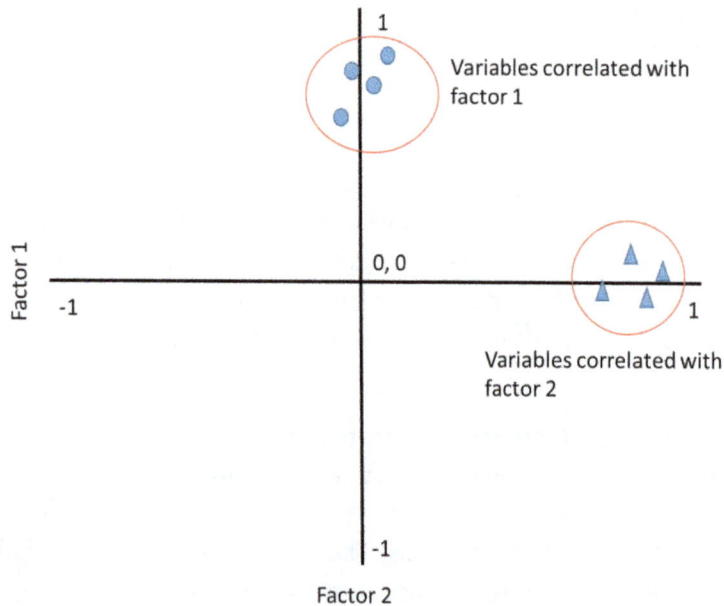

Figure 16.20

Figure 16.20 shows how two factors separate two example features. The features that are correlated with a factor are split so that more of their variances are explained by the factor they are correlated with. When the linear equation for a factor is written, all of the original variables are included, regardless of whether they are strongly correlated with the factor or not. The beta coefficients (loadings) show how much the variable affects the factor. Factor loadings can be represented by a matrix in which the columns are the factors and the rows are the individual loadings. This type of matrix is returned by one of the rotation methods described in the PCA section. The same rotation methods for PCA also apply to factor analysis, and the resulting matrix shows the loadings by factor, just like the loadings by principal components that result from applying rotation to PCA.

$$Factor_1 = b_{1,1}Var1 + b_{2,1}Var2 + b_{3,1}Var3 + b_{4,1}Var4 + b_{5,1}Var5 + b_{6,1}Var6 + \varepsilon$$

$$Factor_2 = b_{1,2}Var1 + b_{2,2}Var2 + b_{3,2}Var3 + b_{4,2}Var4 + b_{5,2}Var5 + b_{6,2}Var6 + \varepsilon$$

$$A = \begin{bmatrix} b_{1,1} & b_{1,2} \\ b_{2,1} & b_{2,2} \\ b_{3,1} & b_{3,2} \\ b_{4,1} & b_{4,2} \\ b_{5,1} & b_{5,2} \\ b_{6,1} & b_{6,2} \end{bmatrix}$$

Figure 16.21

In figure 16.21, the columns of matrix A represent factors. The beta coefficients are the loadings. To get the **factor score** of a particular observation, the linear equations are solved with the loadings substituted for the beta coefficients times the values of their associated variables for the observation. For the purpose of dimension reduction, the factors can replace the original features, and the factor scores can replace the values of the original features. For the second use case listed at the beginning of this section, latent variables are the factors, so if an abstract variable needs to be measured, a factor can be used to represent it.

Factor analysis and PCA are often confused so let us think about how they are different. PCA seeks to find the linear combination of features that maximizes the total variance explained by the linear combination. Factor analysis seeks to find the linear combination of features that maximize the variance that is shared between the features. In a sense, PCA is like factor analysis when all of the variance is assumed to be shared among the features. PCA is preferable to use when the goal is to reduce the dimensions of a model to underlying latent features that explain the most variance (the principal components). Factor analysis is preferable to use when the goal is to find latent factors that explain the observed features, but these factors can also be used to produce a model with fewer dimensions.

Factor Analysis of Mixed Data

When the independent variables are mixed, meaning there are both numeric and categorical variables, in the features set, then neither PCA nor MCA can be used. Factor analysis of mixed data (FAMD) combines PCA and MCA to facilitate dimension reduction on mixed datasets (Husson, Le, & Pages, 2017). The general idea is that the variables are split by type. Categorical variables are crisp encoded and the matrix of standardized residuals is created. The covariance matrix is created for the numeric variables. Then PCA is performed on the complete set of variables. FAMD is computationally intensive, especially if the categorical variables have many categories or if there are a lot of them. For situations where the computational expense of FAMD is not practical, feature selections methods may be preferable.

Linear Discriminant Analysis

Recall from chapter 15 that discriminant function analysis (DFA) can be used for classification or dimension reduction. Chapter 15 promised that dimension reduction would be explored later. Now we will look at linear discriminant analysis (LDA) for the purpose of dimension reduction.

A key difference between LDA and the other dimension reduction techniques in this chapter is that LDA reduces the feature space to a linear combination of features that maximize the separation between classes of the dependent variable (Raschka, 2014). This means that LDA can only be used for dimension reduction when the goal is classification of the target variable. LDA cannot be used for dimension reduction for regression tasks where the goal is to predict a continuous outcome. It is important to note that although LDA seeks to maximize class separation with linear discriminants, it is not guaranteed to be better than PCA, and in fact it is common to see PCA performed for dimension reduction before LDA is used as a classifier (Raschka, 2014). Figure 16.22 illustrates the difference between PCA and LDA.

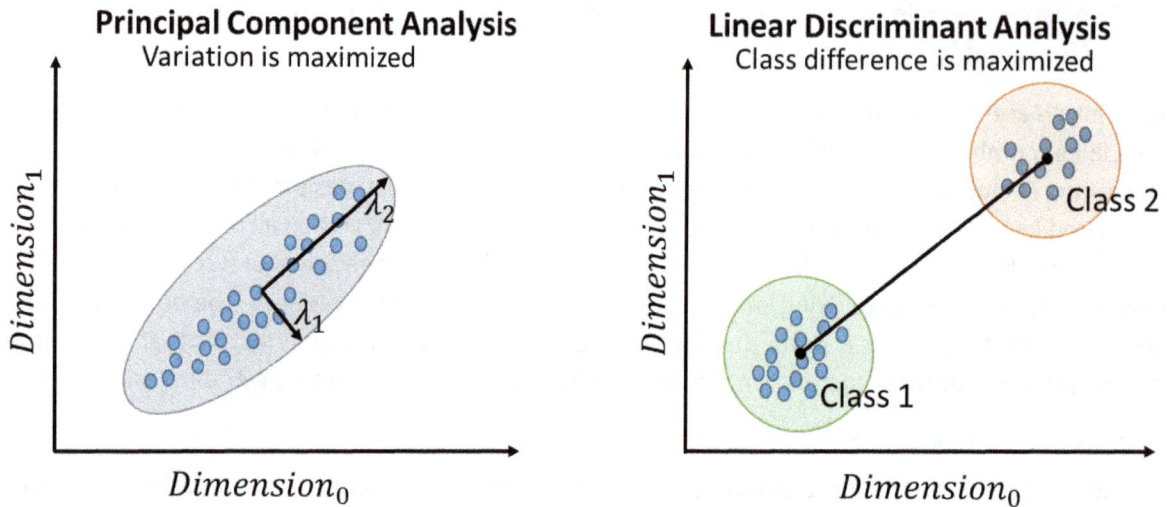

Principal Component Analysis
Variation is maximized

Linear Discriminant Analysis
Class difference is maximized

Figure 16.22

LDA for dimension reduction is carried out by first computing either the scatter matrix or the covariance matrix, and then using either eigendecomposition or SVD to find the linear combinations that maximize the differences between classes. Since the covariance matrix is the normalized version of the scatter matrix, the only difference between decomposing the covariance matrix, versus the scatter matrix, is that the eigenvalues for the decomposition of the covariance matrix are scaled by some constant. This should make sense because normalizing the scatter matrix just scales it to account for different numbers of observations in the classes. The scatter matrix is the quotient of the between class scatter matrix and the within class scatter matrix, as shown in figure 16.23.

$$S = \frac{S_{between\ class}}{S_{within\ class}}$$

Figure 16.23

Recall from chapter 15 that LDA can be carried out using Bayes' discriminant rule or Fisher's discriminant rule. For classification, we used Bayes' rule. For dimension reduction however, we need to use Fisher's rule because we need a matrix to decompose into linear discriminants. The equation in figure 16.23 represents Fisher's discriminant rule. The within-class scatter matrix is computed using the equation in figure 16.24.

$$S_{within\ class} = \sum_{i=1}^{c} \left(\sum_{x \in D_i}^{n} \left(x - \frac{\sum_{x \in D_i}^{n} x}{n_i} \right) \left(x - \frac{\sum_{x \in D_i}^{n} x}{n_i} \right)^T \right)$$

Figure 16.24

In the equation in figure 16.24, c is the number of classes, n is the number of observations in class i, T indicates the transpose of a matrix, and x is the value of an observation. To break the equation down, starting with the inner summation, the fraction that is subtracted from x is the mean vector of the observations in a class. Essentially, the numerator sums the values of the observations in class i, and this sum is divided by the number of observations in class i to get the mean. The difference between each observation and the mean of its class is multiplied by its transpose, and these products for each class are totaled. Therefore, the inner summation represents the class scatter matrix. The outer sum totals the class scatter matrix for every class, resulting in the within class scatter matrix.

The between class scatter matrix is computed using the equation in figure 16.25.

$$S_{between\ class} = \sum_{i=1}^{c} n_i \left(\frac{\sum_{x \in D_i}^{n} x}{n_i} - \frac{\sum x}{n} \right) \left(\frac{\sum_{x \in D_i}^{n} x}{n_i} - \frac{\sum x}{n} \right)^{T}$$

Figure 16.25

In the equation in figure 16.25, the overall mean is subtracted from the class mean. This is multiplied by its transpose and the number of observations in a class, and the result is totaled for all of the classes.

Once the scatter matrix or covariance matrix has been created, it can be decomposed. The researcher can then choose to extract the k eigenvectors with the largest eigenvalues, just like PCA and factor analysis. The loadings and scores are computed in the same way as in PCA and factor analysis.

Recall from chapter 15 that LDA requires making a few assumptions. For the purpose of dimension reduction however, these assumptions can be relaxed. So even if they are violated, LDA can still be used for dimension reduction (Raschka, 2014).

Dimension Reduction with R

In the last chapter, LDA and QDA could not be completed because there was collinearity in the feature set. Since the feature set for the adult dataset is so large (including the dummies), it would be too tedious to manually inspect every correlation pair. Dimension reduction can help though. Since there are both numeric and categorical features in the dataset, the only options for dimension reduction are to use FAMD, dummy encode and use LDA, or use a feature selection method. In this section, we will apply FAMD and LDA. We will also apply PCA to the numeric variables only, just to show how PCA is carried out, because PCA is extremely useful for many problems. Finally, we will return to classification with LDA and QDA, using the reduced dataset.

Let us start with PCA. For comparison, we will work with 2 versions of the dataset: the first version will not dummy encode anything so that categorical variables are left as they are, and the second version will dummy encode categorical variables. So the code below sets up the datasets using some of the cleaning code from the last chapter. At the end of this code block, the numeric variables are separated for PCA.

```
d <- read.table("adult.data", header=F, sep=",")
colnames(d) <- c('age', 'workclass', 'fnlwgt', 'education', 'education_nbr',
                 'marital_status', 'occupation', 'relationship', 'race',
                 'sex', 'capital_gain', 'capital_loss', 'hours_per_week',
                 'native_country', 'salary_bin')
str(d)
head(d)


#Convert variables to desired data types
d$age <- as.numeric(d$age)
d$fnlwgt <- as.numeric(d$fnlwgt)
d$education_nbr <- as.factor(d$education_nbr)
```

```
d$capital_gain <- as.numeric(d$capital_gain)
d$capital_loss <- as.numeric(d$capital_loss)
d$hours_per_week <- as.numeric(d$hours_per_week)

#Since there are 32k rows and <2k rows with nulls, it is safe to discard them
d <- d[(d$workclass!=' ?' & d$occupation!=' ?' & d$native_country!=' ?'),]

#Save the version of the dataset without dummies
d_no_dummies <- d

#Dummy encode categorical variables
d[,which(sapply(d, class)=='factor')] <- sapply(d[,which(sapply(d,
class)=='factor')], substring, 2, 100)  #Gets rid of leading whitespaces that show up
in factor levels
d[,which(sapply(d, class)=='character')] <- lapply(d[,which(sapply(d,
class)=='character')], as.factor)
library(caret)
cols_to_encode <- c('workclass', 'education', 'marital_status', 'occupation',
'relationship', 'race', 'sex', 'native_country')
for (c in cols_to_encode) {
  dummies <- dummyVars(paste0(" ~ ", c), data=d, fullRank=T)
  dummyenc <- data.frame(predict(dummies, newdata=d))
  d <- cbind(d, dummyenc)
}
colnames(d) <- gsub("[.]", "_", colnames(d))
str(d)
#Get rid of redundant columns now that they have been dummy encoded
d$workclass <- NULL
d$education <- NULL
d$education_nbr <- as.numeric(d$education_nbr)
d$marital_status <- NULL
d$occupation <- NULL
d$relationship <- NULL
d$race <- NULL
d$sex <- NULL
d$native_country <- NULL

#Split data into training and test sets
set.seed(14)
train_indices <- createDataPartition(y=d$salary_bin, p=0.7, list=F)
train <- d[train_indices,]
test <- d[-train_indices,]
train_no_dummies <- d_no_dummies[train_indices,]
test_no_dummies <- d_no_dummies[-train_indices,]
train_numeric <- train_no_dummies[,which(sapply(train_no_dummies, class)=='numeric')]
test_numeric <- test_no_dummies[,which(sapply(test_no_dummies, class)=='numeric')]
```

Before PCA is carried out, it is a good idea to do a couple tests to see how useful PCA would be. Bartlett's test can be used to determine whether the correlation matrix of the numeric variables is an identity matrix. If the correlation matrix is an identity matrix (the null hypothesis of Bartlett's test is that it is not), then the matrix can be factorized or decomposed. This is the most important test to perform before attempting PCA, because if the matrix cannot be factorized, it cannot be decomposed. The second test, the KMO statistic, shows whether there is significant correlation among the numeric variables. If there is, then it is likely that the variables share variance, and the greater the shared variance, the more dimensions can be reduced. The third test checks the determinant of the correlation matrix. If the determinant is small, then dimension reduction would be beneficial.

```
#Barlett's test to see if a correlation matrix is an identity matrix
corm <- cor(train_numeric)
library(psych)
cortest.bartlett(corm, n=nrow(train_numeric))
#Low p-value indicates that the correlation matrix is factorable so dimension
reduction can be done

#Create function to find KMO statistic
kmo = function( data ){
  library(MASS)
  X <- cor(as.matrix(data))
  iX <- ginv(X)
  S2 <- diag(diag((iX^-1)))
  AIS <- S2%*%iX%*%S2                       # anti-image covariance matrix
  IS <- X+AIS-2*S2                          # image covariance matrix
  Dai <- sqrt(diag(diag(AIS)))
  IR <- ginv(Dai)%*%IS%*%ginv(Dai)          # image correlation matrix
  AIR <- ginv(Dai)%*%AIS%*%ginv(Dai)        # anti-image correlation matrix
  a <- apply((AIR - diag(diag(AIR)))^2, 2, sum)
  AA <- sum(a)
  b <- apply((X - diag(nrow(X)))^2, 2, sum)
  BB <- sum(b)
  MSA <- b/(b+a)                            # indiv. measures of sampling adequacy
  AIR <- AIR-diag(nrow(AIR))+diag(MSA)  # Examine the anti-image of the correlation
matrix. That is the  negative of the partial correlations, partialling out all other
variables.
  kmo <- BB/(AA+BB)                         # overall KMO statistic
  # Reporting the conclusion
  if (kmo >= 0.00 && kmo < 0.50){test <- 'The KMO test yields a degree of common
variance unacceptable for FA.'}
  else if (kmo >= 0.50 && kmo < 0.60){test <- 'The KMO test yields a poor degree of
common variance of >= 0.5 to < 0.6.'}
  else if (kmo >= 0.60 && kmo < 0.70){test <- 'The KMO test yields a cautionary
degree of common variance of >= 0.6 to < 0.7'}
  else if (kmo >= 0.70 && kmo < 0.80){test <- 'The KMO test yields a satisfactory
degree of common variance of >= 0.7 to < 0.8' }
  else if (kmo >= 0.80 && kmo < 0.90){test <- 'The KMO test yields a high degree of
common variance of >= 0.8 to < 0.9' }
  else { test <- 'The KMO test yields a superbly high degree of common variance.' }

  ans <- list( overall = kmo,
               report = test,
               individual = MSA,
               AIS = AIS,
               AIR = AIR )
  return(ans)
}

#Perform the KMO test to get the KMO statistic
kmo(corm)$overall
kmo(corm)$report
```

```
#Find the determinant of the correlation matrix to see if variable elimination would
be a good idea
det(corm) > 0.00001
#Determinant is sufficiently high so there is no need to eliminate variables
```

Since the p-value of Bartlett's test is significant, the correlation matrix can be factorized. Since the KMO statistic is rather small, it is unlikely that dimension reduction will be helpful. This is no surprise though, for this dataset. We are only looking at a few numeric variables that have little correlation with one another. Any latent features that can be found by factorizing the correlation matrix may only be reflections of the variables themselves. Nevertheless, we will continue for the purpose of example. The determinant is high, suggesting that dimension reduction is not needed, but again, we will persist. These tests tell us that the numeric variables in the adult dataset were most definitely not to blame for the collinearity that was encountered in the last chapter. That means it was likely a combination of one or more of the dummies, or perhaps a dummy and a numeric variable.

Now we can perform PCA, starting off by extracting all of the features and doing no factor rotation.

```
pc1 <- principal(corm, nfactors=length(corm[,1]), rotate="none")
pc1
```

```
#Draw a scree plot
plot(pc1$values, type="b")
```

All 5 of the principal components have loadings > 0.7 (Jolliffe's criterion). The loadings are shown by the SS loadings in the output. This implies that all 5 principal components should be extracted. But since the dataset has > 200 observations, the scree plot was also produced. The scree plot of the SS loadings (the eigenvalues) seems to level off at the 4th component, so let us try extracting 3 principal components.

```
pc2 <- principal(corm, nfactors=3, rotate="none")
pc2
```

```
#Validate PCA with reproduced correlation matrix
factor.model(pc2$loadings)
factor.residuals(corm, pc2$loadings)
#Check how many large residuals there are
residuals <- factor.residuals(corm, pc2$loadings)
residuals <- as.matrix(residuals[upper.tri(residuals)])
sum(abs(residuals) > 0.05) / nrow(residuals)
sqrt(mean(residuals^2))
```

There is a large percentage of residuals, and the mean squared error is 0.19, which is very high (MSE < 0.08 is a good rule of thumb for a good MSE). The large error implies that too little variance is explained by the principal components, meaning we need to extract more. Let us try 4 principal components.

```
pc3 <- principal(corm, nfactors=4, rotate="none")
pc3
```

```
#Check how many large residuals there are
residuals <- factor.residuals(corm, pc3$loadings)
residuals <- as.matrix(residuals[upper.tri(residuals)])
sum(abs(residuals) > 0.05) / nrow(residuals)
sqrt(mean(residuals^2))
```

That is better, but the error is still high. If we add the 5th component though, our feature set will be the same size as the original. There would be no point in dimension reduction. We saw in the initial tests that dimension reduction would probably not be worthwhile for these features, and we are confirming it here. Let us proceed anyway. We will use oblique rotation to get the linear combinations of the original features that produced the principal components. Oblique rotation is preferred to orthogonal rotation here, because there is reason to believe capital gain and capital loss are dependent on one another. A simple scatter plot of these variables shows that if one of them is positive, then the other must be 0: a capital gain means no loss was incurred and vice versa.

```
#Using oblique rotation to examine the factor loadings for each variable
pc4 <- principal(corm, nfactors=4, rotate="oblimin")
print.psych(pc4, sort=TRUE)
#Remove loadings below 0.5 to more clearly see the variables in their factor groups
print.psych(pc4, cut=0.5, sort=TRUE)
```

As the output shows, the first principal component appears to be a linear combination of age and hours per week. The second principal component is a combination of capital gain and capital loss, and the opposite signs indicate that as one increases, the other decreases. We know for a fact that as one increases, the other must be zero, so at least the PCA is on track, even if it is a simplification. The third and fourth principal components are just mirroring two of the original variables. So the principal components are not much better at explaining the variance in the data than the original variables, for this particular dataset.

Supposing PCA was useful for this dataset, let's go ahead and calculate the principal component scores for the test dataset, just as we would if we wanted to use the principal components to make classifications or do a regression. Be sure to use the predict function from the psych package here, and provide the original data that the PCA was fitted to (the data, not the correlation matrix) as the third argument. This ensures that the scores are scaled the same way in both datasets.

```
test_numeric_pcs <- predict.psych(pc4, test_numeric, train_numeric)
```

From here, the principal component scores would be used instead of the original variables for classification or regression.

Now let's move on to FAMD, using the datasets without the dummies. We will start by extracting all of the variables.

```
library(FactoMineR)
famd1 <- FAMD(train_no_dummies[,-15], ncp=ncol(train_no_dummies[,-15]), graph=FALSE)
famd1
```

```
#View the eigenvalues of the latent features (the loadings)
library(factoextra)
famd1_eigenvalues <- get_eigenvalue(famd1)
famd1_eigenvalues
```

```
#Scree plot
fviz_screeplot(famd1)
```

There are no latent features that explain more than 4% of the variance. The scree plot appears to have a knee at feature 5, so let's extract 4 features.

```
#FAMD with 4 extracted features
famd2 <- FAMD(train_no_dummies[,-15], ncp=4, graph=FALSE)

#View the combinations of the original features that produced the latent features
var <- get_famd_var(famd2)
var
var$contrib

#Plot the features in the coordinate plane of the first two latent features
fviz_famd_var(famd2, repel=TRUE)

#Compute FAMD scores for new data - use predict.FAMD from the FactoMineR package
test_no_dummies_lvs <- predict.FAMD(famd2, test_no_dummies[,-15])
```

The plot of the original features on the coordinate plane of the first two latent features (the first two explain the most variance because they are sorted) shows how similar the original features are to one another. It is like a feature cluster plot. The last line predicts the latent features for new data. Now we can use replace the original 14 independent variables with the 4 extracted latent features and retry LDA and QDA. Note that the scores (the values of each observation for the 4 latent features) are stored in the coord element of the FAMD results.

```
train_famd2_res <- famd2$ind
train_reduced <- cbind(salary_bin=train_no_dummies[,15],
as.data.frame(train_famd2_res$coord))
test_reduced <- cbind(salary_bin=test_no_dummies[,15],
as.data.frame(test_no_dummies_lvs$coord))
colnames(test_reduced) <- gsub(" ", ".", colnames(test_reduced))

library(MASS)
lda_model <- lda(formula=salary_bin ~ ., data=train_reduced)
lda_model

#Make predictions for test set
post_lda <- predict(lda_model, test_reduced[,-1])
summary(post_lda)
test_reduced$pred <- post_lda$class
test_reduced$baseline <- '<=50K'   #Baseline assigns everything to majority class
test_reduced$baseline <- factor(test_reduced$baseline, levels=c('<=50K', '>50K'))

#Evaluate the model using the log loss, AUC, and accuracy
library(MLmetrics)
LogLoss(as.integer(test_reduced$pred), as.integer(test_reduced$salary_bin))
AUC(as.integer(test_reduced$pred), as.integer(test_reduced$salary_bin))
Accuracy(as.factor(test_reduced$pred), test_reduced$salary_bin)
Accuracy(as.factor(test_reduced$baseline), test_reduced$salary_bin)
#View the confusion matrix with predicted values on the left
prop.table(table(test_reduced$pred, test_reduced$salary_bin))

qda_model <- qda(formula=salary_bin ~ ., data=train_reduced)
qda_model

#Make predictions for test set
post_qda <- predict(qda_model, test_reduced[,-1])
summary(post_qda)
```

```
test_reduced$qda_pred <- post_qda$class

#Evaluate the model using the log loss, AUC, and accuracy
LogLoss(as.integer(test_reduced$qda_pred), as.integer(test_reduced$salary_bin))
AUC(as.integer(test_reduced$qda_pred), as.integer(test_reduced$salary_bin))
Accuracy(as.factor(test_reduced$qda_pred), test_reduced$salary_bin)
Accuracy(as.factor(test_reduced$baseline), test_reduced$salary_bin)
#View the confusion matrix with predicted values on the left

prop.table(table(test_reduced$qda_pred, test_reduced$salary_bin))
```

The collinearity error has disappeared! LDA shows a classification accuracy of roughly 80%. So by using 4 latent features instead of the original 14, the accuracy only dropped by 3%. This shows the power of dimension reduction. QDA did not do so well though, with an accuracy of 73%, which is worse than the baseline. Since QDA did worse than LDA, the stronger assumptions for LDA must have been worthwhile. QDA trades robustness for statistical power.

The last example for this chapter is to show how LDA can also be used for dimension reduction, instead of just classification. To do this, we will run the LDA function on the dataset without dummies. Note that the LDA function at most computes 1 less linear discriminant than the number of classes. So for 2 classes, only 1 linear discriminant is calculated.

```
#Run LDA for dimension reduction, ignoring the multicollinearity
lda_dim_reduced <- lda(formula=salary_bin ~ ., data=train_no_dummies)

#Explore LDA results from the model that was just built
lda_dim_reduced$prior     #Class prior probabilities
lda_dim_reduced$means     #Class specific means for each predictor
lda_dim_reduced$scaling   #The factor loadings of the predictors on the latent
variables (linear discriminants)
lda_dim_reduced$svd       #Singular values: the ratio of the between group to the
within group standard deviations of the linear discriminants
perc_var_explained <- lda_dim_reduced$svd^2/sum(lda_dim_reduced$svd^2)
perc_var_explained
#Plot the distribution over the linear discriminant (only 1 LD was computed)
plot(lda_dim_reduced)
#Plot the features by their loadings on the linear discriminant
plot(lda_dim_reduced$scaling)
```

After the LDA model is built, the predict function is required to calculate each observation's score for the linear discriminant(s) found.

```
#Calculate the values of each observation using the linear discriminant
train_lda_reduced <- as.data.frame(predict(lda_dim_reduced, train_no_dummies))
head(train_lda_reduced$LD1)  #The observation values of the linear discriminant
```

The resulting values for the linear discriminant can be used in a model where the linear discriminant is the only predictor.

Dimension Reduction with Python

In the last chapter, LDA and QDA could not be completed because there was collinearity in the feature set. Since the feature set for the adult dataset is so large (including the dummies), it would be too tedious to manually inspect every correlation pair. Dimension reduction can help though. Since there are both numeric and categorical features in the dataset, the only options for dimension reduction are to use FAMD, dummy encode and use LDA, or use a feature selection method. In this section, we will apply LDA only, as there is no FAMD capability for Python. We will also apply PCA and factor analysis to the numeric variables.

Let us start with PCA. For comparison, we will work with 2 versions of the dataset: the first version will not dummy encode anything so that categorical variables are left as they are, and the second version will dummy encode categorical variables. So the code below sets up the datasets using some of the cleaning code from the last chapter. At the end of this code block, the numeric variables are separated for PCA.

```python
import pandas as pd
import numpy as np
from sklearn.preprocessing import LabelEncoder, OneHotEncoder, StandardScaler
from sklearn.model_selection import train_test_split
from sklearn.decomposition import PCA, FactorAnalysis
import matplotlib.pyplot as plt
from sklearn.discriminant_analysis import LinearDiscriminantAnalysis

#Set seed for repeatability
seed = 14
np.random.seed(seed)

#Load Iris data and specify dependent variable
d = pd.read_csv('C:/Users/Nick/Documents/Word Documents/Data Science Books/DSILT
Stats Code/12-16 Titanic and Adult Salaries/adult.data',
                names=['age', 'workclass', 'fnlwgt', 'education', 'education_nbr',
'marital_status', 'occupation', 'relationship', 'race',
                        'sex', 'capital_gain', 'capital_loss', 'hours_per_week',
'native_country', 'salary_bin'])
print(d.info())
print(d.head())

#Fixes a bug in printing output from IDLE, it may not be needed on all machines
import sys
sys.__stdout__ = sys.stdout

######
#Data Cleaning

#Since there are 32k rows and <2k rows with nulls, it is safe to discard them
d = d.dropna()
d = d[(d != '?').all(1)]
d = d[(d != ' ?').all(1)]

#Convert variables to desired data types
d[['age', 'fnlwgt', 'capital_gain', 'capital_loss',
'hours_per_week']].apply(pd.to_numeric)
encoder = LabelEncoder()
categorical_vars = ['workclass', 'marital_status', 'occupation', 'relationship',
                    'race', 'sex', 'native_country', 'salary_bin']
categorical_var_mapping = dict()
for cv in categorical_vars:
```

```
    d[cv] = d[cv].str[1:]                     #Gets rid of the leading white spaces in
the text
    d[cv] = encoder.fit_transform(d[cv])  #Encodes as integer
    categorical_var_mapping[cv] = list(encoder.classes_)  #Saves integer to category
mapping

#Get rid of education because it is already encoded as integer in 'education_nbr'
d.drop(['education'], axis=1, inplace=True)

#Save the version of the dataset without dummies
d_no_dummies = d.copy()

print(d.columns)
#Dummy encode categorical variables except for the target (salary_bin)
#Note that pd.get_dummies automatically removes the originals after dummy encoding
d = pd.get_dummies(d, columns=categorical_vars[:6], drop_first=True)
print(d.columns)
print(d.info())

#Perform z-score standardization on the numeric features
#Standardization is only needed if the matrix of the data will be decomposed instead
of the covariance matrix
d[['age', 'fnlwgt', 'education_nbr', 'capital_gain', 'capital_loss',
'hours_per_week']] = StandardScaler().fit_transform(d[['age', 'fnlwgt',
'education_nbr', 'capital_gain', 'capital_loss', 'hours_per_week']])
d_no_dummies[['age', 'fnlwgt', 'education_nbr', 'capital_gain', 'capital_loss',
'hours_per_week']] = StandardScaler().fit_transform(d_no_dummies[['age', 'fnlwgt',
'education_nbr', 'capital_gain', 'capital_loss', 'hours_per_week']])

#Convert the dataframe into numpy array and specify dependent variable
x = d.drop(['salary_bin'], axis=1).values.astype(float)
y = d[['salary_bin']].values
x_nd = d_no_dummies.drop(['salary_bin'], axis=1).values.astype(float)
x_numeric = d_no_dummies[['age', 'fnlwgt', 'education_nbr', 'capital_gain',
'capital_loss', 'hours_per_week']].values.astype(float)

#Split data into training and test sets - be sure to stratify since this is for
classification
x_train, x_test, y_train, y_test = train_test_split(x, y, test_size=0.3, stratify=y,
random_state=seed)
x_nd_train, x_nd_test, y_nd_train, y_nd_test = train_test_split(x_nd, y,
test_size=0.3, stratify=y, random_state=seed)
xn_train, xn_test, yn_train, yn_test = train_test_split(x_numeric, y, test_size=0.3,
stratify=y, random_state=seed)
```

Before PCA is carried out, it is a good idea to do a couple tests to see how useful PCA would be. When we did this with R, we applied Bartlett's test, calculated the KMO statistic, and checked the determinant of the correlation matrix. We will not do any of those tests here, because we are going to use the PCA function from scikit-learn, which uses SVD instead of eigendecomposition. Recall that SVD works regardless of the shape of the matrix, so Bartlett's test is unnecessary. We could look to see if there are significant correlations in the predictors, but the results of PCA will show that anyway, as we saw in the last section. Let's start by extracting all of the principal components and checking the scree plot to see where the knee is.

```
pca_model = PCA(n_components=None, whiten=False, random_state=seed)
pca_dim = pca_model.fit_transform(xn_train)

#Plot first 2 extracted features and the observation class
plt.figure(figsize=(10, 5))
plt.xlabel('Latent Variable 1 (explains most variance)')
plt.ylabel('Latent Variable 2 (explains 2nd most variance)')
plt.title('PCA 2-Dimension Plot with Observation Class')
plt.scatter(pca_dim[:, 0], pca_dim[:, 1], c=yn_train.ravel())
plt.colorbar()
plt.show()

#Get percentage of variance explained (eigenvalue) by each latent variable
(component)
var_explained = pca_model.explained_variance_ratio_
#Calculate cumulative variance explained by each latent variable (component)
cum_var_explained = np.cumsum(np.round(var_explained, decimals=4)*100)
#Plot cumulative explained variance to see how many components should be extracted
plt.plot(cum_var_explained)
plt.xlabel('Latent Variable')
plt.ylabel('Cumulative Variance Explained')
plt.title('Cumulative Variance Explained by Latent Variables')
plt.xticks(np.arange(0, len(var_explained), 1), (np.arange(0, len(var_explained),
1)+1))
plt.show()
```

The scree plot has no discernable knee. So how do we know how many components to extract? One option is to choose our own extraction method. For example, it may be useful to only extract enough principal components to explain 80% of the variance. By choosing some threshold, like 80%, a simple loop can be used to determine how many components are required to reach it.

```
#Choose extracted components based on graph or based on some desired variance
threshold in loop below (default 80%)
#Note that for datasets with many dimensions, the fewer components extracted the
better, even if less variance is explained
var_explained_thresh = 80.0
for idx, cumvar in enumerate(cum_var_explained):
    if (cumvar >= var_explained_thresh):
        pca_extracted_components = idx+1
        break
#pca_extracted_components = 2
pca_features = pca_dim[:, 0:pca_extracted_components]
print('PCA Number of Extracted Features:', pca_features.shape[1])
```

The final PCA model extracts 2 components. The plots show that there is not much benefit to reducing the numeric variables in this dataset. From here, the principal components can be substituted for the original dimensions to build models. The predict function can be used to calculate the principal component scores for new data. Note that scikit-learn has a KernelPCA function that runs the same way as the PCA function. Kernel PCA is the nonlinear version of PCA.

Factor analysis is carried out very similarly to PCA. We will start by extracting all of the factors, and then use the variance explained threshold to pick how many factors to extract for the final model.

```
#Use 'randomized' svd_method if lapack is too slow or dataset is large (randomized
sacrifices a little accuracy)
fact_model = FactorAnalysis(n_components=None, svd_method='lapack',
random_state=seed)
fact_dim = fact_model.fit_transform(xn_train)

#Plot first 2 extracted features and the observation class
plt.figure(figsize=(10, 5))
plt.xlabel('Latent Variable 1 (explains most variance)')
plt.ylabel('Latent Variable 2 (explains 2nd most variance)')
plt.title('Factor Analysis 2-Dimension Plot with Observation Class')
plt.scatter(fact_dim[:, 0], fact_dim[:, 1], c=yn_train.ravel())
plt.colorbar()
plt.show()

#Get percentage of variance explained (eigenvalue) by each latent variable
(component)
#Note - this equation can be found here:
http://stackoverflow.com/questions/29611842/scikit-learn-kernel-pca-explained-
variance
var_explained = np.var(fact_dim, axis=0)/np.sum(np.var(fact_dim, axis=0))
#Calculate cumulative variance explained by each latent variable (component)
cum_var_explained = np.cumsum(np.round(var_explained, decimals=4)*100)
#Plot cumulative explained variance to see how many components should be extracted
plt.plot(cum_var_explained)
plt.xlabel('Latent Variable')
plt.ylabel('Cumulative Variance Explained')
plt.title('Cumulative Variance Explained by Latent Variables')
plt.xticks(np.arange(0, len(var_explained), 1), (np.arange(0, len(var_explained),
1)+1))
plt.show()

#Choose extracted components based on graph or based on some desired variance
threshold in loop below (default 80%)
#Note that for datasets with many dimensions, the fewer components extracted the
better, even if less variance is explained
var_explained_thresh = 80.0
for idx, cumvar in enumerate(cum_var_explained):
    if (cumvar >= var_explained_thresh):
        fact_extracted_components = idx+1
        break
#fact_extracted_components = 2
fact_features = fact_dim[:, 0:fact_extracted_components]
print('Factor Analysis Number of Extracted Features:', fact_features.shape[1])
```

The results of factor analysis are very similar to PCA, and there is not much benefit for this dataset. From here, the factors can be substituted for the original dimensions to build models. The predict function can be used to calculate the factor scores for new data.

The last example for this chapter is to show how LDA can also be used for dimension reduction, instead of just classification. To do this, we will run the LDA function on the dataset without dummies. Note that the LDA function at most computes 1 less linear discriminant than the number of classes. So for 2 classes, only 1 linear discriminant is calculated.

```
#Run LDA for dimension reduction, ignoring the multicollinearity and keeping n
components < the number of original features
#Note if n_components=None, then all of them are kept
lda = LinearDiscriminantAnalysis(n_components=None, solver='svd')
lda_dim_reduced = lda.fit_transform(x_nd_train, y_nd_train.ravel())


#Explore LDA results from the model that was just built
print('Priors:', lda.priors_)          #Class prior probabilities
print('Class means:', lda.means_)   #Class specific means for each predictor
print('Loadings:', lda.coef_)          #The factor loadings of the predictors on the
latent variables (linear discriminants)
print('Explained variance:', lda.explained_variance_ratio_)
print('Mean classification accuracy:', lda.score(x_nd_train, y_nd_train))
#Plot the features by their loadings on the linear discriminant
import seaborn as sns
sns.barplot(lda.coef_[0], d_no_dummies.drop(['salary_bin'], axis=1).columns)
plt.xlabel('Coefficient Loading')
plt.title('Feature Loadings on the First Linear Discriminant')
plt.show()
```

The feature loadings on a linear discriminant are comparable to the loadings on principal components or factors: they show how the features linearly combine to form a latent variable. After the LDA model is built, the predict function can calculate the linear discriminant scores for new data. The resulting values for the linear discriminant can be used in a model where the linear discriminant is the only predictor.

17. Nonlinear Dimension Reduction and Manifold Learning

In the last chapter, we saw how the feature space of a model can be reduced through linear dimension reduction methods. These methods used matrix decomposition to reduce the feature space to a linear combination of the independent variables. Sometimes however there is a nonlinear combination of the independent variables that better models the dependent variable. We will explore nonlinear regression in a later chapter. The difference between regression and modeling the dependent variable on the scores resulting from dimension reduction is that the model produced by dimension reduction consists of latent features instead of the original variables.

Recall from chapter 16 that each feature extends the dimensionality of the data. When data exists in a high dimensional space, it is said to be contained or embedded within a manifold of that space. A **manifold** is a high dimensional space that resembles Euclidean space near each point in the space. **Euclidean space** is either the standard 2D plane that we are used to seeing with x-y coordinates, or it can be a 3D space with x-y-z coordinates. Space with 4 or more dimensions is said to be non-Euclidean. The reason 2D and 3D spaces are called Euclidean is because those were the spaces studied by the Greek mathematician Euclid of Alexandria, who created the postulates that formed the basis for Euclidiean geometry (the geometry that students learn in high school).

So if an n-dimensional manifold, or n-manifold for short, is locally Euclidean around a point or neighborhood of points, intuition should suggest that functions that are locally linear (and globally nonlinear) can model the manifold. The term **embedding** is used to mean that subspaces of dimensions are contained within a higher dimensional manifold. Essentially, it is a fancy way of referring to the locally linear subspaces that exist within a globally nonlinear n-manifold. As an analogy, consider the Earth: it is spherical in shape, but from our limited perspective, it appears flat in every direction. If we want to model a flight path from Los Angeles to San Diego, a straight line does a fairly good job, because the Earth is locally flat over a short distance. If we want to model a flight path from Los Angeles to Sydney however, we need to account for the global curvature of the Earth. We saw in chapter 16 how a kernel function can extend PCA to deal with nonlinearity in the data. The basic idea is that a high dimensional space can be mapped to a lower dimensional space using a similarity function called a kernel. We will see similar methods in this chapter.

The algorithms described in this chapter are often used as a precursor to clustering (grouping data based on similar features), because the lower dimensional mappings they produce naturally group observations together based on how similar their features are in a higher dimensional space. It is important to distinguish that these algorithms are not natively clustering algorithms though, and should not naively be used for clustering. This is because many of these algorithms only preserve local structure, and clustering requires relating the distances between all observations in the global structure. Like the feature extraction methods described in chapter 16 (PCA, MCA, FAMD, and LDA), these nonlinear dimension reduction methods are essentially unsupervised learning algorithms.

Multidimensional Scaling

Multidimensional scaling (MDS) is a way to use the similarities between observations in a high dimensional space to map them to a lower dimensional space (Glen, 2018). MDS works because points that are similar are located near one another in the space of an n-manifold, and points that are spatially close together will remain that way even if the dimensions are reduced. The similarity between observations can be measured by the Euclidean distance between them. A matrix of distances between every two pairs of observations can therefore be called a similarity matrix (or more commonly, a dissimilarity matrix).

Euclidean Distance: $d = \sqrt{(x_2 - x_1)^2 + (y_2 - y_1)^2}$

Figure 17.1

There are two variants of the MDS algorithm: distance scaling and classical scaling. **Classical scaling** involves minimizing a cost function called the **strain function** using eigendecomposition. The classical scaling method is omitted from this book because it produces linear combinations of the independent variables that can be shown to be the equivalent of performing PCA on the dissimilarity matrix. Modern libraries that include functions for MDS, like scikit-learn, use the distance scaling approach (Sklearn User Guide, 2018). The distance scaling approach produces a nonlinear combination of the independent variables, because it entails minimizing a cost function called the **stress function** using an iterative optimization method, such as gradient descent.

The distance scaling MDS algorithm starts by assigning each observation random coordinates in the n-dimensional space. Using these points, the Euclidean distances between every 2 pairs of points are calculated, resulting in a symmetrical dissimilarity matrix. The dissimilarity matrix is compared to the original input matrix via the stress function. Once the stress has been calculated, the coordinates of the observations are adjusted in order to reduce the stress function. This process is repeated with every iteration, as described by the optimization algorithms in chapter 7, until the stress is minimized. The stress function calculates the residual sum of squares from the input distance matrix.

$$min_t Stress_D(t * x_1, ..., t * x_N) = \sqrt{1 - \frac{\left(\sum_{i\neq j=1}^{N} D_{i,j} * \|x_i - x_j\|\right)^2}{\sum_{i\neq j=1}^{N} D_{i,j}^2 * \sum_{i\neq j=1}^{N} \|x_i - x_j\|^2}}$$

Figure 17.2

In the equation in figure 17.2, N is the number of observations, $D_{i,j}$ is the dissimilarity matrix of the input data, t is a scaling factor that makes the stress invariant to scale, and x sub i − x sub j represents the differences between 2 observations in every dimension. The ratio that is subtracted from 1 in the equation is also known as the cosine similarity. **Cosine similarity** is just another way of measuring similarity, much like Euclidean distance. It is defined as the angle between two non-zero vectors, and this angle can be found by computing the inner product of the vectors.

MDS can be performed on the actual values of the dissimilarity matrix, which is referred to as **metric scaling**, or it can be performed on the ranks of the values of the dissimilarity matrix, which is referred to a **nonmetric scaling**. Nonmetric scaling sacrifices information about the magnitude of the distances in order to become more robust to data that do not have an identifiable distribution. Recall from chapters 5 and 14 that ranks were used to calculate nonparametric statistics. The same concept is applied in nonmetric scaling. Nonmetric scaling is preferable to metric scaling when there are missing pairwise distances between observations (such as when observations are missing values) or when the data contains categorical or mixed features.

A scree plot similar to the plot used to determine how many principal components to extract from PCA can be used to determine how many reduced dimensions are appropriate. Refer back to figure 16.8 in chapter 16 for an example. The scree plot for MDS should have stress on the y-axis and the number of dimensions on the x-axis, and the knee will show how many dimensions to reduce the data to. Remember that the point at which the knee occurs should not be included, so if the knee occurs at 3 dimensions, then MDS should reduce the data to 2 dimensions. If there is no discernable knee, then MDS may not be appropriate at all; the number of dimensions of the data may not be able to be reduced (Hastie, Tibshirani, & Friedman, 2009).

After MDS has been applied, the resulting dimensions can be plotted to show how observations are similar to one another in terms of their features. While similar observations will appear to cluster together, the orientation of the points should not be of concern as the axes of an MDS plot are arbitrary.

Isomap

Think back to the analogy of plotting a flight path between two cities over the Earth. If the cities are close, then a straight line is not a bad model, but as the distance between the cities grows, the curvature of the Earth deteriorates the fit of a straight line. The distance between points over a curved space can be calculated using **geodesic distance**. Geodesics applies to curved space in any dimension, not just three. In fact, Euclidean distance is a special case of geodesic distance for two dimensions. The geodesic between two points is a synonym for the shortest path between two points in curved multidimensional space. If we apply this knowledge to the familiar 2D space, we can say that the geodesic between two points is the straight line connecting the points. MDS used Euclidean distance to calculate the similarity between points. Isomap uses the geodesic distance. This is the only difference between the two algorithms.

Isomap stands for isometric mapping (Tenenbaum, de Silva, & Langfor, 2000). Isometric means "of equal value", and if we look at topographic maps like the one in figure 17.3, we can see isolines that demarcate elevations of equal value. The closer these isolines are, the steeper the slope of the terrain represented by the topographic map. Although the isolines in figure 17.3 represent elevation above sea level, the isolines in an n-manifold represent the n features of a dataset. Figure 17.3 was produced by the U.S. Geological Survey.

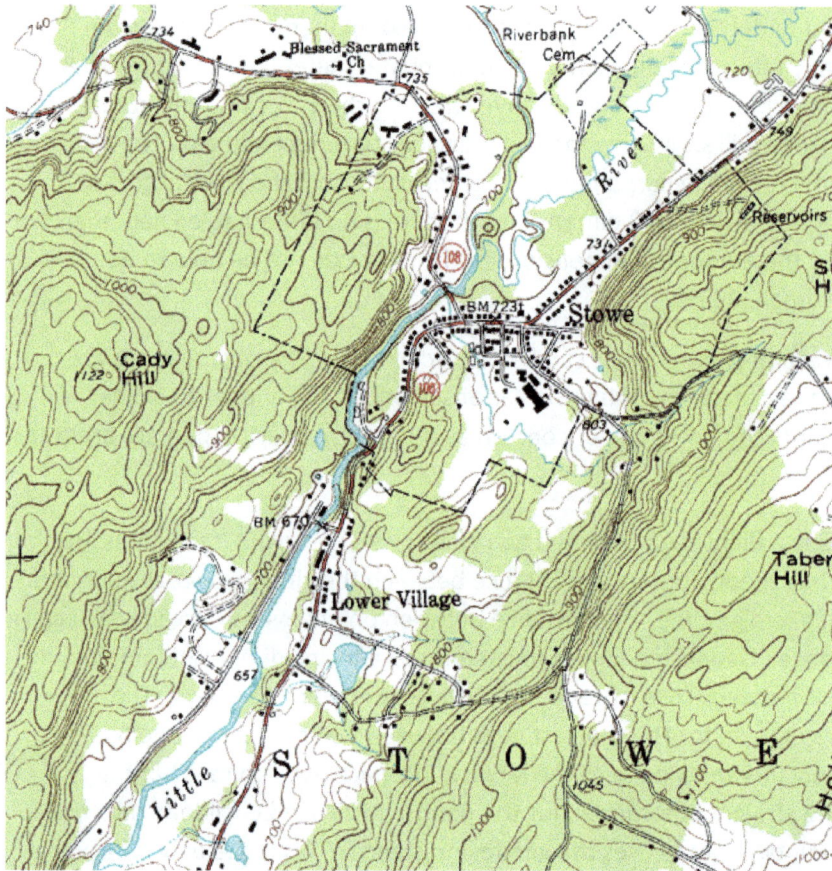

Figure 17.3

Points in an n-manifold can be envisioned as having isolines around them that show regions of space with the same value. The regions form neighborhoods of connected points that can be represented in a graph like the one to the right in figure 17.4. Note that figure 17.4 omits the connections between points in different neighborhoods for the purpose of visual clarity, but the points should all be connected to one another.

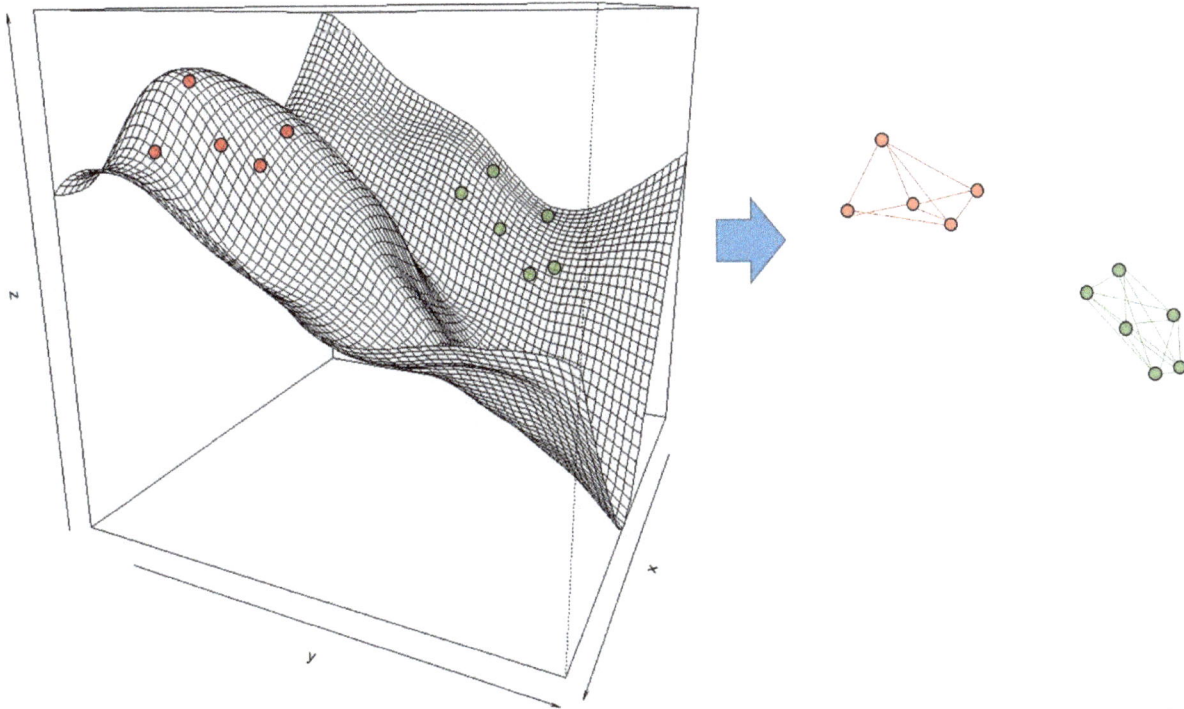

Figure 17.4

In multidimensional space, there are isolines in every dimension. Observations that share a value for a particular feature are on the same isoline for that feature. Therefore, if the geodesic distance between two observations is calculated for every dimension, it is possible to determine how similar the observations are, in terms of all of their features. The isomap algorithm can be carried out like so:

1. Choose a neighborhood size, k. Find the k-nearest neighbors for each point (the k-nearest neighbors algorithm will be explained in a later chapter).
2. Use the Euclidean distance to weight the connections of each point with its k-nearest neighbors.
3. Since each point is only connected to k-nearest neighbors, in order to compute the distance between every two pairs of points, a graph traversal algorithm is needed. We will take an introductory look at graph theory in a later chapter, but the idea is that a path between two points that are not directly connected can be formed by a chain of connections of their neighbors, and their neighbors' neighbors, and so on. The length of the path between two points is the sum of the weighted connections along the path. Figure 17.5 visualizes the concept of a graph path, and the connection weight is the Euclidean distance. Notice how the points that exist in multiple neighborhoods make graph traversal possible. The shortest path between every two pairs of nodes can be found by using a shortest path algorithm. Two common choices for the shortest path algorithm are **Dijkstra's algorithm** and the **Floyd-Warshall algorithm**. It is beyond the scope of this book to describe these algorithms, as they are in the realm of network science.

Path Between Points in Different Neighborhoods, K=2

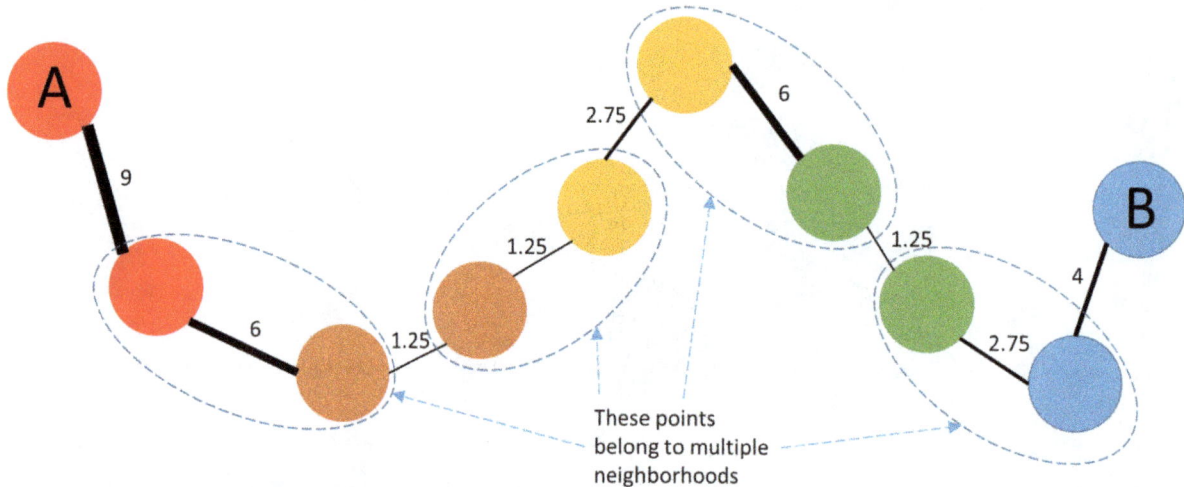

These points belong to multiple neighborhoods

Figure 17.5

In summary, the dissimilarity matrix is formed by using the distance between points as the similarity measure. For points in the same neighborhood, Euclidean distance is used. For points in different neighborhoods, distance is defined by the shortest path between the points. The shortest path between two points is the minimum of the weighted sums of connections between the points. The shortest paths are referred to as geodesic distances.

Once the dissimilarity matrix has been formed, isomap proceeds in the same way as MDS using classical scaling (or PCA): the matrix is decomposed using eigendecomposition, and the top n eigenvectors are used to calculate the coordinates in the lower dimensional space. Just like MDS, isomap can use metric or nonmetric scaling of the distance matrix. Also just like MDS, the reduced dimensions can be plotted and observations that are more similar will appear closer together.

Isomap is sensitive to the choice of the neighborhood size, k. If k is too large, then the neighborhood will consist of points that are not really too similar, but if it is too small, then the neighborhood will leave out points that are similar.

Locally Linear Embedding

Locally linear embedding (LLE) is very similar to isomap, except that instead of using Euclidean distance to weight the connections of each point's k-nearest neighbors, the local weights are determined by the linear combination of the neighbors that best represent each point (Roweis & Lawrence, 2000). LLE can be thought of as using PCA to determine local weights (hence the name locally linear), and stitching together the local PCA's to find the optimal global nonlinear embedding. The local weights determined by PCA produce a linear combination of neighbors to model each point. If the original point were reconstructed from this model, there would be some error due to information loss. This error is modeled by the top equation in figure 17.6.

$$\varepsilon = \sum_{jk} w_j w_k C_{jk}$$

$$C_{jk} = (x - \eta_j)(x - \eta_k)$$

Figure 17.6

In the equations in figure 17.6, w_j is the vector of reconstruction weights (the weights of the linear model for point k), w_k is the vector of original weights for point k, and C_{jk} is the local covariance matrix of nearest neighbors η.

The top equation in figure 17.6 is the cost function for the local embedding part of the LLE algorithm. Since this function is simply the squared error between the model and the ground truth, the optimization method used to minimize it is least squares (OLS). However, the optimization is subject to the constraint that the weights all sum to 1. Therefore, the cost function in figure 17.6 can be minimized by the optimal weights resulting from the top equation in figure 17.7, which can be solved using the Lagrange multiplier (lm) subject to the constraint that the weights sum to 1. It is often more efficient to use the bottom equation in figure 17.7 though, as solving for the weights and rescaling them so that they sum to 1 is equivalent to solving the top equation with the Lagrange multiplier.

$$w_j = \frac{\sum_j C_{jk}^{-1}}{\sum_{lm} C_{lm}^{-1}}$$

$$\sum_j C_{jk} w_k = 1$$

Figure 17.7

When there are more neighbors than dimensions (when k > the number of features), the covariance matrix in the equations in figures 17.6 and 17.7 is nearly singular. This would inflate the weights of the local model, so to correct for this, a regularization parameter r is used to penalize larger weights. We will explore regularization in a later chapter, but the basic idea is that regularization prevents overfitting. The most common parameter used for LLE regularization is Spearman's rho, which was explained in chapter 5. Recall from chapter 5 that Spearman's rank correlation coefficient is a nonparametric correlation coefficient for ranked data. Due to LLE's dependence on regularization, the algorithm may produce distorted embedding if the manifold dimension is larger than 1. Therefore, several methods have been proposed to improve LLE. We will explore one of these methods, modified locally linear embedding, in the next section. LLE is not a useless algorithm though. It can still work very well with the right choices of k and r, and if the data has many features so that k is never larger than the number of input dimensions, then there is no risk of distorted embedding. That is why LLE works particularly well with image data, since images have many input dimensions (each pixel can be thought of as a dimension).

Now that we have optimized the locally linear part of LLE, we can move on to optimizing the global nonlinear embedding. The globally optimal nonlinear embedding can be found by minimizing the cost function in figure 17.8.

$$\Phi(Y) = \sum_i \left| Y_i - \sum_j w_{ij} Y_j \right|^2$$

Figure 17.8

In the equation in figure 17.8, Y represents the lower dimensional vector that is derived from the application of PCA to the higher dimensional vector X (the linear combinations that best represent each observation), i is observation i, j is observation j, and w is the weight between observations i and j.

Once the dissimilarity matrix has been decomposed, the top n-eigenvectors are used to calculate the coordinates in the lower dimensional space. Everything else about this algorithm is the same as described for MDS and isomap.

The greatest benefit to using LLE instead of PCA or MDS is that LLE preserves the neighborhood structure in the context of the global n-manifold. To visualize this, consider reducing the dimensions of an image. The pixels in an image form neighborhood structures. For example, the image in figure 17.9 shows a cat named Walter. Walter's face is near the center of the image, and the neighborhood of pixels that make up his face are arranged in a pattern that resembles the face of a cat, regardless of where the pixels could be translated to. For example, if the pixels were moved to the far right of the image, they would still look the same. PCA and MDS would scatter the information of these pixels throughout the lower dimensional representation of the image. This is not a bad thing, necessarily, it just means the local structure is lost in the lower dimensional representation if the data contains nonlinearity. LLE preserves the local structure of the pixels. This property of LLE makes it better able to represent local structures in lower dimensional space than either PCA or MDS, as long as there are nonlinear structures in the data.

Figure 17.9

LLE is used for facial recognition, the recognition of different configurations of peoples' mouths, the recognition of distinct gait cycles of people walking, hand gesture recognition, and MRI interpretation.

However, due to the problems that arise when k > the number of input dimensions, modifications to the algorithm are often useful.

Modified Locally Linear Embedding

Modified locally linear embedding, or MLLE, uses multiple weight vectors for each point in the manifold, rather than only one weight vector like LLE (Hastie, Tibshirani, & Friedman, 2009). Aside from the modified cost function shown in figure 17.10, MLLE is the same as LLE. Notice that the modified cost function forms a d-dimensional local embedding, meaning each dimension has its own local embedding, t.

$$E(T) = \sum_{i=1}^{N} \sum_{l=1}^{s_i} \left\| t_i - \sum_j w_{ij}^l t_j \right\|^2$$

Figure 17.10

In the equation in figure 17.10, T represents all local embeddings t, N is the number of observations, and s represents all of the singular values of the decomposition of the independent weight vectors w_{ij}^l.

There are other solutions to the local LLE weight matrix problem, such as Hessian Eigenmap Locally Linear Embedding (HLLE), but MLLE is the most computationally efficient. The embedding produced by HLLE is nearly identical to the one produced by MLLE.

Stochastic Neighbor Embedding

So far we have looked at the similarity between observations in an n-manifold in terms of distance. Another way we could look at similarity is in terms of probability. The stochastic neighbor embedding (SNE) algorithm involves converting the Euclidean distances between points in a manifold to conditional probabilities that represent similarities (van der Maaten & Hinton, 2008). Probability will be explained in a later chapter, so this algorithm may not make complete sense until that chapter is presented. Nevertheless, chapter 1 presented the concepts of probability distribution and probability density functions for continuous variables, and those concepts are the foundation of SNE.

The conditional probability of event B occurring, given event A has occurred is shown in the equation in figure 17.11.

$$P(B|A) = \frac{P(A \text{ and } B)}{P(A)}$$

Figure 17.11

The numerator is the probability that both A and B occur, and the denominator is the probability of event A. In the case of SNE, event B is the circumstance of observation x_j being a neighbor to x_i, and event A is x_i. So for SNE, we are interested in determining the conditional probability that x_i would pick x_j as its neighbor, if neighbors were picked from a Gaussian (normal) probability density function centered at x_i. Figure 17.12 visualizes this Gaussian probability density function.

Gaussian Distribution Centered at x_i

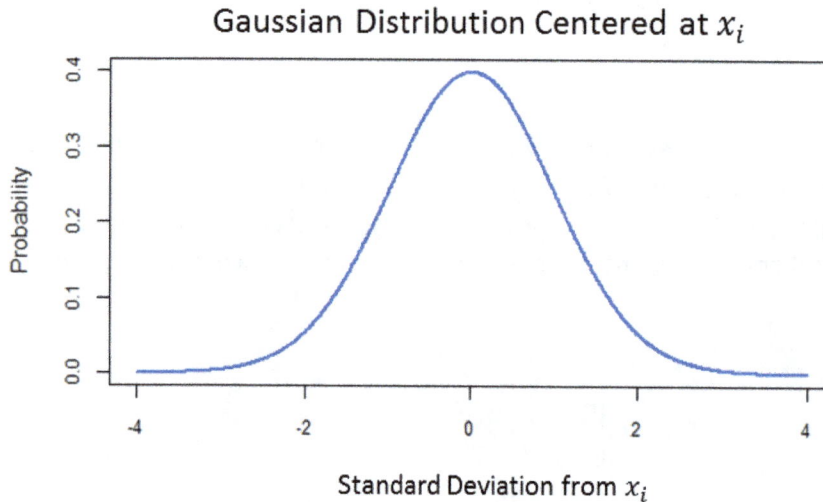

Figure 17.12

The equation in figure 17.13 shows how the similarity of points x_i and x_j is expressed in terms of the conditional probability that x_i would pick x_j as its neighbor, if neighbors were picked from a Gaussian (normal) probability density function centered at x_i. The numerator represents P(x_i and x_j), and the denominator represents P(x_i), just like in the equation in figure 17.11.

$$p(x_j|x_i) = \frac{e^{\left(-\|x_i-x_j\|^2/2\sigma_i^2\right)}}{\sum_{k\neq i} e^{\left(-\|x_i-x_k\|^2/2\sigma_i^2\right)}}$$

Figure 17.13

In the equation in figure 17.13, e is Euler's number and σ_i^2 is the variance of the distribution centered at x_i. This equation shows how points x_i and x_j are similar in the high dimensional space of a manifold. This similarity serves as a frame of reference for the lower dimensional similarity between the transformed points y_i and y_j. The equation in figure 17.14 shows how the similarity of the transformed points y_i and y_j is expressed in terms of the conditional probability that y_i would pick y_j as its neighbor, if neighbors were picked from a Gaussian (normal) probability density function centered at y_i.

$$q(y_j|y_i) = \frac{e^{\left(-\|y_i-y_j\|^2\right)}}{\sum_{k\neq i} e^{\left(-\|y_i-y_k\|^2\right)}}$$

Figure 17.14

The similarity in the higher dimensional space must be compared to the similarity in the lower dimensional space, because a good lower dimensional representation would make q = p. This comparison metric is the cost function to be minimized by an optimization algorithm. For SNE, the cost function is the sum of Kullback-Leibler divergences over all points in the manifold, as shown in figure 17.15. For reference, the **Kullback-Leibler divergence** measures how one probability distribution diverges from a second, expected probability distribution.

$$C = \sum_i \sum_j p(x_j|x_i) \log \frac{p(x_j|x_i)}{q(y_j|y_i)}$$

Figure 17.15

SNE uses a gradient descent algorithm to minimize the cost function in figure 17.15. Since the Kullback-Leibler divergence is not symmetric, the errors are weighted such that representations of points that use distant neighbors are penalized and representations of points that use closer neighbors are rewarded. In this way, SNE maintains local structures in the data mapping, much like LLE.

In order to solve the equations presented so far, the variance of the Gaussian probability density function centered at x_i must be known. Determining this value is not straightforward, because it depends on the density of points in the manifold. Different regions will have different point densities. Dense regions should have smaller variance, whereas sparser regions should have larger variance. Therefore, the variance of the Gaussian distribution of x_i comes from a probability distribution P_i. The optimal variance from distribution P_i is found by using the binary search algorithm, and gradient descent is used to minimize a cost function that holds P_i to a constant perplexity. Perplexity is derived from the concept of entropy from information theory, which is beyond the scope of this book. For the purpose of SNE, perplexity can be thought of as a way to choose the number of neighbors for each point. Perplexity is a hyperparameter that is set manually. Typical values are between 5 and 50 (Wattenberg, Viegas, & Johnson, 2016). It is important to remember that perplexity is not the number of neighbors, but a parameter used in an equation that is optimized to find the number of neighbors.

t-Distributed Stochastic Neighbor Embedding

SNE's optimization is nontrivial, as it requires hyperparameters for perplexity, and each of its internal gradient descent algorithms. Its sensitivity to these hyperparameters make the results it produces unstable (they could change every time the algorithm is run). SNE is also computationally expensive. The t-SNE algorithm attempts to solve these problems. T-SNE works by using a symmetrized version of the SNE cost function, and a t-distribution instead of the Gaussian distribution to represent the similarities between points. Also, for the lower dimensional representation, the t-distribution used to model the similarities has heavy tails that make it much easier to optimize than SNE (van der Maaten & Hinton, 2008).

The cost function for t-SNE shown in figure 17.16.

$$C = \sum_i \sum_j p(x_j x_i) \log \frac{p(x_j x_i)}{q(y_j y_i)}$$

Figure 17.16

The difference between the equations in figurers 17.16 and 17.5 is that 17.16 uses the joint probability distribution for both the high and low dimensional spaces. This makes the Kullback-Leibler divergence symmetrical. For reference, joint probability is the probability of both events A and B occurring. It can be found by the equation in figure 17.17. We will explore joint probability in a later chapter.

$$P(A \ and \ B) = P(A) * P(B|A)$$

Figure 17.17

The t-SNE cost function is minimized by gradient descent. As previously mentioned, the lower dimensional representation uses a heavy tailed t-distributed to model the similarities. Since t-SNE is also using the joint distribution instead of the conditional probability, the equation for the lower dimension is different than in SNE. The equation in figure 17.18 shows how the similarity of the transformed points y_i and y_j is expressed in terms of the joint probability that y_i would pick y_j as its neighbor, if neighbors were picked from a t-distributed probability density function centered at y_i.

$$q(y_i y_j) = \frac{\left(1 + \|y_i - y_j\|^2\right)^{-1}}{\sum_{k \neq l}(1 + \|y_k - y_l\|^2)^{-1}}$$

Figure 17.18

A convenient property of the t-distribution is that the number in figure 17.18 approaches an inverse square law for large distance in the low dimensional map. This makes the mapping nearly invariant to changes in the scale of the map for distant points in the manifold. In general, t-SNE penalizes dissimilar points more heavily than SNE, but it also introduces long range attractional forces that can pull points back together than were incorrectly separated early in the gradient descent optimization process. These properties also make t-SNE good at revealing structures with multiple scales (van der Maaten & Hinton, 2008).

The drawbacks to using t-SNE are that it is computationally expensive, certain implementations are limited to 2-3 lower dimensional embeddings, and global structure is not explicitly preserved unless the points are initialized with PCA (van der Maaten & Hinton, 2008).

Like SNE, the hyperparameters for t-SNE are perplexity and the parameters for gradient descent, such as the learning rate. T-SNE is much less sensitive to the choice of perplexity than SNE, although the results of t-SNE still can vary widely. Larger datasets typically warrant larger perplexity, and common values are between 5 and 50 (van der Maaten & Hinton, 2008).

An important point to make regarding the cluster sizes in t-SNE plots is that they are meaningless. T-SNE scales cluster sizes so that they are even. The distances between clusters is also meaningless, because t-SNE scales the distances in the global geometry of the lower dimensional representation. Therefore, clusters that are farther apart from one another are not necessarily more dissimilar than clusters that are closer together (Wattenberg, Viegas, & Johnson, 2016). These reasons are why t-SNE should not be used for clustering. T-SNE aims to represent similar data points on a lower dimensional embedding. It does not aim to separate points into clusters that maximize the differences between the clusters. Since t-SNE does not preserve global distance structure, distance based clustering methods should never be used after t-SNE, as the results will be erroneous. In general, mixing clustering and t-SNE in any way should be avoided.

Nonlinear Dimension Reduction with Python

Nonlinear dimension reduction is an area of weakness for R. It is much easier to do nonlinear dimension reduction in Python, so all of the examples in this section will be in Python. For this chapter, we will be using the Iris flower dataset, which shows the petal lengths, petal widths, sepal lengths, and sepal widths of 3 different species of Iris flowers. The dataset comes built into R and can be saved to a CSV using R's write.csv function, or it can be downloaded from the UCI Machine Learning repository.

```
import pandas as pd
import numpy as np
from sklearn.preprocessing import LabelEncoder, MinMaxScaler, StandardScaler
from sklearn.manifold import MDS, Isomap, LocallyLinearEmbedding, TSNE
import matplotlib.pyplot as plt

#Set seed for repeatability
seed = 14
np.random.seed(seed)

#Load Iris data and specify dependent variable
alldata = pd.read_csv('C:/Users/Nick/Documents/Word Documents/Data Science
Books/DSILT Stats Code/17-20 and 22 Iris and Motor Trends Cars/iris.csv',
header=True)
print(alldata.info())
print(alldata.head())

#Split data into x and y, and format as numpy arrays
array = alldata.values
x = array[:, 0:4].astype(float)
y = array[:, 4]
#Convert the classes (species) into integers
encoder = LabelEncoder()
encoder.fit(y)
y = encoder.transform(y)

#Scale every feature using min-max normalization and z-score standardization
#Note that y is not scaled because it is categorical, but if it weren't, it would
have to be scaled too
x_norm = MinMaxScaler().fit_transform(x)
x_std = StandardScaler().fit_transform(x)
```

The Iris dataset does not need any data cleaning, so we can jump right into MDS. The code below shows how to apply both metric and non-metric MDS using scikit-learn.

```
#Apply metric MDS, keeping n components < the number of original features
#kernel choices: linear (default), poly (polynomial of degree=degree), rbf, sigmoid,
cosine
mds_model = MDS(n_components=2, metric=True, random_state=seed)
mds_model.fit_transform(x_std)
print(mds_model.get_params())
mds_dim = mds_model.embedding_
print(mds_dim.shape)   #There should be 2 latent variables represented
print('Stress:', mds_model.stress_)

#Plot first 2 extracted features and the observation class
plt.figure(figsize=(10, 5))
plt.xlabel('Latent Variable 1')
plt.ylabel('Latent Variable 2')
plt.title('Metric MDS 2-Dimension Plot with Observation Class')
plt.scatter(mds_dim[:, 0], mds_dim[:, 1], c=y)
plt.colorbar()
plt.show()

#Apply non-metric MDS, keeping n components < the number of original features
```

```
#kernel choices: linear (default), poly (polynomial of degree=degree), rbf, sigmoid,
cosine
mds_model = MDS(n_components=2, metric=False, random_state=seed)
mds_model.fit_transform(x_std)
print(mds_model.get_params())
mds_dim = mds_model.embedding_
print(mds_dim.shape)   #There should be 2 latent variables represented
print('Stress:', mds_model.stress_)

#Plot first 2 extracted features and the observation class
plt.figure(figsize=(10, 5))
plt.xlabel('Latent Variable 1')
plt.ylabel('Latent Variable 2')
plt.title('Nonmetric MDS 2-Dimension Plot with Observation Class')
plt.scatter(mds_dim[:, 0], mds_dim[:, 1], c=y)
plt.colorbar()
plt.show()

#Apply metric MDS for many different choices of dimensions
#Limitation is that nbr dimensions must be < the number of original features
nbr_dim = range(3)
mds_dim_nbr = []
mds_stress_results = []
for nd in nbr_dim:
    mds_model = MDS(n_components=nd+1, metric=True, random_state=seed)
    mds_model.fit_transform(x_std)
    mds_dim_nbr.append(nd+1)
    mds_stress_results.append(mds_model.stress_)
mds_results = {'nbr_dim': mds_dim_nbr, 'stress': mds_stress_results}
mds_results = pd.DataFrame.from_dict(mds_results)
#See which number of dimensions has the lowest stress
plt.plot(mds_results['nbr_dim'], mds_results['stress'])
plt.xlabel('Number of Latent Dimensions')
plt.ylabel('Stress')
plt.title('Scree Plot of Stress by Number of Latent Variables')
plt.show()

#Build MDS model with one less than the number of dimensions at the kneee of the
scree plot
#In this case, that would be 1, so the index is substituted for the second latent
variable
mds_model = MDS(n_components=1, metric=True, random_state=seed)
mds_model.fit_transform(x_std)
mds_dim = mds_model.embedding_
plt.figure(figsize=(10, 5))
plt.xlabel('Index')
plt.ylabel('Latent Variable 1')
plt.title('Metric MDS 1-Dimension Plot with Observation Class')
plt.scatter(mds_dim[:, 0], list(alldata.index), c=y)
plt.colorbar()
plt.show()
```

Notice how well the first dimension of the metric MDS plot separates the species of Iris flowers. This is remarkable separation. This dimensions would make a good predictor for a classifier. Nonmetric MDS is not as useful for this particular dataset. Let us move on to isomap to see how its reduced dimensions compare.

```
#Apply isomap embedding, keeping n components < the number of original features
iso_model = Isomap(n_neighbors=5, n_components=2)
iso_model.fit_transform(x_std)
print(iso_model.get_params())
iso_dim = iso_model.embedding_
print(iso_dim.shape)  #There should be 2 latent variables represented


#Plot first 2 extracted features and the observation class
plt.figure(figsize=(10, 5))
plt.xlabel('Latent Variable 1 (explains most variance)')
plt.ylabel('Latent Variable 2 (explains second most variance)')
plt.title('Isomap 2-Dimension Plot with Observation Class')
plt.scatter(iso_dim[:, 0], iso_dim[:, 1], c=y)
plt.colorbar()
plt.show()


#Apply isomap for many different choices of dimensions
#Limitation is that nbr dimensions must be < the number of original features
nbr_dim = range(3)
iso_dim_nbr = []
iso_reconstruction_errors = []
for nd in nbr_dim:
    iso_model = Isomap(n_neighbors=5, n_components=nd+1)
    iso_model.fit_transform(x_std)
    iso_dim_nbr.append(nd+1)
    iso_reconstruction_errors.append(iso_model.reconstruction_error())
iso_results = {'nbr_dim': iso_dim_nbr, 'error': iso_reconstruction_errors}
iso_results = pd.DataFrame.from_dict(iso_results)
#See which number of dimensions has the lowest reconstruction error
plt.plot(iso_results['nbr_dim'], iso_results['error'])
plt.xlabel('Number of Latent Dimensions')
plt.ylabel('Reconstruction Error')
plt.title('Plot of Error by Number of Latent Variables')
plt.show()
```

Now let's move on to modified LLE to see how it compares.

```
#Apply modified LLE, keeping n components < the number of original features
#method = 'standard' for LLE, 'hessian' for HELLE, or 'modified' for modified LLE
mlle_model = LocallyLinearEmbedding(n_neighbors=5, n_components=2, method='modified',
random_state=seed)
mlle_model.fit_transform(x_std)
print(mlle_model.get_params())
mlle_dim = mlle_model.embedding_
print(mlle_dim.shape)  #There should be 2 latent variables represented


#Plot first 2 extracted features and the observation class
plt.figure(figsize=(10, 5))
plt.xlabel('Latent Variable 1 (explains most variance)')
plt.ylabel('Latent Variable 2 (explains second most variance)')
plt.title('Modified LLE 2-Dimension Plot with Observation Class, 5 neighbors')
plt.scatter(mlle_dim[:, 0], mlle_dim[:, 1], c=y)
```

```
plt.colorbar()
plt.show()

#Try a different number of neighbors
mlle_model = LocallyLinearEmbedding(n_neighbors=15, n_components=2,
method='modified', random_state=seed)
mlle_model.fit_transform(x_std)
mlle_dim = mlle_model.embedding_
plt.figure(figsize=(10, 5))
plt.xlabel('Latent Variable 1 (explains most variance)')
plt.ylabel('Latent Variable 2 (explains second most variance)')
plt.title('Modified LLE 2-Dimension Plot with Observation Class, 15 neighbors')
plt.scatter(mlle_dim[:, 0], mlle_dim[:, 1], c=y)
plt.colorbar()
plt.show()
```

Finally, we can apply t-SNE.

```
#Build TSNE model, learning rate defaults to 1000 but usually best around 200
#Perplexity balances local and global aspects of neighbors, usually best between 5
and 50
tsne_model = TSNE(n_components=2, perplexity=30.0, learning_rate=100.0,
                  n_iter=2000, n_iter_without_progress=30,
                  random_state=seed, method='barnes_hut')
tsne_model.fit_transform(x_std)
print(tsne_model.get_params())
tsne_dim = tsne_model.embedding_
print(tsne_dim.shape)  #There should be 2 latent variables represented
#print('Kullback-Leibler divergence:', tsne_model.kl_divergence_)

#Plot first 2 extracted features and the observation class
plt.figure(figsize=(10, 5))
plt.xlabel('Latent Variable 1')
plt.ylabel('Latent Variable 2')
plt.title('t-SNE 2-Dimension Plot with Observation Class \nperplexity=30,
learning_rate=100')
plt.scatter(tsne_dim[:, 0], tsne_dim[:, 1], c=y)
plt.colorbar()
plt.show()
```

The plots show how each variant of nonlinear dimension reduction has a different take on the data. While the reduced dimensions seem to split the classes of the Iris dataset up fairly well, this is purely coincidental. The goal of dimension reduction is to intelligently reduce the number of features in a dataset.

18. Multivariate Analysis of Variance (MANOVA)

Multivariate analysis of variance (MANOVA) is like ANOVA for situations when there are multiple dependent variables. Consider a couple examples of situations when there would be multiple dependent variables:

- A study aims to measure reduction in high cholesterol and blood pressure, given a plan that involves a diet, exercise, and vitamin supplement. Cholesterol and blood pressure are two dependent variables. The hypothesis to be tested is that both are affected by the diet, exercise, and vitamin supplement.
- A financial researcher wants to study the effect of a new insurance company's initial public offering on the stock prices of Geico, State Farm, and Progressive. In this case, there are three dependent variables: the stock prices of Geico, State Farm, and Progressive. The hypothesis to be tested is that all three are effected by the new company's IPO.

After perusing the above examples, a reasonable question would be to ask why ANOVA could not be carried out multiple times for testing the effects of the independent variable on the individual dependent variables. Recall from chapter 12 that we explored the same question about ANOVA and multiple t-tests. The answer here is similar: combining the tests reduces the probability of type I error. Additionally, it is possible that the dependent variables themselves are related in some way, and ANOVA would not account for this, whereas MANOVA does.

MANOVA predicts several continuous outcome variables based on one or more categorical predictors. Like ANOVA, MANOVA comes in many flavors. If there are 2 categorical independent variables, it is called two-way MANOVA. In general, k independent variables is written as k-way MANOVA. MANOVA is named in a similar manner as ANOVA. So a model with 2 continuous dependent variables, and 3 independent categorical variables from an independent measures design experiment is called a 3-way independent MANOVA. Likewise, a model with 2 continuous dependent variables and 2 independent categorical variables from a repeated measures design experiment is called a 2-way repeated measures MANOVA. Just like how ANOVA becomes ANCOVA when one or more continuous covariate independent variables are added, MANOVA becomes MANCOVA when one or more continuous covariate independent variables are added (Field, Miles, & Field, 2012).

MANOVA produces a matrix of test statistics, rather than only one, and this matrix is derived by comparing the ratios of systematic variance to unsystematic variance (the F-statistic) for several dependent variables. To do this, a matrices of systematic variances and unsystematic variances are created for each dependent variable. The F-statistic calls for them to be divided, but since matrices cannot be divided, the matrices of systematic variances are multiplied by the inverse of the matrices of unsystematic variances. When MANOVA is carried out, any correlations between the dependent variables are found using cross products that represent the total combined error between them. When the test statistic matrix is created, there will be p^2 values in the matrix, where p is the number of dependent variables. Since there are several test statistics produce by MANOVA, there needs to be a way to determine which one should be used to evaluate the hypothesis. Discriminant function variates can be used to calculate the final test statistic. Discriminant function variates are found using discriminant function analysis (DFA), as described in chapter 15. Standard discriminant function analysis involves predicting a categorical dependent variable using several independent variables (it is classification), whereas MANOVA predicts several continuous dependent variables using one or more independent variables (it is regression). In the context of MANOVA, DFA is carried out using Fisher's linear discriminant rule, and it shows how the groups or categories of the independent variables differ based on linear combinations of the dependent variables.

Discriminant Function Variates

Linear combinations of dependent variables are known as **variates**. Therefore, MANOVA shows how groups of the independent variables are predicted by the variates. The linear variates predict which group of the independent variables that an observation belongs to, and the discriminant function of the variates maximizes the differences between the groups. Using DFA for MANOVA is kind of like modeling backwards.

Figurer 18.1 shows an equation for a discriminant function consisting of 2 variates: DV1 and DV2.

$$V_{1i} = b_0 + b_1 DV_{1i} + b_2 DV_{2i}$$

Figure 18.1

Instead of using OLS to estimate the variates, matrix decomposition (either eigendecomposition or SVD) is used. Let us suppose the eigendecomposition is used. If we think about the eigenvalues, we should expect them to be the same when there is no correlation between variables. For example, if two variables are not correlated, their scatterplot will be circular and evenly distributed. Dividing the largest eigenvalue by the smallest would produce 1. If we translate this knowledge to MANOVA, we should expect the eigenvalues of uncorrelated dependent variables to all be the same. Conversely, when variables are correlated, their eigenvalues are very different. Using the example of two correlated variables, we should expect their scatterplot to be more linear the more strongly they are correlated. Dividing the largest eigenvalue by the smallest would produce values that approach infinity as the correlation increases towards 1. If we translate this knowledge to MANOVA, we should expect the eigenvalues of correlated dependent variables to all be very different.

When the eigenvectors and eigenvalues of the cross product matrix for MANOVA are calculated, the ratio of the systematic variance to the unsystematic variance (the test statistic) can be found. To find it, first substitute the values of the eigenvectors back into the discriminant function to get the variate scores for each observation. Then calculate the multiplicative inverse of the cross product matrix of the variates, instead of the cross product matrix of the dependent variables. This will give the eigenvalues that correspond to the F-ratios in ANOVA.

Test Statistics

After the eigenvalues have been calculated, it is necessary to assess how large they are compared to what would be expected by chance alone. The **Pillai-Bartlett Trace** is one way to do this. The equation is the sum of the proportion of explained variance on the discriminant functions. The Pillai-Bartlett Trace is like R^2 for MANOVA.

$$V = \sum_{i=1}^{s} \frac{\lambda_i}{1 + \lambda_i}$$

Figure 18.2

In the equation in figure 18.2, lambda sub i represents the ith eigenvalue. To show how this equation is used, consider the cross product matrix of the variates shown at the top of figure 18.3. The values of this matrix can be substituted into the equation in figure 18.2 to yield the Pillai-Bartlett Trace that has an approximate F-distribution.

$$HE^{-1}_{variates} = \begin{bmatrix} 0.341 & 0.000 \\ 0.000 & 0.069 \end{bmatrix}$$

$$V = \frac{0.341}{1+0.341} + \frac{0.069}{1+0.069} = 0.319$$

Figure 18.3

The Pillai-Bartlett Trace is the most robust test statistic for MANOVA is any of the MANOVA assumptions are violated (we will get to these shortly). But this statistic is unreliable when the number of observations in each group of the independent variables are different, or when the assumption of equal covariance matrices is violated (Field, Miles, & Field, 2012).

Another possible test statistic for MANOVA is the **Hotelling Lawley Trace**, which is the sum of the eigenvalues for each variate as shown in figure 18.4:

$$T = \sum_{i=1}^{S} \lambda_i$$

Figure 18.4

This number compares directly to the F-ratio in ANOVA. It is a robust statistic.

Another test statistic for MANOVA is **Wilk's Lambda**, which is the product of the unexplained variance for each of the variates as shown in figure 18.5:

$$\Lambda = \prod_{i=1}^{S} \frac{1}{1+\lambda_i}$$

Figure 18.5

Wilk's Lambda is equivalent to the SSR / SST for each variate. Large eigenvalues that represent large experimental effects lead to small lambda values, hence statistical significance is found when Wilk's Lambda is small.

Another test statistic for MANOVA is Roy's Largest Root. Roy's Largest Root is like the Hotelling Lawley Trace, but for the first variate only. It is just the eigenvalue for the first variate:

$$\theta = \lambda_{largest}$$

Figure 18.6

Since the eigenvalue for the first variate is the maximum possible between-group difference, it is usually the most powerful test statistic for MANOVA (Field, Miles, & Field, 2012). It is comparable to the F-ratio of univariate ANOVA.

Choosing a Test Statistic

For small samples, or when group differences are concentrated on the first variate (as they most often are), then the order of statistics in terms of power is: Roy's Largest Root, Hotelling's Trace, Wilk's Lambda, and Pillai's trace. However, when the groups differ along more than 1 variate, the power order of these statistics is reversed. It is recommended to use fewer than 10 dependent variables in MANOVA unless the sample size is large (Field, Miles, & Field, 2012).

The Pillai-Bartlett trace is the most robust statistic when any of the MANOVA assumptions are violated, but it is unreliable when sample sizes are different and the assumption of equal covariance matrices is violated (Field, Miles, & Field, 2012).

All four statistics are robust to the multivariate normality assumption being violated, although Roy's Largest Root is affected by platykurtic distributions. Roy's root is also not robust when the homogeneity of covariance matrix assumption is untenable (Field, Miles, & Field, 2012).

Note that all of these test statistics are based on the eigenvalues, so the larger they are, the better the combination of linear variates differentiates the groups.

MANOVA Assumptions

MANOVA assumptions are similar to the assumptions for ANOVA, but extended to the multivariate case.

- **Independence**: Observations should be independent.
- **Random Sampling**: Data should be randomly sampled from the population and measured at the interval level.
- **Multivariate Normality**: In ANOVA, the dependent variable is assumed to be normally distributed within each group. With MANOVA, the dependent variables are collectively assumed to have multivariate normality within groups. This can be tested with the Shapiro-Wilk test (refer to chapter 3).
- **Homogeneity of Covariance Matrices**: In ANOVA, the within group variances are assumed to be homogeneous. In MANOVA, this must be true for each dependent variable, but it is also required that the correlation between any 2 dependent variables is the same for all the groups of the independent variables. This assumptions can be tested by looking at the variance-covariance matrices of the different groups. If they are all equal, then the assumption is met. This assumption can also be tested with Box's M test. Of these two methods, looking at the covariance matrices is the better way. If the homogeneity of covariance matrices assumption is violated, Hotelling's Trace is the most robust statistic.
  - When testing this assumption, if the ratio of the largest variance to the smallest variance for a given variable is < 2, then all is well. Larger samples produce greater variances and covariances, so significant findings can be trusted, even with differences between the covariances in the groups. If the smaller samples produce larger variances and covariances, then the probability values will be liberal however, so the results of MANOVA should entirely not be trusted.

Univariate ANOVA Tests after MANOVA

After MANOVA has been carried out, it will be clear whether the independent variable(s) had any effect on the outcome variables, but it will not be clear which groups of the independent variable(s) differed from one another, nor which outcome the independent variables affected. DFA will provide the answers to these questions, but an alternative to DFA is to use several univariate ANOVA tests: one test per dependent variable. Note however that if the univariate ANOVA tests fail to detect any statistical significance of the independent variables when MANOVA did find statistical significance, it is due to MANOVA's ability to consider correlations between the dependent variables. When this happens, it is useless to continue using univariate tests to clarify which outcomes the treatment effected. DFA should be used instead.

MANOVA with R

Since MANOVA cannot easily be done in Python without significant custom coding, there is no Python section for this chapter. MANOVA and ANOVA are often used in clinical settings where an experiment is carried out using groups of participants, and the results of the experiment are interpreted based on their effects on the target variable. For example, 2 groups of participants could be given 2 different diets to follow for a few weeks, and the target would be body mass index. So the experiment would test for changes in the mean body mass index for each group. That would be a perfect use case for ANOVA. But if the researchers wanted to measure body mass index and cholesterol, then there would be 2 outcome variables so MANOVA would be appropriate.

In this section, we will simulate an experiment using the Iris flower dataset. We will pretend that species, a categorical grouping variable, is a variable that we can control. We will compare the mean sepal length and sepal width across these groups. Our goal will be to determine whether or not species differ in both mean sepal length and mean sepal width. The first task is to verify the multivariate normality assumption using Shapiro's test. Fortunately there is no data cleaning required for the Iris dataset, so we can jump ahead to the assumption testing.

```
d <- iris
str(d)

library(mvnormtest)
#Note that the matrix must be transposed so observations become columns
mshapiro.test(t(as.matrix(d[,c('Sepal.Length', 'Sepal.Width')])))
```

Shapiro's multivariate normality test has a very large p-value, meaning the null hypothesis that the data is normally distributed cannot be rejected. So the normality assumption holds. The other assumption is homogeneity of covariances, which can be tested by applying Levene's test to each condition individually.

```
library(car)
leveneTest(d$Sepal.Length ~ d$Species)
leveneTest(d$Sepal.Width ~ d$Species)
```

There may be a problem with the variance in sepal length being different between species because the p-value is < 0.05. The ratio of the largest group variance to the smallest can be informative.

```
library(dplyr)
group_vars <- group_by(d, Species) %>%
```

```
    summarise(Group_Variance=var(Sepal.Length))
max(group_vars$Group_Variance)/min(group_vars$Group_Variance)
```

Since the ratio is >= 2, and the sample size is small, it looks like the results of MANOVA may not be trustworthy. This could be problematic, but MANOVA is robust to violations in homoskedasticity as long as the group sample sizes are similar. In this case, there are exactly the same number of samples in each group. So we will proceed with MANOVA. It would be best to use Pillai's trace as the test statistic in this case, since it is the most robust to possible violations in the assumptions. The MANOVA function uses Pialli's trace by default.

```
#Perform MANOVA
manova_reg <- manova(cbind(Sepal.Length, Sepal.Width) ~ Species, data=d)
#Default test statistic is Pillai's trace
summary(manova_reg, intercept=TRUE)
summary(manova_reg, intercept=TRUE, test="Wilks")
summary(manova_reg, intercept=TRUE, test="Hotelling")
summary(manova_reg, intercept=TRUE, test="Roy")
```

The results of MANOVA show that there is a significant difference between mean sepal length and mean sepal width between species. ANOVA can be used to determine whether species differ in mean sepal length or mean sepal width.

```
#Use ANOVA to see which species differ in sepal length and width
summary.aov(manova_reg)
```

The results show that species differ in both mean sepal length and mean sepal width. Post hoc tests could be carried out to determine which particular species have different mean sepal lengths and sepal widths, but this exercise will be left to the reader.

19. Regularized Regression

If a model cannot fully describe the relationship between the data it is trained on, then it is said to be **underfitted**. Underfitted models are weak. On the other hand, if a model fully describes the relationship between the data it is trained on but cannot generalize to new data, then it is said to be **overfitted**. Overfitted models are strong only with the data they are trained on, and weak when new observations are presented. In between underfitting and overfitting is a sweet spot of ideal models. Regression models, especially flexible nonlinear regression models such as polynomial or spline regressions, are prone to overfitting. Regularization is a way to penalize models for overfitting. Regularization works by changing the cost function to be optimized. In the context of regression, regularization introduces bias into the model to reduce variance. Recall from chapter 6 that there is a tradeoff between bias and variance (bias describes how close a model is to the ground truth whereas variance describes how sensitive the model is to fluctuations in the data.) Regularized regression models typically have higher bias and lower variance than regression that has been estimated by OLS (James, Witten, Hastie, & Tibshirani, 2013). In this chapter, we will see three ways of regularizing regression.

Before diving into regularization, it should be noted that there are many methods in statistics and machine learning that are designed to prevent overfitting. The quantity of these methods makes overfitting seem like a serious problem, when in fact, it should be preferable to underfitting. If a model is underfitted, then it cannot reveal any insights into the data. At least an overfitted model can describe the data it is trained on. When training models, a good practice is to overfit first, and then try to scale back the fit with regularization. While overfitting can be corrected with model penalization methods or dimension reduction, underfitting can only be corrected by collecting more data or by engineering new features, so it takes more work to correct underfitted models. Some may argue that model tuning can also correct underfitting, and this is true, but only to an extent. Hyperparameter tuning cannot bring a model from zero to hero: only data and features can do that.

Ridge Regression

Regularization penalizes the coefficients of the independent variables in a regression model (James, Witten, Hastie, & Tibshirani, 2013). When predictor coefficients are penalized so that they are estimated to be smaller, it is called **shrinkage**. The reason shrinkage produces more accurate models is that coefficients tend to blow up in magnitude when the model complexity increases. A polynomial regression of the 50th order will have exponentially larger coefficients than a cubic regression, for example. Regularization methods vary based on the type of penalty (lambda) that they use for shrinkage. Ridge regression uses an L2 penalty, because it penalizes the L2-Norm. The **L2-Norm** is simply another name for the least squares error, as introduced in chapter 6. **L2 regularization** adds a penalty that is equivalent to the square of the magnitude of the coefficient. So recall the least squares error cost function (the L2-Norm) as shown in figure 19.1.

$$S = \sum_{i=1}^{n}(y_i - f(x_i))^2$$

Figure 19.1

In the equation in figure 19.1, S is the sum of squared differences between the target variable y and the estimated values f(x sub i). When a penalty term, lambda, is added to the cost function in figure, then it will result in smaller coefficients.

$$w^* = \arg\min_w \sum_j \left(t(x_j) - \sum_i w_i h_i(x_j) \right)^2 + \lambda \sum_{i=1}^{k} w_i^2$$

Figure 19.2

The equation in figure 19.2 shows L2 regularization of least squares. The regularization of the weights is represented by lambda times the sum of the squared weights. In the equation, w star represents the penalized weights of the predictors.

L2 regularization is preferred to L1 regularization when the influence of outliers is important, because squared error gives them more weight. L2 regularization is computationally more efficient than L1 regularization and produces more stable results, meaning that if the data were to shift slightly, the model produced by L2 regularization would not change very much. The drawback to L2 regularization is that since it is sensitive to outliers, it should not be used when outliers should not impact the model (James, Witten, Hastie, & Tibshirani, 2013).

Least Absolute Shrinkage and Selection Operation Regression

Least absolute shrinkage and selection operation (LASSO) shrinks the regression coefficients and has built in feature selection, because some coefficients might shrink so small that they become zero, effectively canceling out any effect they have on the dependent variable (James, Witten, Hastie, & Tibshirani, 2013). In comparison, ridge regression shrinks the coefficients but never reduces any of them to zero. LASSO regression uses an L1 penalty for regularization, because it penalizes the L1-Norm. The **L1-Norm** is simply another name for the least absolute deviation. **L1 regularization** adds a penalty that is equivalent to the absolute value of the magnitude of the coefficient. Consider the least absolute deviation cost function (the L1-Norm) as shown in figure 19.3.

$$S = \sum_{i=1}^{n} |y_i - f(x_i)|$$

Figure 19.3

In the equation in figure 19.3, S is the sum of the absolute differences between the target variable y and the estimated values f(x sub i). When a penalty term, lambda, is added to the cost function in figure 19.3, then it will result in smaller coefficients.

$$w^* = \arg\min_w \sum_j \left(t(x_j) - \sum_i w_i h_i(x_j) \right)^2 + \lambda \sum_{i=1}^{k} |w_i|$$

Figure 19.4

The equation in figure 19.4 shows L1 regularization of least squares. The regularization of the weights is represented by lambda times the sum of the absolute values of the weights. In the equation, w star represents the penalized weights of the predictors.

L1 regularization is more robust to outliers than L2 regularization. It also has built in feature selection, due to the sparse coefficients that it produces. Sparsity in the coefficients of a model means that the coefficients of many independent variables are zero if the variables do not influence the outcome very much. The drawbacks to L1 regularization are that it is computationally inefficient compared to L2 regularization, and it produces unstable results, meaning that if the data were to shift slightly, the model produced by L1 regularization would change drastically (James, Witten, Hastie, & Tibshirani, 2013).

LASSO has a couple flaws. The first is that it can be too aggressive in its feature selection, and if the number of dimensions is large but the number of observations is small, then LASSO saturates once a certain number of features are selected. The second flaw is that when several groups of correlated predictors, LASSO selects one feature from each group and zeros out the coefficients of the others. This can result in underfitting (or too much correction for overfitting). The elastic net was introduced to fix these problems with LASSO.

Elastic Net

A problem with LASSO regression is that it can be too aggressive in its feature selection. Elastic net regression was introduced as an improvement to LASSO that fixed its problem with aggressive feature selection by introducing a L2 penalty. Elastic net is therefore like a blend of ridge and LASSO regressions, because it uses both L1 and L2 penalties. Since it uses both penalties, it is possible to turn an elastic net into either ridge regression or LASSO by changing the penalties accordingly.

Elastic net solves the grouped correlation problem of LASSO by shrinking the coefficients of grouped correlated predictors but not so much that they become zero (much like ridge regression). However, if none of the predictors in a correlated group effect the outcome variable, the coefficients for the entire group are set to zero. The modified weights from the least squares cost function for elastic net is shown in figure 19.5.

$$ w^* = \arg min_w \sum_j \left(t(x_j) - \sum_i w_i h_i(x_j) \right)^2 + \lambda \sum_{i=1}^{k} |w_i| + \lambda \sum_{i=1}^{k} w_i^2 $$

Figure 19.5

In essence, the equation shows that the weights resulting from least square minimization are impacted by both the L1 penalty (lambda times the sum of the absolute values of the weights) and the L2 penalty (lambda times the sum of the squared values of the weights). The constraint is that the L1 and L2 penalty coefficients must sum to 1. Some statistics libraries have alpha and the L1 ratio as hyperparameters for elastic net. Alpha is the sum of the L1 and L2 penalty coefficients. The L1 ratio is the coefficient of the L1 penalty divided by the sum of the coefficients of the L1 and L2 penalties (it is the ratio of L1 penalty to total penalty).

Regularized Regression with R

In this chapter, we will use the motor trends car dataset that shows the miles per gallon (MPG) that different cars get, and several other attributes. This dataset is typically used to predict MPG based on the other features. Since there is no need for data cleaning, we can start right off with ridge regression. The glmnet library that is used in this section covers ridge regression, LASSO, and the elastic net all in one function. The only difference is the shrinkage argument, lambda. Lambda is called alpha in the glmnet package. Finding the best shrinkage parameter is a hyperparameter tuning problem. We will tackle it here by simply trying a bunch of values between 0.01 and 1,000, but it could also be tuned using the methods described in chapter 8.

```
d <- mtcars
str(d)

#Standard linear regression for comparison
linear_reg <- lm(mpg ~ ., data=d)
summary(linear_reg)

library(glmnet)
?glmnet
#glmnet function is the same for ridge, lasso, and elastic net
#alpha=0 for ridge, alpha=1 for lasso, alpha=0.5 for elastic net

#Ridge regression
#Try a bunch of lambda (shrinkage) values between 0.01 and 1,000
lambdas <- 10^seq(3, -2, by=-0.1)
lambdas
#Run the ridge regression using the lambdas
ridge_reg <- glmnet(as.matrix(d[,-1]), d$mpg, alpha=0, lambda=lambdas)
summary(ridge_reg)
#Tune lambda using cross validation
ridge_reg <- cv.glmnet(as.matrix(d[,-1]), d$mpg, alpha=0, lambda=lambdas)
plot(ridge_reg)
#Minimum of the MSE is where the best lambda choice is
best_lambda <- ridge_reg$lambda.min
#Best ridge regression
ridge_reg <- glmnet(as.matrix(d[,-1]), d$mpg, alpha=0, lambda=best_lambda)
coef(ridge_reg)
```

LASSO and elastic net are carried out nearly exactly the same way.

```
#Lasso
#Try a bunch of lambda (shrinkage) values between 0.01 and 1,000
lambdas <- 10^seq(3, -2, by=-0.1)
lambdas
#Run the ridge regression using the lambdas
lasso_reg <- glmnet(as.matrix(d[,-1]), d$mpg, alpha=1, lambda=lambdas)
summary(lasso_reg)
#Tune lambda using cross validation
lasso_reg <- cv.glmnet(as.matrix(d[,-1]), d$mpg, alpha=1, lambda=lambdas)
plot(lasso_reg)
#Minimum of the MSE is where the best lambda choice is
best_lambda <- lasso_reg$lambda.min
#Best ridge regression
lasso_reg <- glmnet(as.matrix(d[,-1]), d$mpg, alpha=1, lambda=best_lambda)
coef(lasso_reg)

#Elastic Net
#Try a bunch of lambda (shrinkage) values between 0.01 and 1,000
lambdas <- 10^seq(3, -2, by=-0.1)
lambdas
#Run the ridge regression using the lambdas
```

```
en_reg <- glmnet(as.matrix(d[,-1]), d$mpg, alpha=0.5, lambda=lambdas)
summary(en_reg)
#Tune lambda using cross validation
en_reg <- cv.glmnet(as.matrix(d[,-1]), d$mpg, alpha=0.5, lambda=lambdas)
plot(en_reg)
#Minimum of the MSE is where the best lambda choice is
best_lambda <- en_reg$lambda.min
#Best ridge regression
en_reg <- glmnet(as.matrix(d[,-1]), d$mpg, alpha=0.5, lambda=best_lambda)
coef(en_reg)
```

The fitted values can be plotted to show how each of the models fit the data. Note that these plots are like the residual plots for standard linear regression. The more linear the plot along the diagonal, the better the model.

```
d$ridge_preds <- predict(ridge_reg, s=best_lambda, newx=as.matrix(d[,-1]))
plot(d$ridge_preds, d$mpg, main='Ridge Regression Predicted vs Actual Values')
d$lasso_preds <- predict(lasso_reg, s=best_lambda, newx=as.matrix(d[,-c(1,12)]))
plot(d$lasso_preds, d$mpg, main='Lasso Predicted vs Actual Values')
d$en_preds <- predict(en_reg, s=best_lambda, newx=as.matrix(d[,-c(1,12,13)]))
plot(d$en_preds, d$mpg, main='Elastic Net Predicted vs Actual Values')
```

A better way to compare the models is to look at the mean squared error (the cost function) that the models minimized. In this case, the ridge regression had the best fit, followed by the elastic net. If ridge regression outperforms LASSO, as it does here, it means that fitting a model by minimizing the squared error (L2) is more effective than minimizing the absolute deviation (L1). Note that the glmnet function minimizes the mean squared error for the L2 norm however, not the squared error. The difference between the textbook definition of the L2 norm (minimizing squared error) and the glmnet version (minimizing mean squared error) is minimal; it just means the coefficients have to be adjusted to account for the differences. The code below performs the transformation.

```
ridge_reg_coef <- solve(t(as.matrix(d[,-c(1,12,13,14)])) %*% as.matrix(d[,-
c(1,12,13,14)]) + ridge_reg$lambda * diag(ncol(d)-4)) %*% t(as.matrix(d[,-
c(1,12,13,14)])) %*% d$mpg
lasso_reg_coef <- solve(t(as.matrix(d[,-c(1,12,13,14)])) %*% as.matrix(d[,-
c(1,12,13,14)]) + lasso_reg$lambda * diag(ncol(d)-4)) %*% t(as.matrix(d[,-
c(1,12,13,14)])) %*% d$mpg
en_reg_coef <- solve(t(as.matrix(d[,-c(1,12,13,14)])) %*% as.matrix(d[,-
c(1,12,13,14)]) + en_reg$lambda * diag(ncol(d)-4)) %*% t(as.matrix(d[,-
c(1,12,13,14)])) %*% d$mpg
#Zero out the least squares penalized coefficients for the features that were removed
by the models due to penalization
ridge_reg_coef * ifelse(coef(ridge_reg)[-1]==0, 0, 1)
lasso_reg_coef * ifelse(coef(lasso_reg)[-1]==0, 0, 1)
en_reg_coef * ifelse(coef(en_reg)[-1]==0, 0, 1)
```

Thanks to StackOverflow for the code to transform the coefficients
https://stackoverflow.com/questions/39863367/ridge-regression-with-glmnet-gives-different-coefficients-than-what-i-compute

Regularized Regression with Python

In this chapter, we will use the motor trends car dataset that shows the miles per gallon (MPG) that different cars get, and several other attributes. This dataset is typically used to predict MPG based on the other features. Before we begin, let us get the data imported and set up a helper function to retrieve output from linear regression, such as p-values and R squared.

```python
import pandas as pd
import numpy as np
from scipy import stats
from sklearn.linear_model import LinearRegression
from sklearn.linear_model import Ridge, Lasso, ElasticNet
import matplotlib.pyplot as plt

d = pd.read_csv('mtcars.csv')
d.info()

def get_lr_output(lr, x, y):
    #Takes a fitted sklearn linear regression as input and outputs results
    #Thanks to stackoverflow: https://stackoverflow.com/questions/27928275/find-p-value-significance-in-scikit-learn-linearregression
    parameters = np.append(lr.intercept_, lr.coef_)
    predictions = lr.predict(x)
    X = pd.DataFrame({"Constant":np.ones(len(x))}).join(pd.DataFrame(x))
    mse = (sum((y-predictions)**2))/(len(X)-len(X.columns))
    sd = np.sqrt(mse*(np.linalg.inv(np.dot(X.T, X)).diagonal()))
    ts = parameters/sd
    p_values = [2*(1-stats.t.cdf(np.abs(i), (len(X)-1))) for i in ts]
    sd = np.round(sd, 4)
    ts = np.round(ts, 4)
    p_values = np.round(p_values, 4)
    parameters = np.round(parameters, 4)
    predictors = pd.Series('intercept')
    predictors = predictors.append(pd.Series(x.columns)).reset_index()[0]
    output_df = pd.DataFrame()
    output_df['predictor'] = predictors
    output_df['coefficient'] = parameters
    output_df['standard_error'] = sd
    output_df['t-statistic'] = ts
    output_df['p-value'] = p_values
    r_squared = round(lr.score(x, y), 4)
    adj_r_squared = round(1-(1-lr.score(x, y))*(len(y)-1)/(len(y)-x.shape[1]-1), 4)
    print(output_df, '\n R-squared:', r_squared, '\n Adjusted R-squared:', adj_r_squared)
    return [output_df, r_squared, adj_r_squared]

#Standard linear regression for comparison
linear_reg = LinearRegression()
linear_reg.fit(d.drop(['mpg'], axis=1), d['mpg'])
linear_reg_output = get_lr_output(linear_reg, d.drop(['mpg'], axis=1), d['mpg'])
linear_reg_output
linear_reg_predictions = linear_reg.predict(d.drop(['mpg'], axis=1))
order = d['disp'].sort_values().index.tolist()
plt.scatter(d['disp'], d['mpg'])
plt.plot(d['disp'][order], linear_reg_predictions[order], color='black')
```

```
plt.xlabel('disp')
plt.ylabel('mpg')
plt.title('Linear Regression Fit')
plt.show()
```

Since there is no need for data cleaning, we can start right off with ridge regression. As with R, we will try different shrinkage parameters between 0.01 and 1,000, but it could also be tuned using the methods described in chapter 8. For consistency, we will call the shrinkage parameters alpha, but they are represented by lambda in the equations earlier in this chapter.

```
#Try a bunch of alpha (shrinkage) values between 0.01 and 1,000
alphas = [10**i for i in np.arange(3, -2, -0.1)]
alphas

#Ridge regression
#Run a ridge regression for every alpha
ridge_results = []
ridge_errors = []
for a in alphas:
    ridge_reg = Ridge(alpha=a).fit(d.drop(['mpg'], axis=1), d['mpg'])
    ridge_predictions = ridge_reg.predict(d.drop(['mpg'], axis=1))
    ridge_results.append(ridge_predictions)
    ridge_mse = np.mean((d['mpg']-ridge_predictions)**2)
    ridge_errors.append(ridge_mse)
print(ridge_errors)

#Plot the fit of a couple different models
order = d['disp'].sort_values().index.tolist()
plt.scatter(d['disp'], d['mpg'])
plt.plot(d['disp'][order], ridge_results[10][order], color='red')
plt.plot(d['disp'][order], ridge_results[25][order], color='blue')
plt.plot(d['disp'][order], ridge_results[45][order], color='green')
plt.xlabel('disp')
plt.ylabel('mpg')
plt.title('Ridge Regressions of Different Shrinkages')
plt.show()

#Run the ridge regression using default alpha and view results
ridge_reg = Ridge().fit(d.drop(['mpg'], axis=1), d['mpg'])
ridge_predictions = ridge_reg.predict(d.drop(['mpg'], axis=1))
ridge_reg_output = get_lr_output(ridge_reg, d.drop(['mpg'], axis=1), d['mpg'])
ridge_reg_output
```

LASSO and elastic net are carried out nearly exactly the same way.

```
#Lasso regression
#Run a lasso regression for every alpha
lasso_results = []
lasso_errors = []
for a in alphas:
    lasso_reg = Lasso(alpha=a).fit(d.drop(['mpg'], axis=1), d['mpg'])
    lasso_predictions = lasso_reg.predict(d.drop(['mpg'], axis=1))
    lasso_results.append(lasso_predictions)
    lasso_mse = np.mean((d['mpg']-lasso_predictions)**2)
    lasso_errors.append(lasso_mse)
```

```
print(lasso_errors)

#Plot the fit of a couple different models
order = d['disp'].sort_values().index.tolist()
plt.scatter(d['disp'], d['mpg'])
plt.plot(d['disp'][order], lasso_results[10][order], color='red')
plt.plot(d['disp'][order], lasso_results[25][order], color='blue')
plt.plot(d['disp'][order], lasso_results[45][order], color='green')
plt.xlabel('disp')
plt.ylabel('mpg')
plt.title('Lasso Regressions of Different Shrinkages')
plt.show()

#Run the lasso regression using default alpha and view results
lasso_reg = Lasso().fit(d.drop(['mpg'], axis=1), d['mpg'])
lasso_predictions = lasso_reg.predict(d.drop(['mpg'], axis=1))
lasso_reg_output = get_lr_output(lasso_reg, d.drop(['mpg'], axis=1), d['mpg'])
lasso_reg_output
#Try a bunch of l1 ratios (determines mix of L1 and L2 regularization) between 0 and
1
l1ratios = list(np.arange(0, 1, 0.01))
l1ratios

#Elastic Net
#Run an elastic net regression for every l1 ratio - ignore the convergence warnings
en_results = []
en_errors = []
for lr in l1ratios:
    en_reg = ElasticNet(l1_ratio=lr).fit(d.drop(['mpg'], axis=1), d['mpg'])
    en_predictions = en_reg.predict(d.drop(['mpg'], axis=1))
    en_results.append(en_predictions)
    en_mse = np.mean((d['mpg']-en_predictions)**2)
    en_errors.append(en_mse)
print(en_errors)

#Plot the fit of a couple different models
order = d['disp'].sort_values().index.tolist()
plt.scatter(d['disp'], d['mpg'])
plt.plot(d['disp'][order], en_results[20][order], color='red')
plt.plot(d['disp'][order], en_results[45][order], color='blue')
plt.plot(d['disp'][order], en_results[90][order], color='green')
plt.xlabel('disp')
plt.ylabel('mpg')
plt.title('Elastic Nets of Different L1 to L2 Ratios')
plt.show()

#Run the elastic net regression using default l1_ratio of 0.5 and view results
en_reg = ElasticNet().fit(d.drop(['mpg'], axis=1), d['mpg'])
en_predictions = en_reg.predict(d.drop(['mpg'], axis=1))
en_reg_output = get_lr_output(en_reg, d.drop(['mpg'], axis=1), d['mpg'])
en_reg_output
```

Examining the output, we can see that the results we found using R are validated: ridge regression had the best fit as measured by R squared, followed by elastic net. If ridge regression outperforms LASSO, as it does here, it means that fitting a model by minimizing the squared error (L2) is more effective than minimizing the absolute deviation (L1). Note that unlike the glmnet function, which minimizes mean squared error for the L2 norm, the functions from scikit-learn minimize the textbook definition.

20. Nonlinear Regression and Classification

This chapter is a bit of a misnomer because several of the models explained hitherto are nonlinear (logistic regression, QDA, MANOVA). But the models presented in this chapter are all nonlinear and they do not really fit into the other chapters.

Polynomial Regression

Polynomial regression is nonlinear regression using a polynomial model. For example, a simple quadratic equation is shown in figure 20.1.

$$y = ax^2 + bx + c$$

Figure 20.1

If x is the only independent variable, then a standard linear regression can be converted to a quadratic regression as in figure 20.2.

$$y = bx + c \quad \rightarrow \quad y = ax^2 + bx + c + \varepsilon$$

Figure 20.2

In these equations, c is the constant, a and b are the coefficients for x^2 and x, respectively, and epsilon is the model error. The same conversion can be done for regression with multiple independent variables x and z:

$$y = bx + hz + c \quad \rightarrow \quad y = ax^2 + bx + gz^2 + hz + (ax * gz) + c + \varepsilon$$

Figure 20.3

Notice the inclusion of an interaction term (ax*gz) in the polynomial model with 2 independent variables in figure 20.3. Interaction terms look for combined effects of multiple variables. Although most statistics texts state that interactions are required for polynomial models, it is actually up to the researcher whether or not to include them in the model. Interaction terms would be beneficial, especially for low order polynomials, if it is thought that two predictors should have a combined effect on the outcome. Interaction terms would not yield any benefit if there are many predictors in the model however, as the interpretation of such interaction terms becomes nearly impossible as the number of predictors increases (Hastie, Tibshirani, & Friedman, 2009).

If interaction terms are included in the model, there should be interaction terms for every combination of predictors, up to the degree of the polynomial minus 1. So in the example in figure 20.3 above, the degree of the polynomial is two so the interaction term between the independent variables for the 1st degree model (the original linear model) is included. If the degree of the polynomial is increased higher, interactions will need to be added for higher degrees as well. For example, if we increase the degree to three, the equation in figure 20.3 becomes:

$$y = dx^3 + ax^2 + bx + iz^3 + gz^2 + hz + (bx * hz) + (ax^2 * hz) + (gz^2 * bx) + c + \varepsilon$$

Figure 20.4

As another example, if we keep the degree at three but add a 3rd predictor, the equation for the cubic polynomial would be:

$$y = dx^3 + ax^2 + bx + iz^3 + gz^2 + hz + lw^3 + mw^2 + nw + (bx * hz) + (bx * nw) + (hz * nw)$$
$$+ (ax^2 * hz) + (ax^2 * nw) + (gz^2 * bx) + (gz^2 * nw) + (mw^2 * bx) + (mw^2 * hz)$$
$$+ c + \varepsilon$$

Figure 20.5

The higher the degree of the polynomial, the closer the fit of the nonlinear model, but at some point the model overfits. It is up to the researcher to balance bias and variance to find the best model, using methods like feature selection or looking at the AIC or BIC.

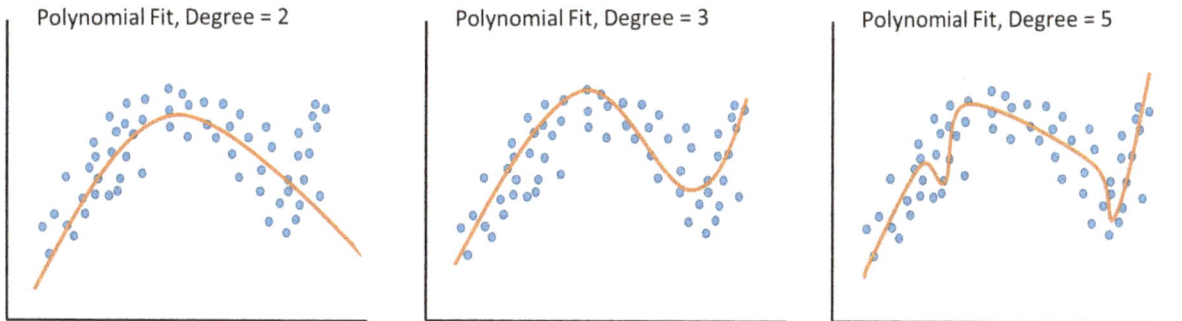

Figure 20.6

Isotonic Regression

Iostonic regression, a.k.a. monotonic regression, is a method for modeling data that is always increasing (just like how monotonic functions are either always increasing or always decreasing). An example of data with an increasing trend is plotted in the sections at the end of this chapter. The modeling process involves finding the line that best fits the data in terms of least squares, such that each consecutive observation is greater than the value of the observation before it. The data is therefore ranked in ascending order, and can be described by the directed graph G = (N, E), where N is the number of observations and E is the set of independent variables. The cost function for isotonic regression is shown in figure 20.7.

$$min \sum_{i=1}^{n} w_i(x_i - a_i)^2 \; subject \; to \; x_i \le x_j \in E_{i,j}$$

Figure 20.7

In the equation in figure 20.7, w is the weight, n is the number of observations, and E is the set of edges between nodes i and j.

Exponential and Logarithmic Regression

Exponential regression fits exponential growth or decay curves. Exponential regression models take the form in figure 20.8.

$$y = \beta_0 + \beta_1 e^{\beta_2 x} + \varepsilon$$

Figure 20.8

To model an exponential curve, the natural logarithm is taken for both sides of the equation, and a linear model is fitted with OLS. Before interpreting the results, it is important to either first transform the resulting model back into exponential form by exponentiating from Euler's number, e, or interpret the results in terms of the logarithmic relationship.

Logarithmic regression fits logarithmic curves. Logarithmic regression models take the form in figure 20.9.

$$\ln(y) = \beta_0 + \beta_1 \ln(x) + \varepsilon$$

Figure 20.9

To model a logarithmic curve, an OLS linear model is fitted to the natural logarithm of the variables.

Smoothing Functions

Smoothing is the estimation of a trend in the dependent variable as a function of the predictors. The simplest smoothing function is a moving average. A **moving average** is the average of a set of data points over a sliding window. For example, a MA(6) smoothing function is the average of 6 consecutive data points when the points are ordered. The problem with moving averages is that they require a certain number of points to return a value. For example, the MA(6) smoothing function is undefined for data points 1 through 5 and only returns a value from point 6 onward. A **local regression (LOESS)** is a slightly more complex smoother that solves the problem of moving averages being undefined at the beginning of a dataset. LOESS fits a weighted regression within each nearest neighbor window of data points when the points are ordered. More distant neighbors to a point are given lower weights, while closer neighbors are weighted higher. The steps to create a LOESS smoothing function are as follows:

1. Select a span parameter that determines the number of nearest-neighbors. For example, a span of 0.4 means that 40% of the data points are included in each neighborhood. 20% of this span will come from the left of a point, and 20% will come from the right.
2. Calculate the weights according to the equation in figure 20.10.

$$w_i = \begin{cases} \left(1 - \left(\frac{x_i - x}{h}\right)^3\right)^3, & x_i \text{ is in the neighborhood of } x \\ 0, & elsewhere \end{cases}$$

Figure 20.10

In the equation in figure 20.10, h is the width of the neighborhood (span*N).

3. Regress Y on the weighted x-values.

Another smoothing function is a **spline**. We saw how splines can be computed to estimate missing values in data in chapter 4. Splines also make good smoothing functions. Splines are fitted by minimizing the penalized sum of squares (Hastie, Tibshirani, & Friedman, 2009). There are several types of splines that could be fitted, but recall from chapter 4 that the cubic spline is the most common. It has been shown that the cubic spline with knots at every data point is the function that minimizes the penalized sum of squares. Since it is not practical to have knots at every data point for large datasets, a different type of spline is needed: the regression spline. **Regression splines** fit p knots and optimize the placement of the knots by using a penalization parameter to control the smoothness of the spline. Figure 20.11 shows the differences between cubic spline and LOESS smoothing methods.

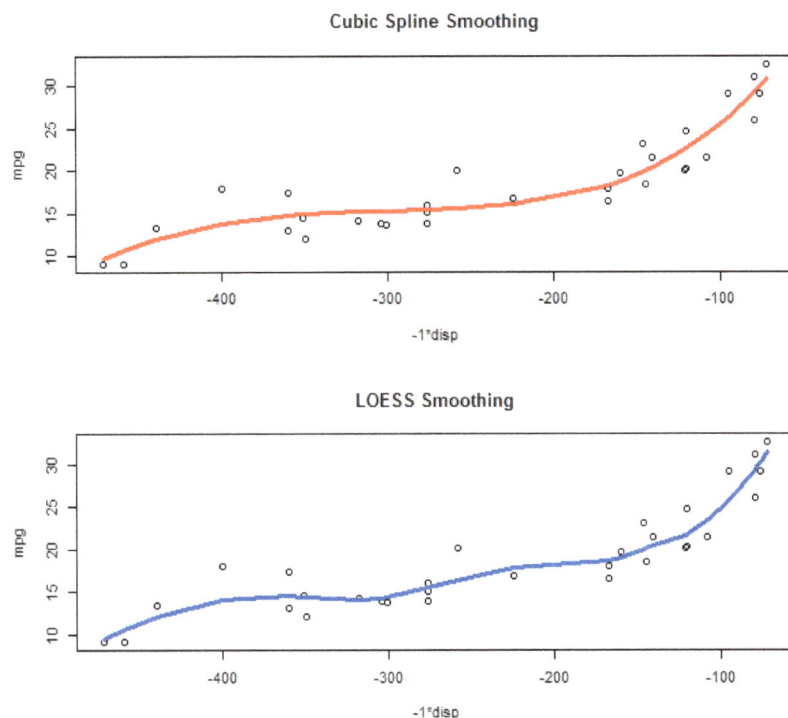

Figure 20.11

Generalized Additive Models

Sometimes the relationship between the outcome and the predictors is nonlinear and so complicated that it cannot be well described by a single nonlinear equation. One way of dealing with these relationships is to break the data into partitions and produce a separate model for each partition. Each partition should be small enough so that a simple linear or nonlinear model can accurately describe the relationship between the outcome and the predictors for the partition. Often, the best estimator for a partition is a smoothing function, like a cubic spline. When several linear or nonlinear smoothing functions are stitched together, they can produce very good global models. This technique is called **additive modeling**.

Generalized additive models (GAMs) are additive techniques where the effects of the predictors are captured by smoothing functions. GAMs are arguably more powerful than any the regression or classification models we have explored so far, because they are nonparametric, they can model complex linear and nonlinear relationships, they have built in regularization to prevent overfitting, and they are nearly as easy to explain as linear regression (Larsen, 2015). The last point about ease of explanation is important because many nonparametric modeling techniques in machine learning are difficult to explain. Models that are difficult to explain are called black boxes. GAMs are not black boxes, but they can often perform just as well if not better than black box models (Larsen, 2015). The form of a GAM is shown in figure 20.12.

$$g\big(E(Y)\big) = \alpha + f_1(x_1) + f_2(x_2) + \cdots + f_m(x_m) + \varepsilon$$

Figure 20.12

In the equation in figure 20.12, Y is the dependent variable, E(Y) is the expectation of Y, and g(Y) is the link function that links E(Y) to the predictors x_1 through x_m. The functions f_1 through f_m are the smoothing functions of the data for the partitions.

GAMs can also be used for binary classification by using the logistic function or logit. The form of a GAM for binary classification is shown in figure 20.13.

$$g(E(Y)) = \log\left(\frac{P(Y=1)}{P(Y=0)}\right)$$

Figure 20.13

In the equation in figure 20.13, the logit link function, g(Y), is defined as the log of the ratio of the probability that the outcome is 1 to the probability that the outcome is 0.

The cost function for GAMs is the penalized log likelihood function as shown in figure 20.14.

$$log\ likelihood = \sum_{i=1}^{N}[y_i\ln(P(y_i)) + (1 - y_i)\ln(1 - P(y_i))]$$

$$penalty = \sum_{j=1}^{p}\lambda_j\int\left(s_j^n * x_j\right)^2 dx$$

$$GAM\ cost\ function = (2 * log\ likelihood) - penalty$$

Figure 20.14

In these equations, s is the smoothing function. The cost function for GAMs used for binary classification is shown in figure 20.15.

$$log\ likelihood = \sum_{i=1}^{N}[y_i\ln(\hat{p}_i) + (1 - y_i)\ln(1 - \hat{p}_i)]$$

$$\hat{p}_i = \frac{1}{1 + e^{\left(-\hat{a}-\Sigma_{j=1}^{p}(s_j * x_{ij})\right)}} = P(Y = 1|x_1, ..., x_p)$$

Figure 20.15

The bottom equation in figure 20.15 is the logistic function (see chapter 9). The penalty used in a GAM depends on the smoothing functions chosen, which depend on the smoothing parameters represented by the lambdas in figure 20.16. The higher this regularization penalty, the smoother the model. This is how GAMs prevent overfitting.

$$penalty = \sum_{j=1}^{p}\lambda_j\int\left(s_j^n * x_j\right)^2 dx$$

Figure 20.16

The penalty shown in figure above is the sum of the squared second derivatives of the smoothing functions. The smoother a curve, the more linear it will be, meaning its squared second derivative (the slope of the slopes of the smoothing function) will be closer to 0. Therefore, the sum of these second derivatives is a good measure of the global smoothness of the GAM.

Since the parameters (lambdas) of the smoothing functions must be selected, they are hyperparameters. They can be tuned through cross validation. We will explore cross validation in a later chapter. Typically, either cubic splines or regression splines are used as the smoothing functions, but polynomial regressions or LOESS could also be used (Hastie, Tibshirani, & Friedman, 2009). Since GAMs are models of models, then can be thought of as an ensemble of regressions.

GAMs are fitted using the **local scoring algorithm**, which is an extension of the **backfitting algorithm**. Conceptually, backfitting starts by fitting a global function of local smoothing functions to minimize the error of each observation from the global mean. At this point, each local smoothing function is given a coefficient. In the next iteration, the coefficients are updated based on the error of the current configuration. This process is repeated until it converges. Since the algorithm uses knowledge of the previous iteration to inform its computations in the current iteration, it is called backfitting. The local scoring algorithm follows a similar procedure for fitting GAMs. It starts by setting all smoothing functions to 0. Then the variables are iterated over to determine the smoothing functions:

The estimated log-odds for observation i is found by the equation in figure 20.17. Recall that the log odds is derived from the logistic function (the bottom equation in figure 20.15).

$$\log odds = \hat{\alpha} + \sum_{j=1}^{p} (s_j * x_{ij})$$

Figure 20.17

The pseudo dependent variable z_i is created and smoothed against x_i using the weights w_i defined in the equation in figure20.18.

$$z_i = \log odds + \frac{y_i - \hat{p}_i}{w_i}$$

$$w_i = \hat{p}_i(1 - \hat{p}_i)$$

Figure 20.18

Since the weights for x_i have now changed, the log odds, p hat and w must be updated. Then the process is repeated for every predictor until the smoothing functions converge.

Although GAMs are nonparametric, they are still subject to the problem of multicollinearity. Therefore, it is good practice to perform dimension reduction if multicollinearity is present in the predictors, before producing a GAM. GAMs also struggle with overfitting, despite having built in regularization. So to correct for an overfitted GAM, increase the penalization by increasing lambda. Another criticism of GAMs is that while they are not black boxes, they are not as easy to understand as other methods. In a business setting, it is often desirable to have easily explainable models. Therefore, as powerful as GAMs are, they are often overlooked in favor of regression and classification trees, which are easier to explain, less computationally intensive, and often equally as powerful.

Regression and Classification Trees

We just saw how GAMs split data into partitions where simple models can be fitted to the partitions. GAMs are somewhat like a combination of step functions fitted to a dataset. Another regression method based on partitioning is called the regression tree. A regression tree is based on **recursive partitioning**, or partitioning partitions until simple models can be fitted to the partitions, much like a GAM. We can imagine the terminal nodes or leaves of a regression tree to be the smallest partitions. Partitions of different sizes will be on different branches of the tree. Figure 20.19 shows how a dataset of 100 observations could be partitioned.

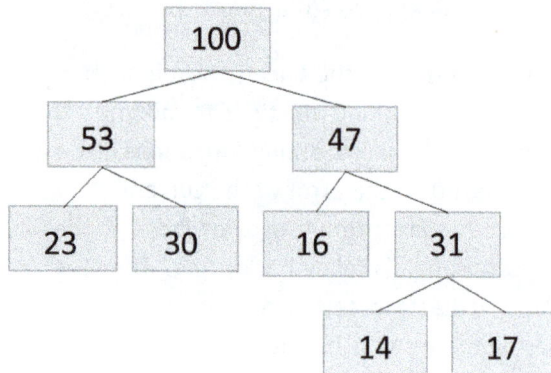

Figure 20.19

Whereas GAMs use a smoothing function to model each data partition, regression trees use a simple average. In other words, the model for each leaf in a regression tree is the mean of the dependent variable for all the observations in the leaf. This makes regression trees much faster to compute than GAMs. Regression trees produce global models of the dependent variable that look like piecewise functions.

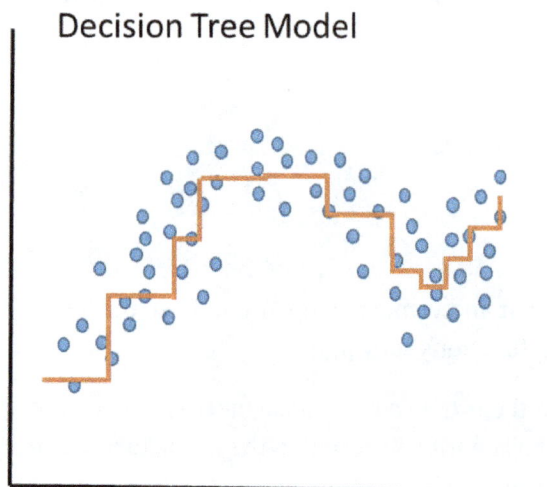

Figure 20.20

Since the models for the tree nodes are simple means, the power of a regression tree must come from optimal partitions of the data. The best split of the data into partitions is determined by maximizing information gain, which is the same as minimizing the error of the tree. To understand what this means, think about the dichotomous keys used to identify species of plants and animals. They start by asking a question about a particular feature, like "does it have a yellow stripe down the spine?" The answer can be either yes or no, resulting in one partition of creatures with a yellow stripe down the spine and another partition of creatures without the stripe. This process continues until the terminal nodes have only one species. In the case of regression and classification trees, the process continues until some stopping criterion is met. We will define the stopping criterion in a moment.

It is worth briefly mentioning model trees. **Model trees** are a variation of regression trees that use linear regressions to model the data in every leaf instead of simple averages. Standard deviation is used as the splitting criterion. Model trees often outperform regression tees, but are less commonly used. The reason for this is that regression trees are faster and far surpass model trees when they are used in an ensemble model, such as a random forest. Ensemble models are beyond the scope of this book, but it should be noted that ensembles of regression and classification trees are almost always preferred to using a single tree.

Figure 20.21 shows a sample dichotomous key represented as a decision tree. Regression and classification trees are collectively referred to as decision trees.

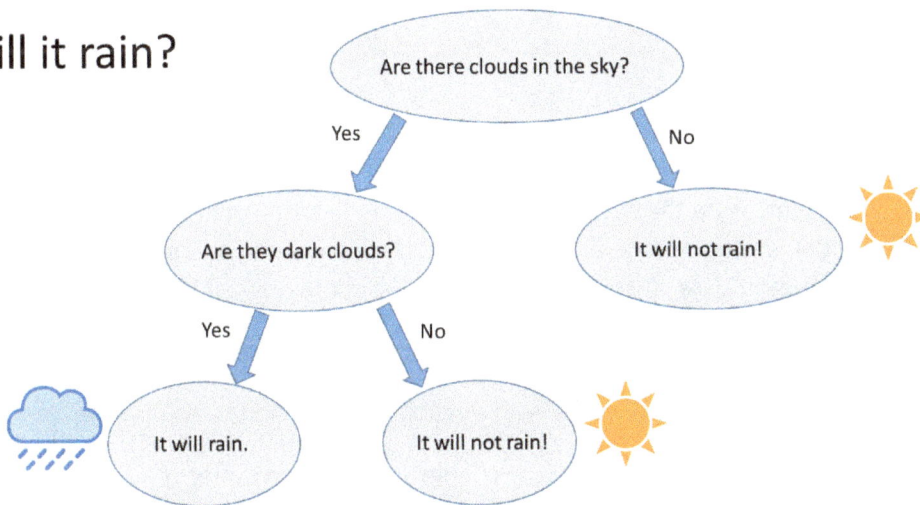

Figure 20.21

Decision trees classify observations based on their known features. In the diagram of a decision tree in figure 20.21, each node (leaf) is a group of observations, and the features that lead to the leaves are called branches. One feature in the tree would be that there are no clouds in the sky. That feature leads to the leaf on the far right of figure 20.21 that consists of cloudless days when it did not rain. Decision trees are grown with as few branches as possible.

The goal for a regression tree is to minimize the error of the tree, and the error of the tree is the sum of squared errors for each node. This can be expressed in terms of the variance of each node, as in figure 20.22.

$$S = \sum_{c=1}^{L} n_c \sigma_c^2$$

Figure 20.22

In the equation in figure 20.22, L is the total number of leaves in the tree, n sub c is the number of observations in leaf c, and sigma squared sub c is the variance of leaf c. The goal for a classification tree is a bit different, as the outcome is categorical instead of numeric. Instead of minimizing the error in terms of node variance, we can minimize the error in terms of Gini impurity. **Gini impurity** indicates how homogenous a collection of observations is. Gini is measured on a scale of 0 to 1, where 0 is perfect purity or perfect homogeneity (Hastie, Tibshirani, & Friedman, 2009). To understand how Gini works, think about the probability of finding an observation belonging to class k in any given group. That probability is the number of observations in class k, n sub k, divided by the total number of observations in the group, n.

$$p_k = \frac{n_k}{n}$$

Figure 20.23

If there are two classes, j and k, then impurity can be found by the equation in figure 20.24.

$$i(t) = 1 - p_j^2 - p_k^2$$

Figure 20.24

The equation in figure 20.24 shows the Gini impurity at node t. The probabilities of j and k can never be more than 1, so squaring them and subtracting them from 1 produces a value between 0 and 1 that shows impurity at node t. To measure the change in impurity of node t after splitting it into new nodes t sub j and t sub k, subtract the products of the probability of each class, p sub j and p sub k, and the impurity of the new node for each class, i(t sub j) and i(t sub k), from the original impurity, i(t):

$$\Delta i(t) = i(t) - p_j i(t_j) - p_k i(t_k)$$

Figure 20.25

Using the information about a node's impurity and the change in its impurity after splitting into partitions, the best split can be determined based on which split would result in the greatest reduction in impurity. It is possible for the same predictor to be used to split nodes at two different levels of the tree. For example, if one of the predictors is age group (age that has been binned into groups), splitting a node into sub nodes in which one sub node contains only people in the 18-24 age group might reduce impurity the most at the first level, and splitting the other sub node on the 70+ age group at level two might reduce impurity the most for that group.

It is possible to have two or more observations that are identical in all their features, but have different values of the dependent variable. These observations would incorrectly be grouped together, and there is nothing that can be done about it because there simply is not enough information to differentiate them. When this happens, model will have a small error.

It is also possible for two observations to be in the same leaves all the way down until the very last branch, which results in two terminal nodes with 1 observation in each node. When a decision tree is grown all the way out, perfect splits like this are common, but the resulting model will overfit the data. The process of deciding which level to stop growing branches to prevent overfitting is called **pruning**.

Pruning can be tuned through cross validation (Hastie, Tibshirani, & Friedman, 2009). We will explore cross validation in a later chapter.

To see how impurity informs node splits, consider the example shown in figure 20.26.

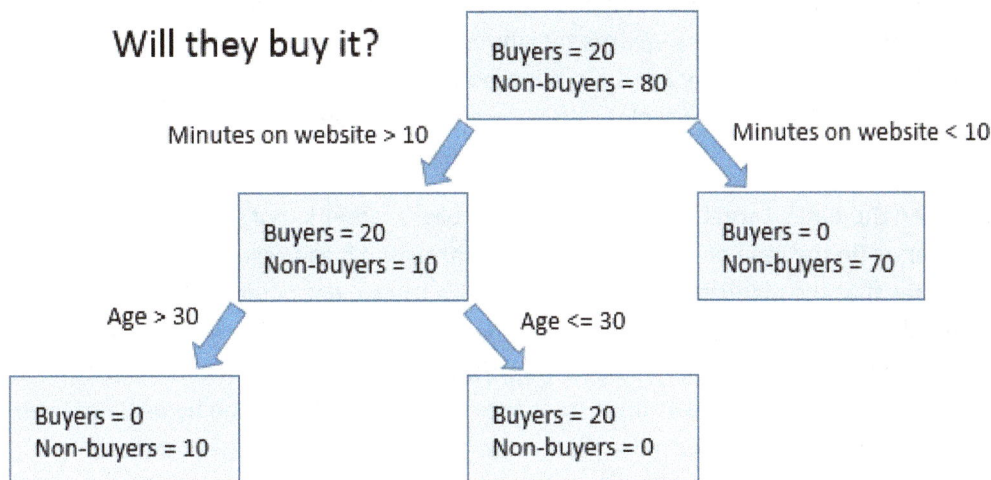

Will they buy it?

Buyers = 20
Non-buyers = 80

Minutes on website > 10

Minutes on website < 10

Buyers = 20
Non-buyers = 10

Buyers = 0
Non-buyers = 70

Age > 30

Age <= 30

Buyers = 0
Non-buyers = 10

Buyers = 20
Non-buyers = 0

Figure 20.26

The result of a classification tree is a piecewise decision boundary. Instead of using the mean of the dependent variable for each node like regression trees, classification trees use the majority vote of their nodes' class predictions to predict the dependent variable.

Decision trees (both regression and classification trees) are easy to interpret because the models they produce follow logical rules in the partitioning of the data. They do not require one-hot encoding of categorical variables, because they can handle splitting categorical variables as they are. They are also robust to multicollinearity and null or missing values, they have built in feature selection making them useful for datasets with many features, and they scale well to large datasets. For these reasons, tree based learning is a very flexible and widely used approach. The drawbacks to decision trees are that they are not very accurate, they are unstable (meaning they change drastically when there are changes to the data), and they are prone to overfitting if not properly pruned. Decision trees are also limited in their ability to handle highly complex relationships between the dependent and independent variables, because the more complex the relationship, the larger the tree must be, and larger trees are unwieldy (Hastie, Tibshirani, & Friedman, 2009). Many of these drawbacks are overcome by using ensembles of trees, but that is beyond the scope of this book.

K-Nearest Neighbors

The k-nearest neighbors (KNN) algorithm can be used for regression or classification. As its name implies, it involves choosing a value k that decides how many neighbors each data point should have. Each data point and its observations are evaluated over the feature space. In regression, the predicted outcome for an observation is the mean of the outcome for its k neighbors, just like how the predicted outcome for a node in a regression tree is the mean of the outcome for that node. In classification, the predicted outcome for an observation is the class that the majority of its neighbors have, just like how the predicted outcome for a node in a classification tree is the majority vote of the outcome for that node. An optional approach to KNN is the weight the neighbors depending on their distance to the observation whose value is being predicted. We saw this in many of the manifold learning techniques explored in chapter 17. Weighting the neighbors is not always necessary though, as KNN often does well without any weighting (James, Witten, Hastie, & Tibshirani, 2013).

The KNN algorithm does not actually produce a model. Instead, the predicted values of new observations are based on the values of their k-nearest neighbors. For this reason, KNN is referred to as a lazy algorithm. Since KNN does not produce a model, it is a poor choice of algorithm to use when it is desirable to identify relationships between variables. KNN's use is limited purely to identifying similar observations.

The choice of k depends on the data. Larger datasets usually require a larger k, but a larger k produces muddier decision boundaries (in the case of classification). K controls the bias-variance tradeoff for KNN. If k is large, it ensures that the algorithm is able to generalize to new data, but if k is too large, the algorithm could miss small patterns. A rule of thumb is to set k between 3 and 10, or set it to the square root of the number of training records for large datasets (James, Witten, Hastie, & Tibshirani, 2013). The fewer outliers there are in the data, the less important the choice of k will be. The hyperparameter optimization methods described in chapter 8 can be used to tune k, using either RMSE as the cost function for regression or cross-entropy as the cost function for classification. Cross validation can also be used to choose k, or cross validation can be combined with the hyperparameter optimization methods from chapter 8. We will see how cross validation is carried out in a later chapter.

Once k has been chosen, the neighbors are determined based on a distance metric. Euclidean distance is used most often. For categorical features, the overlap metric or Hamming distance is used most often. Other choices include the Mahalanobis distance, Manhattan distance, cosine similarity, and even the correlation coefficient. The table in figure 20.27 shows the differences between various distance metrics. Note that the most common distance metrics, Euclidean and Manhattan, are akin to calculating the RMSE and MAE, respectively.

Distance Metric	Equation	Use Case
Minkowski	$$D(X,Y) = \left(\sum_{i=1}^{n} |x_i - y_i|^p \right)^{\frac{1}{p}}$$ Where n is the number of observations and i is the ith observation.	Used for continuous variables measured on the same scale. As p increases, larger values become more influential in the distance and smaller values are neglected.
Euclidean (Minkowski Distance of order p = 2, a.k.a. the L2 Norm)	$$D(X,Y) = \sqrt{\sum_{i=1}^{n} |x_i - y_i|^2}$$ Where n is the number of observations and i is the ith observation. Note that x_i and y_i are points in k dimensional space with k coordinates. For 2D space (k=2), the equation expands to	Used for continuous variables measured on the same scale. Euclidean gives more weight to outliers than Manhattan, so it is better to use when outliers are very rare.

	the sum of $(x_1 - y_1)^2 + (x_2 - y_2)^2$, which is the form presented in high school algebra. The vectorized version of the above equation is shown below. $$D(X,Y) = \sqrt{(X-Y)^T(X-Y)}$$	
Manhattan (Minkowski Distance of order p = 1, a.k.a. the L1 Norm)	$$D(X,Y) = \sum_{i=1}^{n} |x_i - y_i|$$ Where n is the number of observations and i is the ith observation.	Used for continuous variables measured on the same scale. Research by Aggarwal, Hinneburg, & Keim, 2001 has shown that Manhattan is preferable to Euclidean in high dimensional space. Manhattan also gives less weight to outliers than Euclidean.
Mahalanobis	$$D(X,Y) = \sqrt{\sum_{i=1}^{n} \frac{(x_i - y_i)^2}{\sigma_i^2}}$$ Where n is the number of observations and i is the ith observation. The vectorized version of the above equation is shown below. $$D(X,Y) = \sqrt{(x-y)^T S^{-1}(x-y)}$$ Where S is the covariance matrix between X and Y, and S^{-1} is the inverse of the covariance matrix: $$S = \frac{1}{n}\sum_{i=1}^{n}(x_i - \mu)(x_i - \mu)^T$$	Used for continuous variables measured on the same or different scales, where the variances of the variables are not equal or when they covary. Minkowski, Euclidean, and Manhattan distance metrics do not take covariance into account, and they assume that the variances are equal. If the variables do not covary, then the set of points equidistant from a point is a sphere. If they do covary, then the set of points equidistant from a point is stretched, like the ellipse formed by correlated variables in a 2D space. Mahalanobis distance is good for detecting outliers and is commonly used with learning algorithms that make use of radial kernels. Mahalanobis distance requires knowing the covariance matrix, because it is inverted in the calculation of the distance (so it is computationally intensive too).
Hamming	$$D_{i,j} = \frac{q+r}{k}$$ Where q is the number of binary variables that equal 1 for the ith observation and 0 for the jth observation, r is the number of binary variables that equal 0 for the ith observation and 1 for the jth observation, and k is the total	Used for categorical variables that have been binary encoded. Hamming distance can also be formulated as an algorithm to determine the similarity between strings, by finding the minimum number of changes that must be made to one string to get it to equal the other.

	number of binary variables. So the distance is the ratio of the number of unmatched binary variables to the total number of binary variables.	
Gower	$$S_{i,j} = \frac{\sum_{k=1}^{n} S_{i,j,k}}{\sum_{k=1}^{n} w_k}$$ Where w sub k = 0 if the comparison is invalid (either of the values to be compared is missing) and 1 if the comparison is valid. $S_{i,j,k} = 0$ if $w_k = 0$ or if the variables being compared are categorical and their values are not equal. If $w_k = 1$, then $S_{i,j,k} = 1$ if the variables being compared are categorical and their values are equal, or $S_{i,j,k} = 1 - \lvert a_{i,k} - a_{j,k} \rvert / R(k)$ if the variables being compared are numeric. R(k) is the range of the kth variable.	Used for mixed categorical and continuous variables.
Cosine Similarity	$$\cos(\theta) = \frac{\sum_{i=1}^{n} X_i Y_i}{\sqrt{\sum_{i=1}^{n} X_i^2} \sqrt{\sum_{i=1}^{n} Y_i^2}}$$	Used for continuous variables measured on the same or different scales, where the overall similarity is more important than the absolute distances. I.e. comparing 2 stock prices where 1 is trading at \$5/share and the other at \$500/share. Cosine similarity equals Pearson correlation when the vectors X and Y are normalized by subtracting their means. It is often used to compare similarity between texts or documents in terms of their subject matter. The result ranges from -1 (completely different) to 1 (exactly the same), unless the values of X and Y are always positive, in which case it will range from 0 to 1.

Figure 20.27

Since KNN relies on calculating the distance between every pair of observations, the algorithm cannot accept any missing values. Nominal variables must also be treated using a distance metric that handle them, or by transforming them in some way to make them numeric. For example, a categorical variable could be one-hot encoded or label encoded. Recall from chapter 4 that if label encoding is used, the intervals between the categories must be the same. If they are not, then label encoding should not be used to transform the variable for use in KNN.

The benefits of using KNN are that it is insensitive to outliers and makes no assumptions about the data. The drawbacks are that it is computationally expensive because it requires calculating the distance between every two pairs of observations, it is sensitive to irrelevant or excessive features, and it is sensitive to the scale of the data. Since scale affects KNN, it is best practice to scale all of the features using standardization or normalization before applying KNN. Recall from chapter 4 that standardization is preferable to normalization when outliers should be given more weight, so take this into consideration when deciding which scaling method to use. In some datasets, like the Wisconsin Breast Cancer dataset, outliers are very informative. Since KNN is nonparametric and reasonably accurate, it is often selected as the first choice for classification problems when little is known about the data. It is often a useful benchmark for regression as well.

Data Reduction with KNN

A unique use for KNN is data reduction. When modeling huge datasets, it is often the case that a small subset of the observations are influential and the rest just "fall in line". For example, suppose we again want to model students' favorite colors on age and gender. Let us suppose our dataset consists of 2 billion observations. Many of these observations will be duplicates and therefore provide no additional value to a model being fitted to the data. But even if every observation were unique (say we measure age at the millisecond level), there would still be many observations that would not provide any additional value. If the models learns that 10 year old boys like the color red, then observations of one boy who is 10.320057 years old and likes the color red and another boy who is 10.320857 years old and likes the color red are providing the model with the same information. But if the models sees a boy who is 10.340027 years old who likes the color green, then this observation is important because it presents a contradiction to what the model has learned so far. As another example, think about human faces in photographs. If we wanted to detect human faces in a dataset of 1 million images, then the images that actually contain human faces are much more informative to the model than the images that do not. Informative observations that influence the model are called **prototypes**.

Data reduction is especially useful for classification. KNN can perform data reduction by first selecting the class outliers (the observations that KNN classifies incorrectly). The rest of the data is then split into prototypes and absorbed data, where the absorbed data are the observations that can accurately be classified by a model trained only on the prototypes.

Data reduction is dependent on prototype selection. Prototype selection is carried out the same way as feature selection. There is forward selection where observations are added incrementally and backward selection where observations are removed incrementally. The process of finding prototypes is computationally expensive and often defeats the purpose (reducing training time) of doing data reduction. Data reduction is an area that is still under active research and there are not any better options available at this time.

Nonlinear Regression and Classification with R

Let us continue using the motor trends car dataset that we started using in the last chapter. To start off, we can use polynomial regression of orders 3, 5, and 10 to model mpg on one numeric variable (disp =

displacement in cubic inches). When the results are plotted, it is obvious how higher order polynomials more closely fit the data.

```
d <- mtcars
str(d)

#Polynomial regression of orders 3, 5, 10 with 1 predictor
poly_reg3 <- lm(mpg ~ disp + poly(disp, 3, raw=T), data=d)
poly_reg5 <- lm(mpg ~ disp + poly(disp, 5, raw=T), data=d)
poly_reg10 <- lm(mpg ~ disp + poly(disp, 10, raw=T), data=d)

summary(poly_reg3)
summary(poly_reg5)
summary(poly_reg10)

plot(poly_reg3)
plot(poly_reg5)
plot(poly_reg10)

#Plot the fitted lines
x_order <- order(d$disp)
poly_reg3_preds <- predict(poly_reg3)
poly_reg5_preds <- predict(poly_reg5)
poly_reg10_preds <- predict(poly_reg10)
plot(d$disp, d$mpg, main='Polynomial Regression of Orders 3, 5, and 10', xlab='disp',
ylab='mpg')
lines(d$disp[x_order], poly_reg3_preds[x_order], col='orange', lwd=3)
lines(d$disp[x_order], poly_reg5_preds[x_order], col='green', lwd=3)
lines(d$disp[x_order], poly_reg10_preds[x_order], col='purple', lwd=3)
```

Polynomial regression for more than one numeric predictor is carried out as follows:

```
#Polynomial regression of order 3 with many predictors
poly_reg3 <- lm(mpg ~ disp + poly(disp, 3, raw=T) + hp + poly(hp, 3, raw=T) + drat +
wt, data=d)
summary(poly_reg3)
plot(poly_reg3)
```

To carry out isotonic regression, we need to simulate data that is increasing. We can do this by multiplying displacement by -1 and adding random noise to mpg. Then the regression can be fitted and plotted.

```
#Simulate a function of mpg that is increasing
dsim <- data.frame(mpg=d$mpg+rnorm(1), disp=d$disp*(-1))

#View scatter plot of x and mpg
plot(dsim$disp, dsim$mpg)

#Fit isotonic regression
iso_reg <- isoreg(dsim$disp, dsim$mpg)
iso_reg
iso_reg$yf   #Fitted values
plot(iso_reg, xlab='-1*disp', ylab='mpg')
```

The next code block shows how to apply a 6 value centered moving average, cubic spline smoothing, and LOESS smoothing. The plots show the difference between each smoother.

```
#Smoothing with moving average of a 6 value window (the observation and 6 values
around it)
library(zoo)
smooth_ma <- rollmean(dsim$mpg, k=6, align='center', na.pad=T)
smooth_ma
plot(row.names(dsim), dsim$mpg, main='Moving Average Smoothing', xlab='Row Index',
ylab='mpg')
lines(row.names(dsim), smooth_ma, col='darkgreen', lwd=3)


#Smoothing with cubic spline
library(splines)
cubic_spline <- lm(mpg ~ bs(disp, knots=c(min(dsim$disp), mean(dsim$disp),
max(dsim$disp))), data=dsim)
cubic_spline
newd <- data.frame(disp=dsim$disp)
cubic_spline_vals <- predict(cubic_spline, newdata=newd)
x_order <- order(dsim$disp)  #Store the order for the x-axis
plot(dsim$disp, dsim$mpg, main='Cubic Spline Smoothing', xlab='-1*disp', ylab='mpg')
lines(dsim$disp[x_order], cubic_spline_vals[x_order], col='red', lwd=3)


#Smoothing with LOESS
smooth_loess <- loess(mpg ~ disp, data=dsim, span=0.5)  #smoothing span of 50% of the
data
smooth_loess
smoothed_points <- predict(smooth_loess)
x_order <- order(dsim$disp)  #Store the order for the x-axis
#plot(dsim$disp, dsim$mpg, main='LOESS Smoothing', xlab='-1*disp', ylab='mpg')
lines(dsim$disp[x_order], smoothed_points[x_order], col='blue', lwd=3)
```

For GAMs, we can start by building a GAM to regress mpg on disp and wt. Note that in the model specification, the "s" denotes a smoothing spline. This model is essentially fitting additive splines.

```
library(gam)
gam_model <- gam(mpg ~ s(disp, df=6) + s(wt, df=6), data=d)  #Note s = smoothing
spline
summary(gam_model)
plot(gam_model, se=T, main="GAM Smoothing for Variable")
```

When categorical variables of two or more categories are added to the GAM, they are not fitted with splines.

```
#Adding a binary variable to the GAM
gam_model_b <- gam(mpg ~ s(disp, df=6) + s(wt, df=6) + vs, data=d)
summary(gam_model_b)
plot(gam_model_b, se=T, main="GAM Smoothing for Variable")


#Adding a categorical variable to the GAM
d$hp_bin <- cut(d$hp, 5)
table(d$hp_bin)
gam_model_b_c <- gam(mpg ~ s(disp, df=6) + s(wt, df=6) + vs + hp_bin, data=d)
summary(gam_model_b_c)
plot(gam_model_b_c, se=T, main="GAM Smoothing for Variable")
```

```
#Compare models
anova(gam_model, gam_model_b, gam_model_b_c)
```

We can create a GAM to perform classification too. This is essentially a logistic regression of a categorical variable on additive splines, plus any categorical variables. Note that the I() function in R is used to cut a numeric variable into bins. In this case, we are cutting mpg into bins of < 20 and >= 20 to treat it as a categorical variable.

```
#Performing classification (logistic regression) with the GAM
gam_model_class <- gam(I(mpg < 20) ~ s(disp, df=6) + s(wt, df=6) + vs + hp_bin,
data=d, family=binomial)
summary(gam_model_class)
plot(gam_model_class, se=T, main="GAM Smoothing for Variable")
#This models the conditional probabilities for mpg being < 20 and >=20
#Note the y-axis of these plots is the logit
```

Regression and classification trees are most easily fitted using the rpart package. The method argument determines whether the tree is for regression or classification. Below, we use a regression tree to predict mpg using all of the other variables in the motor trends cars dataset. Then we use a classification tree to predict species using all of the other variables in the Iris flower dataset.

```
#Regression tree - use rpart method='anova'
library(rpart)
reg_tree <- rpart(mpg ~ ., data=mtcars, method='anova',
                  control=rpart.control(minsplit=20, cp=0.001))
summary(reg_tree)
rsq.rpart(reg_tree)
plot(reg_tree)
text(reg_tree)
```

```
#Classification tree - use rpart method='class'
library(rpart)
class_tree <- rpart(Species ~ ., data=iris, method='class',
                    control=rpart.control(minsplit=20, cp=0.001))
summary(class_tree)
plot(class_tree)
text(class_tree)
```

We have not been following best practices by splitting the datasets we have been fitting. The examples here are just showing how to apply the various modeling techniques described in the chapter. The reader should note that if these techniques are applied to data in the wild, it is important that the models are trained and tested on separate datasets.

KNN regression and classification can be performed as follows, using the FNN and class libraries.

```
#Split data into training and test sets
library(caret)
set.seed(14)
train_indices <- createDataPartition(y=mtcars$mpg, p=0.7, list=F)
train <- mtcars[train_indices,]
test <- mtcars[-train_indices,]

#Normalize numeric variables
```

```
normalize <- function(x) {
  return ((x - min(x)) / (max(x) - min(x)))
}
train_norm <- as.data.frame(sapply(train, normalize), row.names=row.names(train))
test_norm <- as.data.frame(sapply(test, normalize), row.names=row.names(test))

#KNN Regression, k=3
library(FNN)
knn_reg <- knn.reg(train=train_norm, test=test_norm, y=train_norm$mpg, k=3,
algorithm='kd_tree')
knn_reg
plot(mtcars$mpg, knn_reg$pred, xlab="actual mpg", ylab="predicted mpg")
print(c("RMSE:", sqrt(mean(knn_reg$residuals^2))))

#Split data into training and test sets
library(caret)
set.seed(14)
train_indices <- createDataPartition(y=iris$Species, p=0.7, list=F)
train <- iris[train_indices,]
test <- iris[-train_indices,]

#Normalize numeric variables
normalize <- function(x) {
  return ((x - min(x)) / (max(x) - min(x)))
}
train_norm <- as.data.frame(sapply(train[,-which(colnames(train) %in% c('Species'))],
normalize), row.names=row.names(train))
test_norm <- as.data.frame(sapply(test[,-which(colnames(train) %in% c('Species'))],
normalize), row.names=row.names(test))
#Label encode target
train_norm$Species <- as.integer(train$Species)
test_norm$Species <- as.integer(test$Species)

#KNN Classification, k=3
library(class)
knn_classifier <- class::knn(train=train_norm, test=test_norm, cl=train$Species, k=3,
prob=F)
knn_preds <- as.integer(knn_classifier)
library(MLmetrics)
LogLoss(knn_preds, test_norm$Species)
AUC(knn_preds, test_norm$Species)
Accuracy(knn_preds, test_norm$Species)
#View the confusion matrix with predicted values on the left
prop.table(table(knn_preds, test_norm$Species))
```

The number of neighbors has been fixed at 3, and the scaling method chosen was normalization.

Nonlinear Regression and Classification with Python

Let us continue using the motor trends car dataset that we started using in the last chapter. To start off, we can use polynomial regression of orders 3, 5, and 10 to model mpg on one numeric variable (disp = displacement in cubic inches). When the results are plotted, it is obvious how higher order polynomials more closely fit the data.

```
import pandas as pd
```

```python
import numpy as np
from scipy import stats
from sklearn.linear_model import LinearRegression
import matplotlib.pyplot as plt
from sklearn.isotonic import IsotonicRegression
from sklearn.tree import DecisionTreeClassifier, DecisionTreeRegressor

d = pd.read_csv('mtcars.csv')
d.info()

def get_lr_output(lr, x, y):
    #Takes a fitted sklearn linear regression as input and outputs results
    #Thanks to stackoverflow: https://stackoverflow.com/questions/27928275/find-p-
value-significance-in-scikit-learn-linearregression
    import numpy as np
    parameters = np.append(lr.intercept_, lr.coef_)
    predictions = lr.predict(x)
    X = pd.DataFrame({"Constant":np.ones(len(x))}).join(pd.DataFrame(x))
    mse = (sum((y-predictions)**2))/(len(X)-len(X.columns))
    sd = np.sqrt(mse*(np.linalg.inv(np.dot(X.T, X)).diagonal()))
    ts = parameters/sd
    p_values = [2*(1-stats.t.cdf(np.abs(i), (len(X)-1))) for i in ts]
    sd = np.round(sd, 4)
    ts = np.round(ts, 4)
    p_values = np.round(p_values, 4)
    parameters = np.round(parameters, 4)
    predictors = pd.Series('intercept')
    predictors = predictors.append(pd.Series(x.columns)).reset_index()[0]
    output_df = pd.DataFrame()
    output_df['predictor'] = predictors
    output_df['coefficient'] = parameters
    output_df['standard_error'] = sd
    output_df['t-statistic'] = ts
    output_df['p-value'] = p_values
    r_squared = round(lr.score(x, y), 4)
    adj_r_squared = round(1-(1-lr.score(x, y))*(len(y)-1)/(len(y)-x.shape[1]-1), 4)
    print(output_df, '\n R-squared:', r_squared, '\n Adjusted R-squared:',
adj_r_squared)
    return [output_df, r_squared, adj_r_squared]

linear_reg = LinearRegression()
linear_reg.fit(d.drop(['mpg'], axis=1), d['mpg'])
linear_reg_output = get_lr_output(linear_reg, d.drop(['mpg'], axis=1), d['mpg'])

def polynomial_regression(x, y, deg=1, x_exclude_from_poly=[]):
    #Takes dataframe input and outputs a fitted polynomial regression of order deg
    #The list passed as x_exclude_from_poly has features that should not have
polynomial terms
    #PolynomialFeatures transforms a formula like a+b+c to a^2+ab+ac+b^2+bc+c^2 for
example of 2nd degreee
    #Example: polynomial_regression(df[['a', 'b', 'c']], df['d'], deg=2,
x_exclude_from_poly=['c'])
    from sklearn.preprocessing import PolynomialFeatures
    from sklearn.linear_model import LinearRegression
    import numpy as np
```

```
    x_poly = x.drop(x_exclude_from_poly, axis=1)
    x_remainder = x[[col for col in x.columns if col not in x_poly.columns]]
    poly = PolynomialFeatures(degree=deg)
    x_poly = poly.fit_transform(np.asarray(x_poly).reshape(-1,len(x_poly.columns)))
    x_poly = pd.DataFrame(x_poly[:,1:])  #Gets rid of first column of 1's
    colnames = poly.get_feature_names()[1:]
    x_poly.columns = colnames
    x = pd.concat([x_poly, x_remainder], axis=1)
    linear_reg = LinearRegression().fit(x, y)
    return linear_reg, x

#Polynomial regression of orders 3, 5, 10 with 1 predictor
poly_reg_three, x_poly_three = polynomial_regression(d[['disp']], d['mpg'], deg=3)
poly_reg_three_output = get_lr_output(poly_reg_three, x_poly_three, d['mpg'])
poly_reg_five, x_poly_five = polynomial_regression(d[['disp']], d['mpg'], deg=5)
poly_reg_five_output = get_lr_output(poly_reg_five, x_poly_five, d['mpg'])
poly_reg_ten, x_poly_ten = polynomial_regression(d[['disp']], d['mpg'], deg=10)
poly_reg_ten_output = get_lr_output(poly_reg_ten, x_poly_ten, d['mpg'])

print(poly_reg_three_output)
print(poly_reg_five_output)
print(poly_reg_ten_output)


#Plot the fitted lines
order = d['disp'].sort_values().index.tolist()
plt.scatter(d['disp'], d['mpg'])
plt.plot(d['disp'][order], poly_reg_three.predict(x_poly_three)[order],
color='orange')
plt.plot(d['disp'][order], poly_reg_five.predict(x_poly_five)[order], color='green')
plt.plot(d['disp'][order], poly_reg_ten.predict(x_poly_ten)[order], color='purple')
plt.xlabel('disp')
plt.ylabel('mpg')
plt.title('Polynomial Regression of Orders 3, 5, and 10')
plt.show()
```

Polynomial regression for more than one numeric predictor is carried out as follows:

```
#Polynomial regression of order 3 with many predictors
poly_reg_three_complex, x_poly = polynomial_regression(d[['disp', 'hp', 'drat',
'wt']], d['mpg'], deg=3, x_exclude_from_poly=['drat', 'wt'])
poly_reg_three_complex_output = get_lr_output(poly_reg_three_complex, x_poly,
d['mpg'])
print(poly_reg_three_complex_output)
```

To carry out isotonic regression, we need to simulate data that is increasing. We can do this by multiplying displacement by -1 and adding random noise to mpg. Then the regression can be fitted and plotted.

```
#Simulate a function of mpg that is increasing
dsim = d[['mpg', 'disp']].copy()
dsim['mpg'] = dsim['mpg']+np.random.normal()
dsim['disp'] = dsim['disp']*(-1)
dsim.head()
#Convert df to numpy array
dsim_x = np.asarray(dsim['disp'])
```

```
dsim_y = np.asarray(dsim['mpg'])

#View scatter plot of x and mpg
plt.scatter(dsim['disp'], dsim['mpg'])
plt.xlabel('disp')
plt.ylabel('mpg')
plt.title('Scatterplot of x and y')
plt.show()

#Fit isotonic regression
iso_reg = IsotonicRegression()
print(iso_reg.get_params())
iso_fitted_values = iso_reg.fit_transform(dsim_x, dsim_y)
iso_predictions = iso_reg.predict(dsim_x)
print('R squared:', iso_reg.score(dsim_x, dsim_y))

#Plot the fitted line
order = dsim['disp'].sort_values().index.tolist()
plt.scatter(dsim['disp'], dsim['mpg'])
plt.plot(dsim['disp'][order], iso_fitted_values[order], color='brown')
plt.xlabel('disp')
plt.ylabel('mpg')
plt.title('Isotonic Regression')
plt.show()
```

The next code block shows how to apply a 6 value centered moving average and LOESS smoothing. The plots show the difference between each smoother.

```
#Smoothing with moving average of a 6 value window (the observation and 6 values
around it)
rolling_window = pd.Series(dsim['mpg']).rolling(window=6, center=True)
smooth_ma = rolling_window.mean()
print(smooth_ma)
plt.scatter(list(dsim.index), dsim['mpg'])
plt.plot(list(dsim.index), smooth_ma, color='darkgreen')
plt.xlabel('Row Index')
plt.ylabel('mpg')
plt.title('Moving Average Smoothing')
plt.show()

#Smoothing with LOESS
import statsmodels.api as sm
loess = sm.nonparametric.lowess(dsim_y, dsim_x)
loess_x = list(zip(*loess))[0]
loess_y = list(zip(*loess))[1]
plt.scatter(dsim['disp'], dsim['mpg'])
plt.plot(loess_x, loess_y, color='blue')
plt.xlabel('-1*disp')
plt.ylabel('mpg')
plt.title('LOESS Smoothing')
plt.show()
```

For GAMs, we can start by building a GAM to regress mpg on disp and wt. The Python implementation of GAMs in Pygam uses the Bayesian method, which will be described in a few chapters. Like the R implementation, Bayesian GAMs are essentially fitting additive splines.

```
from pygam import LinearGAM, LogisticGAM
gam_model = LinearGAM().fit(d[['disp', 'wt']], d['mpg'])
print(gam_model.summary())
gam_predictions = gam_model.predict(d[['disp', 'wt']])
gam_mse = np.mean((gam_predictions-d['mpg'])**2)
print('MSE:', gam_mse)

#Plot the predictions with confidence intervals
plt.plot(list(d.index), gam_predictions, 'r--')
plt.plot(list(d.index), gam_model.prediction_intervals(d[['disp', 'wt']], width=.95),
color='b', ls='--')
plt.scatter(list(d.index), d['mpg'], facecolor='gray', edgecolors='none')
plt.xlabel('Row Index')
plt.ylabel('mpg')
plt.title('GAM Prediction with 95% Condidence Interval')
plt.show()

#Plot with simulated posterior
for response in gam_model.sample(d[['disp', 'wt']], d['mpg'], quantity='y',
n_draws=50, sample_at_X=d[['disp', 'wt']]):
    plt.scatter(list(d.index), response, alpha=0.03, color='k')
plt.plot(list(d.index), gam_predictions, 'r--')
plt.plot(list(d.index), gam_model.prediction_intervals(d[['disp', 'wt']], width=.95),
color='b', ls='--')
plt.xlabel('Row Index')
plt.ylabel('mpg')
plt.title('GAM Prediction with 95% Condidence Interval')
plt.show()

#Plot the partial dependencies of the predictors with confidence intervals
plt.rcParams['figure.figsize'] = (12, 8)
fig, axs = plt.subplots(1, len(list(d[['disp', 'wt']].columns)))
titles = list(d[['disp', 'wt']].columns)
for i, ax in enumerate(axs):
    partial_dep, confidence = gam_model.partial_dependence(d[['disp', 'wt']],
feature=i+1, width=0.95)
    print(partial_dep)
    order = d[['disp', 'wt']][titles[i]].sort_values().index.tolist()
    ax.plot(d[['disp', 'wt']][titles[i]].values[order], partial_dep[order])
    ax.plot(d[['disp', 'wt']][titles[i]].values[order], confidence[0][:, 0][order],
c='grey', ls='--')
    ax.plot(d[['disp', 'wt']][titles[i]].values[order], confidence[0][:, 1][order],
c='grey', ls='--')
    ax.set_title(titles[i])
plt.show()
#The strength & direction of the relationship corresponds to the slope of the line
#Nonlinear lines should have smoothing applied (they are already smoothed in this
example)

#Try different hyperparameters
spline_exp = [3, 3]      #Type of spline to fit to each variable
nbr_splines = [10, 20]   #This must be > spline order
gam_model = LinearGAM(spline_order=spline_exp, n_splines=nbr_splines).fit(d[['disp',
'wt']], d['mpg'])
print(gam_model.summary())
```

```
gam_predictions = gam_model.predict(d[['disp', 'wt']])
gam_mse = np.mean((gam_predictions-d['mpg'])**2)
print('MSE:', gam_mse)
```

When categorical variables of two or more categories are added to the GAM, they are not fitted with splines.

```
#Add binary and categorical predictors to the GAM
from sklearn.preprocessing import LabelEncoder
encoder = LabelEncoder()
d['h_bin'] = encoder.fit_transform(pd.cut(d['hp'], 5))
gam_model = LinearGAM().fit(d[['disp', 'wt', 'vs', 'h_bin']], d['mpg'])
print(gam_model.summary())
gam_predictions = gam_model.predict(d[['disp', 'wt', 'vs', 'h_bin']])
gam_mse = np.mean((gam_predictions-d['mpg'])**2)
print('MSE:', gam_mse)
```

We can create a GAM to perform classification too. This is essentially a logistic regression of a categorical variable on additive splines, plus any categorical variables. In this case, we are cutting mpg into bins of < 20 and >= 20 to treat it as a categorical variable.

```
#Performing classification (logistic regression) with the GAM
d['mpg_bin'] = encoder.fit_transform(pd.cut(d['mpg'], [0, 20, 100]))
gam_model = LogisticGAM().gridsearch(d[['disp', 'wt', 'vs', 'h_bin']], d['mpg_bin'])
print(gam_model.summary())
print('Classification Accuracy:', gam_model.accuracy(d[['disp', 'wt', 'vs',
'h_bin']], d['mpg_bin']))
```

Note that the y-axis in the resulting plot is the logit. The Bayesian GAM is modeling the conditional probabilities of an observation belonging to each class (<20 mpg and >=20mpg).

Regression and classification trees are most easily fitted using the rpart package. The method argument determines whether the tree is for regression or classification. Below, we use a regression tree to predict mpg using all of the other variables in the motor trends cars dataset. Then we use a classification tree to predict species using all of the other variables in the Iris flower dataset.

```
#Regression tree
reg_tree = DecisionTreeRegressor(criterion='mse',
min_samples_split=20).fit(d.drop(['mpg', 'mpg_bin'], axis=1), d['mpg'])
print(reg_tree.get_params)
for i, f in enumerate(d.drop(['mpg', 'mpg_bin'], axis=1).columns):
    print('Importance of', f, reg_tree.feature_importances_[i])
reg_tree_predictions = reg_tree.predict(d.drop(['mpg', 'mpg_bin'], axis=1))
print('MSE:', reg_tree.score(d.drop(['mpg', 'mpg_bin'], axis=1), d['mpg']))

iris = pd.read_csv('iris.csv')
iris.info()

#Classification tree
class_tree = DecisionTreeClassifier(criterion='gini',
min_samples_split=20).fit(iris.drop(['Species'], axis=1), iris['Species'])
print(class_tree.get_params)
class_tree_predictions = class_tree.predict(iris.drop(['Species'], axis=1))
```

```
class_tree_prob_predictions = class_tree.predict_proba(iris.drop(['Species'],
axis=1))
print('Classification Accuracy:', class_tree.score(iris.drop(['Species'], axis=1),
iris['Species']))
```

We have not been following best practices by splitting the datasets we have been fitting. The examples here are just showing how to apply the various modeling techniques described in the chapter. The reader should note that if these techniques are applied to data in the wild, it is important that the models are trained and tested on separate datasets.

KNN regression and classification can be performed as follows.

```
from sklearn.model_selection import train_test_split
from sklearn.preprocessing import LabelEncoder, MinMaxScaler
from sklearn.neighbors import KNeighborsRegressor, KNeighborsClassifier
from sklearn import metrics

#KNN regression
x = d.drop(['mpg'], axis=1).values
y = d[['mpg']].values
normalizer = MinMaxScaler()
x = normalizer.fit_transform(x)
y = normalizer.fit_transform(y)
x_train, x_test, y_train, y_test = train_test_split(x, y, test_size=0.3,
random_state=14)

knn_reg = KNeighborsRegressor(n_neighbors=3)
knn_reg.fit(x_train, y_train)
preds = knn_reg.predict(x_test)
plt.scatter(y_test, preds)
plt.xlabel('actual mpg')
plt.ylabel('predicted mpg')
plt.show()
print('RMSE:', np.sqrt(metrics.mean_squared_error(y_test, preds)))

#KNN classification, k=3
encoder = LabelEncoder()
categorical_vars = ['Species']
categorical_var_mapping = dict()
for cv in categorical_vars:
    iris[cv] = encoder.fit_transform(iris[cv])  #Encodes as integer
    categorical_var_mapping[cv] = list(encoder.classes_)  #Saves integer to category
mapping

x = iris.drop(['Species'], axis=1).values
y = iris[['Species']].values
normalizer = MinMaxScaler()
x = normalizer.fit_transform(x)
x_train, x_test, y_train, y_test = train_test_split(x, y, test_size=0.3,
random_state=14)

knn_class = KNeighborsClassifier(n_neighbors=3)
knn_class.fit(x_train, y_train.ravel())
preds = knn_class.predict(x_test)
```

```
pred_probs = knn_class.predict_proba(x_test)
print(metrics.log_loss(y_test, pred_probs, labels=np.array([0,1,2])))
print(metrics.accuracy_score(y_test, preds))
#View the confusion matrix with predicted values on the left
print(metrics.confusion_matrix(y_test, preds))
```

The number of neighbors has been fixed at 3, and the scaling method chosen was normalization.

21. Multilevel Models

Multilevel linear models are designed to model data with hierarchical structure. The classic example is modeling some aspect of student behavior. Students can be grouped into classes, which can be grouped into schools, which can be grouped into counties, and so on. A researcher could expand the scope of hierarchy arbitrarily, but there is usually some limitation in data collection that prevents this. It would not make sense, for example, to include the county level in a model if the data is limited to only a handful of schools. In this example, the county level would only be useful if there were data for every school in multiple counties.

A reasonable question to ask would be why hierarchies could not simply be added as a categorical variable. For example, if student grade point average (GPA) were being modeled on time spent studying, why could not class be added as a covariate and ANCOVA be performed? There is no reason why that could not be done, but multilevel models offer advantages to ANCOVA and other generalized linear models when hierarchy is present in the data. Specifically, multilevel models do not require the assumptions of homogeneity of variance among different groups of the dependent variable, nor do they require the assumption that observations are independent. This allows multilevel models to be used for repeated measures design experiments. Additionally, by adding a categorical variable to represent level, there is an implicit assumption that the effects of the other variables are the same at every level, which is often a flawed assumption. For example, suppose race and state were added to the student GPA model. Now there would be hierarchy levels for class and state. If those features were simply used as categorical variables, then the implicit assumption would be that the effects of time spent studying and race on GPA were uniform regardless of class or state. This assumption would be flawed.

Another reasonable question to ask would be why the data could not simply be aggregated to the higher level. For example, if student GPA were being modeled on time spent studying and class, why could not time spent studying and GPA be averaged by class, and a model built off of the simplified dataset? The problem with this approach is that it discards information about the within-group variance, and inflates the parameters of the aggregated variables. Likewise, disaggregating higher level data to a lower level is also a bad idea. It is also bad to create new variables aggregated at the higher level and include them with the lower level variables. For example, if the average time spent studying by class were added as a new variable, then the model of GPA would be have individual level time spent studying, class, and class average time spent studying as independent variables. The obvious problem with this is that average time spent studying by class would be repeated for each student, or more generally, the aggregate variable at the higher level would be duplicated for every unique observation at the lower level. This would inflate the fit of the model and the coefficients, because class and class average time spent studying would account for overlapping variance in the model. If there were other numeric variables in the model, this could also introduce collinearity. So it is best not to re-aggregate data for different levels, as it would produce erroneous regression results.

Multilevel models allow the coefficients of predictors to vary by each level of the hierarchies in the data. This makes them very flexible. For example, if income were being predicted by age, occupation, and city, then it is likely that the coefficients of age and occupation should vary by city, as cities with higher costs of living should also have higher incomes. Each hierarchical grouping in the data is called a level. So a 1 level model is used for data that are all on the same level – this would be no different than a general linear model. A level 2 model is used for data that have 2 levels, like student and class. A level 3 model is used for data that have 3 levels, like student, class, and school. The dependent variable in multilevel models must be examined for the lowest level of the hierarchy.

Multilevel models allow the assumption of independence between observations to be tossed aside, which means that multilevel models can handle data with multiple measurements of the same observations (Field, Miles, & Field, 2012). For example, repeated measures design experiments, which often cannot be modeled using other methods, can successfully be modeled using a multilevel model. Time series data can also be analyzed using a multilevel model. Missing data is less problematic in multilevel models because the missing values can be estimated by the model with decent accuracy.

Random Effects and Fixed Effects Models

Multilevel models can be random effects models, fixed effects models, or mixed effects models (a combination of random and fixed effects).

A **random effects model** is a model that allows either the intercept or the slope (coefficient) of one or more predictors to fluctuate. The fluctuation can be over time or for different levels of a hierarchy. The type of fluctuation depends on the data. If it is panel data or time series data, then the fluctuation is over time. We will explore time series models in a later chapter. In this chapter, we will focus on data with a hierarchical structure, as in the case of multilevel models, so the fluctuation of a random effects model is over the different levels of the hierarchy. Random effects models assume that observation or individual specific effects are uncorrelated with the independent variables. Random effects models can either be random slope (the intercept is fixed), random intercept (the coefficients of the predictors are fixed), or random slope and intercept (the intercept and the coefficients of the predictors are allowed to vary). A random effects model takes the form shown in figure 21.1.

$$Y_{ij} = \mu + U_i + W_{ij}$$

Figure 21.1

In the equation in figure 21.1, Y sub ij is the outcome for observation j in the ith level of a hierarchy, mu is the population mean of Y, U sub i is the level specific random effect, and W sub ij is the individual specific effect. U sub i shows the difference between the average of Y for level i versus the average of Y for the entire population. W sub ij shows the deviation of the jth observation's value of Y from the average of the ith level of the hierarchy.

A **fixed effects model** is a model that holds both the intercept and the slopes (coefficients) of the predictors constant. Fixed effects models assume that observation or individual specific effects are correlated with the independent variables. Fixed effects models are used when the group means are non-random, that is; when the samples in the dataset are selected non-randomly. In panel data, the group means are individual specific means, since the same individual has multiple measurements over time. The simplest way of fixing effects is to include dummy variables for each individual or observation in the dataset, minus one to prevent multicollinearity. By doing this, it is easier to quantify the correlation between the individual specific effects and the independent variables, because the individuals become binary independent variables themselves.

It is important to note that in an experimental setting, the results of fixed effects models cannot generalize beyond the experiment, because by definition, fixed effects models assume that all of the conditions that could possibly be measured are included in the experiment. Random effects models do not have this restriction.

A **mixed effects model** is a blend of random and fixed effects. It is essentially a random effects model with dummies added for different categorical representations. For example, a random effects model predicting GPA from time spent studying, where the coefficient of time spent studying is allowed to vary by class (class is the hierarchy in the dataset), would become a mixed effects model if dummies for race and gender were added.

To build a multilevel model, it is best to start with a fixed model and randomize 1 parameter at a time. The models can be compared by subtracting the log-likelihood of the new model from the old. The caveat is that models can only be compared if they are estimated by MLE and if the same effects are used in both models. If the researcher chooses to scale the data using normalization before building a multilevel model, normalization could be carried out as normal, or it could be applied at the group level, meaning the observations in each group are normalized independently. When the results of a model that is fitted to normalized data are interpreted, it should be noted that the intercept shows the value of the outcome when the independent variables equal their average values.

When to Use Multilevel Models

Sometimes it is unclear whether or not a multilevel model is appropriate for a dataset. Maybe it is uncertain whether or not hierarchies exist in the data, and if they do, it is possible they would not have significant effects that could reduce the performance of standard linear models. To determine if a multilevel model is needed, it is necessary to determine if there is significant variation between the groups or levels of the hierarchy. This can be done by looking at the intra-class correlations.

The **intra-class correlation (ICC)** is the correlation between two observations within the same level of the hierarchy. The higher the ICC, the lower the variance is within the level, which means that the variance between different levels is high. What constitutes a high ICC depends on the data and the context, so it is not possible to use a one-size fits all heuristic, but like any correlation the ICC ranges from 0 meaning no correlation to 1 meaning perfect correlation. When the ICC is large, it is a good indication that 1) multilevel modeling should be used, and 2) random intercepts should be used in the multilevel model. If the model has random intercepts and fixed slopes, then the ICC is the same as the variation within each level. The variation within a level is called the **variance partition coefficient (VPC)**. The ICC and VPC diverge when random slopes are introduced into the model, because the ICC is a function of the variables with random slopes. If the independent variables with random slopes are continuous, there could be an infinite number of ICC's. If the independent variables with random slopes are categorical, then there could be as many ICC's as categories. In these situations, most statistical packages calculate a single ICC based on a value of zero for the coefficients of all the independent variables with random slopes. This means that the ICC that is computed in by those packages is the ICC for a fixed slope model only.

A simpler way to think of the ICC is to consider exam scores for students in different schools. It is plausible that students within the same school should have scores that correlate with one another more closely than with scores for students in other schools. This is intra-class correlation. The level 1 units (the students) have exam scores that correlate within the level 2 units (the schools). We will explore this exact problem at the end of this chapter.

After it is known that a multilevel model is required, it must be decided whether a fixed effects model should be used or a random effects model. If the data does not come from an experiment that dictates through its design which model would be more appropriate, then one way to do choose is to simply create both types of models and compare them. For example, first create a generalized least squares model with only a fixed intercept as the predictor (i.e. a fixed intercept model), and then create a second model with only a random intercept as the predictor (i.e. a random intercept model). Then compare the results to see which model is a better fit.

Multilevel Models with R

Multilevel models are much easier to build with R than with Python, so there is no Python section for this chapter. For this chapter, we will use a dataset from the University of Bristol Centre for Multilevel Modeling that contains exam scores for 1,905 students from 73 schools in England (University of Bristol Centre for Multilevel Modeling, 2018 and Goldstein, et. al., 1993). The dataset can be downloaded here: http://www.bristol.ac.uk/cmm/learning/support/datasets/. The dataset contains the following fields: school number, student number for the school, a binary flag for gender where 0 = male and 1 = female, the score for an essay on the exam, and the final exam score. Multilevel models in R can be built with the lme4 or nlme packages. We will use the lme4 package.

```
library(lme4)

#Data has exam scores (both the total and the score for an essay on the exam), for
1905 students from 73 schools in England
#Data from http://www.bristol.ac.uk/cmm/learning/support/datasets/
d <- read.table('exam_scores.DAT', header=F, sep='')
colnames(d) <- c('school', 'student', 'female_flag', 'essay_score', 'exam_score')
d <- d[d$school!='\032',]  #Gets rid of junk row
d$school <- as.integer(d$school)
head(d)
```

Let us start by constructing a standard linear regression of exam score on school, the femal flag, and essay score.

```
#Linear regression of exam_score on school, female_flag, and essay_score
linear_reg <- lm(exam_score ~ school + female_flag + essay_score, data=d)
summary(linear_reg)
preds <- predict(linear_reg, d[,-c(2,5)])
rmse <- sqrt(mean((preds-d$exam_score)^2))
rmse  #15.1
```

The RMSE is 15.1, meaning the average error of a standard linear regression is about 15 points. This will serve as a baseline for comparing the next few models we will build. There is an hierarchy in this dataset: students belong to schools. Instead of using a multilevel model to account for the hierarchical structure of the data, what would happen if we simply dummy encoded school?

```
#Linear regression of exam_score on school dummies, female_flag, and essay_score
d$school <- as.factor(d$school)
dummies <- model.matrix(~d$school)[,-1]
colnames(dummies) <- substr(colnames(dummies), 3, 100)
d <- cbind(d[,-1], dummies)
d <- d[,-1]  #Drop student
linear_reg <- lm(exam_score ~ ., data=d)
```

```
summary(linear_reg)
preds <- predict(linear_reg, d[,-3])
rmse <- sqrt(mean((preds-d$exam_score)^2))
rmse  #12.5
```

The fit has improved over the standard linear regression, however we know that there is a hierarchical structure to this data. That means that the level 1 units (the students) within the same level 2 units (the schools) will be correlated. By dummy encoding, we are ignoring this correlation. Let us ignore the hierarchy for a moment longer though. Another way we could model this data is to split it into several models. What if we try building a separate regression for each school?

```
#Many linear regressions (1 for each school) of exam_score on female_flag, and
essay_score
d <- read.table('exam_scores.DAT', header=F, sep='')  #Read data back in to reset
everything
colnames(d) <- c('school', 'student', 'female_flag', 'essay_score', 'exam_score')
d <- d[d$school!='\032',]  #Gets rid of junk row
d$school <- as.integer(d$school)
schools <- unique(d$school)
linear_regs <- by(data=d, INDICES=d$school, FUN=function(x) {lm(exam_score ~ school +
female_flag + essay_score, data=d)})
```

We can get the error of each model by iterating through the list. This exercise will be left to the reader, but a quick inspection will reveal that there has not been much improvement in the mean error over the use of school dummies. So we will proceed with multilevel modeling to see how we can improve fit by accounting for ther hierarchy in the data. The first model we will try is a varying intercept model. In this model, the intercept will be allowed to vary by the group in the hierarchy (so each school will have a different intercept).

```
#Use multilevel model to account for exam_score differences between schools and
within schools
#start with varying intercept model - allows intercept to vary by the group variable
school
ml_reg <- lmer(exam_score ~ (1 | school) + female_flag + essay_score, data=d)
summary(ml_reg)
preds <- predict(ml_clust_reg, d[,-c(2,5)])
rmse <- sqrt(mean((preds-d$exam_score)^2))
rmse  #12.5
```

The average RMSE across all of the models is 12.5. We still have not improved over the use of school dummies. One way we could improve the fit is through feature engineering. It is plausible that there are different types of students in each of the schools. Some students will study hard and score well on both the essay and the exam as a whole. Others may poor essay writers but still score high on the exam overall. Others will likely have low scores in both the essay and overall. If we apply an unsupervised learning method like k-means clustering, we can see if there are any natural groupings of students.

```
#Create artificial grouping variable using k-means
kclusters <- kmeans(as.matrix(d[,c('female_flag', 'essay_score')]), centers=5)
library(fpc)
plotcluster(d[,c('school', 'essay_score')], kclusters$cluster)
d$cluster <- kclusters$cluster  #Assumes the row ordering is the same
```

Lookin at the plot, k-means has clearly separated students by essay score. We can use the cluster number as a new feature in the regression. Do not be distracted by k-means. It is nothing fancy, because all it has accomplished is finding natural bins for students based on their essay scores. Other binning methods may be efficient as well, but by using k-means, which clusters based on Euclidean distance, we can be sure that the binnings are good fits. We can also think of student bins as an artificial hierarchy. It does not make much sense to do this, but it will help show how the multilevel model specification would change if there were a legitimate second level to the hierarchy. So we will use the cluster number as an artificial level 3, above the school level. Let us now rebuild the varying intercept model, using nesting for cluster and school. That is, we will consider student exam scores within the same cluster to be correlated across schools, and student exam scores within the same school to be correlated.

```
#Varying intercept model with mixed effect term for varying intercepts by cluster and
schools within clusters
ml_clust_reg <- lmer(exam_score ~ (1 | cluster/school) + female_flag + essay_score,
data=d)
summary(ml_clust_reg)
preds <- predict(ml_clust_reg, d[,-c(2,5)])
rmse <- sqrt(mean((preds-d$exam_score)^2))
rmse  #11.6
```

The RMSE has been reduced to 11.6, which is a marked improvement over standard linear regression, and a slight improvement over the varying intercept model without cluster. What would happen if we got rid of the nesting? In other words, what if we build the same varying intercept model but assume that student exam scores are correlated within clusters, and separately correlated within schools?

```
#Alternatively, do a varying intercept model without nesting
ml_clust_reg <- lmer(exam_score ~ (1 | cluster) + (1 | school) + female_flag +
essay_score, data=d)
summary(ml_clust_reg)
preds <- predict(ml_clust_reg, d[,-c(2,5)])
rmse <- sqrt(mean((preds-d$exam_score)^2))
rmse  #12.4
```

The fit has decreased, implying that nesting cluster and school hierarchies more accurately captures the variance in student exam scores. Now let's try a varying slope model where the intercepts are fixed for school and cluster, but the slopes are allowed to vary by each level.

```
#Varying slope model to explore the effect of essay_score (a student level variable)
as it varies across clusters and schools
ml_clust_reg <- lmer(exam_score ~ (1 + essay_score | cluster) (1 + essay_score |
school) + female_flag, data=d)
summary(ml_clust_reg)
preds <- predict(ml_clust_reg, d[,-c(2,5)])
rmse <- sqrt(mean((preds-d$exam_score)^2))
rmse  #12.4
```

This model was not any better. So we will try one more thing: allow both the slope and intercept to vary across cluster and school, and consider the nested hierarchy of cluster/school.

```
#Same thing but with mixed effect term for varying intercepts by cluster and schools
within clusters, and varying slopes by essay_score by cluster and schools within
clusters
ml_clust_reg <- lmer(exam_score ~ (1 + essay_score | cluster/school) + female_flag,
data=d)  #Fails to converge
summary(ml_clust_reg)
preds <- predict(ml_clust_reg, d[,-c(2,5)])
rmse <- sqrt(mean((preds-d$exam_score)^2))
rmse  #11.6
```

This model is on par with the nested varying intercept model. If we plot the observed versus predicted values and color by absolute error, we can see which observations the model got wrong.

```
#Plot error by observation to see where the model messed up
library(ggplot2)
abs_dev <- abs(preds-d$exam_score)
d$abs_dev <- abs_dev
ggplot(d, aes(x=row.names(d), y=exam_score)) + geom_point(aes(colour=abs_dev)) +
scale_colour_gradient(low='white', high='black') + ggtitle('Absolute Error by
Observation')
```

According to the plot, the model had trouble with students who scored very low on the exam, as there seem to be more observations with large absolute errors below a score of 60.

22. Resampling

By sampling a dataset many times, it is possible to get more information about it than otherwise possible. Resampling can be thought of as wringing out a wet rag: the more time the rag is twisted, the more water can be squeezed out, up to a point. Resampling can be used for validating models, tuning hyperparameters, or producing nonparametric estimates of sample statistics like the mean or variance.

Recall from chapter 2 that the Central Limit Theorem states that as the size of a sample approaches the population size, the sampling distribution of the sample mean approaches the population mean. This is true for all sample statistics. So the sample variance also approaches the population variance as the sample size increases. If we cannot increase the sample size, or cannot resample the population, we can instead resample the sample we already have to get more information out of it. When a sample is resampled, the new samples form a sampling distribution, allowing the distribution of a sample statistic to be estimated. In other words, resampling allows us to estimate the sampling distribution of a sample statistic. The same concept applies to models: when a model is fitted on several samples, it tends to be able to generalize to the larger population better, because resampling allows the sampling distribution of the model's error to be estimated. Therefore, models that are trained using the resampling methods described in this chapter are more robust than models trained on a single sample (James, Witten, Hastie, & Tibshirani, 2013). In fact, it is common practice in data science to use resampling any time that a model is fitted.

The two most popular resampling methods for estimating sampling distributions (and therefore by the Central Limit Theorem, also the population parameters), are bootstrapping and jackknifing. Of the two, bootstrapping is far more common. The most popular resampling method for hyperparameter tuning and model validation is cross-validation.

Bootstrapping

Bootstrapping is carried out by drawing many samples from the same dataset, with replacement. When sampling with replacement is performed, a sample that is drawn in one iteration is "put back" into the dataset so that is has a chance of being drawn again in the next sample. This process is carried out B times, where B is some large arbitrary number. As an analogy, think about drawing marbles from a bag. The bag represents a sample of marbles from the global marble population. If we decide to draw B samples from the bag, all of the marbles we remove for one sample are put back into the bag for the next sample. The number of marbles drawn per sample should be roughly 2/3 of the total number of marbles in the bag. The consensus in the data science community is that bootstrapping should ideally draw roughly 2/3 samples out of the total sample size N, or more specifically, 0.632*N (James, Witten, Hastie, & Tibshirani, 2013). The reason 63.2% of the sample size is desirable is beyond the scope of this book, but it should be obvious that the sample should be smaller than the original or else we would just be copying the sample. The practice of using 0.632*N is referred to as the 0.632 rule.

The standard error of the population parameter estimate of the bootstrapped samples can be found by equation 22.1

$$SE_B(\hat{\alpha}) = \sqrt{\frac{1}{B-1} \sum_{r=1}^{B} \left(\hat{\alpha}r - \frac{1}{B} \sum_{r'=1}^{B} \hat{\alpha}r' \right)^2}$$

Figure 22.1

In the equation in figure 22.1, alpha hat is the population parameter to be estimated, B is the number of bootstrapped samples, and r is one particular iteration or bootstrap.

Bootstrapping depends on the assumption that the observations in the dataset are independent, so bootstrapping should not be used for datasets from repeated measures experiments.

Jackknifing
Jackknifing is carried out just like bootstrapping except that the samples are drawn without replacement. Referring back to the marble analogy, the marbles are not put back into the bag after being drawn, but instead removed so that no future samples can contain them. The sampling distribution resulting from jackknifing is a t-distribution with N-1 degrees of freedom. Just like bootstrapping, jackknifing relies on the assumption of independence of the observations. Unlike bootstrapping however, jackknifing is mainly used to estimate the variance of a sampling distribution, or the log of the variance if the sampling distribution shows that it is not normally distributed. It is not as consistent for other parameters. It can be shown that the jackknife estimates are approximations of the bootstrapped estimates.

Cross Validation
When models are fitted to a dataset, it is best practice to split the dataset so that part of it is used for training the model, and the other part is used for testing it. This is called the train/test split and the resulting datasets are called the training and test sets. It is important that the split be carried out randomly so that the test set serves as a good benchmark for how the model performs against new data and so that the training set contains enough variation that the model can be trained well. A common ratio for splitting the data is 80% training and 20% test, but there are no rules. To use the test set to validate a model, the data in the test set is fed into the model to produce a predicted outcome. The predictions are then compared to the true outcome of the test set to determine error.

In addition to the train/test split, it is also good practice to carve out a validation dataset. The **validation set** is a randomly sampled dataset pulled out of the training set, and it is used to approximate the test set error. If bootstrapping is used to sample several validation sets, then a sampling distribution of the test set error is formed. The technique of using bootstrapping to form several validation sets is called **cross validation**. As we saw in the previous section, bootstrapping the validation set error allows us to approximate the test set error fairly well. The validation set is particularly useful for tuning a model's hyperparameters, as the hyperparameters that minimize the validation set error are the ones that should also minimize the test set error (James, Witten, Hastie, & Tibshirani, 2013). In most academic papers, the results of the test set are reported to judge how good a model is. Therefore, the use of a validation set prevents contamination of the results that could have occurred if the test set were used to tune a model instead. There are two drawbacks to using a validation set:

1. The validation set error is highly variable depending on which observations are used to form the validation set.
2. Validation error may overestimate test error because it takes away observations that could be used to train and improve the model.

Let us think about the first drawback. If we are trying to predict a categorical outcome, and we draw a validation set from the training set that pulls out all of the observations from a particular class, then the training set will have no samples belonging to that class. Any model we build would get 100% of the predictions for that class wrong, because it would not have been trained to recognize samples belonging to the class. On the other hand, if we draw a validation set that does not include at least one observation from each class, then the validation set error would be optimistically lower than it should be. So it should be clear that the validation set must be sampled so that it has one or more observations from each class, and the training set should also include one or more observations from each class. The same can be said for the test set: the way we divide up the data should leave at least one observation from each class in each set. The practice of sampling such that every group or class is represented in the sample is called **stratified sampling**.

The second drawback to using a validation set is harder to overcome. The size of the validation set certainly plays a role in how the test error is estimated, but it is up to the researcher to choose a validation set size that is large enough to estimate the test error but small enough to not take away too many observations from the training set that could improve the model.

There are many types of cross validation. We will explore the following two, as they are the most common:

1. Leave one out cross validation
2. K-fold cross validation

Leave one out cross validation creates n splits of the training and validation sets, with only 1 observation in each validation set. The number of splits n, is the number of observations in the training set. So if there are 100 observations, then leave one out cross validation produces 100 validation sets, each comprised of 1 observation. The remaining 99 observations would form the training set. The validation errors are averaged together to approximate the test set error.

Leave one out cross validation always produces the same estimate of the test set error, which makes it useful for use in academia, where replication is important. It also greatly reduces the risk of overestimating the test set error. The drawback to using this method is that it is time consuming because the model has to be fitted n times, and the larger the dataset, the longer it will take.

Rather than using n validation sets, we could use k validation sets, where k < n. This is called **k-fold cross validation**. K-fold cross validation randomly splits the training set into k groups of equal size. One of the groups (folds) is used as the validation set and the others are used as the training set. This process is repeated k times until every group or fold has been used as the validation set once. The resulting k validation set errors are averaged to produce the estimate of the test set error. Typical sizes for k are 5 and 10. Like leave one out, k-fold cross validation also greatly reduces the risk of overestimating the test set error. Unlike leave one out however, k-fold cross validation is much less computationally intensive. The cost of computational speed is slightly more bias in the test set error estimate. In general however, k-fold usually produces a better estimate of the test set error than leave one out, because the estimates have lower variance and the bias, although usually higher than leave one out, is not often too much higher (James, Witten, Hastie, & Tibshirani, 2013).

Figure 22.2

Figure 22.2 shows how 5-fold cross validation is carried out.

When to Use Bootstrapping vs Cross Validation

A common question is whether k-fold cross validation or bootstrapping is better for model validation. The answer is neither: the choice of resampling method depends on what the researcher hopes to achieve in terms of the bias-variance tradeoff. Bootstrapping is used to produce estimates of model parameters, while cross validation is used to make sure models are capturing all of the signal in the data. If the goal is to reduce variance in a model's parameter estimates, then bootstrapping is the better choice. Bootstrapping can fail for small samples or when there are outliers in the training data, and these failures result in higher bias for bootstrapped estimates. If the goal is to reduce the bias of a model, then cross validation is better. Cross validated models can have high variance if not enough data is held out in the folds (that's why leave one out cross validation has higher variance than k-fold). Instead of choosing between bootstrapping and cross validation, a third option is to repeat k-fold cross validation many times. This is called repeated cross validation. **Repeated k-fold cross validation** has the benefit of reducing bias without causing variance to grow too high.

Resampling to Deal with Class Imbalance

When a classification task must be performed and there is a large class imbalance, the class with the most observations will overwhelm any model fitted to the data. Class imbalances actually occur quite often, because it is common for one class to be a small percentage of the total number of observations. Some examples are detecting fraudulent transactions, predicting voluntary terminations, and identifying fake reviews on websites like Amazon. Class imbalances must be corrected before models can be fitted. For example, if 99% of the observations belong to one class, then a model that simply predicts all observations belong to the majority class will be 99% accurate. By resampling the data to balance the classes, features that truly separate the minority class from the majority will be given stronger influence in the model, which will make it more able to detect observations that actually belong to the smaller class, even if it means sacrificing some predictive accuracy overall.

One way to resample the data to balance the classes of the dependent variable is to downsample. **Downsampling** involves replacing the training set with random samples taken, with replacement, from the training data, and ensuring that the minority class is better represented in the new training set. The minority class can be equally represented, but it does not have to be. Simply tipping the ratio of the majority to minority class from a severe imbalance of 90:10 to 30:10 may be sufficient. The choice of the ratio is up to the researcher, and it often depends on the circumstances and type of problem to be modeled. What constitutes an imbalance is also subjective, although many imbalances, like fraudulent transactions, are so severe that the minority class may make up 1% or fewer of the observations.

As an example of downsampling, suppose there is a dataset of 100 observations. Suppose that 90 of the observations belong to class A and 10 belong to class B. A balanced downsample would be 20 observations with 10 from each class. The 20 downsampled observations might contain duplicates of some of the 10 observations belonging to class B, because the sampling was done with replacement, or they might contain no duplicates. A model could then be fitted to this downsampled training data and validated on the larger validation set that retains the original class proportions. The key is to only downsample the training set. The validation set should retain the original class proportions so that it can produce error estimates in line with what would be expected for new data. So if k-fold cross validation is being used, then the fold that is held out as the validation set should not be resampled.

The disadvantage with downsampling is that it is only useful for very large datasets, because it reduces the sample size used for training. If the dataset is small, then an alternative sampling technique to deal with class imbalance is **synthetic minority oversampling technique (SMOTE)**. SMOTE involves iterating over the observations in the minority class and creating synthetic observations that are similar to them (Blagus & Lusa, 2013). The algorithm starts by finding the k-nearest neighbors for one observation belonging to the minority class. One of the k-nearest neighbors is randomly chosen to assist in creating a new synthetic observation. The synthetic observation is created by multiplying the feature vector between the selected neighbor and the observation by a small number (between 0 and 1) and adding it to the feature vector of the observation. The resulting synthetic data point will have similar features to the observation belonging to the minority class. This process is carried out for every observation in the minority class, as many times as necessary to bring the class ratio up to some desired ratio. The ratio does not have to be 1:1. In fact, it is usually better to maintain some imbalance. The purpose of resampling to remove severe class imbalances is to prevent the model from being overwhelmed by one class. As long as there is a significant number of observations from each class, even if the ratio is uneven, it should not negatively impact any models fitted on the data. Like downsampling, SMOTE should be performed only on the training set, and the class imbalance in the validation set should be left alone.

Resampling with R

Bootstrapping can be done using the boot package in R. Let us try bootstrapping an estimate of the mean sepal length for a random sample of the Iris flower dataset and compare it to the mean of the entire dataset.

```
d <- iris
str(d)

#Take a sample of the dataset for comparison of bootstrap estimate
ds <- d[sample(nrow(d), size=50),]

library(boot)
#Bootstrap estimate of mean (k=1 for 1 statistic)
average <- function(data, indices) {
  return(mean(data[indices]))
```

```
}
bootresults <- boot(data=ds$Sepal.Length, statistic=average, R=1000)
plot(bootresults)
#Compare bootstrapped estimate to the sample mean and the population mean
mean(ds$Sepal.Length)
mean(d$Sepal.Length)
#Bootstrapped CI
boot.ci(bootresults, type="bca")
```

The bootstrapped estimate is close to the overall mean sepal length, and both lie within the boostrapped confidence interval. That is how one parameter can be boostrapped. Now let us bootstrap the coefficients of a linear regression of mpg on wt and disp, from the Motor Trends cars dataset.

```
#Bootstrap cofficient estimates of regression (k>1 for many statistics)
linear_reg_coef <- function(formula, data, indices) {
  d <- data[indices,]
  linear_reg <- lm(formula, data=d)
  return(coef(linear_reg))
}
bootresults <- boot(data=mtcars, statistic=linear_reg_coef, R=1000,
formula=mpg~wt+disp)
plot(bootresults, index=1)   #Intercept
plot(bootresults, index=2)   #wt coef
plot(bootresults, index=3)   #disp coef
#Bootstrapped CI
boot.ci(bootresults, type="bca", index=1)   #Intercept
boot.ci(bootresults, type="bca", index=2)   #wt coef
boot.ci(bootresults, type="bca", index=3)   #disp coef
```

The bootstrapped confidence intervals form useful ranges for showing the variability of the coefficient estimates. Now let us bootstrap a Naïve Bayes classifier with the caret package to show how the bootstrapping process we coded above can be greatly simplified through the use of the caret package. Ignore the fact that we are using Naïve Bayes right now, as the algorithm will be explained in a later chapter.

```
#Bootstrapping a model with caret
library(caret)
train_control <- trainControl(method="boot", number=100)
nb_class <- train(Species~., data=d, trControl=train_control, method="nb")
nb_class
```

Moving on to cross validation, the caret package makes 10-fold cross validation extremely easy. Let us fit a classification tree to the Iris dataset to predict species class.

```
#k-fold cross validation
library(caret)
library(rpart)
train_control <- trainControl(method="cv", number=10, savePredictions=TRUE)
class_tree <- train(Species~., data=d, trControl=train_control, method="rpart")
```

The same process can be used to tune a model's hyperparameters. Let us use 10-fold cross validation to tune the Laplace correction parameter (fL) of a Naïve Bayes classifier to predict species class for the Iris dataset. To do this, we need to specify a hyperparameter grid, as caret will grid search the values in the grid we supply.

```
#k-fold cross validation with hyperparameter tuning
hyperparam_grid <- expand.grid(.fL=c(0), .usekernel=c(FALSE))
#Train & tune
nb_class <- train(Species~., data=d, trControl=train_control, method="nb",
tuneGrid=hyperparam_grid)
nb_class
```

Now let us try hyperparameter tuning using repeated 10-fold cross validation. We will repeated the 10-fold cross validation 3 times to tune the alpha and lambda regularization parameters of regularized regression.

```
#Repeated k-fold cross validation with hyperparameter tuning for regularized
regression
library(glmnet)
glmnet_grid <- expand.grid(alpha=c(0, 0.1, 0.2, 0.4, 0.6, 0.8, 1), lambda=seq(0.01,
0.2, length=20))
#glmnet_ctrl <- trainControl(method="cv", number=10)          #k-fold without
repeating
glmnet_ctrl <- trainControl(method="repeatedcv", number=10, repeats=3)  #k-fold with
3 repeats
glmnet_fit <- train(Species~., data=d, method="glmnet", preProcess=c("center",
"scale"), tuneGrid=glmnet_grid, trControl=glmnet_ctrl)
glmnet_fit
```

Caret conveniently lists the tuned hyperparameters as alpha = 0.1 and lambda = 0.03.

Resampling with Python

Bootstrapping can be done using the resample function from sklearn utils. Let us try bootstrapping the classification accuracy of a decision tree classifier fitted to the Iris flower dataset.

```
import numpy as np
import pandas as pd
from sklearn.utils import resample
from sklearn.preprocessing import LabelEncoder
from sklearn.tree import DecisionTreeClassifier
from sklearn.metrics import accuracy_score
from sklearn.linear_model import LinearRegression
import matplotlib.pyplot as plt

d = pd.read_csv('iris.csv')
le = LabelEncoder()
d['Species'] = le.fit_transform(d['Species'])
d = d.values

n_iterations = 1000
n_size = int(len(d)*0.5)

#Bootstrap estimate the classification accuracy of a decision tree
scores = list()
for i in range(n_iterations):
    train = resample(d, n_samples=n_size)
    x_train = train[:,0:3]
    y_train = train[:,4].reshape(-1,1)
    test = np.array([x for x in d if x.tolist() not in train.tolist()])
```

```
    x_test = test[:,0:3]
    y_test = test[:,4].reshape(-1,1)
    dtc_model = DecisionTreeClassifier()
    dtc_model.fit(x_train, y_train)
    predictions = dtc_model.predict(x_test)
    score = accuracy_score(y_test, predictions)
    scores.append(score)
plt.hist(scores)
plt.show()
alpha_level_of_sig = 0.95
p_val_lower = ((1.0-alpha_level_of_sig)/2.0)*100
lower_ci = max(0.0, np.percentile(scores, p_val_lower))
p_val_upper = (alpha_level_of_sig+((1.0-alpha_level_of_sig)/2.0))*100
upper_ci = min(1.0, np.percentile(scores, p_val_upper))
print('95% Confidence Interval:', lower_ci*100, upper_ci*100)
```

The confidence interval has been bootstrapped, and the plot shows the bootstrapped model accuracy score. We can do the same thing for a linear regression, but we can also get bootstrapped estimates of the predictor coefficients. Let's regress sepal length on the other variables.

```
#Bootstrap estimate of regression coefficients
#Regress Spepal.Length on the other vars
coefs = list()
for i in range(n_iterations):
    train = resample(d, n_samples=n_size)
    x_train = train[:,1:4]
    y_train = train[:,0].reshape(-1,1)
    test = np.array([x for x in d if x.tolist() not in train.tolist()])
    x_test = test[:,1:4]
    y_test = test[:,0].reshape(-1,1)
    lr_model = LinearRegression()
    lr_model.fit(x_train, y_train)
    coefs.append(lr_model.coef_)
coefsarr = np.array(coefs).reshape((1000, 3)).T
plt.hist(coefsarr[0])   #Plot of coefficients for first var only
plt.show()
alpha_level_of_sig = 0.95
p_val_lower = ((1.0-alpha_level_of_sig)/2.0)*100
lower_ci = max(0.0, np.percentile(coefsarr[0], p_val_lower))
p_val_upper = (alpha_level_of_sig+((1.0-alpha_level_of_sig)/2.0))*100
upper_ci = min(1.0, np.percentile(coefsarr[0], p_val_upper))
print('95% Confidence Interval:', lower_ci*100, upper_ci*100)
```

The histogram shows the bootstrapped estimate of the coefficient for the first predictor, and the bootstrapped confidence interval for the coefficient is printed. The bootstrapped confidence intervals form useful ranges for showing the variability of the coefficient estimates.

Moving on to cross validation, we can use k-fold cross validation to get the average score of a model over k-folds. This is useful for situations where multiple types of models need to be compared. Think of it like the National Hockey League (NHL) playoffs: rather than judging NHL teams on the outcome of 1 game, we can judge them better based on the outcomes of 7 games.

```
from sklearn.model_selection import KFold, RepeatedKFold, train_test_split,
GridSearchCV, cross_val_score
```

```
#K-fold cross val to get the average accuracy of a model over the folds
#This is useful for robust model assessment so different models can be compared

#Specify the model
class_tree = DecisionTreeClassifier(criterion='gini', min_samples_split=20)
#Set up k-fold
kf = KFold(n_splits=10, shuffle=True, random_state=14)
#Fit the model to each fold and track accuracy
scores = list()
for train_index, test_index in kf.split(d):
    x_train, x_test = d[train_index, 0:3], d[test_index, 0:3]  #Columns 0-3 are
predictors
    y_train, y_test = d[train_index, 4], d[test_index, 4]       #Column 4 is target
    class_tree.fit(x_train, y_train)
    predictions = class_tree.predict(x_test)
    score = accuracy_score(y_test, predictions)
    scores.append(score)
print('Mean accuracy score over 10 folds:', np.mean(scores))
```

K-fold cross validation can also be used to tune a model's hyperparameters. We can set up a hyperparameter grid with values to choose from. Then the GridSearchCV function will try every combination of values that we provide, fitting each to 10 folds, and save the results of each model. The model with the highest average accuracy score is selected as the model with the best hyperparameters. We can then retrieve that model's parameters and performance. Alternatively, we can fit the tuned model over another 10 folds to get its average accuracy score. This is called **nested cross validation**, because we already tuned the model using an inner cross validation loop, and now we are getting its average score from an outer cross validation loop.

```
#Split dataset into training and test sets (CV will further split up the training
set)
x_train, x_test, y_train, y_test = train_test_split(d[:, 0:3], d[:, 4],
test_size=0.20, random_state=14)
#Specify the model
class_tree = DecisionTreeClassifier(random_state=14)
#List the hyperparameters to be tuned
param_grid = [{'criterion': ['gini', 'entropy'], 'min_samples_split': [2, 5, 10,
20]}]
#Perform grid search 10-fold cross validation on the hyperparameters listed, using
accuracy as the evaluation metric
gs = GridSearchCV(estimator=class_tree, param_grid=param_grid, scoring='accuracy',
cv=10, return_train_score=False)
gs.fit(x_train, y_train)
print('Hyperparameters of best model (the model with the highest mean accuracy over
10 folds):', gs.best_params_)
print(pd.DataFrame.from_dict(gs.cv_results_))  #Shows the full results that can be
inspected to verify the best_params_ and mean score
#Optional outer cross validation loop: this is nested cross validation, as the outer
CV is run over the model tuned with inner CV
scores = cross_val_score(gs, x_train, y_train, scoring='accuracy', cv=10)
print('Mean accuracy of optimally tuned classification tree over 10 folds: %.3f +/-
%.3f' % (np.mean(scores), np.std(scores)))
```

Finally, we can do the same hyperparameter tuning process using repeated cross validation.

```
#Split dataset into training and test sets (CV will further split up the training
set)
x_train, x_test, y_train, y_test = train_test_split(d[:, 0:3], d[:, 4],
test_size=0.20, random_state=14)
#Specify the model
class_tree = DecisionTreeClassifier(random_state=14)
#List the hyperparameters to be tuned
param_grid = [{'criterion': ['gini', 'entropy'], 'min_samples_split': [2, 5, 10,
20]}]
#Set up k-fold
rkf = RepeatedKFold(n_splits=10, n_repeats=3, random_state=14)
#Perform grid search 10-fold cross validation on the hyperparameters listed, using
accuracy as the evaluation metric
gs = GridSearchCV(estimator=class_tree, param_grid=param_grid, scoring='accuracy',
cv=rkf, return_train_score=False)
gs.fit(x_train, y_train)
print('Hyperparameters of best model (the model with the highest mean accuracy over
10 folds):', gs.best_params_)
print(pd.DataFrame.from_dict(gs.cv_results_))  #Shows the full results that can be
inspected to verify the best_params_ and mean score
#Optional outer cross validation loop: this is nested cross validation, as the outer
CV is run over the model tuned with inner CV
scores = cross_val_score(gs, x_train, y_train, scoring='accuracy', cv=10)
print('Mean accuracy of optimally tuned classification tree over 10 folds: %.3f +/-
%.3f' % (np.mean(scores), np.std(scores)))
```

23. Probability and Bayesian Statistics

In every chapter in this book so far, we have been exploring statistics through the lens of frequentist statistics. An alternative perspective is Bayesian statistics. Bayesian statistics does not build on top of frequentist statistics or vice versa, but rather, the Bayesian/frequentist approaches are like considering whether a glass of water is half full or half empty. The glass is both! It is simply a matter of perspective. At a time when machine learning, deep learning, and big data are all the rage, the 200+ year old ideas in Bayesian statistics still appear in state of the art models and work with datasets of any size.

Bayesian vs Frequentist Statistics

As we have seen throughout this book, the frequentist approach expresses probability in terms of the lifetime of an experiment. Hypothesis tests are frequentist in nature. If we want to know the probability of getting heads on a coin flip, we can say that out of 1,000 flips, if a coin lands on heads 500 times, then we can be confident that the coin has a 50% probability of landing on heads. This confidence is dependent on the stopping criterion (sample size) though. If another researcher flips the same coin 10,000 times and find that it lands on heads 4,000 times, then that researcher would conclude that the coin has a 40% probability of landing on heads. So does the coin land on heads 50% of the time or 40%? There is no way to know which is right. This problem is the reason many statisticians despise the p-value, because it is used to draw conclusions about a hypothesis test, and yet it is dependent on sample size.

The Bayesian approach is probabilistic in nature. It involves using prior beliefs as a baseline for what we should expect to happen, and when new information is introduced, we update our beliefs. The Bayesian approach is much more intuitive because we use it every day. Suppose John lives in a two story house and has lost his phone. John's first thought upon realizing that his phone is lost is likely "where have I been with my phone?" The places John has been with his phone form his prior belief, giving him some ideas for where his phone could be now. If John has his girlfriend Jill call his phone, and he hears it ringing upstairs, there is new information introduced. John updates his prior belief with this new information: "since I was in the bedroom earlier, and my phone is ringing from upstairs, then my phone is probably in the bedroom." A frequentist would approach this problem differently. If John were a frequentist, he would have thought about all the previous times he had lost his phone and the places he found it. Then he would guess that his phone's location is probably where he has found it most often in the past. But if John had never lost his phone before, or if he had only lost it in somebody else's house, then he would not have a good sample to build a "model" of where his phone could be. This is why the Bayesian approach can be more useful than the frequentist approach. While the frequentist approach depends on sample size, the Bayesian approach depends on prior beliefs.

Probability

Bayesian statistics is based on a fundamental theorem called Bayes Theorem. Before we can define Bayes Theorem though, we must review probability. The tables in figures 23.1 and 23.2 summarize the basic rules of probability for independent and dependent events, respectively (Glen, 2018). **Independent events** are events that do not impact the probability of each other: the occurrence of A has no impact on the probability of B and vice versa. Two coin tosses are independent events because neither toss has any influence on the outcome of the other. **Dependent events** are events that do impact the probability of each other: the occurrence of A changes the probability of B and vice versa. Dependent events often happen in order, but they do not have to. Failing to pay an electric bill increases the probability of the power going out, so the events are dependent and happen in order. Buying peanut butter increases the probability of buying jelly, and vice versa, so they are dependent events, and these events can happen in either order.

colspan Probability Rules of Independent Events			
Rule Description	Notation	Visual Aid	Notes
Probability of event A	$P(A) = \dfrac{number\ of\ times\ A\ occurs}{number\ of\ total\ things\ that\ can\ occur}$		Probability is always between 0 and 1, where 1 is absolute certainty.
Probability of A not occurring	$P(A^c) = 1 - P(A)$		This is called the **compliment**.
Probability of A and B	$P(A\ and\ B) = P(A) * P(B)$		Only the shared region.
Probability of A and B if A and B have no common outcomes	$P(A\ and\ B) = 0$		Events are **mutually exclusive** or **disjoint** (they cannot both occur).
Probability of A or B	$P(A\ or\ B) = P(A) + P(B) - P(A\ and\ B)$		Subtract shared region so that it is not included twice.
Probability of A or B if A and B have no common outcomes	$P(A\ or\ B) = P(A) + P(B)$		Probability of either event, and the events are mutually exclusive.

Figure 23.1

Probability Rules of Dependent Events			
Rule Description	Notation	Visual Aid	Notes
Probability of event B, given A has occurred	$P(B\|A) = \dfrac{P(A\ and\ B)}{P(A)}$		This is called **conditional probability**. It is like asking which portion of A is shared by B.
Probability of A and B	$P(A\ and\ B) = P(A) * P(B\|A)$		
Probability of B	$P(B) = P(A\ and\ B) + (P(A^c) * P(B\|A^c))$		P(B) is the sum of the conditional probability of B given that A has occurred and the conditional probability of B given that A has not occurred.
Probability of event A, given B has occurred	$P(A\|B) = \dfrac{P(B\|A)P(A)}{P(B)}$		**Bayes' Theorem** (see the next section)

Figure 23.2

Probability confuses many statistics students when it is required to solve word problems, so here a few examples to show how these simple rules can apply to a variety of scenarios.

1. Suppose Lebron James makes 90% of his free throws. If he shoots 10, what is the probability that he misses all of them?

    Each shot is an independent event.

    P(miss all 10) = P(miss 1) + P(miss 2) + ... + P(miss 10)

    P(miss 1 and miss 2, etc.)   ← requires multiplication rule

    P(miss all 10) = 0.1^10

P(miss all 10) = 1e-10 (pretty much 0)

2. Suppose Lebron James makes 90% of his free throws. If he shoots 10, what is the probability that he makes at least 5 of the 10?

> P(makes at least 5) + P(misses all 10) = 1
>
> P(makes at least 5) = 1 − P(misses all 10)
>
> P(makes at least 5) = 1 − 1e-10
>
> P(makes at least 5) = 0.9999 (pretty much 1)
>
> Note that this question could also be answered by using the binomial distribution, which will be defined later in this chapter.

3. Suppose a 6 sided die is tossed and a coin is flipped. What is the probability that the die is even and the coin is tails?

> The die toss and coin flip are independent events.
>
> P(even and tails) = P(even) * P(tails)
>
> P(even and tails) = (3/6) * (1/2) = ¼

4. Suppose a card is drawn from a deck. What is the probability that the card is either a spade or a queen?

> P(spade or queen) = P(spade) + P(queen) − P(spade and queen)
>
> P(spade or queen) = 13/52 + 4/52 − 1/52
>
> P(spade or queen) = 16/52

5. Suppose a card is drawn from a deck and it happens to be a king. The card is not placed back into the deck. If another card is drawn, what is the probability that it too is a king?

> Drawing a king changes the probability of drawing a king on the next pick, so the events are dependent.
>
> P(king2 | king1) = (4-1)/(52-1) = 3/51

6. Suppose a group of students is asked to rate their agreement with the statement "My favorite color is red." If their responses are measured on a Likert scale, they would look like this:

	Strongly Agree	Agree	Neutral	Disagree	Strongly Disagree	Total
Male	15	5	1	7	3	31
Female	6	5	7	8	2	28
Total	21	10	8	15	5	59

What is the probability that a student's favorite color is red, given that they are male?

The question is looking for conditional probability.

Of the 31 male students, 20 agreed or strongly agreed that their favorite color is red.

P(favorite color is red | male) = 20/31

The diagram in figure 23.3 shows how to know when to use the rules for independent or dependent events.

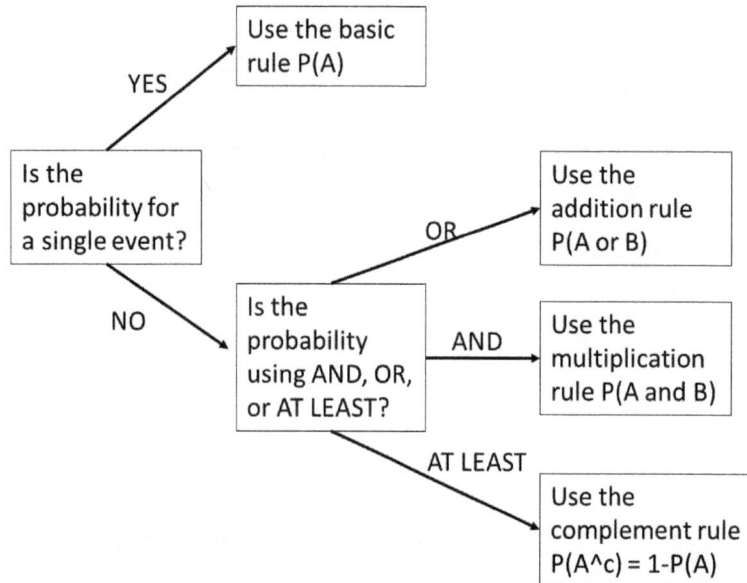

Figure 23.3

Probability rules are easy to apply when it is easy to count the number of times an event can happen. But counting occurrences is not always easy. Let us refer back to word problem number 3 with the example of the die and the coin. When we want to know the probability of the die being even and the coin being tails, we are looking at events out of every combination of results. We used the multiplication rule and split these probability up before, but another way to look at it would be to count the number of combinations of die and coin values. Since there are 6 possible die values and 2 possible coin values, we can use the equation in figure below to determine that there are 12 possible combinations of the die and coin. We can use the same formula to determine that there are 3 possible combinations of even die values and tails. Therefore, 3/12 possible **combinations** is 0.25, which matches the result we got before. It is easy for us to count the possible combinations of events with only a few possible outcomes, but it is much harder as the number of outcomes increases. For example, suppose we randomly sampled 6 letters from the English alphabet with replacement. How would we know how many combinations there could be? The equations in figure 23.4 show the counting rules for combinations and permutations. **Permutations** are ordered combinations of events. Note that the words repetition and replacement are interchangeable. If we sample with replacement, then the probability rules in the table below must allow repetition.

Combination and Permutation Rules		
Rule Description	Notation	Notes
Sequence of choices where there are p ways to make the first choice, q ways to make the second choice, r ways to make the third choice, and so on.	$p * q * r * ...$	General rule
Sequence of r objects chosen from n total objects, where objects are chosen with replacement so they can be chosen again.	n^r	**Permutation with repetition/replacement**: there are always n objects to choose from and order matters.
Sequence of r objects chosen from n total objects, so there are n-r ways to make the second choice, n-2r ways to make the third choice, and so on.	$nPr = \dfrac{n!}{(n-r)!}$	**Permutation without repetition**: the n objects are distinct and order matters.
Sequence of n objects where there are n1 of the 1st type of object, n2 of the 2nd type of object, and nk of the kth type of object.	$\dfrac{n!}{n_1! * n_2! * ... * n_k!}$	**Permutation without repetition and non-distinct n objects**, where order matters. (E.g. how many ways can a sequence be arranged?)
Assortment of r objects chosen from n total objects: it is like asking how many assorted ways can r objects out of n be arranged.	$\dfrac{n(r+n-1)!}{r!(n-1)!}$	**Combination with repetition/replacement**: there are always n objects to choose from and order does not matter.
Assortment of r objects chosen from n total objects, so there are n-r ways to make the second choice, n-2r ways to make the third choice, and so on.	$nCr = \dfrac{n!}{r!(n-r)!}$	**Combination without repetition**: the n objects are distinct and order does not matter.

Figure 23.4

Here are a few examples of probability word problems that require counting rules (combinations or permutations) to solve.

1. Suppose a bag of marbles contains 10 blue marbles and 30 red. If 3 marbles are drawn without replacement, what is the probability of all 3 marbles being blue?

    P(3 blue) = number of ways to draw 3 blue / total number of possible ways to draw 3 marbles

    Order does not matter, so combinations without replacement should be used.

P(3 blue) = 10C3 / 40C3 = 3/247 = 1.2%

2. To win the Powerball lottery, it is necessary to match all 5 balls and the Powerball. The balls are numbered from 1 to 55, and the Powerball can be numbered from 1 to 42. The 5 balls can be matched in any order. What is the probability of winning the Powerball lottery with one ticket?

P(win lottery) = number of ways to win / total number of possible outcomes

Since the question asks for only 1 ticket, there is only 1 way to win.

Each ball is an independent event.

Total number of possible outcomes = number of ways the 5 balls can be arranged * number of possible values of the Powerball

Since the order of the 5 balls does not matter, it is a combination where 5 numbers are drawn from 55 total numbers without replacement (the balls are not put back after being drawn, meaning the number can only appear once). This requires combination without repetition. Since the number of possible values of the Powerball is 42, no calculation is necessary for the Powerball part of the equation.

Total number of possible outcomes = 55C5 * 42

Total number of possible outcomes = 146,107,962

P(win lottery) = 1/146,107,962 or nearly 1 in 150 million

Figure 23.5 below shows guidelines that help indicate when to use counting rules.

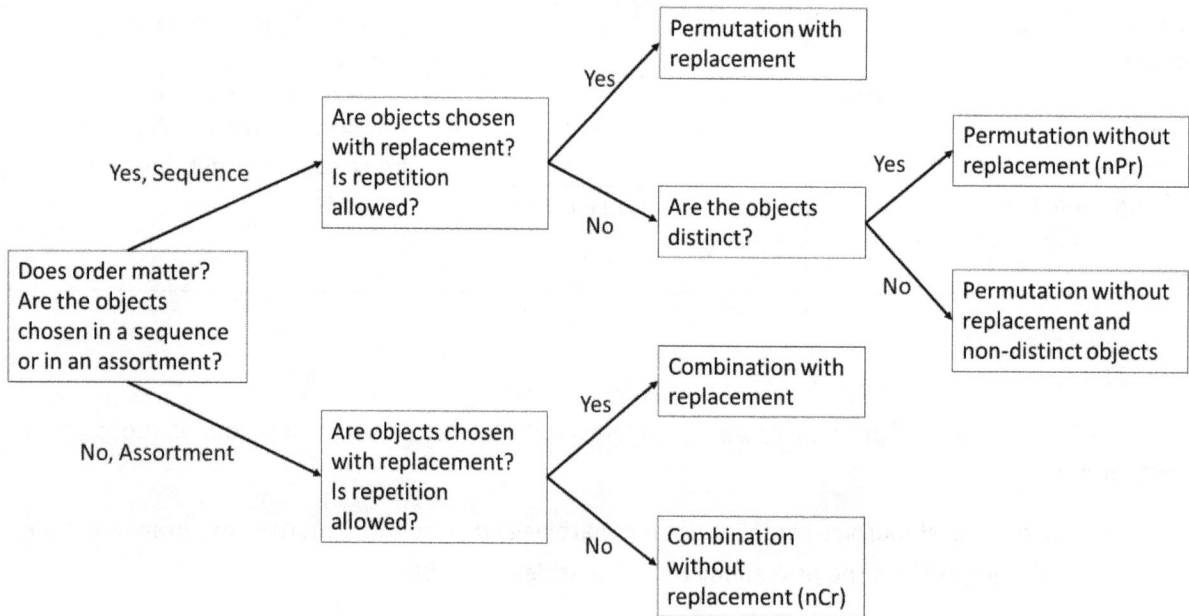

Figure 23.5

More complex ways of using combinations, permutations, and factorials exist. For example, consider the complexity of determining how many possible passwords are possible given a set of criteria.

Problems like these are solved using the principles of a branch of mathematics called combinatorics, which is beyond the scope of this book.

Bayes' Theorem

Bayes' theorem was presented in figure 23.2. It is shown again in figure 23.6 for reference.

$$P(A|B) = \frac{P(B|A)P(A)}{P(B)}$$

Figure 23.6

In the equation in figure 23.6, P(A) is the prior probability of event A. The **prior probability** is the probability event A is expected to occur on its own, independent of event B. In the equation, P(B|A)/P(B) is the support event B provides for A. P(A|B) is the posterior probability of A, given that B has occurred. The **posterior probability** is the updated belief in the probability of event A occurring, given event B has occurred, as determine by updating the prior belief P(A) with the support from event B. Bayes' theorem is important because it provides a way to swap between P(A|B) and P(B|A).

Notice that the numerator of Bayes' theorem is equivalent to P(A and B) in the dependent probability table in figure 23.2. Sometimes the numerator in figure 23.6 is replaced by P(A and B), since both representations show the joint probability of A and B. **Joint probability** is the probability that both events A and B are true. If there were 24 events instead of just A and B, joint probability would be the probability that all 24 are true.

As an example of how Bayes' theorem can be applied, let us consider coin flipping. Coin flipping is a very academic example with little practical use in the real world, but it the procedure we will follow can be directly transferred to any problem dealing with binary outcomes in the real world. Since the outcome is binary, and coin flips are independent events, a series of coin flips (a series of independent events) produces outcomes that form a binomial distribution. The **binomial distribution** is the discrete probability distribution of the number of positive outcomes in a sequence of n independent experiments. The probability of exactly k positive outcomes in the sequence of n trials sampled with replacement is given by the probability mass function in figure 23.7.

$$P(k, n, p) = P(x = k) = \frac{n!}{k!\,(n-k)!}p^k(1-p)^{n-k}$$

Figure 23.7

In the equation in figure 23.7, x is a binomially distributed random variable (a binary variable), p is the probability of a positive outcome for a single trial k, and n is the total number of trials. A quick aside: $\frac{n!}{k!(n-k)!}$ is called the binary coefficient, which is the number of combinations of k items that can be sampled from a set of n total items (nCk) and arranged in Pascal's triangle.

When there is only one trial, in other words when n=1, the binomial distribution takes on a special form called the **Bernoulli distribution**, and the one trial used to form the Bernoulli distribution is called a **Bernoulli trial**. Thus, the binomial distribution can be thought of as the discrete probability distribution of the number of positive Bernoulli trials. In the case of a coin flip, a single flip would be a Bernoulli trial.

The mean and variance of the binomial distribution are shown in figure 23.8.

$$mean = np$$

$$variance = np(1 - p)$$

Figure 23.8

The mean and variance of the Bernoulli distribution are shown in figure 23.9.

$$mean = p$$

$$variance = p(1 - p)$$

Figure 23.9

As we have seen throughout this book, every probability distribution has a likelihood function that estimates the parameters of the distribution, given a dataset. Maximizing the likelihood function gives the best estimates for the parameters, and that is the point of MLE (refer back to chapter 7 for a refresher on MLE). For the Bernoulli distribution, the maximum likelihood estimator of the probability of a positive outcome p is the sample mean. For the binomial distribution, the likelihood function to be maximized to estimate the PMF is given in the equation in figure 23.10.

$$L(p) = \prod_{i=1}^{n}(p^{x_i} * (1 - p)^{1-x_i})$$

Figure 23.10

Notice that the equation in figure 23.10 does not contain the PMF of the binomial distribution from figure 23.7, rather, it is the product of a series of Bernoulli trials. That is because maximizing the PMF for a binomial distribution is equivalent to maximizing $p^x * (1 - p)^{n-x}$ for a fixed sample x, because $p^x * (1 - p)^{n-x}$ is the Bernoulli likelihood function of a single Bernoulli trial and the equation in figure 23.10 is the Binomial likelihood function of a series of Bernoulli trials.

The Bernoulli distribution and the binomial distribution can be generalized to the **categorical distribution** and the **multinomial distribution**, respectively (Barber, 2012). While the Bernoulli and binomial distributions deal with binary outcomes, the categorical and multinomial distributions deal with any number of outcomes. The probability mass functions that define these distributions are shown in figure 23.11.

PMF Categorical Distribution: $f(x|p) = \prod_{i=1}^{k} p_i^{x_i}$

PMF Multinomial Distribution: $f(x_1, \ldots, x_k, n, p_1, \ldots, p_k) = \begin{cases} \frac{n!}{x_1! * \ldots * x_k!} p_1^{x_1} * \ldots * p_i^{x_k}, & when \ \sum_{i=1}^{k} x_i = n \\ 0, & otherwise \end{cases}$

Figure 23.11

Getting back to coin flipping, a likelihood function we might want to estimate is the probability of the coin landing on heads, P(H). Therefore, theta = P(H). For a Bernoulli trial, since the maximum likelihood estimator is the sample mean, P(H) for one trial is 0.5, which makes sense. For some number of trials n, would the probability change? We should not expect it to, but suppose we did not know that the probability of heads was exactly 0.5. In other situations in the real world, we might not know the probability of each category of a binary outcome. For example, if we were selling lemonade, and at the end of the day more women bought lemonade than men, it might be because women are more likely to buy lemonade. So suppose we do not have former knowledge that heads has a probability of 0.5. Bayes' theorem requires a prior probability, but if we do not know what it could be, one option would be to sample it from a uniform (Gaussian) distribution and only assume that the distribution has a particular mean and standard deviation. This would make it a **Gaussian prior**. Since our goal is to estimate the posterior distribution, we need to make sure it matches the shape of the prior distribution. Therefore, if the prior has a Gaussian distribution, the posterior should as well. When the prior and posterior have the same distribution, they are called **conjugate distributions**, and the prior is referred to as the **conjugate prior**. If the conjugate prior has a Gaussian distribution, then the likelihood function is Gaussian too. Prior probabilities are more often drawn from a beta distribution, though, rather than a Gaussian distribution (Barber, 2012). The **beta distribution** is symmetrical, just like the Gaussian distribution, and as the number of parameters approaches infinity, the beta distribution approaches the Gaussian distribution, but the beta distribution is shaped slightly differently. For binary outcomes that form the Bernoulli and binomial distributions, the conjugate prior comes from a beta distribution with parameters alpha and beta:

$$P(p) = \frac{p^{\alpha-1}(1-p)^{\beta-1}}{B(\alpha,\beta)}$$

$$\text{where the Beta function } B(\alpha,\beta) = \frac{(\alpha-1)!(\beta-1)!}{(\alpha+\beta-1)!}$$

Figure 23.12

The equation in figure is the probability density function of the beta distribution, from which conjugate priors are sampled. Alpha and beta are the parameters of the prior, so they are hyperparameters. Alpha and beta can be derived from the mean and standard deviation of the distribution of the prior.

$$mean = \frac{\alpha}{\alpha+\beta}$$

$$standard\ deviation = \sqrt{\frac{\alpha\beta}{(\alpha+\beta)^2(\alpha+\beta+1)}}$$

Figure 23.13

The equations in figure 23.13 show the mean and standard deviation of the beta distribution of the conjugate prior. Thus, if the mean and standard deviation of the prior are known, then alpha and beta can be found. Once a prior is sampled from the beta distribution, it is possible to calculate the posterior for a dataset x. This is done by combining Bayes' theorem with the likelihood function from figure 23.10.

$$P(p|x,n) = \frac{P(x,n|p)P(p)}{P(x,n)} = \frac{(p^x * (1-p)^{n-x})(p^{\alpha-1}(1-p)^{\beta-1})}{B(\alpha,\beta)P(x,n)}$$

Figure 23.14

If we assume that the conjugate prior distribution (the beta distribution) has mean 0.6 and standard deviation 0.1, then we can update the posterior as trials are carried out. For example, if there were 7 heads after 10 trials, then the posterior distribution will shift towards a mean of 0.7. The figure below shows how the posterior probability of heads, P(H) changes as more trials are carried out. After 1,000 trials (1,000 coin flips), the posterior is pretty close to 0.5.

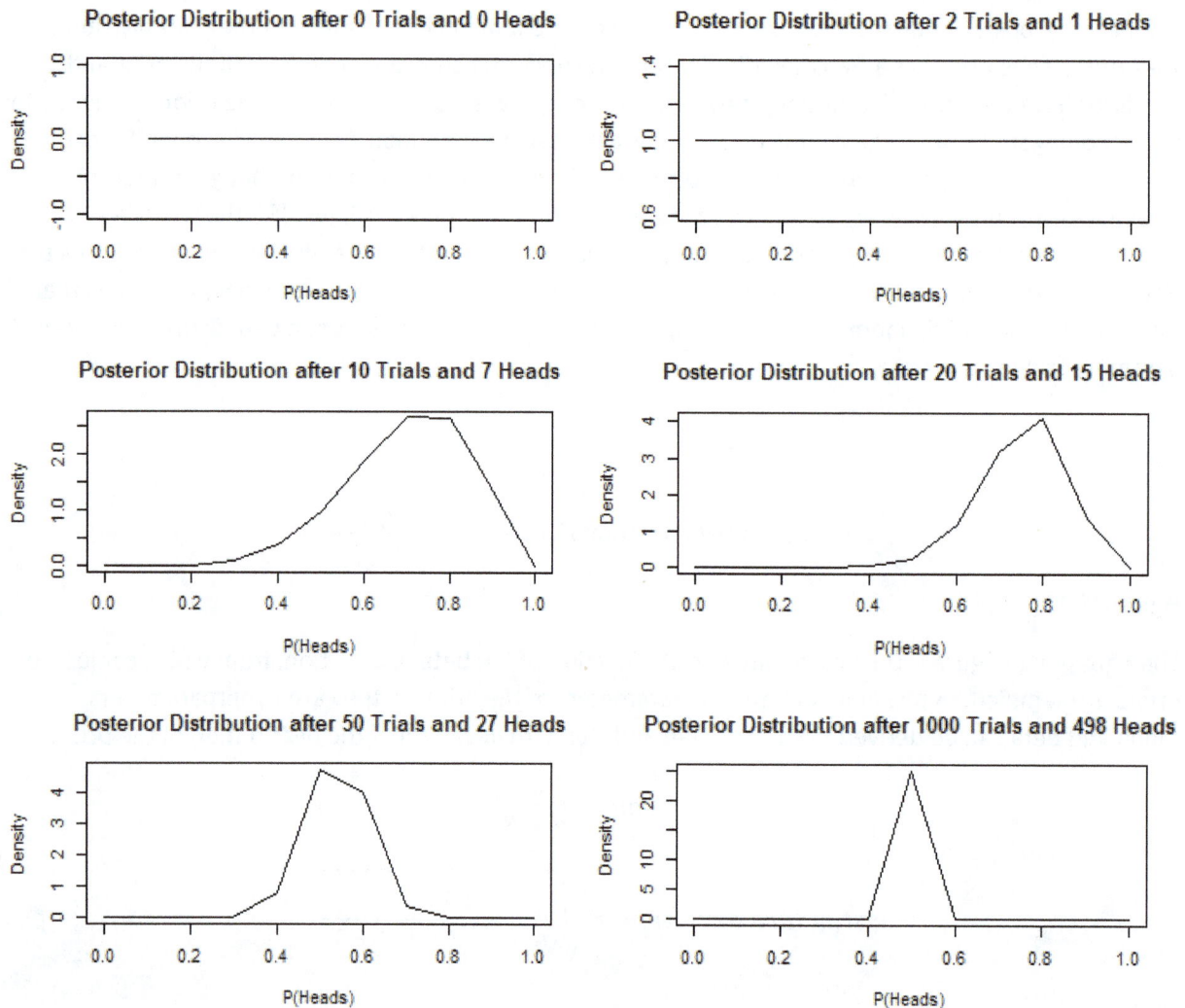

Figure 23.15

Introduction to Graph Theory

We have seen how joint probabilities of dependent events can be represented, but it quickly becomes difficult to mathematically represent joint probabilities of many events. Graph representations allow probabilities of dependent events to be represented much more cleanly than a massive equation. A **graph** is simply a collection of circles (nodes or vertices) with lines (links or edges) connecting them. A graph can also be called a **network**. If the links between nodes are directional one way, the graph is called a directed graph. If the links are either undirected or directed both ways, the graph is called an undirected graph (Tsvetovat & Kouznestov, 2011). Figure 23.16 shows the difference between these types of graphs.

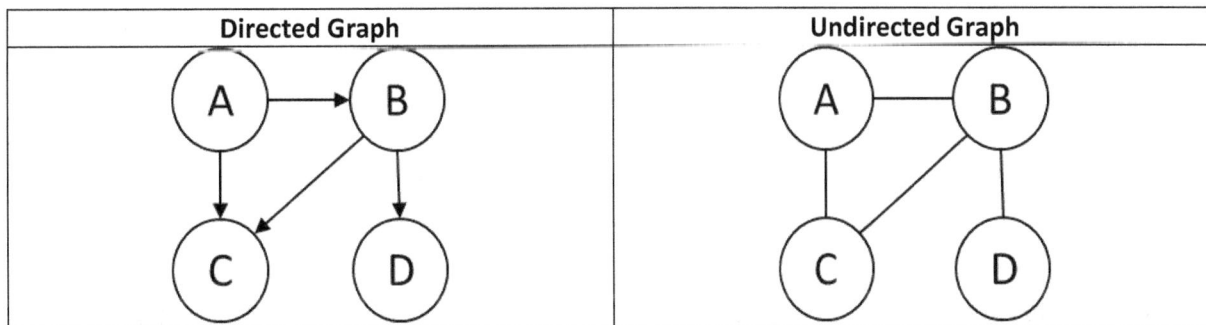

Figure 23.16

The links between nodes can have weights, which enable functions to be calculated over the graph. If a node has other nodes the point to it through a sequence, such as node C receiving links from A and B in the directed graph in figure 23.16, then the nodes pointing to it are called the node's **ancestors**.

In the directed graph in figure 23.16, the following statements are true:

- A and B are ancestors of node C.
- Since A and B are the direct ancestors of node C, meaning they have a direct path that points to C, A and B are the **parents** of node C, and node C is the **child** of A and B.
- **Ancestors** are any nodes in a sequence that point to a node. Node B is a parent to D, but A is not. However, since node A is a parent of B, A is an ancestor to node D.
- **Descendants** are any nodes in a sequence that receive a link from ancestors. Node D is a child of B and a descendent of A.

A **cycle** is a directed path that starts and returns to the same node. A cycle is essentially a loop. There are no loops in figure 23.16, but if a link from D to A were formed, then there would be a cycle from A to B to D and back to A. Since the directed graph in figure 23.16 has no cycles, it is called a **directed acyclic graph (DAG)**. DAG's are very useful for probability modeling (Barber, 2012).

In the undirected graph in figure 23.16, the following statement is true:

- Nodes A, B, and C are all connected to one another, so they form a clique. A clique is a fully connected subset of nodes. All of the members of the clique are neighbors.

Graphs can be encoded as an edge list or as an adjacency matrix. An **edge list** shows what node to node links exist in the graph. For example, the edge list for the directed graph in figure 23.16 is: L = {[A,B], [A,C], [B,C], [B,D]}. The edge list for an undirected graph has edges in both directions. For example, the edge list for the undirected graph in figure 23.16 is: L = {[A,B], [A,C], [B,A], [B,C], [B,D], [C,A], [C,B], [D,B]}. An **adjacency matrix** is a binary matrix showing which nodes are connected. For a directed graph, cell A_{ij} = 1 if there is a link from node i to node j. An undirected graph has a symmetrical adjacency matrix. The adjacency matrices for the graphs in figure 23.16 is shown in figure 23.17.

Direct Graph Adjacency Matrix	Undirected Graph Adjacency Matrix
$A = \begin{array}{c} A \\ B \\ C \\ D \end{array} \begin{bmatrix} 0 & 1 & 1 & 0 \\ 0 & 0 & 1 & 1 \\ 0 & 0 & 0 & 0 \\ 0 & 0 & 0 & 0 \end{bmatrix}$	$A = \begin{array}{c} A \\ B \\ C \\ D \end{array} \begin{bmatrix} 0 & 1 & 1 & 0 \\ 1 & 0 & 1 & 1 \\ 1 & 1 & 0 & 0 \\ 0 & 1 & 0 & 0 \end{bmatrix}$

Figure 23.17

A graphical model of joint probabilities is any model that can be used to represented probabilistic relationships. There are many types of graphs, so there are many types of graphical models. We will only explore two in this book: Bayesian networks and Markov networks.

Bayesian Networks

Bayesian networks, a.k.a. belief networks or Bayesian belief networks, are graphical models that extend Bayes' theorem to model the join probabilities of any number of random variables. The graphical models in Bayesian networks take the form of DAG's. The key to using Bayesian networks is to specify which variables are independent (Barber, 2012). One a model is specified, Bayesian inference (or simply, inference) can be applied. Bayesian networks take the form of the equation in figure 23.18.

$$P(x_1, \ldots, x_D) = \prod_{i=1}^{D} P(x_i | pa(x_i))$$

Figure 23.18

In the equation in figure 23.18, x is a variable or event, D is the number of variables, x_i is the ith variable, and pa(x_i) represents the parental variables of the variable x_i. Parental variables are the parent nodes in the DAG. Applying the equation in figure 23.18 requires knowing the conditional probabilities $P(x_i | pa(x_i))$ for every possible state of the parent nodes. For a large number of variables, this task is intractable, so instead the conditional probabilities are parameterized in a lower dimensional feature space using a statistical model. The model could be any of the models described in this book, because once the parameters of the conditional probabilities are estimated, they can be inserted into the equation in figure 23.18.

Let us consider a couple examples with only a handful of variables to see how Bayesian networks can be used for inference. Suppose Traci walks outside one morning and notices that her grass is wet. She wants to know whether her grass is wet because it rained last night or if she left her sprinkler on. She looks over to her neighbor John's yard and sees that his grass is also wet. This causes her to infer that it must have rained last night. Traci's inference is correct, but we can use Bayesian networks to show how correct she is, in terms of how probable her inference is correct.

Let $R \in \{0,1\}$, where R =1 means that it rained last night,

$S \in \{0,1\}$, where S =1 means that Traci left the sprinkler on all night,

$J \in \{0,1\}$, where J =1 means that Jack's grass is wet,

and $T \in \{0,1\}$, where T =1 means that Traci's grass is wet.

A Bayesian model would represent the probability distribution of the joint set, P(R, S, J, T). By using the equation in figure 23.18 to model the situation, and then expanding it out to chained conditional probabilities, the model becomes:

$$P(R,S,J,T) = P(T|J,R,S) * P(J,R,S)$$

$$P(R,S,J,T) = P(T|J,R,S) * P(J|R,S) * P(R|S) * P(S)$$

Now we can make a few independence assumptions. For example, we can assume that Traci's yard being wet depends only on the occurrence of rain or the sprinkler being left on. Therefore, P(T|J, R, S) can be reduced to P(T|R, S). We can also assume Traci's sprinkler cannot get John's yard wet, so P(J|R, S) reduces to P(J|R), and we can assume that the occurrence of rain is not influenced by the sprinkler, so P(R|S) reduces to P(R). Now the model is:

$$P(R,S,J,T) = P(T|R,S) * P(J|R) * P(R) * P(S)$$

A graphical representation of this model is shown below.

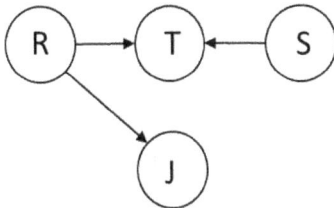

All that is required to solve the problem are the prior probabilities. Let us suppose that Traci lives somewhere where the average ratio of rainy days to total days in the year is 0.2, and the probability that Traci uses her sprinkler on any given day is 0.1. If it rains, both Traci's and John's yards will be wet, so P(J=1|R=1) = 1 and P(T=1|R=1, {S=0 or S=1}) = 1. We can assume John also uses a sprinkler so there is a small probability that his grass could be wet even if it did not rain: P(J=1|R=0) = 0.2. We can also assume there is a small chance that the grass does not look wet even if Traci did leave the sprinkler on: P(T=1|R=0, S=1) = 0.9. Given these values, we can use Bayes' theorem to create the formula below, which shows the probability that Traci left the sprinkler on, given that her grass is wet and John's grass is wet.

$$P(S = 1|T = 1, J = 1) = \frac{P(S = 1, T = 1, J = 1)}{P(T = 1, J = 1)}$$

Expanding this equation out yields:

$$P(S = 1|T = 1, J = 1) = \frac{\sum_{J,R} P(T = 1, J, R, S = 1)}{\sum_{J,R,S} P(T = 1, J, R, S)}$$

$$P(S = 1|T = 1, J = 1) = \frac{\sum_{R} P(T = 1, J = 1, R, S = 1)}{\sum_{R,S} P(T = 1, J = 1, R, S)}$$

$$P(S = 1|T = 1, J = 1) = \frac{\sum_{R} P(J = 1|R) * P(T = 1|R, S = 1) * P(R) * P(S = 1)}{\sum_{R,S} P(J = 1|R) * P(T = 1|R, S) * P(R) * P(S)}$$

$$P(S = 1|T = 1, J = 1) = \frac{0.034}{0.214}$$

$$P(S = 1|T = 1, J = 1) = 0.159$$

Therefore, the posterior belief that Traci left the sprinkler on changes from the prior probability of 0.1 to 0.159, given that Traci's grass is wet and John's grass is wet. Since the probability that Traci left the sprinkler on is only about 16%, it is reasonable for her to infer that rain caused her grass to be wet.

Markov Networks

Whereas Bayesian networks are always DAG's, Markov networks are undirected graphs that can be either cyclic or acyclic (Barber, 2012). Markov networks take the form of the equation in figure 23.19.

$$P(x_1, \dots, x_n) = \frac{1}{Z} \prod_{c=1}^{C} \phi_z(\chi_z)$$

Figure 23.19

$$Z = \sum_{a,b,d} \phi(a,b)\phi(b,d)$$

Figure 23.20

In the equation in figurer 23.19, Z is a constant that ensures normalization called the partition function. The **partition function**, Z, is defined in figure 23.20, as the sum of the products of the potentials for a set of variables a, b, and d. A potential is represented by phi. A **potential** is a non-negative function of the variable. Phi(a,b) is an example of a joint potential. A probability distribution can be redefined as the normalized potential. Normalization means that the sum of the values in the potential sum to 1, which is also true for a probability distribution, so the potential is just a non-negative function that acts as a substitute for the probability distribution.

Potentials are important because a Markov network is defined as a product of potentials on the maximal cliques of an undirected graph (refer back to figure 23.19). In the equation in figure 23.19, c=1,...,C are the maximal cliques of an undirected graph. Recall from chapter 4 that normalization is a way to scale data to make it possible to directly compare observations measured on different scales on a common scale. In the case of a Markov network, the product of potentials over all of the cliques is normalized so that they are transformed to a common scale that sums to 1 (essentially, they are scaled to a common probability distribution) (Barber, 2012).

When the clique potentials in a Markov network are all positive, they form a Gibbs distribution or Boltzmann distribution. A **Gibbs distribution**, or **Boltzmann distribution**, is a probability distribution of potentials in a system with many possible states, where a **state** is defined as a set of conditions within the system. For example, in thermodynamics, the state of a system is defined in terms of its energy and temperature. In layman's terms, this is like saying that the state of a room in building is defined by its temperature. Since there is a probability associated with each temperature, the set of possible temperatures forms a Gibbs distribution.

As an example of a Markov network, consider the Boltzmann machine. The **Boltzmann machine** is a Markov network on binary variables that takes the form of the equation in figure 23.21.

$$P(x) = \frac{1}{Z(w,b)} e^{\sum_{i<j} w_{ij} x_i x_j + \sum_i b_i x_i}$$

Figure 23.21

In the equation in figure 23.21, e is Euler's number, w is the weight of a link in the network, and b is the bias. The Boltzmann machine is a fully connected undirected graph, with a link between nodes i and j for $w_{ij} \neq 0$, as shown in figure 23.22. The Boltzmann machine is useful in machine learning because it has been used as a basic model of distributed memory. It is beyond the scope of this book to dive further into the Boltzmann machine, because it is a type of neural network. But it is worth noting here that all neural networks are graphical models.

Boltzmann Machine: A Fully Connected Undirected Graph

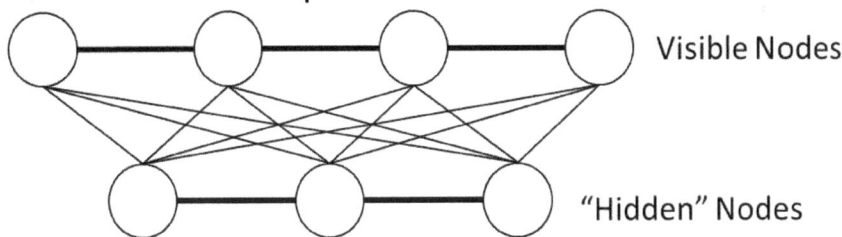

Figure 23.22

Not all Markov networks are fully connected, but all Markov networks do have some common properties.

1. **Global Markov Property** – Any two subsets of nodes (variables) in a Markov network are independent if they are separated by a subset S. So if A and B are connected only through S, or in other words if every link path from A to B passes through S, then A and B are independent.
2. **Local Markov Property** – All linked nodes (variables) in a Markov network are dependent. This property makes Markov networks useful in studying physical phenomena, as local dependencies can give rise to global collaborative phenomena.
3. **Pairwise Markov Property** – Any two non-adjacent nodes (variables) in a Markov network are independent.

These properties are ordered by their strength, so the global property is the strongest. In addition to these properties, Markov networks exhibit the general **Markov property**. To understand what this means, think about a collection of ordered random variables. A collection of ordered random variables is called a stochastic process. **Stochastic processes** are indexed or ordered in some way, such as by time. The condition of a stochastic process at one particular index or point in time is called a state, as was previously defined. If the conditional probability distribution of future states of a stochastic process depends only on the current state, and not any other previous states, then the process is said to exhibit the Markov property. So the conditional probabilities of future states of Markov networks, which are also called **Markov random fields** if the networks represent stochastic processes, depend only on the current state (Rice, 2014). Stochastic processes that exhibit the Markov property are called **Markov**

processes. A random walk, as we will see in a later chapter, is an example of a Markov process. Another example is the Markov chain. A **Markov chain** is a model that describes a sequence of possible events in which the probability of an event at time t depends only on the state of the process at time t-1. A Markov chain is the simplest type of Markov network, and it consists of a Markov processes with discrete time points. Figure shows how all of these Markovian concepts are related, since there are so many definitions that they become entangled in a paragraph of text.

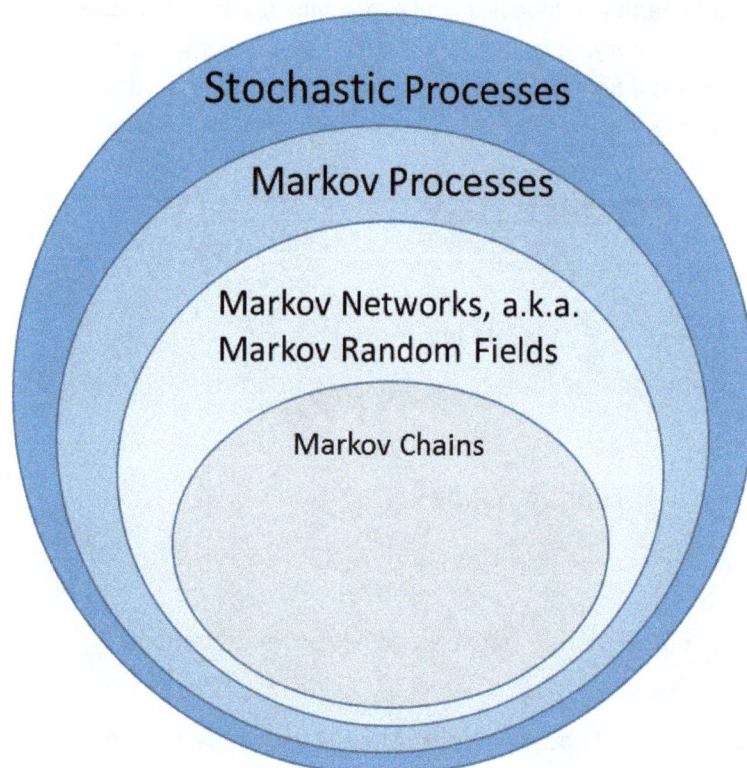

Figure 23.23

We have defined Markov chains as exhibiting the Markov property, but technically, when the current state depends only on the previous state, the model is said to be a **first order** Markov model. It is possible to have a 2nd order Markov model where the current state depends on the previous 2 states, and an nth order Markov model where the current state depends on the previous n states (Barber, 2012). The Markov property for an nth order Markov model is defined below. The strict Markov property, which consists of only first order models, limits n to 1.

$$P(x_i|x_{i-1}, x_{i-2}, ..., x_1) = P(x_i|x_i, ..., x_{i-n})$$

Figure 23.24

The part of the equation to the right of the equals sign defines an nth order Markov chain. From now on in this book, whenever a Markov chain is referred to, it will be assumed to be a first order model, and so it is defined by the equation below.

$$P(x) = P(x_i|x_{i-1})$$

Figure 23.25

A reasonable question at this point is why Markov networks are modeled as undirected graphs if they exhibit the Markov property, where the current state depends only on the previous state. Why should not the graph be directed as a long chain of nodes where every node is connected only to the node before it and the node after it? Indeed, Markov chains are usually depicted this way, but it is just as accurate to depict them as undirected because the joint density of the states in a Markov chain can be decomposed into a product of conditional densities, which can also be represented by the factorized decomposition of an undirected graph. The top equation in figure 23.26 shows how the join density of a directed Markov chain of n elements can be decomposed into a product of conditional densities. The bottom equation in figure 23.26 shows how the undirected version of the same graph can be factorized.

$$P(x) = P(x_n|x_{n-1}) * \ldots * P(x_2|x_1) * P(x_1)$$

$$P(x) = \Phi_{n,n-1}(x_n, x_{n-1}) * \ldots * \Phi_{2,1}(x_2, x_1)$$

Figure 23.26

In general, any directed graph can be transformed into an undirected graph and vice versa. Undirected graphs yield more flexible models, making Markov networks slightly more powerful than Bayesian networks (Barber, 2012). However, it is very important to consider whether the transformation is necessary, as computations for undirected graphs are much more intensive. When a directed graph sufficiently captures the problem to be solved, there should be no need to convert it. For example, consider voting preferences in groups of friends. Friends typically influence one another, so it is reasonable to use an undirected graph to capture this influence, because the direction is uncertain and likely goes both ways. However, bank to bank wire transactions, for example, are directional: a remitter sends money to a beneficiary. So it would be sufficient to model a network of wire transactions as a directed graph.

From now on in this book, Markov models will be modeled as directed graphs, because it is more intuitive to see how sequences of events interact through directed graphs. The directed graph in figure 23.27 shows an example of a Markov chain with two states.

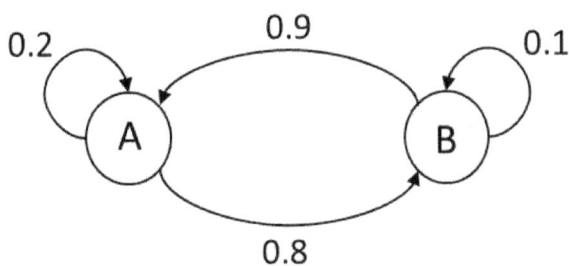

Figure 23.27

The links in the Markov chain are weighted to show the transition probabilities between the two states. There are 4 possible transitions, each with its own probability. When a node loops back on itself, it is said to be sticky. This is a very useful property in a variety of situations. For example, consider the stock market. If we imagine node A in figure to be a bull market and node B to be a bear market, it is reasonable to think that a day in a bull market might follow other days in a bull market, and likewise for a bear market (Halls-Moore, 2016). We will see this concept in a later chapter when we look at volatility clustering in financial time series. The probabilities of the state transitions in the Markov chain shown in figure 23.27 can be represented by a transition matrix as shown in figure 23.28.

	A	B
A	0.2	0.8
B	0.9	0.1

Figure 23.28

If this Markov chain were used to calculate a sequence of 20 states, we should expect the sequence to have more A's than B's. For example, if the process starts on B, then it has a 90% chance of transitioning to state A on the next step in the sequence and a 10% chance of staying in state B. If it transitions to A, then it has an 80% chance of transitioning back to B in the 3rd step in the sequence, and a 20% chance of staying in state A. If we changed the transition probabilities, we could produce processes with different behaviors. Some processes might settle into a steady state or converge into an equilibrium. The equation in figure 23.29 shows how to determine if some state pi is an equilibrium distribution of a Markov chain.

$$\pi_j = \sum_{i \in S} \pi_i P(x_{n+1} = j | x_n = i)$$

Figure 23.29

The equation in figure 23.29 is showing that equilibrium distribution for state j equals the sum of the weighted probabilities that the next state in the sequence is j, given the current state is i. Not all Markov chains have a reachable steady state.

Bayesian Regression

In the section of this chapter about Bayes' theorem, we saw how the posterior probability could be represented by a likelihood function that produces the best model parameters when maximized through MLE. That same concept underlies Bayesian regression. If we start with a standard linear regression model shown in the top equation in figure 23.30, it can be rewritten in terms of a probabilistic model as shown by the middle equation in figure 23.30. The middle equation models the dependent variable y (the normally distributed posterior probability) as a function of x with parameters alpha, beta, and standard deviation sigma. The script N denotes that y is normally distributed with mean mu and standard deviation sigma. The bottom equation in figure 23.30 shows the likelihood function to be maximized (Tim, 2016).

$$y = \alpha + \beta x + \varepsilon$$

$$y = \mu + \varepsilon, \text{ where } \mu = \alpha + \beta x \text{ and } y \sim \mathcal{N}(\mu, \sigma)$$

$$\arg max \prod_{i=1}^{N} \mathcal{N}(y; \alpha + \beta x, \sigma)$$

Figure 23.30

The equation at the bottom of figure 23.30 is stating that the maximum likelihood is the product over all of the observations of the density function of a normal distribution evaluated at y points, with parameters alpha, beta*x, and sigma. If the equation at the bottom of figure 23.30 were optimized, it would be no different than the frequentist approach where the likelihood function is maximized (by MLE). Instead of doing this, the Bayesian approach adds prior distributions for the estimated parameters alpha, beta, and sigma to the model, resulting in the equation shown in figure 23.31. Note that the equation in the bottom of figure 23.31 essentially stating that the posterior is proportional to the likelihood function times the prior, as shown in the top of figure 23.31.

$$posterior \propto likelihood * prior$$

$$f(\alpha, \beta, \sigma | Y, X) \propto \prod_{i=1}^{N} \mathcal{N}(y; \alpha + \beta x, \sigma) * f(\alpha) f(\beta) f(\sigma)$$

Figure 23.31

The result of optimizing this equation is a set of posterior probabilities for the parameters alpha, beta, and sigma. If conjugate priors were used, then the posterior distribution is known (if a conjugate prior was assumed, then the posterior follows the same distribution as the prior) (Salvatier, Wiecki, & Fonnesbeck, 2018). In the coin flipping example from the Bayes' theorem section of this chapter, we assumed that the conjugate prior had a mean of 0.6 and a standard deviation of 0.1. In many real world applications however, the conjugate prior and posterior distributions are unknown, making the estimation of the final model difficult. Fortunately, there are Markov methods that can produce good estimates of the posterior distribution, and we will explore these in a later chapter.

Once the final model is chosen, either by using the Markov methods that will be described later, or by assuming a conjugate prior and optimizing the parameters, the tests for significance for the Bayesian model can be carried out. Whereas the frequentist approach gives every parameter estimate a p-value that indicates whether or not the null hypothesis can be rejected, the Bayesian approach gives a Bayes factor. **Bayes factor** is independent of sample size and is calculated as the ratio of the posterior odds to the prior odds, which is equivalent to the ratio of the probability of observing the data given model 1 to the probability of observing the data given model 2.

$$K = \frac{\dfrac{P(M1 = null | z, N)}{P(M2 = alt | z, N)}}{\dfrac{P(M1 = null)}{P(M2 = alt)}}$$

Figure 23.32

Model 1 represents the null hypothesis and model 2 represents the alternative hypothesis. The null hypothesis is that a parameter has a probability of 1 at only one value, and zero probability everywhere else. This would mean the probability distribution is a vertical line at a certain value. The alternative hypothesis is that all values of the parameter are possible. This would mean the probability distribution is flat.

A Bayes factor, K, < 0.1 means that the null hypothesis can be rejected. A broader table of interpretation that is cited in most statistics texts is provided in figure 23.33 below.

K	Strength of Evidence Supporting the Null Hypothesis (Model 1)
1-3	Not worth mentioning
3-20	Positive
20-150	Strong
>150	Very strong

Figure 23.33

An alternative to the p-value for the frequentist approach is the confidence interval. A comparable metric for the Bayesian approach is the **credible interval**. The narrowest credible interval contains the highest probability density for the parameter and includes the mode of the distribution. This interval is called the **high density interval (HDI)**. The HDI is formed from the posterior distribution and includes a range of 95% of the most plausible values of the parameter. The HDI is susceptible to change when the parameter is transformed, so critics of the HDI recommend using confidence intervals because they are based on percentiles and so are invariant to transformations in the parameter. Conversely, critics of the confidence interval argue that it can be wrong 5% of the time (for a 95% interval) because it excludes 5% of the possible trials. Conceptually, the confidence intervals in the frequentist approach give the probability that the parameter equals some value. In contrast, the credible interval or HDI in the Bayesian approach gives the probability that the parameter is some value drawn from a posterior distribution. The arguments for and against credible intervals come down to the fundamental differences between the Bayesian and frequentist approaches. It is up to the researcher to determine which method is better for any given problem.

One last note to make about Bayesian regression is that it is more robust to outliers than frequentist regression. When there are outliers in the dataset, the frequentist approach is to use robust regression methods or try to transform the data to remove the outliers. The Bayesian approach deals with outliers by choosing priors with wider tails, like the t-distribution. This makes the probability of observing an outlier slightly greater (Salvatier, Wiecki, & Fonnesbeck, 2018).

Classification with Naïve Bayes

If variables are considered to be events, then the goal of classification is to determine how the probability of one event relates to the probability of another. In other words, the goal is to find the joint probability of two events. To find joint probability, it is necessary to know whether the two events are independent or dependent.

- If the events are independent, then the probability of both occurring is the product of their probabilities. For example, if 20% of all emails are spam, and 5% of all emails contain the word Viagra, then .2*.5 = .01, or 1% of all emails are spam messages containing the word Viagra.
- If the events are dependent, then the probability of both occurring is described by Bayes Theorem.

The Naïve Bayes classifier uses Bayes' Theorem to find the posterior join probability of the outcome variable and the independent variables. The independent variables are assumed to be independent of one another and equally important, which is a naïve assumption, so that is where the naïve part of the name Naïve Bayes comes from. This assumption is rarely ever true, but Naïve Bayes is still useful because probabilities are used to determine classification, and probabilities ranging from 51% to 99% are rounded to the same class (Raschka, 2015).

The strengths of the Naïve Bayes classifier are that is it simple to execute and trains quickly, it does well with noisy data and can easily handle missing data, and it requires few training samples to be fitted, yet still performs well with large samples. Its weaknesses are that it relies on the faulty assumption of independent and equally important features, it is not ideal for data sets with many numeric features, and the estimated probabilities are less reliable than the predicted classes. Naïve Bayes is often a very good classifier for text data. For example, it is commonly used to train email spam filters (Raschka, 2015). The equation that represents the class probabilities of the Naïve Bayes classifier is shown in figure 23.34.

$$P(y_c|x_1, \dots, x_N) = \frac{1}{Z}P(y_c) \prod_{i=1}^{N} P(x_i|y_c), \text{ where } Z = \sum_c P(y_c)P(x|y_c)$$

Figure 23.34

The equation in figure 23.34 states that the probability of y belonging to class c, given the evidence of the features, is equal to the product of the probabilities of each piece of evidence conditioned on the class c, times the prior probability of class c, times a scaling factor 1/Z, which converts the final product to a probability between 0 and 1. It was previously stated that Naïve Bayes was based on Bayes' Theorem, and yet the equation in figure 23.34 is much different in appearance than Bayes' Theorem. The reason for this is that when the independence assumption is made between the individual features, as well as between the classes of the dependent variable, Bayes Theorem simplifies to the form shown in figure 23.34. Turning the equation in figure 23.34 into a classifier for y = y_c produces the equation in figure 23.35.

$$\hat{y} = \arg\max P(y_c) \prod_{i=1}^{N} P(x_i|y_c)$$

Figure 23.35

Thus, the estimated class if found by the maximum class probability. Notice that if any one of the features has a prior probability of 0, meaning it is not observed in the dataset, then the entire product in the equations 23.34 and 23.35 would be 0, regardless of the probabilities of other features. This can happen when a categorical feature has a new level appear that has never been observed belonging to a given class before. To overcome this problem, a Laplace estimator (a very small, nonzero number) is added to the frequency of each level of each class. Most of the time, the estimator is set to 1, ensuring that every class-feature combination is observed at least once.

Naïve Bayes, like any classifier, can be used to predict numeric outcomes, assuming the numeric outcome is binned into categories. In general however, binning numeric outcomes is a bad idea because it eliminates some information by grouping ranges of values together.

Bayesian Networks and Classification with R
This code sections for this chapter will be a bit different than previous chapters. Since different aspects of Bayesian statistics are easier with R, and other easier with Python, we will split some of the concept examples up. The bnlearn library for R is unquestionably the easiest way to visualize Bayesian networks,

so we will start with that. The library comes with a built in dataset containing data that is well
structured for analysis in a graphical model: the coronary dataset. This dataset has a field to indicate
whether an individual is a smoker, a field indicating whether an individual does mental work, another for
physical work, a binned blood pressure variable, a binned protein ratio variable, and a binary indicator
for whether the individual has heart disease. The variables in this dataset are ripe for dependencies that
can be represented graphically. A hill climbing algorithm can automatically detect these dependencies,
and we can plot the resulting network.

```
library(bnlearn)
set.seed(14)
d <- coronary
colnames(d) <- c('smoker', 'mental_work',
                'physical_work', 'blood_pressure',
                'protein_ratio',
                'heart_disease')
head(d)
#Find a good network structure using hill climbing algorithm
net <- hc(d)
plot(net)
```

The resulting plot shows causal relationships using arrows. Some of them look ok, but others defy logic.
For example, the link from mental_work to heart_disease does not make sense. We can manually
remove this link. We can also add a link from smoking to heart_disease to prevent heart_disease from
being orphaned and because it seems reasonable for smoking to effect heart_disease.

```
#Remove link from mental_work to heart_disease, bc heart disease does not depend on
mental work
head(net$arcs)  #Shows how links are structured
net$arcs <- net$arcs[-which((net$arcs[,'from'] == 'mental_work' & net$arcs[,'to'] ==
'heart_disease')),]
#Add link from smoker to heart_disease
set.arc(net, from='smoker', to='heart_disease')
amat(net)['smoker', 'heart_disease'] <- 1  #Update adjacency matrix
plot(net)
```

The remaining links seem plausible. We can view the conditional probability tables for each
node/variable to see how each variable is effected by others. The conditional probability tables for a
particular variable can be called by referencing the variable as if it were a column in a dataframe. Let us
examine the conditional probabilities for blood pressure and protein ratio.

```
#Create conditional probability tables for each node
cpts <- bn.fit(net, data=d)
print(cpts$blood_pressure)
print(cpts$protein_ratio)
```

Blood pressure is only conditional on smoking. This can be verified by comparing the plot of the
network, which shows a directional link from smoking to blood pressure, to the conditional probability
table, which shows how each bin of blood pressure depends on each value of the smoker variable.
Similarly, the protein ratio is conditional on smoking and mental work. The conditional probabilities
show $P(B|A)$, so for blood pressure for example, $P(\text{blood\_pressure} < 140 \mid \text{non-smoker}) = 0.54$.

Using this Bayesian network, we can perform Bayesian inference. Let's use Bayes' theorem to find the probability that the protein ratio is <3 if smoker = no. In other words, we will find P(protein_ratio < 3 | smoker = no). This is equivalent to the equation to find P(A|B) in figure 23.6 and the bottom row of figure 23.2.

```
#Perform inference
cpquery(cpts, even=(protein_ratio=="<3"), evidence=(smoker=="no"))
```

So there is a 63% probability that protein ratio < 3 if an individual is a non-smoker. We can combine events too. For example, what is the probability that a non-smoker with blood pressure > 140 has a protein ratio < 3?

```
cpquery(cpts, event=(protein_ratio=='<3'), evidence=(smoker=='no' &
blood_pressure=='>140'))
```

Roughly 62%. Lastly, what is the probability that an individual has heart disease, given that they are a smoker?

```
cpquery(cpts, event=(heart_disease=='pos'), evidence=(smoker=='yes'))
```

Roughly 15%. Now we will turn our attention to the old faithful dataset. This dataset contains the eruption times of the geyser called "Old Faithful" in Yellowstone National Park. We can use the bnlearn package to determine the direction and strength of the relationship between the number of people waiting and eruptions. Common sense tells us there should be no directional relationship at all, but let's see if we can prove that using Bayesian inference. First we will discretize the dataframe and use bootstrap resampling to average multiple directed acyclic graphs. In this example, we will bootstrap 500 networks (R=500).

```
library(bnlearn)
set.seed(14)

#Look at skewed data
hist(faithful$eruptions)

#Discreteize dataframe of entirely continuous variables
dfaithful <- discretize(faithful, metho='hartemink', breaks=3, ibreaks=25,
idisc='quantile')

#Bootstrap resampling to average multiple DAGs, where R = number of network
structures
boot <- boot.strength(dfaithful, R=500, algorithm='hc',
algorithm.args=list(score='bde', iss=10))
```

Now we can look at the strength in terms of edge frequency, and the direction of the edge.

```
boot[boot$strength > 0.85 & boot$direction >= 0.5,]
```

Since the edge strengths are identical, the direction of the link is uniform. In other words, there is no directional relationship, just as we suspected.

```
avg.boot <- averaged.network(boot, threshold=1)
plot(avg.boot)
```

The plot shows an undirected graph.

The last technique we will explore with R is classification with Naïve Bayes. Let's use the Iris dataset, since it is very familiar to us by this point.

```
library(e1071)

#Naive Bayes
nb_model <- naiveBayes(Species ~ ., data=iris)
nb_model
preds <- predict(nb_model, iris[,-which(colnames(iris) %in% c('Species'))])
table(preds, iris$Species)
```

The confusion matrix shows that Naïve Bayes made 6 errors, resulting in an accuracy of 96%. Note that we broke good practice by not using a test set, but this simple example shows how the algorithm can be deployed.

Bayesian Regression and Classification with Python

This code sections for this chapter will be a bit different than previous chapters. Since different aspects of Bayesian statistics are easier with R, and other easier with Python, we will split some of the concept examples up. We skipped over combinations, permutations, and a direct application of Bayes' theorem with R since we could apply inference using bnlearn. So in this section, we will define custom functions for combinations, permutations, and Bayes' theorem. The combination and permutation functions are simple:

```
import math

def permutation(n, k):
    return math.factorial(n) / math.factorial(k)

def combination(n, k):
    return permutation(n, k) / math.factorial(k)
```

These functions are self explanatory, so we will skip any examples and move on to Bayesian inference using the textbook coin flipping example. This example was adapted from a tutorial by Folkman, 2015. A fair coin has probability of heads = 0.5 and probability of tails = 0.5. Each flip is a Bernoulli trial, as described earlier in this chapter, and Bernoulli trials are drawn from a Binomial distribution. If we use numpy to simulate 1,000 samples from a binomial distribution with a probability of 0.5, we should expect the mean of those samples to be about 0.5.

```
import numpy as np
from scipy import stats
import matplotlib.pyplot as plt

#Simulate 1000 coin flips (Bernoulli trials) from a binomial dist
bernoulli_trials = np.random.binomial(n=1, p=0.5, size=1000)
print(np.mean(bernoulli_trials))
```

The mean is indeed close to 0.5. The probability mass function (PMF) gives the probability of a positive hit for a single Bernoulli trial. We can consider a positive hit to be whatever we want, but it is common practice to make heads positive.

```
print(stats.bernoulli.pmf(1, 0.5))
```

So the probability of a single head is about 0.5. No surprise. Let's generate 1,000 random binary numbers to simulate coin flipping data.

```
d = np.random.randint(2, size=1000)
```

For independent events, the probability of observing the data, d, is the product of the PMF for every Bernoulli trial in the sequence. We can look at the probability of observing the random data, d, on a scale of 0-1.

```
#For indpendent events, the probability of observing the data is the product of the
PMF
print(np.product(stats.bernoulli.pmf(d, 0.5)))

#Look at the probability of the data on an x scale of 0-1
x = np.linspace(0, 1, 100)
prob = [np.product(stats.bernoulli.pmf(d, p)) for p in x]
print(prob)
```

So far we have done nothing extraordinary. Now we will look at the prior distribution of a fair coin.

```
#Plot distribution of a fair coin (prior)
fair_flips = bernoulli_flips = np.random.binomial(n=1, p=.5, size=1000)
p_fair = np.array([np.product(stats.bernoulli.pmf(fair_flips, p)) for p in x])
p_fair = p_fair / np.sum(p_fair)
plt.plot(x, p_fair)
plt.title('Prior Probability')
plt.show()
```

For comparison, let's simulate an unfair coin by setting p=0.8.

```
#Look at the probability of an unfair coin (sample)
d = np.random.binomial(n=1, p=0.8, size=1000)
prob = np.array([np.product(stats.bernoulli.pmf(d, p)) for p in x])
print(prob)
```

Suppose our observed data has a probability of 0.8 but the prior is 0.5. If this were true, then there were 80% heads in the observed data when we expected 50% heads from the prior. Using Bayes' theorem, we can apply Bayesian inference to update our beliefs about the posterior probability distribution of the observed data. When we plot the posterior, we should expect it to have shifted towards a mean of 0.8. Let us simulate 100 Bernoulli trials with 80% heads, and then 1,000 Bernoulli trials with 80% heads. The more observations there are, the greater the update to the posterior distribution. So we should expect the posterior to have shifted more for the 1,000 trials.

```
#Plot the posterior distribution after applying Bayes' Theorem
def bayes(n_sample, n_prior=100, observed_p=0.8, prior_p=0.5):
    x = np.linspace(0, 1, 100)
    sample = np.random.binomial(n=1, p=observed_p, size=n_sample)
    observed_dist = np.array([np.product(stats.bernoulli.pmf(sample, p)) for p in x])
    prior_sample = np.random.binomial(n=1, p=prior_p, size=n_prior)
    prior = np.array([np.product(stats.bernoulli.pmf(prior_sample, p)) for p in x])
    prior = prior / np.sum(prior)
    posterior = [prior[i] * observed_dist[i] for i in range(prior.shape[0])]
```

```
    posterior = posterior / np.sum(posterior)

    fig, axes = plt.subplots(3, 1, sharex=True, figsize=(8,8))
    axes[0].plot(x, observed_dist)
    axes[0].set_title("Sampling Distribution")
    axes[1].plot(x, prior)
    axes[1].set_title("Prior Distribution")
    axes[2].plot(x, posterior)
    axes[2].set_title("Posterior Distribution")
    plt.tight_layout()
    plt.show()

    return posterior

bayes(100)
bayes(1000)
```

Everything we expected to see has been shown in the plots produced by the bayes function that we defined.

Now let's move on to Markov Chains. Markov Chains are used to simulate a series of dependent events that are related by fixed probabilities. This makes them good at generating fake text. Markov chains, and even HMMs, are used to generate fake tweets that appear reasonably normal, for example. An example of using a HMM to create fake bible verses can be found at: https://github.com/nlinc1905/Markov-Model-Bible-Verse-Creation. We will use a Markov chain to generate fake speeches for President Trump. The dataset that we will use for training consists of campaign speeches made by Donald Trump, altered from its original form for ease of use. The dataset can be downloaded from the GitHub repository for this book, but it was curated by McDermott, 2016. When we load the data, we need to split it into a corpus of documents, where each document is a word. Then we can create a generator to produce word pairs from the corpus. The generator can be used to create a dictionary where the keys are the unique words used in the campaign speeches, and the values are the list of words that appear after the key word.

```
import numpy as np

#Trump speech generator
d = open('trump_speeches.txt', encoding='utf8').read()
corpus = d.split()
#Function to generate all pairs of words
#Keys are all unique words and all of the words that appear after the key are stored
as a list of values
def word_pairs(corpus):
    for i in range(len(corpus)-1):
        yield (corpus[i], corpus[i+1])
pairs = word_pairs(corpus)
words = {}
for w1, w2 in pairs:
    if w1 in words.keys():
        words[w1].append(w2)
    else:
        words[w1] = [w2]
```

All that is required to initialize the Markov chain text generator is a random first word. We can set the stopping point to whatever we choose. In this example, we will generate 40 words. The Markov chain chooses a random word from the list of words following a given word to be the next word in the sequence. It is essentially a memoryless process where the next word depends only on the current word in the sequence.

```
#Choose a word to initialize the Markov chain and the length of the chain in number
of words
first_word = np.random.choice(corpus)
nbr_words = 40
def markov_chain_text(initial_state, word_pair_dict, stop_time):
    state = initial_state
    chain = [state]
    for w in range(stop_time):
        state = np.random.choice(word_pair_dict[chain[-1]])
        chain.append(state)
    return ' '.join(chain)

print(markov_chain_text(first_word, words, nbr_words))
```

The result is a 40 word speech generated from the typical speaking pattern of Donald Trump during his campaign speeches.

Let us consider another Markov chain example. Sticking with the political theme, we will build a Markov chain to model voters voting for the democrat and republican parties in a sequence of presidential elections. This type of situation resembles the graph in figure 23.27, where there are 2 states: democrat and republican, and transition probabilities between each state. To model this situation, we need to make a few assumptions. First, we will assume that the initial split of the voting population is 55% republican and 45% democrat. Next, we will assume the following transition probabilities:

- 80% of republican voters will vote republican in the next election
- 20% of republican voters will vote democrat in the next election
- 70% of democrats will vote democrat in the next election
- 30% of democrats will vote republican in the next election

If we manually calculate the split of the voting population in the next election, we can see how the Markov chain can be expected to transition from state t to state t+1.

```
#Assume the following split of voting population for this elect (time t)
rep_1 = 0.55
dem_1 = 0.45
#Assume the following transition probabilities
rep_to_rep = 0.8
rep_to_dem = 0.2
dem_to_dem = 0.7
dem_to_rep = 0.3
#Calculate split of voting population at next election (t+1)
rep_2 = rep_1*rep_to_rep + dem_1*dem_to_rep
dem_2 = dem_1*dem_to_dem + rep_1*rep_to_dem
print('Republican vs Democrat Split for Next Election:', rep_2, 'vs', dem_2)
```

Note that the operation to calculate the split at time t+1 can be made more efficient by vectorizing the code with a dot product. We can define a function to calculate the split of the voting population for any future election, assuming the transition probabilities remain constant. To enable vectorization of the code, we will put the initial split and the transition probabilities into matrix form.

```python
#Define initial state and transition matrix to make calculations simpler in the
future
state_1 = np.array([rep_1, dem_1])
transm = np.array([[rep_to_rep, rep_to_dem], [dem_to_rep, dem_to_dem]])
#Define function to generalize the calculations above to any future time
def markov_chain(initial_state, transition_matrix, stop_time):
    state = initial_state
    chain = [state]
    for t in range(stop_time):
        state = np.round(np.dot(state, transm), 4)
        chain.append(state)
    return chain
print(markov_chain(state_1, transm, 10))
```

The proportion of voters appears to converge around a republican to democrat ratio of 60:40. Recall that the steady state is when the initial state * transition matrix = the initial state, meaning no more changes in the state occur. So rather than continuing for an indefinite time, we can adjust the Markov chain function to stop if a steady state has been reached.

```python
def markov_chain(initial_state, transition_matrix, stop_time):
    state = initial_state
    chain = [state]
    for t in range(stop_time):
        if (np.round(np.dot(state, transm), 4) == state).all():
            print('Steady state reached at time t =', t)
            break
        else:
            state = np.round(np.dot(state, transm), 4)
        chain.append(state)
    return chain
print(markov_chain(state_1, transm, 20))
```

We can see that the steady state is reached at time t=9. Note that in the real world, the transition probabilities are likely to change between elections, so this is not a good model of reality. Nevertheless, it serves to show how Markov chains work.

We skipped Bayesian regression with R because it is easier to do in Python using the PyMC3 library. Recall from earlier in the chapter that the conjugate prior and posterior distributions are usually unknown, making the analytical approach to performing Bayesian regression very difficult. It is for this reason that Bayesian models never received as much attention in classical statistics, because it was usually intractable and impractical to use them. With modern computing power, big data, and the Markov Chain Monte Carlo algorithms that will be described in the next chapter however, Bayesian models are seeing a surge in popularity. To perform Bayesian regression here, we will use a Markov Chain Monte Carlo (MCMC) sampling algorithm to estimate the conjugate distributions. Ignore the details for now, as they will be explained in a later chapter.

To utilize PyMC3 to build a Bayesian regression model, we will define a custom function. This function takes the regression formula in the same format as R (e.g. y~x1+x2), which is unusual for Python, but it is made possible through the Patsy library, which PyMC3 uses in the backend. Using the formula specification, our function will automatically choose the prior distributions for each predictor variable. Note that we could define them ourselves, but as we will see, a normal distribution works the majority of the time. Even for skewed data, using a normally distributed prior is not a bad thing, because the MCMC algorithm is highly robust. The power of the MCMC algorithm makes Bayesian regression rival other powerful machine learning algorithms despite having a simle formulation.

Our function will return the model and the trace. The trace is a collection of sampled posterior distributions for each predictor. We will explore these in more detail shortly. First, let's define the function:

```python
import pymc3 as pm
import numpy as np
import pandas as pd
import matplotlib.pyplot as plt
import seaborn as sns

def glm_mcmc(df, formula, nbr_iters, likelihood=None, mcmc_sampler=None,
target_accept=0.8, tune=500):
    '''
    This function is a wrapper for GLM Bayesian regression, using MCMC.  The
    default MCMC method is NUTS (No U-Turn Sampler), but others are available:
        Binary variables --> BinaryMetropolis
        Discrete variables --> Metropolis
        Continuous variables --> NUTS

    Parameters:
        df = pandas dataframe
        formula = regression formula as specified in R or Patsy
        nbr_iters = number of iterations to sample over in MCMC
        likelihood = the family of likelihood functions for the data (e.g.
pm.glm.families.Normal())
        mcmc_sampler = the MCMC sampling method (e.g. pm.NUTS, pm.Metropolis),
default NUTS
        target_accept = adapts the step size to make acceptance probability closer to
target, higher target --> smaller steps, range(0,1)
        tune = number of tuning samples to use in each iteration, higher tune -->
more probability space explored

    If acceptance probability is higher than the target, increase the tune
    parameters.  For more info about this, see:
        https://discourse.pymc.io/t/warning-when-nuts-probability-is-greater-than-
acceptance-level/594

    If there are divergences after tuning, increase the target_accept
    parameter.  For more info about this, see:

https://docs.pymc.io/notebooks/Diagnosing_biased_Inference_with_Divergences.html

    A great intro to GLM Bayesian regression with PYMC3 can be found here:
        https://docs.pymc.io/notebooks/GLM-linear.html
```

```
    '''
    bayesian_model = pm.Model()  #Initializes a Bayesian model container
    with bayesian_model:
        if likelihood is None:
            #Priors and observed sampling dist (likelihood) are automatically set
            pm.glm.GLM.from_formula(formula, df)
        else:
            pm.glm.GLM.from_formula(formula, df, family=likelihood)
        if mcmc_sampler is None:
            mcmc_sampler = pm.NUTS(target_accept=target_accept)
        else:
            mcmc_sampler = mcmc_sampler(target_accept=target_accept)
        #Perform inference to estimate the posterior, using nbr_iters MCMC sample
posteriors
        trace = pm.sample(draws=nbr_iters, step=mcmc_sampler,
                          init='auto', n_init=200000,
                          cores=None, tune=tune, progressbar=True,
                          random_seed=14)
    return bayesian_model, trace
```

One of the shortcomings of Patsy is that it has no ability to mimick the "." capability in R that allows a dependent variable to be regression on all of the other variables in a dataframe. We can define a custom function to overcome this shortcoming:

```
def patsy_formula(df, dependent_var, *excluded_cols):
    '''
    This function generates the R style formula for Patsy, omitting the
    dependent variable and any other specified variables from the right
    side of the equation.  This is useful because Patsy does not have
    a '.' operator.
    '''
    df_columns = list(df.columns.values)
    df_columns.remove(dependent_var)
    for col in excluded_cols:
        df_columns.remove(col)
    return dependent_var + ' ~ ' + ' + '.join(df_columns)
```

Now let's import the Motor Trends cars dataset, specify a Patsy formula, build a Bayesian regression model for mpg on all other variables, and then view the results (the trace).

```
#GLM of mpg on disp
d = pd.read_csv('mtcars.csv')
formula = patsy_formula(d, 'mpg')

#Bayesian regression
bayesian_reg, trace = glm_mcmc(d, formula, nbr_iters=2000)
print(pm.summary(trace))  #Regression results
#SD is the error term
pm.autocorrplot(trace)
plt.show()
```

For this regression model, we used 2,000 MCMC samples. The trace shows the estimated posterior distributions for each predictor over all of the iterations. Since the MCMC sampling algorithm requires time to converge, it is customary to omit the first half of the iterations when viewing the trace. That way, only the latter half, the half that converged, will be viewed and interpreted. Whereas frequentist regression assumes a normal distribution for the predictor coefficients, Bayesian regression shows the entire sampling distribution for each coefficient.

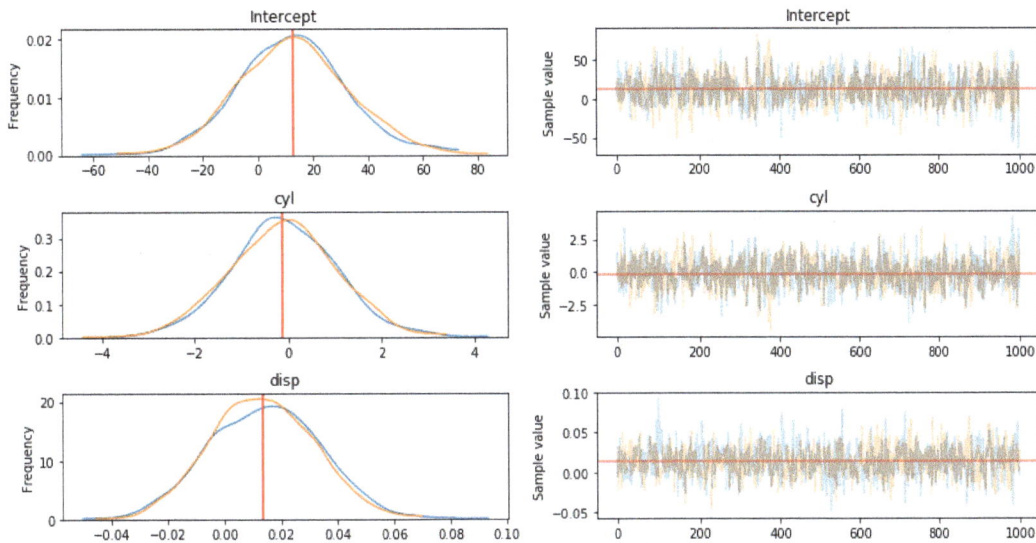

The plots on the left of the traceplot are the distributions across the iterations of MCMC. If the distributions for a coefficient align, then the algorithm is very confident in the estimated distribution of the coefficient. The reported value of the coefficient is usually the mean of this sampled distribution. Dpending on its shape, it may be better to report the median. The plots to the right of the distribution plots show the convergence of the MCMC algorithm. If these plots appear stationary (they have a fairly consistent mean and look like white noise), then the results are reliable. If the plots are non-stationary, then different prior distributions can be chosen, or a larger number of iterations can be used. Unfortunately, since Bayesian regression is sampling coefficient values for each predictor thousands of times, it is much slower than traditional least squares frequentist regression. But the benefit is that the estimation process for the coefficient values is more transparent.

```
#First 1000 estimates are called "burn-in"
pm.traceplot(trace[1000:], lines={k: v['mean'] for k,v in
pm.summary(trace[1000:]).iterrows()})
plt.show()
```

Now we can look at the RMSE to evaluate the model. We can also calculate the R^2 manually to see how well the model fits the data.

```
ppc = pm.sample_ppc(trace[1000:], samples=500, model=bayesian_reg)
rmse = np.sqrt(np.sum((ppc['y'].mean(0).mean(0).T - d['mpg'])**2) / d.shape[0])

def get_r2(df, ppc, target):
    sse_model = np.sum((ppc['y'].mean(0).mean(0).T - df[[target]])**2)[0]
    sse_mean = np.sum((df[[target]] - df[target].mean())**2)[0]
    return 1-(sse_model / sse_mean)
print('R Squared:', get_r2(d, ppc, 'mpg'))
```

The posterior predictions can be obtained through the traces. The target variable's credible interval can also be obtained.

```
#Get the posterior predictions
traces = pm.trace_to_dataframe(trace)[['Intercept'] + list(d.drop(['mpg'],
axis=1).columns)]
x = d.drop(['mpg'], axis=1)
x.insert(0, 'intercept', 1)
likelihoods = np.dot(x, traces.T)
likelihoods_sd = np.tile(pm.trace_to_dataframe(trace)[['sd']].T, (d.shape[0], 1))
likelihood = np.random.normal(likelihoods, likelihoods_sd)
print(likelihood.shape)  #Should have d.shape[0] rows and as many columns as MCMC
samples

#Get credible intervals for the target variable
dfp = pd.DataFrame(np.percentile(likelihood,[2.5, 25, 50, 75, 97.5], axis=1).T,
                   columns=['0_25','25','50','75','97_5'])
```

Finally, we can plot the target variable over the observation index and include the credible intervals to
see the predicted values versus the true values. This is akin to viewing the predicted versus actual
values and confidence intervals for frequentist regression.

```
pal = sns.color_palette('Purples')
fig, ax = plt.subplots(1, 1, figsize=(7, 7))
ax.fill_between(d.index, dfp['0_25'], dfp['97_5'], alpha=0.7, color=pal[1], label='CR
95%')
ax.fill_between(d.index, dfp['25'], dfp['75'], alpha=0.5, color=pal[4], label='CR
50%')
ax.plot(d.index, dfp['50'], alpha=0.5, color=pal[5], label='Median')
plt.plot(d.index, d['mpg'], 'bo')
plt.legend()
plt.title('Bayesian Regression Predicted Credible Intervals vs Actual')
plt.show()
```

For another example, let's do a LASSO regularized Bayesian regression. This can be achieved by using a
Laplacian as the prior distribution, because a Laplacian prior will hold the distribution of coefficients for
each predictor near zero, and only the predictors with significant effects will escape the pull towards
zero. We will need to modify our Bayesian regression function to accommodate the Laplacian priors.
Also note that the MCMC sampler can be switched from NUTS to Metropolis, because it seems to
converge better for this type of regression.

```
from scipy.stats import norm, laplace

#Sense check: compare Laplacian to normal dist
def plot_laplace_vs_normal(norm_sd=1., b=1.):

    dstrb = pd.DataFrame(index=np.linspace(-10, 10, 1000))
    dstrb['normal'] = norm.pdf(dstrb.index.values, loc=0, scale=norm_sd)
    b0 = max(b*.5,0)
    b2 = min(b*2,10)
    dstrb['laplace b={}'.format(b0)] = laplace.pdf(dstrb.index.values, loc=0,
scale=b0)
    dstrb['laplace b={}'.format(b)] = laplace.pdf(dstrb.index.values, loc=0, scale=b)
    dstrb['laplace b={}'.format(b2)] = laplace.pdf(dstrb.index.values, loc=0,
scale=b2)
    dstrb.plot(style=['--','-','-','-'], figsize=(12,4))
    plt.show()
```

```
plot_laplace_vs_normal()

def lasso_mcmc(df, formula, nbr_iters, likelihood=None, mcmc_sampler=None,
target_accept=0.8, tune=500):
    '''
    This function is a wrapper for LASSO Bayesian regression, using MCMC.  The
    default MCMC method is Metropolis, but others are available:
        Binary variables --> BinaryMetropolis
        Discrete variables --> Metropolis
        Continuous variables --> NUTS (No U-Turn Sampler)

    Parameters:
        df = pandas dataframe
        formula = regression formula as specified in R or Patsy
        nbr_iters = number of iterations to sample over in MCMC
        likelihood = the family of likelihood functions for the data (e.g.
pm.glm.families.Normal())
        mcmc_sampler = the MCMC sampling method (e.g. pm.NUTS, pm.Metropolis),
default NUTS
        target_accept = adapts the step size to make acceptance probability closer to
target, higher target --> smaller steps, range(0,1)
        tune = number of tuning samples to use in each iteration, higher tune -->
more probability space explored

    If acceptance probability is higher than the target, increase the tune
    parameters.  For more info about this, see:
        https://discourse.pymc.io/t/warning-when-nuts-probability-is-greater-than-
acceptance-level/594

    If there are divergences after tuning, increase the target_accept
    parameter.  For more info about this, see:

https://docs.pymc.io/notebooks/Diagnosing_biased_Inference_with_Divergences.html

    A great intro to GLM Bayesian regression with PYMC3 can be found here:
        https://docs.pymc.io/notebooks/GLM-linear.html
    '''
    bayesian_model = pm.Model()  #Initializes a Bayesian model container
    with bayesian_model:

        priors = {"Intercept": pm.Laplace.dist(mu=0, b=0.1),
                  "Regressor": pm.Laplace.dist(mu=0, b=0.1)
                  }

        if likelihood is None:
            #Priors and observed sampling dist (likelihood) are automatically set
            pm.glm.GLM.from_formula(formula, df, priors=priors,
                                    family=pm.glm.families.Normal())
        else:
            pm.glm.GLM.from_formula(formula, df, priors=priors,
                                    family=likelihood)
        if mcmc_sampler is None:
            mcmc_sampler = pm.Metropolis(target_accept=target_accept)
        else:
```

```
            mcmc_sampler = mcmc_sampler(target_accept=target_accept)
        #Perform inference to estimate the posterior, using nbr_iters MCMC sample
posteriors
        trace = pm.sample(draws=nbr_iters, step=mcmc_sampler,
                          init='auto', n_init=200000,
                          cores=None, tune=tune, progressbar=True,
                          random_seed=14)
    return bayesian_model, trace


lasso_reg, trace = lasso_mcmc(d, formula, nbr_iters=2000)
print(pm.summary(trace))
pm.traceplot(trace[1000:], lines={k: v['mean'] for k,v in
pm.summary(trace[1000:]).iterrows()})
plt.show()
#So using Laplacian priors was a horrible idea


#Model PPC and RMSE for evaluation
ppc = pm.sample_ppc(trace[1000:], samples=500, model=lasso_reg)
rmse = np.sqrt(np.sum((ppc['y'].mean(0).mean(0).T - d['mpg'])**2) / d.shape[0])


print('R Squared:', get_r2(d, ppc, 'mpg'))  #R^2 shows the model does not fit the
data at all


#Get the posterior predictions
traces = pm.trace_to_dataframe(trace)[['Intercept'] + list(d.drop(['mpg'],
axis=1).columns)]
x = d.drop(['mpg'], axis=1)
x.insert(0, 'intercept', 1)
likelihoods = np.dot(x, traces.T)
likelihoods_sd = np.tile(pm.trace_to_dataframe(trace)[['sd']].T, (d.shape[0], 1))
likelihood = np.random.normal(likelihoods, likelihoods_sd)
print(likelihood.shape)  #Should have d.shape[0] rows and as many columns as MCMC
samples


#Get credible intervals for the target variable
dfp = pd.DataFrame(np.percentile(likelihood,[2.5, 25, 50, 75, 97.5], axis=1).T,
                   columns=['0_25','25','50','75','97_5'])


#Plot target variable over index and the credible intervals from Bayesian regression
pal = sns.color_palette('Purples')
fig, ax = plt.subplots(1, 1, figsize=(7, 7))
ax.fill_between(d.index, dfp['0_25'], dfp['97_5'], alpha=0.7, color=pal[1], label='CR
95%')
ax.fill_between(d.index, dfp['25'], dfp['75'], alpha=0.5, color=pal[4], label='CR
50%')
ax.plot(d.index, dfp['50'], alpha=0.5, color=pal[5], label='Median')
plt.plot(d.index, d['mpg'], 'bo')
plt.legend()
plt.title('Bayesian Regression Predicted Credible Intervals vs Actual')
plt.show()
```

The Bayesian regression function that we defined could also be modified to perform classification through logistic regression by defining the logit as the link function. This will be left as an exercise for the reader.

Finally, we can move on to Naïve Bayes classification with Python. This is a lot easier than Bayesian regression.

```
from sklearn.naive_bayes import GaussianNB
import pandas as pd
from sklearn.metrics import confusion_matrix

d = pd.read_csv('iris.csv')

#Train model with default hyperparameters
nb_model = GaussianNB()
nb_model.fit(d.drop(['Species'], axis=1), d['Species'])
preds = nb_model.predict(d.drop(['Species'], axis=1))
pred_probs = nb_model.predict_proba(d.drop(['Species'], axis=1))
print(confusion_matrix(d['Species'], preds))
```

The confusion matrix shows that Naïve Bayes made 6 errors, resulting in an accuracy of 96%. Note that we broke good practice by not using a test set, but this simple example shows how the algorithm can be deployed.

24. Markov Models

In the chapter about Bayesian statistics, Markov networks were introduced. Markov networks fall into a broader class of Markov models. The simplest Markov model, the Markov chain, was introduced in chapter 23. In a Markov chain, every state of the Markov process is represented by a node, and the transition probabilities are the only parameter of the model. Sometimes there are hidden states in a Markov process. Hidden Markov models can be used to model processes with hidden states.

Hidden Markov Models

Hidden Markov models (HMMs) are Markov processes with hidden or unobserved states (Barber, 2012). HMMs are the simplest form of dynamic Bayesian network (a Bayesian network that changes over time – any model that includes the word "dynamic" is simply a model with time states). Although an HMM has unknown states, each state has a known sequence of outputs drawn from a probability distribution of possible outputs, so it is still possible to make inferences using HMMs. HMMs can be thought of as processes with hidden components that produce observed data. They can answer questions like: given some output sequence, what path was taken to produce it? HMMs are used for speech recognition, gene sequencing, and have even been proposed as models of cognition (Barber, 2012). There are three types of problems that HMMs are well suited for:

1. Evaluation problems – Find the conditional probability of a sequence of observations, given an HMM. These types of problems can be solved using the forward algorithm.
2. Decoding problems – Given an HMM and a sequence of observations, find the optimal (most likely) sequence of the hidden states. These types of problems can be solved using the Viterbi algorithm.
3. Learning problems – Find the HMM that maximize the probability of a sequence of observations. These types of problems are like curve fitting, but instead of a curve, an HMM is fitted. These types of problems can be solved using the expectation maximization algorithm, or the forward-backward algorithm.

The algorithms for solving these problems are shown in figure 24.1.

$$P(O|\lambda) = \sum_{i=0}^{N-1} \alpha_{T-1}(i)$$

$$\gamma_t(i) = \frac{\alpha_t(i)\beta_t(i)}{P(O|\lambda)}$$

Figure 24.1

The top equation in figure 24.1 solves evaluation problems. It shows that the probability of a sequence of observations, O, given model lambda, equals the sum from observation i=0 to N-1 total observations of the probability of the partial observation sequence up to the final time -1, $\alpha_{T-1}(i)$. The bottom equation in figure 24.1 solves decoding problems. It shows that the most likely state sequence for time t, $\gamma_t(i)$, equals the ratio of the product of forward pass, $\alpha_t(i)$, and the backward pass, $\beta_t(i)$, to the probability of sequence O, given model lambda. Learning problems do not have an equation given, because they can be solved by combining the equations shown.

The algorithms in figure 24.1 are examples of dynamic programming. Dynamic programming involves breaking a problem into simpler sub-problems and solving the sub-problems. The solutions to the sub-problems are stored, or "memorized", so that when they appear again, the solution can be recalled without re-computation. The sub-problems can then be recombined to yield an optimal solution to the larger problem. A deeper explanation of dynamic programming is required to implement HMMs from scratch, but that is beyond the scope of this book. The examples we will see in the R and Python sections of this chapter do the heavy lifting behind the scenes. For further information on HMMs and dynamic programming, see Read, 2011, and Stamp, 2018.

The general form of a HMM is shown in figure 24.2.

$$P(y, x) = P(y)P(x|y)$$

Figure 24.2

In the equation in figure, P(y) is the probability of a sequence of events. In other words, P(y) is a Markov chain. P(y) is the hidden part of an HMM. P(x|y) is called the emission model, or the sequence of observed events. P(y) and P(x|y) are defined by the equations in figure 24.3.

$$P(y) = \prod_{i=1}^{N+1} P(y_i|y_{i-1})$$

$$P(x|y) = \prod_{i=1}^{N} P(x_i|y_i)$$

Figure 24.3

The probabilities of x sub i and y sub i are drawn from the transition matrix. As an example of an HMM, consider a prisoner trying to determine if the weather outside is sunny, rainy, or cloudy. The prisoner cannot observe the state of the weather directly. Instead, the prisoner only sees the prison guards walking around. Sometimes the guards have umbrellas, and other days they do not. The directed graph to the left of figure 24.4 illustrates the underlying model in this scenario and the graph to the right shows what the prisoner can observe.

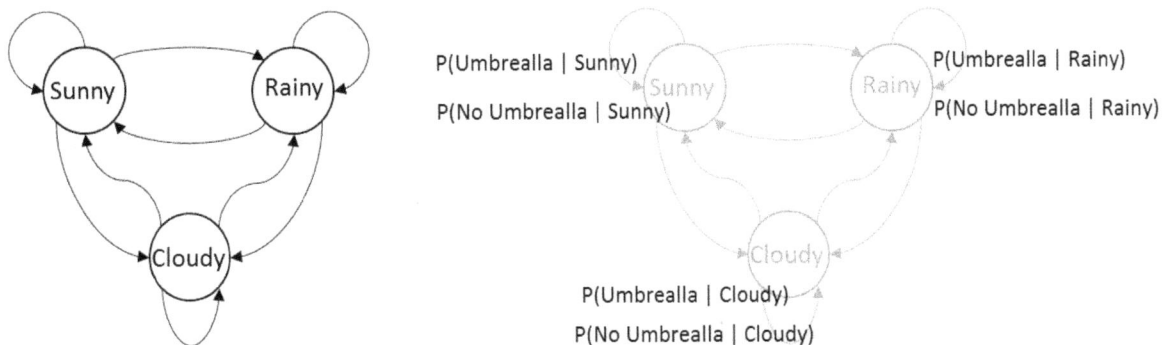

Figure 24.4

After t days, the prisoner has observed the binary sequence O = {o1, o2, ..., ot}, where each o can be either "umbrella" or "no umbrella". This sequence of observations is produced by a Markov process with unknown states that can be represented as Q = {q1, q2, ..., qt}, where each q can be "sunny", "rainy", or "cloudy". The prisoner wants to know the probability of the hidden sequence, given the observed sequence, P(q1, ..., qt|o1, ..., ot). According to Bayes' theorem, the probability for a single day is given by:

$$P(q_i|o_i) = \frac{P(o_i|q_i)P(q_i)}{P(o_i)}$$

Figure 24.5

So for a sequence of length t, the conditional probability of the unobserved sequence on the observed sequence is given by:

$$P(q_1, ..., q_t|o_1, ..., o_t) = \frac{P(o_1, ..., o_t|q_1, ..., q_t)P(q_1, ..., q_t)}{P(o_1, ..., o_t)}$$

Figure 24.6

Recall from chapter 23 that the Markov property holds that the probability of an event at time t depends only on the event at time t-1. Taking this into consideration, the probability of a sequence of hidden states $P(q_1, ..., q_t)$ can be represented as:

$$P(q_1, ..., q_t) = \prod_{i=1}^{t} P(q_i|q_{i-1})$$

Figure 24.7

If we assume that the observations are independent (the guards' possession of umbrellas one day is independent of their possession of umbrellas the next day), then the probability of the observations given a sequence of hidden states is given by:

$$P(o_1, ..., o_t|q_1, ..., q_t) = \prod_{i=1}^{t} P(o_i|q_i)$$

Figure 24.8

Putting all of these together yields the final HMM:

$$P(q_1, ..., q_t|o_1, ..., o_t) \propto \prod_{i=1}^{t} P(o_i|q_i) * \prod_{i=1}^{t} P(q_i|q_{i-1})$$

Figure 24.9

The parameters of this HMM are the transition probabilities, $P(q_i|q_{i-1})$, the observation probabilities, $P(o_i|q_i)$, and the initial state probabilities, $P(q_i)$.

A **trellis** is a representation of the possible state sequences y of an HMM for a single observation sequence x. By running a shortest path algorithm on a trellis, where paths are computed as the sum of the transition probabilities, it is possible to determine the most likely state sequence y. By repeating this process, a sequence of most likely states can be formed (Hoover, 2017). The visualization at the top of figure 24.10 shows an example of a trellis derived from the first 4 time steps of the HMM in figure 24.4. Note that the edge weights between nodes in this trellis are not shown, because no probabilities were assigned to the state transitions in figure 24.4. If there were probabilities for the state transitions, they would be the edge weights in the trellis.

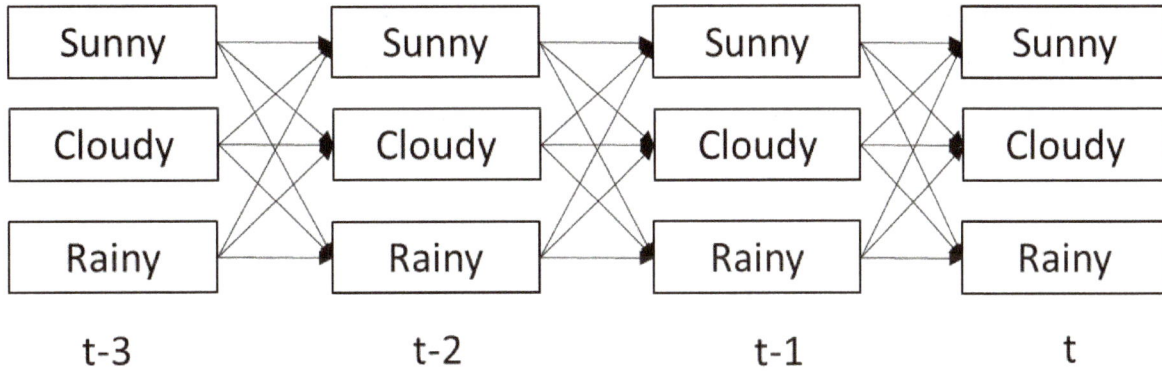

Figure 24.10

Kalman Filter

HMMs model discrete or continuous observations with a discrete/categorical state space of the hidden variable(s). They can be extended to allow continuous state spaces of the hidden variables, where the hidden and observed variables are linearly related and follow normal distributions. Under these conditions, a Kalman Filter can be used to estimate the joint probability distribution of the variables for each time frame (Protopapas, 2014). An example of a dynamical system that meets these criteria is the locomotion of a robotic vehicle. Suppose a robotic vehicle is driving from point A to point B, but in order to get to point B it must pass through a box. The robot's velocity and position are measurable when it can be observed, but when it goes into the box, these variables are hidden. Logically, we can imagine that the robot would likely travel in a straight path through the box, as shown in figure 24.11, but we cannot be sure. A Kalman filter is a model that can calculate the robot's most probable velocity and position (the hidden variables) while it is in the box, assuming it cannot be observed while it is in the box.

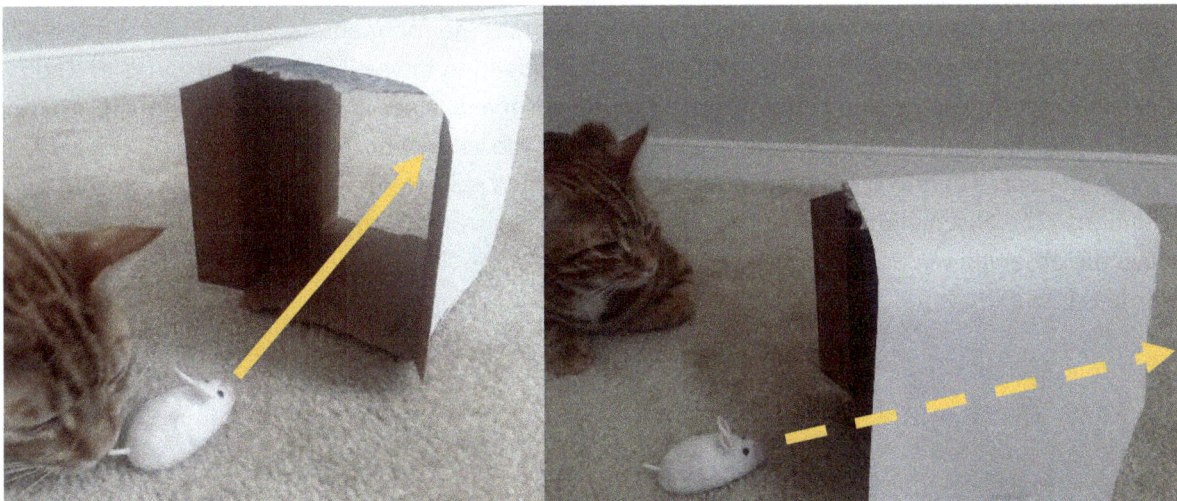

Figure 24.11

The Kalman filter is extremely useful and has been called one of the most important discoveries of the 20th century (Grewal & Andrews, 2001). It is used to control manufacturing processes, track objects in computer vision, and for the navigation of aircraft and spacecraft. The Kalman filter played a key role in the Apollo missions (Grewal & Andrews, 2001). It is even used to predict likely future courses for dynamical systems that cannot be controlled, like the flow of rivers during floods and stock prices.

The Kalman filter was originally designed for variables that were linearly related, but it can also be used for variables that are nonlinearly related. The nonlinear version of the Kalman filter is called the extended Kalman filter.

The relationship between position and velocity in a dynamical system can be represented by normal probability distributions as shown in figure 24.12.

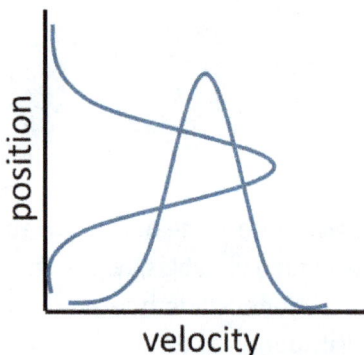

Figure 24.12

Figure 24.12 assumes that position and velocity are uncorrelated. It is more realistic to assume that they are correlated, because the position at time t depends on the position at time t-1 and the velocity at time t-1 (Babb, 2015). The covariance matrix can show how these two variables trend together at each state in a dynamical system. In a simple model, the covariance matrix can be used to predict the position and velocity at a future time, allowing it to produce the precise position and velocity of hidden states. But this assumes that there are no external forces that could affect the robot's position and velocity. What if there is a cat hiding in the box, and as soon as the robot enters, the cat bats it around? For this reason, there needs to be an uncertainty variable added to the estimate of each variable. There also needs to be some level of uncertainty added to the observed values of the variables though, because in the real world, sensors are not always reliable. The value reported by a sensor could be off by a small amount. Therefore, even the observed position and velocity are drawn from probability distributions. The observations give an indication about which states of the system were most likely to produce the observed values of the variables at each time frame. The Kalman filter combines all of these ideas into a single process:

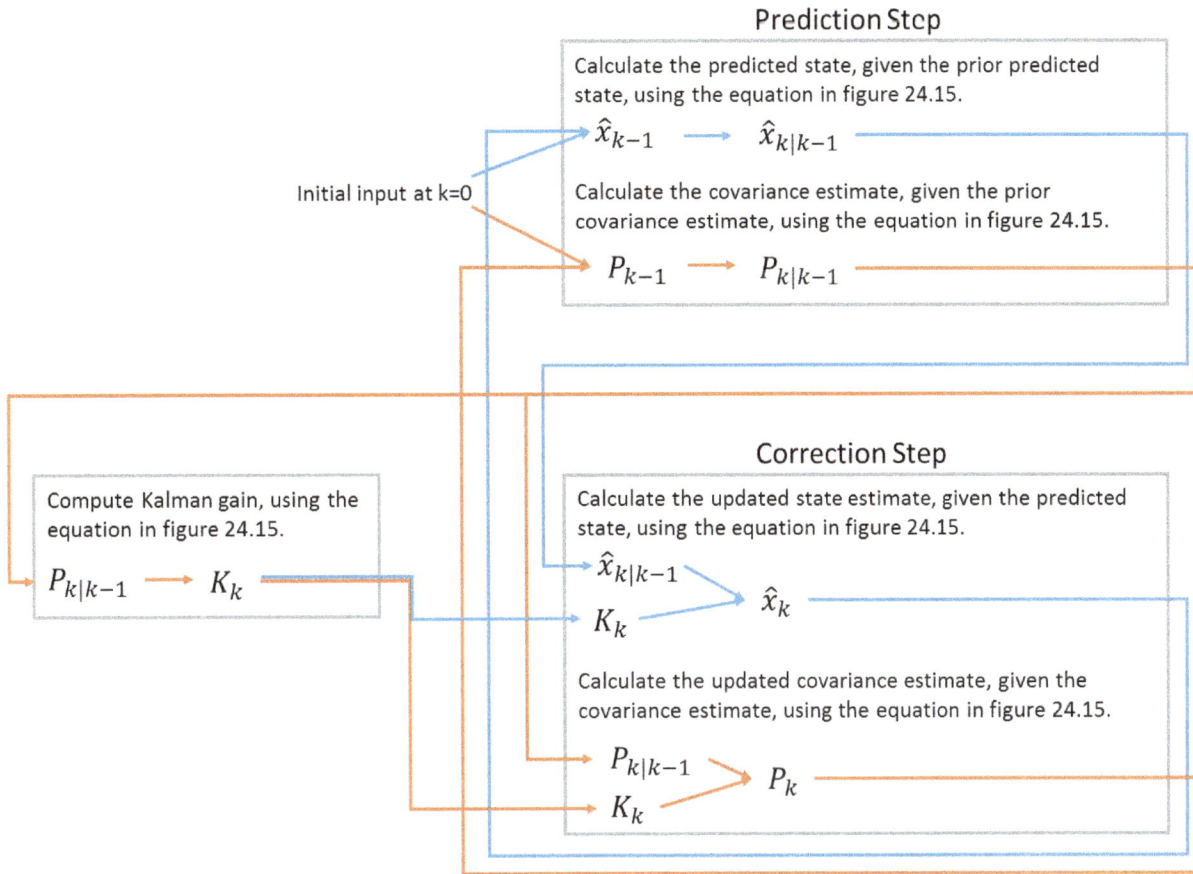

Figure 24.13

The first step in the process is to build a mathematical model of the problem. For simplicity, a linear relationship is assumed in the equation below.

$$x_k = A_k x_{k-1} + B_k u_k + w_k$$

$$z_k = H_k x_k + v_k$$

Figure 24.14

The top equation in figure 24.14 describes the observation values. We will call x sub k the signal value. So the top equation shows that the signal value, x sub k, equals the state transition model, A sub k, that is applied to the previous state x sub k-1, plus the control matrix, B sub k, that is applied to the control vector u sub k, plus the process noise, w sub k, drawn from a normal distribution with covariance Q sub k. The control input model (control matrix * control vector) accounts for external influences, like a cat batting around a robotic mouse.

The bottom equation in figure 24.14 describes the measurement noise associated with signal readings. We will call z sub k the measurement value. So the bottom equation shows that the measurement value, z sub k, equals the matrix mapping of the true state space into the observed state space, H sub k, applied to the signal x sub k, plus observation noise, v sub k, drawn from a normal distribution with covariance R sub k.

In many cases, the matrices A, B, and H are just constants (Czerniak, 2018). The noise functions are assumed to be normally distributed, which is a poor assumption for real world processes. Nevertheless, the Kalman filter converges on good estimations even when the noise parameters are poorly estimated. The second step in the Kalman filter process is to determine the initial values. Since the Kalman filter entails making iterative predictions and corrections, there are two update cycles that must be defined: the time update (prediction) and the measurement update (correction).

Time Update (Prediction, or Prior)	Measurement Update (Correction, or Posterior)
$$\hat{x}_{k|k-1} = A_k\hat{x}_{k-1} + B_ku_k$$ This equation gives the predicted state estimate, based on the previously predicted state estimate.	$$\hat{x}_k = \hat{x}_{k|k-1} + K_k\left(z_k - H\hat{x}_{k|k-1}\right)$$ This equation gives the updated state estimate. K sub k is the optimal Kalman gain, found by: $$K_k = P_{k|k-1}H_k^T\left(R_k + H_kP_{k|k-1}H_k^T\right)^{-1}$$
$$P_{k|k-1} = A_kP_{k-1} + A_k^T + Q_k$$ This equation gives the predicted covariance estimate.	$$P_k = (I - K_kH_k)P_{k|k-1}(I - K_kH_k)^T + K_kR_kK_k^T$$ This equation gives the updated covariance estimate.

Figure 24.15

The third step in the Kalman filter process is to iterate over each time frame, and make predictions and corrections. This step carries out the process displayed in the flow chart in figure 24.13.

The trickiest thing about implementing a Kalman filter is that its formulation depends on the dynamical system that is represented (Czerniak, 2018). There are implementations of Kalman filters out there, but they require preparing the input from a dynamical system beforehand. A Kalman filter cannot simply be applied to a dataset like many of the algorithms in this book. For example, estimating the internal temperature of a combustion engine based on sensor readings from outside the engine requires a different setup than estimating the position of a robot inside a box. At the end of this chapter, we will implement a Kalman filter in Python for estimating the position of a robot that follows a random walk, including the estimation of its position when it disappears inside a box.

Recall the smoothing methods introduced in chapter 20. The Kalman filter can actually be used as a smoother too. The Kalman filter is a great smoother because it is robust to outliers and can adapt to constantly changing conditions. A Kalman filter can be used to smooth a random walk and is unaffected by non-stationarity, which makes it a great tool for financial time series and stock prices, as we will see in a later chapter. The disadvantages to using a Kalman filter are that it assumes the variables are sampled from a normal distribution, and that there is a linear relationship between them. When these assumptions are violated, the extended Kalman filter (for nonlinearity) or a particle filter (for non-normal distributions) should be used instead.

Markov Chain Monte Carlo Algorithms

Recall from chapter 23 that Bayesian models produce posterior distributions. If a conjugate prior is used, then the posterior distribution has a known shape and the parameters of the model can be inferred. It is not possible to use conjugate priors for all Bayesian models, because models with many parameters become intractable to solve analytically. Many models require the posterior distribution to be approximated instead. Chapter 23 briefly mentioned that this could be done through Markov chain

Monte Carlo (MCMC) algorithms. Monte Carlo methods are a class of algorithms that rely on random sampling to obtain numerical estimates. The idea behind MCMC is that repeated random samples form good estimates of the posterior distributions of a Bayesian model's parameters. But the samples are not drawn completely at random or else the approximation would never approach anything close to the true posteriors. Instead, each consecutive sample is drawn from a distribution that is dependent on the previous sample. This means that the samples all form a Markov chain, and since the starting point does not matter in a Markov chain, MCMC are powerful sampling methods. As an analogy for MCMC, imagine trying to cross a stream by stepping across rocks. While randomly choosing rocks might work, the likelihood of a rock tipping over and dumping a person into the stream could be high. A better option would be to step on the rocks that look like they are steady, and with each step, the options for the next possible rock depend on the current rock because they must be close enough to step to. This is like MCMC.

Figure 24.16

The Monte Carlo method of random sampling was named after the Monte Carlo Casino in Monaco, by a physicist named Nicholas Metropolis. Metropolis developed an algorithm for MCMC called the Metropolis algorithm, which was generalized a few years later by the statistician Wilfred Keith Hastings. Thus, the first MCMC algorithm we will explore is called the Metropolis-Hastings algorithm.

Metropolis-Hastings Algorithm

The Metropolis-Hastings (MH) algorithm works by generating samples that approximate a distribution P(x), using a function f(x) that is proportional to the probability density. Since it is a MCMC technique, the algorithm is performed iteratively so that the samples form a Markov Chain (each consecutive sample depends on the sample before it). The twist is that each sample that is chosen for the next iteration has some probability of being rejected, and if the candidate sample is rejected, the current sample is reused for the next iteration. The probability of accepting a candidate sample for the next iteration is determined by comparing the evaluation of function f(x) for the current sample and the candidate sample to the distribution P(x), through the application of an acceptance function alpha.

Samples are drawn for some number of iterations chosen by the researcher, so the number of iterations is the stopping criterion for the algorithm. The output of the algorithm is the sequence of selected samples (Halls-Moore, 2016).

The very first sample in the MH algorithm is chosen from the prior distribution of each random variable. After that, all of the candidate samples are chosen from the proposal distribution q. The most common choice for a proposal distribution is the normal distribution. In general, symmetric proposal distributions meet the following property:

$$Q(x^i|x^{i-1}) = q(x^{i-1}|x^i)$$

Figure 24.17

In the equation in figure 24.17, x^i is the candidate sample and x^{i-1} is the current sample. When the proposal distribution is normal, it is also symmetric by definition, and the candidate randomly perturbs the Markov Chain by some random value: it is essentially a random walk (previous value + random change). So when the proposal distribution is normal, the algorithm can be referred to as a random-walk MH algorithm.

The proposal distribution does not need to be symmetrical however. Sometimes it might be preferable to choose an asymmetric distribution that skews toward higher or lower values. For example, the log-normal density skews towards higher values. This might be necessary if there is a constraint that must be imposed, such as the samples not being negative (Yildirim, 2012a).

The acceptance function, alpha, calculates the ratio between the probability of the data, given the proposed sample, and the probability of the data, given the current sample: P(data | proposed sample) / P(data | current sample). The probability of the data, given each sample, can be found using Bayes theorem, because the likelihood of the sample is multiplied by the prior probability to yield the posterior probability of the data:

$$Posterior\ Probability\ of\ Proposed\ Sample = \frac{P(x|\mu_{proposed\ sample})P(\mu_{proposed\ sample})}{P(x)}$$

$$Posterior\ Probability\ of\ Current\ Sample = \frac{P(x|\mu_{current\ sample})P(\mu_{current\ sample})}{P(x)}$$

Figure 24.18

Therefore, the ratio of the probability of the data, given the proposed sample, to the probability of the data, given the current sample, is found by the equation below.

$$\frac{\frac{P(x|\mu_{proposed\ sample})P(\mu_{proposed\ sample})}{P(x)}}{\frac{P(x|\mu_{current\ sample})P(\mu_{current\ sample})}{P(x)}} = \frac{P(x|\mu_{proposed\ sample})P(\mu_{proposed\ sample})}{P(x|\mu_{current\ sample})P(\mu_{current\ sample})}$$

Figure 24.19

Notice how the denominator, P(x), is canceled out in the reduced form. This denominator was what made it intractable to analytically solve for the posterior probability. Now that is has been removed, it is possible to approximate the posterior probability using the ratio in the equation, because the ratio is proportional to the posterior distribution. Let us think about why the function is proportional to the posterior distribution: the ratio, which essentially compares the posterior probability of two samples in a Markov Chain, implies that regions of high posterior probability will be sampled more often than regions of low posterior probability. The equation in figure 24.19 is a high level version of the acceptance function to make it easy to see the concept. The actual acceptance function is given below.

$$\alpha\left(x^{candidate}|x^{i-1}\right) = \min\left\{1, \frac{q(x^{i-1}|x^{candidate})P(x^{candidate})}{q(x^{candidate}|x^{i-1})P(x^{i-1})}\right\}$$

Figure 24.20

Once the acceptance function has been computed, it is compared to a random number drawn uniformly between 0 and 1, and if the random number is smaller than the acceptance ratio, then the proposed sample is accepted (Yildirim, 2012a).

Gibbs Sampling

Gibbs sampling is a special case of the MH algorithm, where the acceptance probability is always 1; the candidate sample is always accepted. This can be advantageous when dealing with multivariate distributions, because the sample points are multidimensional, and datasets with more features (dimensions) have more complex behavior. For multivariate distributions, the MH algorithm can take a very long time to find a good sequence of samples, due to the probability that a sample could be reused in successive iterations. By always accepting the candidate sample, Gibbs sampling can be faster. The catch is that Gibbs sampling might go down the wrong rabbit hole; it could hastily pick samples that are not good estimates of the posterior and end up with a suboptimal sequence of samples. The MH algorithm is less likely to do this (Yildirim, 2012b).

Gibbs sampling is more commonly used than the MH algorithm, especially for statistical inference. One example where Gibbs sampling is used is to predict which politician a voter will most likely vote for. The process starts by calculating the joint distribution of the random variables, and using it to derive an initial sample of the posterior conditionals for each random variable. The key point is that the posterior distribution itself is not sampled from, but rather the conditionals of the random variables that make up the posterior are sampled. This process repeats for each iteration until some chosen number of iterations have passed. Therefore, like the MH algorithm, the stopping criterion is some number of iterations that is chosen by the researcher. Since there is no acceptance function, samples from higher regions of probability in the posterior distribution are automatically accepted (Yildirim, 2012b).

For each iteration in the sampling process, the random variables that are included in the posterior conditionals are sampled one at a time. For example, the samples for random variables 1 through D in the ith iteration is represented below.

$$x_1^i \sim P\left(X_1 = x_1|X_2 = x_2^{i-1}, X_3 = x_3^{i-1}, \ldots, X_D = x_D^{i-1}\right)$$

$$\ldots$$

$$\ldots$$

$$x_D^i \sim P\left(X_D = x_D|X_1 = x_1^i, X_2 = x_2^i, \ldots, X_D = x_{D-1}^i\right)$$

Figure 24.21

Markov Models with Python

Markov models are easier to build with Python, so there is no R section for this chapter. We will also skip implementations of the MCMC sampling algorithms, since we put them into practice in the last chapter on Bayesian regression. So let us start with HMMs. Recall that there are 3 types of problems that HMMs can solve, and the algorithms used in the model depend on the type of problem to be solved:

1. Evaluate likelihood of data given model parameters and observed states
2. Decode optimal sequence of hidden states given model parameters and observations
3. Learn the model parameters given observations

Consider the following problem, which is an example of an evaluation problem:

There are 2 states: healthy and sick. There are 3 symptom observation types: dizziness, chills, no-symptoms. Suppose that we know whether a person is healthy or sick, and can produce a sequence of states (healthy/sick). The sequence of symptoms corresponding to this sequence of states is hidden or unknown. There are probabilities for a person to transition from one state to the other. These form a transition matrix. There is some probability that a healthy person could show all 3 symptoms, and some probability that a sick person could show all 3 symptoms. These probabilities form an emission matrix. There is also a baseline probability that a person is healthy or sick, and we can use this as the starting condition. The challenge is therefore to find the probability of a sequence of symptom observations, given a sequence of hidden states (healthy/sick). If we draw this problem out, it looks like this:

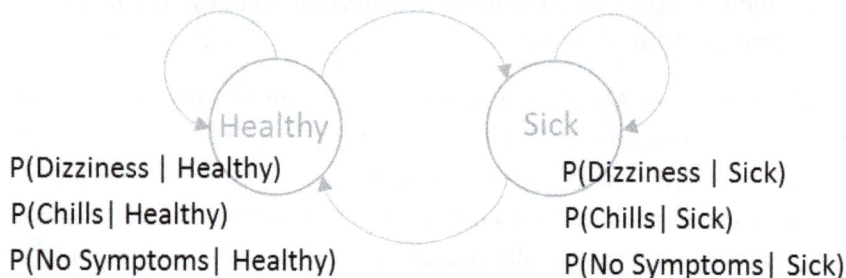

The gray arrows form the transition matrix. The black text are the possible emissions from each state, so they form the emission matrix. We will make a few assumptions about the probabilities and manually assign them to start the problem. Then we will convert the more readable probabilities into numpy matrices for vectorization. To build a HMM, we need the hmmlearn library.

```python
import numpy as np
from hmmlearn import hmm
import math

states = ('healthy', 'sick')
observations = ('no-symptoms', 'cold', 'dizzy')

start_probability = {'healthy': 0.8, 'sick': 0.2}
startm = np.array([0.8, 0.2])
transition_probability = {
```

```
    'healthy' : {'healthy': 0.8, 'sick': 0.2},
    'sick' : {'healthy': 0.4, 'sick': 0.6},
    }
transm = np.array([[0.8, 0.2],
                   [0.4, 0.6]])
emission_probability = {
    'healthy' : {'no-symptoms': 0.6, 'cold': 0.3, 'dizzy': 0.1},
    'sick' : {'no-symptoms': 0.1, 'cold': 0.3, 'dizzy': 0.6},
    }
emism = np.array([[0.6, 0.3, 0.1],
                  [0.1, 0.3, 0.6]])
```

The HMM model is defined as follows:

```
hmm_model = hmm.MultinomialHMM(n_components=len(states), algorithm='viterbi')
hmm_model.startprob_ = startm
hmm_model.transmat_ = transm
hmm_model.emissionprob_ = emism
```

This is a type 1 problem: evaluate the likelihood of a sequence of obervations. This probability of a given symptom for any state can be found by directly looking at the emission matrix. So we will instead look at the probability of various sequences of symptoms, regardless of state.

```
y = np.array([[0]])
print('Probability of first observation in a sequence being',
      observations[0], 'regardless of state is', math.exp(hmm_model.score(y)))
y = np.array([[1]])
print('Probability of first observation in a sequence being',
      observations[1], 'regardless of state is', math.exp(hmm_model.score(y)))
y = np.array([[2]])
print('Probability of first observation in a sequence being',
      observations[2], 'regardless of state is', math.exp(hmm_model.score(y)))
y = np.array([[1, 1, 0]])
print('Probability of', observations[1], observations[1], observations[0],
      'regardless of state is', math.exp(hmm_model.score(y)))
y = np.array([[0, 0, 2]])
print('Probability of', observations[0], observations[0], observations[2],
      'regardless of state is', math.exp(hmm_model.score(y)))
```

Now using the same scenario, let's find the most likely series of hidden states(healthy/sick), given an observed series of symptoms. This is a type 2 problem: decode the series of hidden states, given model parameters and observations.

```
y = np.array([[2]]).T
print('Given observation', observations[2],
      'the most likely hidden state is', states[hmm_model.decode(y)[1][0]],
      'with probability', math.exp(hmm_model.decode(y)[0]))
y = np.array([[0, 1, 0]]).T
print('Given observations', observations[0], observations[1], observations[0],
      'the most likely sequence of hidden states is',
states[hmm_model.decode(y)[1][0]],
      states[hmm_model.decode(y)[1][1]], states[hmm_model.decode(y)[1][2]],
      'with probability', math.exp(hmm_model.decode(y)[0]))
y = np.array([[2, 1, 2]]).T
print('Given observations', observations[2], observations[1], observations[2],
```

```
        'the most likely sequence of hidden states is',
states[hmm_model.decode(y)[1][0]],
        states[hmm_model.decode(y)[1][1]], states[hmm_model.decode(y)[1][2]],
        'with probability', math.exp(hmm_model.decode(y)[0]))
```

Now consider the following scenario, which is an example of a learning problem:

There are 2 states: bear market and bull market. These states are hidden - we do not know whether the stock market is in a pessimistic bearish cycle or an optimistic bullish cycle. However, we can observe the daily returns of the S&P500. There are probabilities for the market to transition from one state to the other. These form the transition matrix. We do not know what these probabilities are though. We also do not know the emission probabilities. We may not even know the starting condition probabilities. The challenge is therefore to find the most likely model, given a series of observations. This is a type 3 problem: learn the model parameters given observations.

The Baum-Welch algorithm (an implementation of expectation maximization forward-backward) can be used to estimate the maximum likelihood model that generated an observed sequence of observations. There is an implementation in hmmlearn that forms the backbone of hmmlearn's GaussianHMM function. The Gaussian HMM is similar to the multinomial HMM from the previous examples, except it assumes the emission probabilities are normally distributed.

We will now collect 13 years (Feb 1, 1993 – Jan 1, 2016) of S&P 500 data, using the SPY ticker symbol.

```
import datetime as dt
import numpy as np
import pandas as pd
import pandas_datareader.data as web
from matplotlib import cm, pyplot as plt
from matplotlib.dates import YearLocator, MonthLocator
import warnings
from hmmlearn.hmm import GaussianHMM

start_dt = dt.datetime(1993, 2, 1)
end_dt = dt.datetime(2016, 1, 1)

#d = web.DataReader('SPY', 'morningstar', start_dt, end_dt)
#d.to_csv('spy.csv')  #Write to csv so future runs don't need to call the API
d = pd.read_csv('spy.csv')
d = d[['Date', 'Close']]
date_index = pd.to_datetime(d['Date'])
dts = pd.Series(d['Close'])
dts.index = date_index
log_returns = pd.Series(np.log(dts/dts.shift(1)).dropna())
#Note that log returns = percent change
#print(d.pct_change().dropna())
log_returns.plot()
plt.title('Log Returns for SPY 1993-2016')
plt.show()
dts.plot()
plt.title('Closing Price of SPY 1993-2016')
plt.show()

#Convert log returns to a numpy array
```

```
log_returns_arr = np.column_stack([log returns]).reshape(-1, 1)
```

Now we can fit a Guassian HMM for 2 components (hidden states) and plot the predicted states of the fitted model.

```
#Ignore the litany of deprecation warnings
warnings.filterwarnings("ignore")

#Create a HMM with Gaussian emissions and assume there are 2 hidden states (bear and
bull markets)
hmm_model = GaussianHMM(n_components=2, covariance_type="full",
n_iter=1000).fit(log_returns_arr)
hidden_states = hmm_model.predict(log_returns_arr)

#Stack plots for each of the hidden states
#This plotting code comes from hmmlearn's documentation:
#http://hmmlearn.readthedocs.io/en/latest/auto_examples/plot_hmm_stock_analysis.html
fig, axs = plt.subplots(hmm_model.n_components, sharex=True, sharey=True)
colors = cm.rainbow(np.linspace(0, 1, hmm_model.n_components))
for i, (ax, color) in enumerate(zip(axs, colors)):
    mask = hidden_states == i
    ax.plot_date(dts[1:].index[mask], d['Close'][1:][mask], ".", linestyle='none',
c=color)
    ax.set_title("Hidden State #%s" % i)
    ax.xaxis.set_major_locator(YearLocator())
    ax.xaxis.set_minor_locator(MonthLocator())
    ax.grid(True)
plt.show()
```

There were quite a few periods where the market flip flopped between states quickly, making it look like the graphs overlap. But if the graph is expanded in size, it is clear that they do not overlap. The model seemed to capture the bear markets in state 0 and bull markets in state 1, but is that really what these states represent? The model used log returns to predict the hidden states, assuming normally distributed emissions. It is more likely that the model captured volatility clustering, and that state 0 represents times of increased volatility that typically coincides with bear markets. Consider for example how state 0 dominates the dot com bubble of the late 1990s, 2002-2003, and the great recession in 2008-2009.

The HMM model we have just built captures regime change. Regime change is a time series segmentation method that assumes different segments of a time series are generated by a system with distinct parameters. We will explore time series segmentation in a later chapter.

Now that we have thoroughly explored HMMs, we can turn our attention to the Kalman filter. First we will see how a Kalman filter can smooth 2 dimensional data by fitting it to a random walk. We will allow both x and y to vary randomly. The random walk can be thought of as the path of a robot.

```
import numpy as np
import random
import matplotlib.pyplot as plt
from pykalman import KalmanFilter

random.seed(14)
```

```python
def randomWalk2d(n_steps=1000, walkType="financial"):
    x = np.zeros(n_steps)
    y = np.zeros(n_steps)
    if walkType == "random":
        for s in range(1, n_steps):
            val = random.randint(1,4)
            if val == 1:
                x[s] = x[s - 1] + 1
                y[s] = y[s - 1]
            elif val == 2:
                x[s] = x[s - 1] - 1
                y[s] = y[s - 1]
            elif val == 3:
                x[s] = x[s - 1]
                y[s] = y[s - 1] + 1
            else:
                x[s] = x[s - 1]
                y[s] = y[s - 1] - 1
    else:
        for s in range(1, n_steps):
            val = random.randint(1,3)
            if val == 1:
                x[s] = x[s - 1] + 1
                y[s] = y[s - 1]
            elif val == 2:
                x[s] = x[s - 1]
                y[s] = y[s - 1] + 1
            else:
                x[s] = x[s - 1]
                y[s] = y[s - 1] - 1
    return x,y

x, y = randomWalk2d()

#Plot random walk
plt.title("Random Walk ($n = " + str(1000) + "$ steps)")
plt.plot(x, y)
plt.show()

robotpath = np.array((x,y)).T
```

We now need to set up the parameters for the Kalman filter. Refer back to Kalman filter section for help understanding which parameters need to be specified.

```python
#Parameter set up
init_state_mean = [robotpath[0, 0],
                   0,
                   robotpath[0, 1],
                   0]
init_state_cov = np.ones((4, 4))
obs_mat= [[1, 0, 0, 0],
          [0, 0, 1, 0]]
obs_cov = np.eye(2)
trans_mat = [[1, 1, 0, 0],
             [0, 1, 0, 0],
```

```
         [0, 0, 1, 1],
         [0, 0, 0, 1]]
trans_cov = np.eye(4)*1e-6
```

The Kalman filter is fitted to the random walk all at once, producing state means and state covariances.

```
kf = KalmanFilter(n_dim_obs=2, n_dim_state=4,
                  initial_state_mean=init_state_mean,
                  initial_state_covariance=init_state_cov,
                  observation_matrices=obs_mat,
                  observation_covariance=obs_cov,
                  transition_matrices=trans_mat,
                  transition_covariance=trans_cov)

state_means, state_covs = kf.filter(robotpath)
```

The Kalman filter can be further smoothed using expectation maximization. Then we can plot the two versions of the filter for comparison. We can also plot the change in each variable (x and y) to see how the position of the robot is affected by each variable individually. It is clear that x is increasing more rapidly than the y values in this random walk.

```
#Smoothing with expectation maximization
kf1 = kf.em(robotpath, n_iter=5)
smoothed_state_means, smoothed_state_covariances = kf1.smooth(robotpath)


#Plot the fitted Kalman filter
plt.plot(#x, y, 'bo',
         #state_means[:, 0], state_means[:, 2], 'ro',
         x, y, 'b--',
         state_means[:, 0], state_means[:, 2], 'r--',)
plt.title('Kalman Filter Fitted to a 2D Random Walk')
plt.show()
plt.plot(#x, y, 'bo',
         #state_means[:, 0], state_means[:, 2], 'ro',
         x, y, 'b--',
         smoothed_state_means[:, 0], smoothed_state_means[:, 2], 'r--',)
plt.title('Kalman Filter Fitted to a 2D Random Walk and Smoothed with Expectation
Maximization')
plt.show()


#Plot the change in each variable over time
times = range(robotpath.shape[0])
plt.plot(#times, robotpath[:, 0], 'bo',
         #times, robotpath[:, 1], 'ro',
         times, state_means[:, 0], 'b--',
         times, state_means[:, 2], 'r--',)
plt.title('Split Plot of the Change in x and y throughout the Random Walk')
plt.show()
plt.plot(#times, robotpath[:, 0], 'bo',
         #times, robotpath[:, 1], 'ro',
         times, smoothed_state_means[:, 0], 'b--',
         times, smoothed_state_means[:, 2], 'r--',)
plt.title('Split Plot of the Smoothed Change in x and y throughout the Random Walk')
plt.show()
```

Now let us try deleting some points from the random walk to see how the Kalman filter adapts to missing states. This will simulate a robot going somewhere where its coordinates cannot be observed. Doing this presents a problem however. The x values were allowed to vary randomly when the random walk was created. In the code below, we will make the assumption that x always increase by 1 for the points that are missing. This is a faulty assumption, but unless we can replicate the same random number generation process that was used to create the random walk, it is the best we can do.

```python
#Remove observations 400-500
x_h = x.copy()
y_h = y.copy()
x_h[range(400, 500)] = np.nan
y_h[range(400, 500)] = np.nan
robotpath_obscured = np.array((x_h, y_h)).T

#Parameter set up
init_state_mean = [robotpath_obscured[0, 0],
                   0,
                   robotpath_obscured[0, 1],
                   0]
init_state_cov = np.ones((4, 4))
obs_mat= [[1, 0, 0, 0],
          [0, 0, 1, 0]]
obs_cov = np.eye(2)
delta_x = x_h[1] - x_h[0]   #If x were time, this would be more meaningful, but here
it is just 1.0
#Note that this trans_mat is the same as before, but we're using delta_x to show how
it would change if dt were not 1.0
trans_mat = [[1, delta_x, 0, 0],
             [0, 1, 0, 0],
             [0, 0, 1, delta_x],
             [0, 0, 0, 1]]
trans_cov = np.eye(4)*1e-6

kf = KalmanFilter(n_dim_obs=2, n_dim_state=4,
                  initial_state_mean=init_state_mean,
                  initial_state_covariance=init_state_cov,
                  observation_matrices=obs_mat,
                  observation_covariance=obs_cov,
                  transition_matrices=trans_mat,
                  transition_covariance=trans_cov)
```

In this case, we will apply the Kalman filter iteratively as information about each state presents itself. This serves as an example for how the Kalman filter can be implemented for online learning.

```python
#Make sure the tuples have the same number of dimensions as the initial state mean
and initial state cov matrix (4, and (4,4) in this case)
state_means_h = np.zeros((x_h.shape[0], 4))
state_covs_h = np.zeros((x_h.shape[0], 4, 4))

#Apply the KF iteratively to interpolate missing states
#Note that both x and y are missing, but we are assuming x always increments by 1
(delta_x). so we really only need to interpolate y
for t in range(robotpath_obscured[:,0].shape[0]):
    if np.isnan(robotpath_obscured[t,0]):
```

```
            x_h[t] = x_h[t-1]+1  #Increments x[t] by x[t-1]+1 when x[t] is missing
            state_means_h[t], state_covs_h[t] = (
                kf.filter_update(state_means_h[t-1],
                                  state_covs_h[t-1],
                                  observation=x_h[t])  #Only x is available
            )
    else:
            state_means_h[t], state_covs_h[t] = (
                kf.filter_update(state_means_h[t-1],
                                  state_covs_h[t-1],
                                  observation=robotpath_obscured[t])  #Both x and y are
available
            )


#Plot the fitted Kalman filter
plt.plot(x, y, 'b--', label="Random Walk")
plt.plot(state_means[:, 0], state_means[:, 2], 'g--', label="KF with No Missing
Values")
plt.plot(state_means_h[:, 0], state_means_h[:, 2], 'r--', label="KF with Missing
Values")
plt.grid()
plt.legend(loc="upper left")
plt.xlabel('x')
plt.ylabel('y')
plt.title('Random Walk with Hidden States')
plt.show()
```

The plot of the fitted filter shows the problem with our assumption about the unit increase in x. It caused the filter to predict very large y values! Notice that the filter relies on the previous state to calculate the next state mean, and since the states leading up to the point at which the data disappears are seeing large increases in y, the filter expects these increases to continue. When the values reappear, the filter realizes how wrong it is and drops way back down to correct itself.

Let us move on to an example where the Kalman filter performs better. A common real world application of the Kalman filter is to smooth sensor readings. Every sensor has some element of random noise in its readings. Here, we will just exaggerate the noisiness of a voltmeter (a sensor for measuring the voltage of electricity). We will simulate noisy voltmeter readings for 100 time intervals and use a Kalman filter to smooth the values, attempting to predict the true voltage value.

```
class Voltmeter:
    def __init__(self, _truevoltage, _noiselevel):
        self.truevoltage = _truevoltage
        self.noiselevel = _noiselevel
    def GetVoltage(self):
        return self.truevoltage
    def GetVoltageWithNoise(self):
        return random.gauss(self.GetVoltage(), self.noiselevel)

voltmeter = Voltmeter(1.25, 0.25)

random.seed(14)

#Simulate voltmeter readings for 100 time steps
```

```python
measuredvoltage = []
truevoltage = []
for i in range(100):
    measured = voltmeter.GetVoltageWithNoise()
    measuredvoltage.append(measured)
    truevoltage.append(voltmeter.GetVoltage())

#Parameter set up
init_state_mean = np.array((measuredvoltage[0]))
init_state_cov = np.ones((1, 1))
obs_mat= np.matrix([1])
obs_cov = np.eye(1)
trans_mat = np.matrix([1])
trans_cov = np.eye(1)*1e-6

kfv = KalmanFilter(n_dim_obs=1, n_dim_state=1,
                   initial_state_mean=init_state_mean,
                   initial_state_covariance=init_state_cov,
                   observation_matrices=obs_mat,
                   observation_covariance=obs_cov,
                   transition_matrices=trans_mat,
                   transition_covariance=trans_cov)

observations = np.array((measuredvoltage)).reshape(100,1)
state_means_v, state_covs_v = kfv.filter(observations)

#Plot the change in each variable over time
times = range(len(measuredvoltage))
plt.plot(times, measuredvoltage, 'b--', label='Measured Voltage')
plt.plot(times, truevoltage, 'g--', label='True Voltage')
plt.plot(times, state_means_v, 'r--', label='Kalman Filter')
plt.title('Noisy Voltage Measurements Smoothed by Kalman Filter')
plt.legend(loc="upper left")
plt.show()
```

The plot shows that the Kalman filter very closely predicts the true voltage.

25. Time Series Analysis

Recall from chapter 1 that panel data consists of repeated observations of the same variables or features over a period of time, and that time series data is a special case of panel data where there is only one entity being measured over time. Time series data consists of ordered observations about one entity or variable over time. The analysis of time series data is categorized by the goal of the analysis:

- **Forecasting** – Time series forecasting seeks to predict the current or future value of the dependent variable based on previously observed values. Forecasting is very similar to regression, except that previous values of the dependent variable can be included as predictors. An example of forecasting that we will look at in this chapter is predicting future stock prices.

- **Classification** – Time series classification seeks to classify the dependent variable based on a time series of predictors. An example is classifying heartbeat patterns as irregular or normal.

- **Signal Processing** – Signal processing involves distinguishing between signal and noise, and extracting the true signal out of the transmission of a wave. It is like dimension reduction where the dimensions that are discarded are meaningless noise. Signal processing covers topics like robotic navigation, voice detection and recognition, seismology, and interpreting output of scientific instruments like a telescope or particle accelerator. Signal processing is its own discipline, so it will be explored in a separate chapter.

- **Anomaly Detection** – Anomaly detection aims to detect breaks from normal patterns. It can be used to detect fraud and money laundering in financial data, and detect intrusions in computer networks. Anomaly detection can also be used to detect events in time series data that could signal a break in a pattern. Such a break might require a new forecasting models to be trained for each segment of time around the break, such as in quantitative stock trading. Anomaly detection with time series data is closely related to the outlier detection methods described in chapter 4.

- **Segmentation** – Segmentation aims to break time series data into segments where each segment has unique characteristics. It can be thought of as unsupervised clustering of time series data. Segmentation is most often used as an intermediate step in forecasting, classification, signal processing, or anomaly detection. An example of time series segmentation that we will look at in this chapter is segmenting booms and recessions in financial data.

A common question about panel data and time series analysis is how to handle forecasts and classifications for different lengths of time, or different levels of measurement. For example, suppose we want to predict employee job satisfaction scores that are measured on a weekly basis. So every employee has 1 score per week. Regression/forecasting could predict the score 6 weeks or 6 months into the future (6 months of weekly forecasts). It would not make sense to try to predict the score at the month level, if the data is measured at the week level. A month level prediction could be derived by summarizing the weekly data, such as by averaging, but a model fitted to data measured at one level can only make predictions on that level. Now suppose that the goal were classification instead of regression, and the scores were binned so that scores 0.5 or greater were "good" and scores less than 0.5 were "bad". The same principle holds: a classifier that is fitted to weekly data can only predict weekly classes. A model trained at the week level could not predict a daily score, for instance.

Another common question about panel data and time series data is how to add more covariates to the model. The independent variables of time series models do not need to be restricted to lags of the dependent variable. The models we are about to explore specifically apply to models of a dependent variable on its lags, but it is perfectly acceptable to add more regressors. For example, a model of sales over time could include lags of sales, categorical or dummy predictors like quarter and month, as well as continuous predictors like dollars spent on marketing.

One more common question about modeling panel data and time series data is how different levels of data aggregation should be mixed: can weekly averages be combined with monthly totals, for instance? The answer is that different levels of aggregation should not be mixed at all. To understand why, consider aggregating employee salary up to the manager level, such as by averaging. There would be one average salary per manager that would be shared by all the employees under that manager. Average salary would therefore be a wasted feature because it would not yield any information about the dependent variable that would not already be captured by the manager variable by itself. If the data is time series data with different levels or a hierarchical structure, then a combination of time series and multilevel modeling is possible.

Serial Correlation/Autocorrelation

Serial correlation, a.k.a. autocorrelation, is when a variable is correlated with itself at different time intervals. For example, today's price of General Electric's stock partially depends on the price it was yesterday. Serial correlation can occur at any time interval: if GE's stock price today depends on its price x days ago, then GE's stock price exhibits serial correlation. Serial correlation also applies to the errors of a model. If today's stock price is correlated with yesterday's stock price, then any model for today's stock price will have errors that correlate with the model for yesterday's stock price. Recall from chapter 6 that serial correlation violates the assumptions of linear regression. To test for serial correlation, the residuals plot of a regression model can be inspected, or the Durbin-Watson test can be applied, or the correlogram of a variable and its lags (or the residuals) can be inspected. So far, we have only used the residuals plot and the DW statistic to test for serial correlation. In this chapter, we will explore the use of correlograms.

Serial correlation usually occurs at several time intervals. For example, GE's stock price could be correlated with its price yesterday, its price 7 days ago, and its price 30 days ago. The time intervals for which a variable is serially correlated are often repetitive, such as retail sales spikes during the yearly holiday shopping season. The phenomenon of serial correlation following a cycle is called **seasonality**. Seasonality is not limited to a yearly cycle; it can occur every 2 years or every 20. Some statistics texts restrict seasonality to cycles that occur yearly (or seasonally) however, and refer to the phenomenon of longer cycles as cyclicality. In this book, we will refer to all cyclical serial correlation as seasonality.

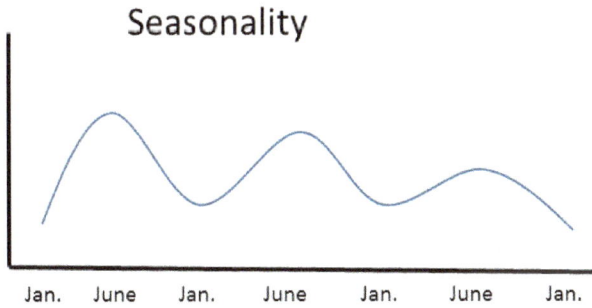

Seasonality

Jan. June Jan. June Jan. June Jan.

Figure 25.1

In addition to having seasonal patterns, time series data can often follow a long term **trend**. For example, as the American population grows, so does consumption. So a model of retail sails might show both seasonality and a long term increasing trend. There are bound to be disruptions to any trend. When disruptions are soon erased and the time series returns to the trend, the trend is called a **deterministic trend**. Deterministic trends exhibit **mean-reversion**, which means that they revert to the average in the absence of any perturbation. When disruptions or perturbations permanently shift the time series, the trend is called a **stochastic trend**. Stochastic is another word for random. If a time series does not have any trend, then its mean and variance remain constant over time. This type of time series is referred to as a **stationary process**. If a time series has a trend, then its mean and variance are different between different periods of time. This type of time series is referred to as a **non-stationary process**. If the removal of a trend from a time series causes it to become stationary, then it is called a **trend stationary process**. If a time series can be made stationary by subtracting one or more of its lags from every value, then it is called a **difference stationary process** (recall that a lag is the value of a variable at the previous time step).

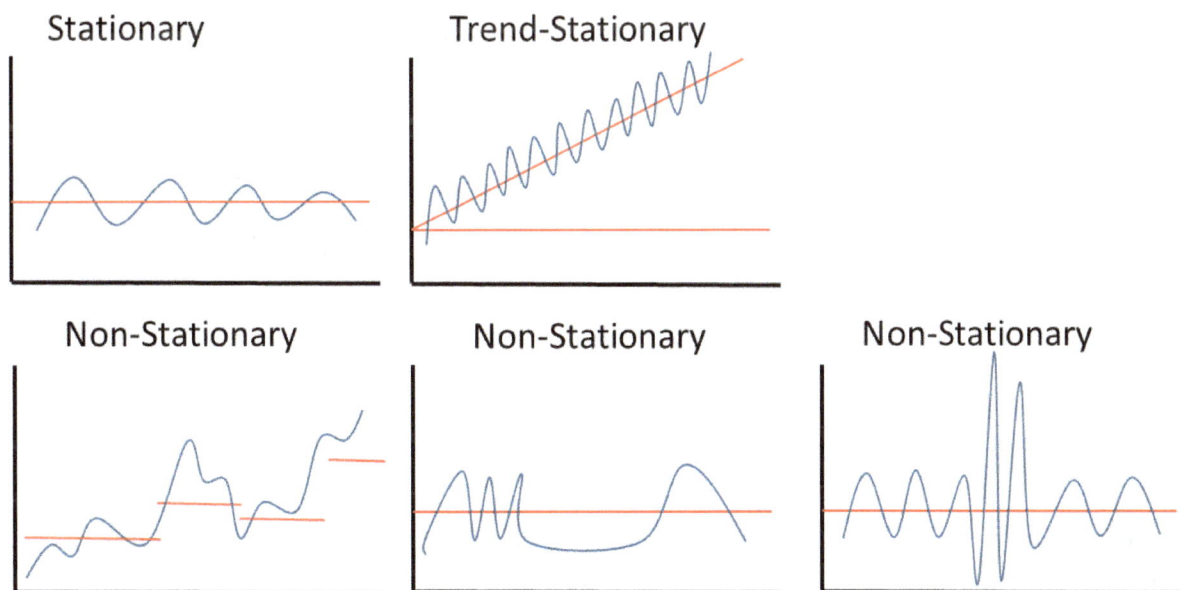

Stationary   ## Trend-Stationary

Non-Stationary   ## Non-Stationary   ## Non-Stationary

Figure 25.2

Time Series Decomposition

When the trend and seasonality are removed from time series data, the remainder is called the residual, or noise. The process of removing trend and seasonality is called decomposition. Classical decomposition was developed in the 1920's and is still the default method taught in most statistics courses (Wooldridge, 2009). Classical decomposition involves using a moving average to estimate the trend in a time series. The order of the moving average (the size of the sliding window) must be found by trial and error and by manual inspection. If seasonality is expected to occur once per week, a moving average of order 7 is probably best. Similarly, if seasonality is expected to occur once per quarter, a moving average of order 4 is appropriate. To perform classical decomposition, it is necessary to choose either an additive model or a multiplicative model to represent the time series.

Additive Model	Multiplicative Model
$y_t = S_t + T_t + R_t$	$y_t = S_t * T_t * R_t$

Figure 25.3

In figure 25.3, S represents the seasonal component, T represents the trend, and R represents the residual noise. Additive models are used when the variations around the trend do not vary, whereas multiplicative models are used when the variations around the trend are proportional to the level of the time series. For example, if the seasonality has larger swings as the time series increases, then a multiplicative model is a better choice. The multiplicative model can be converted to an additive model by taking the log of the equation. If an additive model is chosen, then the model is subtracted from the time series to produce the de-trended series. If a multiplicative model is chosen, then the time series is divided by the model to produce the de-trended series. Seasonality is then estimated by averaging the de-trended values for each seasonal interval (such as a month). Finally, the seasonality estimate is either subtracted from the de-trended series (for the additive model) or the de-trended series is divided by the seasonality estimate (for the multiplicative model). Since classical decomposition relies on the moving average for smoothing, it is undefined for the first few and last few observations. Classical decomposition is unable to handle changes in seasonality over the long term, and if there are only a few periods in the time series, it is not robust enough to handle the larger variation of a small dataset.

Seasonal trend decomposition using locally weighted regression and scatterplot smoothing (LOESS) is a more modern decomposition method that overcomes the weaknesses of classical decomposition. Recall from chapter 17 that LOESS is a more robust smoothing method than the moving average. Seasonal trend decomposition using LOESS is abbreviated STL, and it allows the seasonal component of a time series to vary over time. STL is also robust to outliers, making it better able to model small datasets than classical decomposition. STL only works with additive models though, so if it is suspected that a multiplicative model is needed, the log of the multiplicative model can be taken to transform it into an additive model. Then STL can be applied to produce the predicted log values of y, and the equation can be exponentiated to transform it back to the original scale.

The residual of a decomposed time series can be used to inform forecasting, anomaly detection, and segmentation, as we will soon see. The residual of a decomposed time series is essentially the signal minus any seasonal or trend effects.

Random Walks

A perfect time series model will eliminate all serial correlation from the residuals. Perfect models are impossible, so it is safe to say that any model of time series data will produce serially correlated residuals. Since a perfect model eliminates serial correlation of the residuals, one way to choose good time series models is to look for models that produce very little serial correlation in the residuals (the differences between the forecasted and actual values). But in order to judge how small serial correlation should be for a good time series model, there needs to be a basis for comparison. It turns out that independently sampled values from a Gaussian (normal) distribution simulate serially uncorrelated residuals. Since independently sampled values from a Gaussian distribution are just white noise, Gaussian white noise is a good baseline for comparison for a time series model's residuals. If the model is good, the serial correlation of its residuals will be close to the serial correlation of Gaussian white noise (close to 0), so its correlogram will look like the one in figure 25.4.

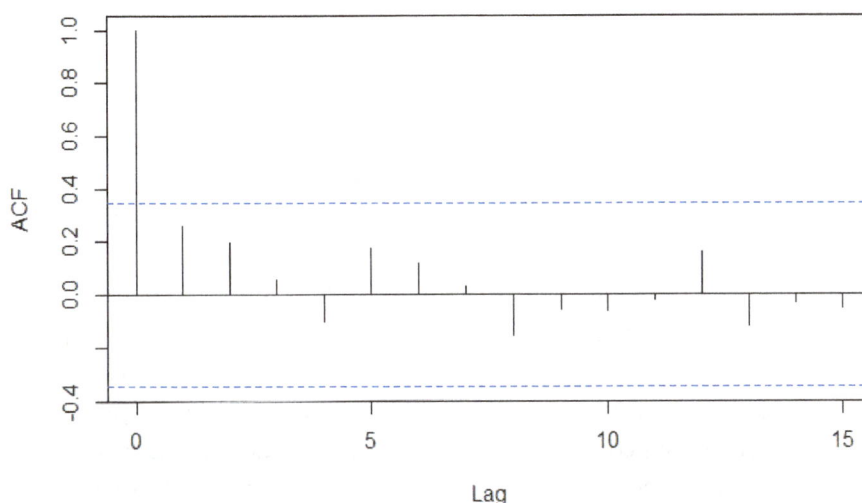

Figure 25.4

A correlogram shows how a variable is correlated with its lags, where the y-axis is correlation and the x-axis has the lags. The 95% confidence interval can be overlaid, as it is in figure 25.4, and any correlations outside of this interval can be considered significant. The correlogram is the plot of the results of the **autocorrelation function (ACF)** or **partial autocorrelation function (PACF)**. The ACF simply calculates the correlation of each observation with its k lags. The PCF calculates the correlation of each observation with its k lags, while accounting for intervening correlations. An intervening correlation might make the ACF correlations appear larger. For example, if a variable is correlated with its first lag only, then the effect of the first lag correlation carries over to other lags (if A is correlated with B, which is correlated with C, then A appears to be correlated with C more strongly than it really is) (Halls-Moore, 2016).

Since random white noise has no serial correlation, we can specify a basic time series model in terms of a model plus random noise. The simplest time series model is just the first lag. So the simplest model of time series data is the first lag plus random noise. This type of model is called a random walk, and figure 25.5 shows the equation and a graphical example.

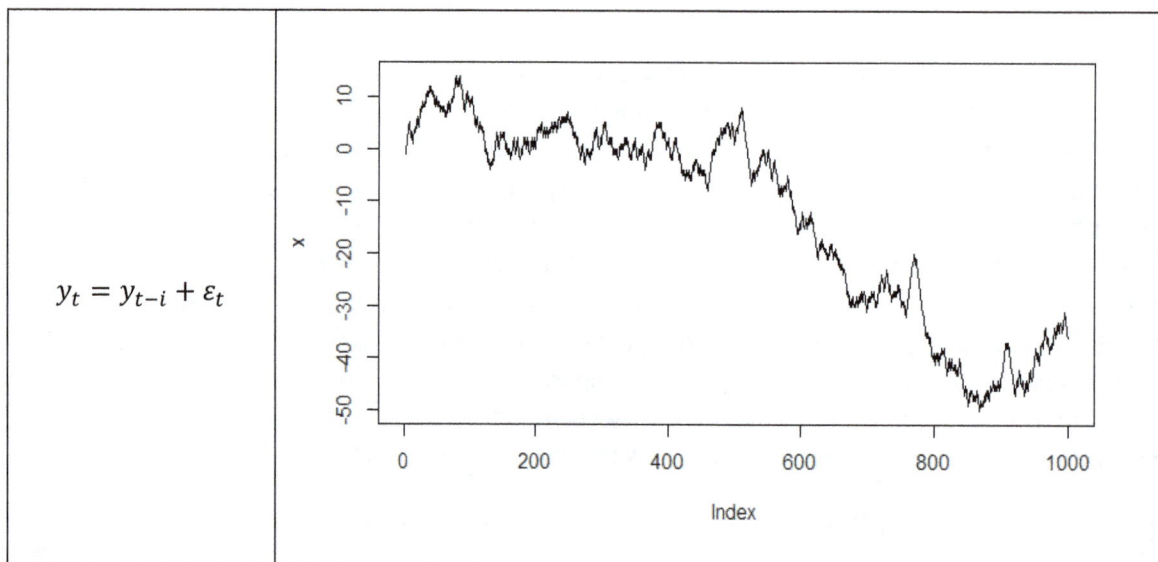

$$y_t = y_{t-i} + \varepsilon_t$$

Figure 25.5

If a constant were added to the right side of the equation in figure 25.5, then the model would be called a random walk with drift. The random error term could be positive or negative, so the predicted value is the previous value plus some random change. The mean of a random walk is 0, but its covariance depends on time (a random walk's covariance = variance * time), meaning random walks are non-stationary processes. Since the covariance increases as time increases, it is not possible to create a model that can predict the long term future values of random walks. Random walks have often been used to describe stock prices. If a random walk is found to be a good model for a particular stock, then it is nearly impossible to predict the future prices of that stock. Random walks are not always good models for stocks however. Sometimes stocks are correlated with more distant lags. For example, today's price might be correlated with the price the stock was 14 days ago. If this were true, then models that use longer lags would be better fits than a random walk. One kind of model that uses longer lags is the autoregressive model.

Autoregressive Models and Moving Averages

Autoregressive (AR) models are specified in terms of their order, p, which represents the number of lags in the model. An AR(3) model is an autoregressive model with three lags. An AR(1) model is an autoregressive model with one lag, or in other words, an AR(1) model is a random walk with a lag coefficient (alpha) of 1. Figure 25.6 defines an AR(p) model.

$$y_t = \sum_{i=1}^{p} \alpha_i y_{t-i} + \varepsilon_t$$

Figure 25.6

As with the random walk, if a constant were added to the right side of the equation, the AR model would have drift. AR models can be either stationary or non-stationary, depending on the alpha coefficients. If a stationary AR model is found to be a good model of a stock's price, then it is possible to predict the stock's future prices. To determine if an AR model is stationary, find the roots of the equation in figure 25.7. If the absolute values of all the roots are > 1, then the process is stationary.

$$1 - \alpha_1 z - \cdots - \alpha_p z^p = 0$$

Figure 25.7

In the equation in figure 25.7, z is some variable, alpha is the coefficient of the ith lag, and p is the number of lags. In order for an AR(p) model to be stationary, the absolute values of all the roots of the solution to the equation in figure must be > 1.

Another way to test for stationarity in an AR model is the **unit root test**, a.k.a. the **Dickey-Fuller test**. The unit root test tests for the presence of a unit root in an AR model. A **unit root** simply means that if one of the equations in figure 25.8 is solved and produces root of 1, then the AR model has a unit root and is non-stationary. In order to be stationary, the absolute values of all of the roots must be < 1 (Wooldridge, 2009). Note that this is the exact opposite of what was stated must be true for the equation in figure 25.7. That is because the equation in figure 25.7 is rearranged so that the roots must be > 1 for the process to be stationary.

There are three equations for each version of the Dickey-Fuller test, as shown in figure 25.8. The top equation tests for a unit root, the middle equation tests for a unit root with drift, and the bottom equation tests for a unit root with drift and a deterministic time trend.

$$\Delta y_t = \beta y_{t-1} + u_t$$

$$\Delta y_t = \alpha_0 + \beta y_{t-1} + u_t$$

$$\Delta y_t = \alpha_0 + \alpha_1 t + \beta y_{t-1} + u_t$$

Figure 25.8

Each version of the test has the null hypothesis that there is a unit root (so the null is that the series is non-stationary), and a separate critical value.

The Dickey-Fuller test can also be carried out after the serial correlation in a time series has been removed. To remove the serial correlation, the Dickey-Fuller test is augmented by adding the lagged values of the dependent variable to the equations in figure 25.8 above. This produces the **Augmented Dickey-Fuller test**, as shown in the equations in figure 25.9.

$$\Delta y_t = \beta y_{t-1} + \sum_{s=1}^{m} (\gamma_s \Delta y_{t-s}) + u_t$$

$$\Delta y_t = \alpha_0 + \beta y_{t-1} + \sum_{s=1}^{m} (\gamma_s \Delta y_{t-s}) + u_t$$

$$\Delta y_t = \alpha_0 + \alpha_1 t + \beta y_{t-1} + \sum_{s=1}^{m} (\gamma_s \Delta y_{t-s}) + u_t$$

Figure 25.9

In the equations in figure 25.9, alpha sub 0 is the intercept (a constant), and beta and gamma are the coefficients. In statistical packages, the Augmented Dickey-Fuller test produces a p-value and the test statistic, which follows a chi-square distribution. Either the test statistic or the p-value can be used to determine whether the null, that the series is non-stationary, can be rejected.

The fit of an AR model depends on its order p, which can be chosen by picking the order that optimizes the AIC or BIC. Refer back to chapter 9 for definitions of the AIC and BIC. Once p is known, MLE is used to optimally estimate the alpha parameters of an AR model. A p-value or confidence interval computed for the AR model can then indicate whether the estimates of the model parameters (the alpha coefficients) are likely to be due to chance or not.

Now that we have defined AR models, we can move on to **moving average (MA) models**. In chapter 20, we saw how moving averages act as smoothing functions. In the context of time series, moving averages are smoothing functions that model white noise error terms: the model the difference between the value of the dependent variable at time t and the AR model of its lags (if the dependent variable has an AR model component). Time series MA models are always left moving averages: they are the average of prior noise terms. Whereas an AR model of order p is a linear combination of previous time steps, a MA model of order q is a linear combination of previous white noise terms. AR models weight perturbations from a trend in a decreasing order so that more recent disruptions are weighted more heavily. MA models weight all perturbations equally, but are limited to only the past q disruptions. A MA(q) model is represented by the equation in figure 25.10.

$$y_t = \mu + \varepsilon_t + \sum_{i=1}^{q} \beta_i \varepsilon_{t-i}$$

Figure 25.10

In the equation in figure 25.10, mu is the mean of the series y over all time points, epsilon is the white noise error term, and beta is the coefficient of the ith error term. If the MA modeled only the white noise error terms, then the mean, mu, would be assumed to be zero and the variance of the errors would be assumed to be constant. The correlogram of a MA(q) model should have a significant spike at lag q, but only at lag q, except for the margin of error. For example, if the margin of error or confidence interval is 5%, then it would still be ok to model a time series with a MA(q) process if there was a spike at lag q and a small spike at some other non-sequential lag. Just like AR processes, the order q for MA processes can be determined by optimizing the AIC or BIC.

Moving averages are unlikely to be able to model a time series by themselves, as they are aimed at explaining "shocks" to a time series caused by random events. AR processes are unlikely to be able to model a time series by themselves either, as they are aimed at explaining the momentum and past behavior of the series. If these models were combined, they might more accurately describe time series, and that is exactly what ARMA and ARIMA models do.

Autoregressive moving averages (ARMA) of order (p, q) combine the ability of moving averages to explain random shocks and the AR model's ability to explain past behavior. They are represented by the equation in figure 25.11.

$$y_t = c + \sum_{i=1}^{p} (\alpha_i y_{t-i}) + \sum_{i=1}^{q} (\beta_i \varepsilon_{t-i}) + \varepsilon_t$$

Figure 25.11

In the equation in figure 25.11, c is a constant, the first sum is the AR(p) component, the second sum is the MA(q) component, and epsilon is the residual error. The individual AR and MA components (the summations) of the ARMA model should be recognizable as coming from figures 25.6 and 25.10. The ARMA model is essentially just a linear combination of the AR and MA models.

ARMA models should be fitted and estimated using the Box-Jenkins method (Box, Jenkins, & Reinsel, 1976). The **Box-Jenkins method** is a procedural approach to fitting ARMA models to ensure that the optimal p and q are selected. The procedure is carried out as follows:

1. Ensure that the time series is stationary. To do this, simply plot the ordered time series to visually look for non-stationarity. A plot of the ordered time series is called a **sequence plot**. Alternatively, plot the correlogram of the ACF. Non-stationarity can be identified from the ACF correlogram by looking for significantly autocorrelated lags with slow correlation decay. If the series is non-stationary, differencing or some other transformation method should be applied to convert the series to a stationary series. If a series cannot be converted, ARMA models cannot be used.

2. Look for seasonality that needs to be modeled. Seasonality should be obvious when step 1 is carried out, as both the sequence plot and the ACF correlogram will show it. In the correlogram, look for significantly correlated lags that show periodicity (e.g. every 3 lags are correlated). Seasonality does not necessarily need to be differenced out, because the AR component will model it.

3. Choose p and q by either minimizing the AIC or BIC, or by examining the ACF and PACF correlograms. The chart below can be followed to help choose the type of model based on the appearance of the ACF or PACF correlogram (Nau, 2018).

Situation	Response
ACF shows significant lags that decay to 0	Look at the PACF to identify the order of p, and use an AR(p) model
ACF shows significant lags that decay to 0 and PACF shows only 1 correlated lag, AND the time series has been differenced already	Use an ARIMA(p, d, 0) model, where p is the correlated lag in the PACF
ACF has alternating positive and negative correlations that decay to 0	Look at the PACF to identify the order of p, and use an AR(p) model
ACF has one or more correlated lags	Use a MA(q) model, where q is where the ACF zeros out
ACF has one correlated lag and the PACF shows significant lags that decay to 0	Use an ARIMA(0, d, q) model, where q is the correlated lag in the ACF
ACF shows decay but only after the first few lags	Use an ARMA(p, q) model, where p is picked from the PACF and q is where the ACF zeros out or gets small
ACF shows no correlated lags	AR and MA models cannot be used – the time series has no patterns that can be modeled
ACF shows periodic correlated lags	Look at the PACF to identify the order of p, based on the lowest order lag, and use an AR(p) model

ACF does not decay at all	Non-stationary time series – AR and MA models cannot be used unless differencing can transform the series into a stationary series
ACF shows a positively correlated first lag and PACF shows a sharp cutoff	Use an AR(p) model, where p is where the PACF cuts off
ACF shows a sharp cutoff and/or the first lag is negatively correlated	Use a MA(q) model, where q is where the ACF cuts off
If the sum of the AR terms is near 1	Reduce the number of AR terms by one and increase the order of differencing by 1
If the sum of the MA terms is near 1	Reduce the number of MA terms by one and reduce the order of differencing by 1

Figure 25.12

The chart above should make it obvious that time series modeling using AR/MA models is as much an art as it is a science. If heuristics like the ones shown in figure 25.12 are not appealing, then choosing the p and q by optimizing the AIC or BIC is a suitable alternative (Nau, 2018). After the appropriate model has been selected, it is fitted using MLE. After fitting the model, the ACF and PACF correlograms of the residuals should show no significant lags, as this would mean that the model explained the time series very well. Another way to evaluate ARMA models is to use the Ljung-Box test. The **Ljung-Box** test computes the tests statistic shown in figure 25.13. It tests the null hypothesis that the time series is independently distributed (the correlations are 0). The alternative hypothesis is that the time series exhibits serial correlation. The Ljung-Box test can be used as an alternative to the Durbin-Watson test to look for serial correlation (Halls-Moore, 2016). In the context of ARMA modeling, it can be used to assess the fit of a model by testing the null that the model residuals are not correlated, because a good model should show no serial correlation in the residuals.

$$Q = n(n+2) \sum_{k=1}^{h} \frac{\hat{\rho}_k^2}{n-k}$$

Figure 25.13

In the equation in figure 25.13, n is the sample size, rho hat sub k is the sample autocorrelation at lag k, and h is the number of lags being tested. The test statistic for the Ljung-Box test follows a chi-square distribution with h degrees of freedom.

Autoregressive integrated moving averages (ARIMA) or order (p, d, q) combine the ability of moving averages to explain random shocks and the AR model's ability to explain past behavior, just like ARMA models. The difference between ARMA and ARIMA models are that ARMA models only combine past behavior and random noise. ARIMA models combine past behavior and random noise on the differenced values of a time series. This means that ARIMA models replace the values of the time series with the difference between the values and their lags. The order of the differencing is given by d, thus ARIMA models have order (p, d, q) for the AR component, the differencing component, and the MA component.

Differencing is the process of subtracting the lagged values of a time series from the non-lagged values. The first difference is the first lag subtracted from the values in the series. First differencing shows how the series changes from one period to the next.

$$First\ Difference\ =\ y_t - y_{t-1}$$

Figure 25.14

Recall that when time series are decomposed, the seasonality and trend are subtracted (or differenced) from the series to produce the residual. Decomposition smooths a time series by removing repetitive or predictable patterns. Similarly, differencing is a way to smooth a time series. It is sometimes desirable to smooth a time series that is non-stationary (its variance changes over time), because differencing may cause the series to become stationary. Such time series are called difference stationary. To understand how this works, look at figure 25.15.

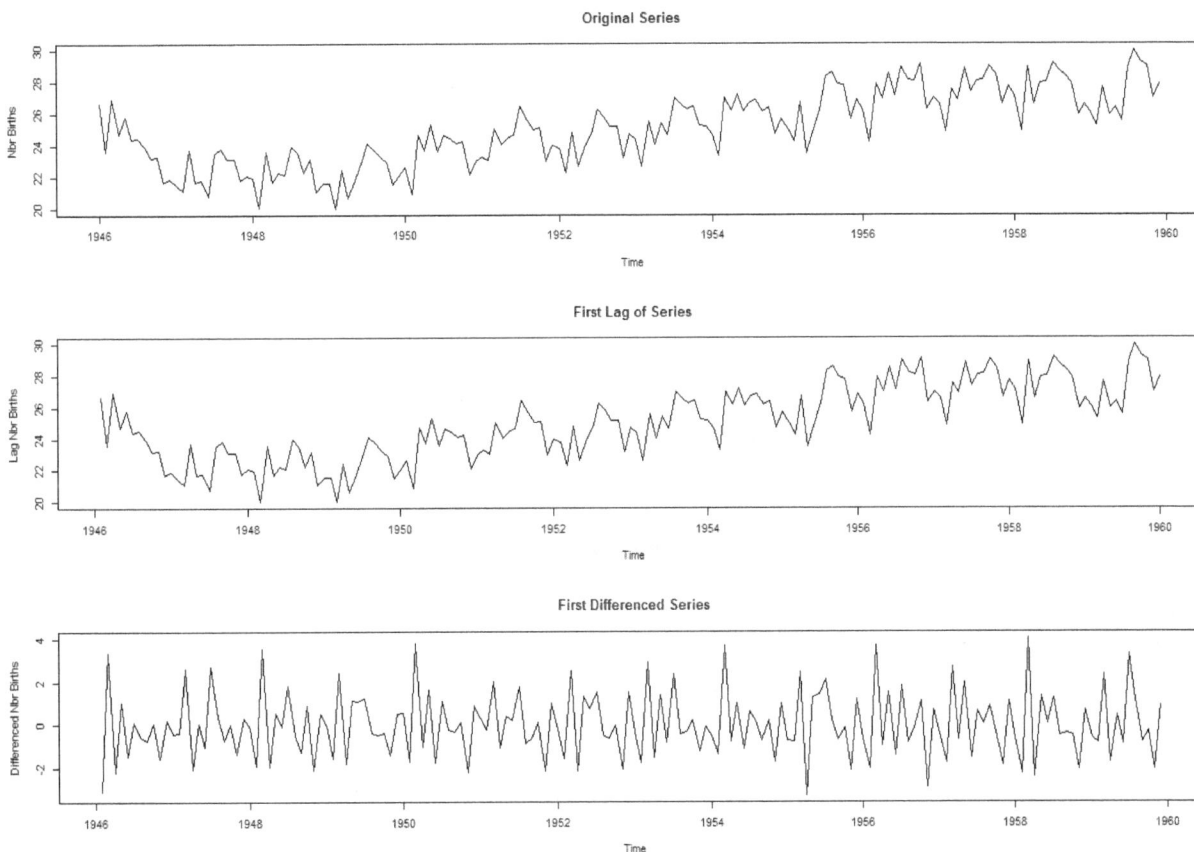

Figure 25.15

The top graph in figure 25.15 is a non-stationary time series. The middle graph is the first lag of the series in the top graph. When this middle graph is subtracted from the top, the result is the graph on the bottom, which is a stationary time series.

Working with stationary time series is always easier than working with non-stationary time series, because statistics like the mean, variance, covariances, and correlations with other variables are only predictive of future values if the series is stationary. Consider why this is true: if a series is non-stationary, then its mean and variance change over time by definition, which prevents the mean and all models that minimize mean squared error from being good models of future behavior. Therefore, transformations like differencing that convert a non-stationary time series to a stationary one are very useful.

ARIMA models of order (p, d, q) are represented in figure 25.16.

$$y_t = c + \mu + \sum_{i=1}^{p}(\alpha_i y_{t-i}) + \sum_{i=1}^{q}(\beta_i \varepsilon_{t-i}) + \varepsilon_t$$

Figure 25.16

The variables in equation in figure 25.16 are the same as in the equation for ARMA models in figure 25.11. The constant, c, is optional. The difference between ARIMA and ARMA is that the y's in the ARIMA equation in figure 25.16 are the differenced values of time series Y. If d = 0, then y = Y. If d = 1, then $y_t = Y_t - Y_{t-1}$. If d = 2, then $y_t = (Y_t - Y_{t-1}) - (Y_{t-1} - Y_{t-2})$.

Like ARMA models, ARIMA models are estimated using the Box-Jenkins method. The table in figure 25.12 can be used to determine the optimal p, d, and q. Alternatively, p, d, and q can be chosen by trying different orders and choosing the ones that optimize the AIC or BIC. After the appropriate model has been selected, it is fitted using MLE. After fitting the model, the ACF and PACF correlograms of the residuals should show no significant lags, as this would mean that the model explained the time series very well. Another way to evaluate ARIMA models is to use the Ljung-Box test.

As previously mentioned, a requirement for the Box-Jenkins method is that the time series is stationary. If the time series is non-stationary, then differencing, de-trending using a smoothing method must be performed to transform it into a stationary series. If de-trending is required, it would be carried out by subtracting the smoothed trend from the time series. Although seasonal differencing is not required by the Box-Jenkins method, it may be done if the researcher believes it would improve the fit of the model. Alternatively, if seasonal differencing is not performed, the seasonal terms are included in the model. For example, a time series with quarterly seasonality would have an AR(4) term included in the model.

ARIMA models can be converted to AR models, MA models, or ARMA models by modifying the p, d, and q orders. For example, an AR(1) model is the same as an ARIMA(1, 0, 0) model, a MA(1) model is the same as an ARIMA(0, 0, 1) model, and an ARMA(1, 1) model is the same as and ARIMA(1, 0, 1) model. An ARIMA(0, 1, 0) model is a random walk. An ARIMA(0, 1, 1) model is an example of exponential smoothing, because it is an exponentially weighted moving average of past values.

One of the drawbacks to AR, MA, ARMA, and ARIMA models is that none of them take stochastic volatility into consideration.

Stochastic Volatility

So far, the time series models we have looked at have assumed that the variance of the time series is constant over time. This assumption is often wrong. It is common for the variance of a time series to change over time. When this occurs, the variance can be envisioned as having a random distribution. If the variance has a random distribution, then the volatility (variance) of a time series is itself a stochastic process and can be modeled as such. Recall from chapter 3 that when variance is non-constant, the data is said to exhibit heteroskedasticity. Heteroskedasticity can occur at regular intervals, such as periods of higher volatility occurring in conjunction with seasonal effects. In financial time series, like stock prices, it is common for volatility to be correlated with periods of increasing volatility. In other words, if a stock's price suddenly drops lower than usual on a particular day, it could cause a ripple effect of several shareholders selling the stock and further reducing the price. In this example, the volatility of the stock would be clustered. Periods of higher volatility would be clustered together with other periods of higher volatility (Halls-Moore, 2016). This phenomenon is called **volatility clustering**. If the heteroskedasticity of a time series is dependent on periods of increased volatility, it is called **conditional heteroskedasticity**.

At this point, a reasonable question would be to ask what the difference between a heteroskedastic and non-stationary time series is, since we have defined both concepts in terms of non-constant variance. Heteroskedasticity is when the only variance changes over time. Non-stationarity is when either the mean or variance, or both, and other statistical properties like covariance or correlation with other variables, change over time. All heteroskedastic time series are non-stationary, but only some non-stationary time series are heteroskedastic.

If there is volatility clustering in a time series, in other words, if there is conditional heteroskedasticity, then one way to model the volatility clustering is to use a model that forecasts volatility based on previous values. In other words, use an autoregressive model for conditional heteroskedasticity. **Autoregressive conditional heteroskedasticity (ARCH)** models of order (q) take the form of the equation in figure 25.17.

$$\varepsilon_t = c + \sqrt{\alpha_0 + \sum_{i=1}^{q} \alpha_i \varepsilon_{t-i}^2}$$

Figure 25.17

In the equation in figure 25.17, epsilon represents error (the residuals), c is an optional constant, alpha sub 0 is the intercept and alpha sub i is the coefficient of the squared lagged residual.

ARCH should only be applied to a time series that has already been fitted by another time series model that has removed trends and seasonality (i.e. any serial correlation is removed) and has reduced the residuals to appearing like white noise. For example, if a model is fitted by an ARMA or ARIMA model, or if its trend and seasonality have been removed by decomposition, only the residuals will be left unexplained. When the residuals are viewed in a correlogram, none should be significant. This would mean that the residuals resemble white noise, since none of the lags of white noise are statistically significant either. When the residuals resemble white noise, they can be squared and then re-plotted in the correlogram to see if any of the squared residuals are significant. If they are, then ARCH can be used.

Since we have seen how adding a moving average component to an autoregressive model produces the more powerful ARMA and ARIMA models, a reasonable question would be to ask why a moving average could not be added to the ARCH model. The answer is that it can, and this is called generalized ARCH. **Generalized autoregressive conditional heteroskedasticity (GARCH)** models of order (p, q) are shown by the equation in figure 25.18.

$$\sigma_t = c + \sqrt{\alpha_0 + \sum_{i=1}^{q} \alpha_i \varepsilon_{t-i}^2 + \sum_{i=1}^{p} \beta_i \sigma_{t-i}^2}$$

Figure 25.18

In the equation in figure 25.18, epsilon again represents the residuals, alpha sub 0 is the intercept, alpha sub i is the coefficient of the squared lagged residual, and beta sub j is the coefficient of the lagged variance.

Rather than using the moving average, it is possible to use an exponentially weighted moving average (EWMA) that gives more weight to more recent residuals. The drawback to using EWMA instead of GARCH is that the researcher must define the weight decay factor, and if the decays is chosen arbitrarily, the results could be unstable (they would vary depending on the choice of decay). The combination of ARIMA and GARCH, or ARIMA + GARCH, is a powerful time series model.

A common question about GARCH is how to decide how many residuals to include. Should the stopping point be 1 residual or 100? This number can be determined by starting with 1 residual and applying the Ljung-Box test until the p-value is less than some level of significance, such as 10%. At that point, no more residuals should be included (Christopher, 2016).

Quantitative Trading: Mean Reversion and Cointegration

The models described in this chapter so far can apply to any time series, but we have been leaning towards applying them to stocks. A full treatment of quantitative finance is beyond the scope of this book, but in this section, we will digress a bit to see how the models we have explored can be used in basic stock trading strategies.

Mean reversion generally means that observations tend to gravitate toward the mean. In the context of financial time series, mean reversion is the idea that a stock's price tends to gravitate toward its average price over time. When the price is below its mean, the stock is considered a good buy because the assumption is that it will rise when it reverts to its mean. When the price is above the mean, the stock is considered a good sell, because the assumption is that it will fall when it reverts to its mean. This concept is why many technical trading platforms provide 50 and 100 day moving averages in their stock charts (Halls-Moore, 2016). When the price rises above the x day moving average, it could be in an upswing, whereas if it falls below this moving average, the stock could be in a downswing. The shorter the span of the moving average, the more sensitive it is to shocks in the price. Exponentially weighted averages are more price sensitive too, because they give more weight to recent price changes. It is therefore common for trading platforms to show several moving averages, and if a moving average with a shorter span crosses over the moving average with a longer span, it may signal a shifting trend in the stock price (Nolan & Lang, 2015).

Figure 25.19

The candlestick chart in figure 25.19 shows Apple Inc.'s stock price from April 1, 2013 to June 27, 2014. The 20 day moving average is blue, the 50 day moving average is turquoise, and the 200 day moving average is fuchsia. Days when the price fell are shown in red.

We know that moving averages are far from perfect models, so trading solely based on moving averages is risky behavior. But mean reversion is still useful. Consider tracking the price of a stock like Alphabet Inc. (Google's holding company). If there is a week of bad press that pushes Alphabet's stock price down, it is reasonable to expect this shock to disappear and that the price will revert to its mean (assuming the shock pushes the price below the mean). Now consider tracking the stock price of a very similar company, like Facebook. Broad market factors should impact Alphabet and Facebook the same way, but company specific shocks may only affect one or the other. As long as the long-term share prices of these companies are influenced by similar market factors, we can follow a pairs trading strategy. **Pairs trading** is a strategy that exploits the tendency for the stock prices of similar companies to behave similarly, and when a shock affects only one of the pairs, there is a temporary deviance from this trend that creates an opportunity for profit (Nolan & Lang, 2015). Suppose for example that Alphabet's and Facebook's stock prices trend together, and that a week of great press for Facebook temporarily disrupts this trend. If a short position is taken on Facebook (meaning money is borrowed to sell a stock whose price is expected to drop in the future) and a long position is taken on Alphabet (meaning money is borrowed to buy a stock whose price is expected to rise in the future), when the spread between the stocks converges back to the trend, profit is made. The profit would come from Alphabet if it rises to the trend. If instead, Facebook fell to return to its balance with Alphabet, the profit would come from shorting Facebook. The yellow box in figure 25.20 roughly illustrates this situation. Note that figure 25.20 does not represent real stock prices for these companies, and in order to compare them on a chart like this, they would have to be transformed to the same scale.

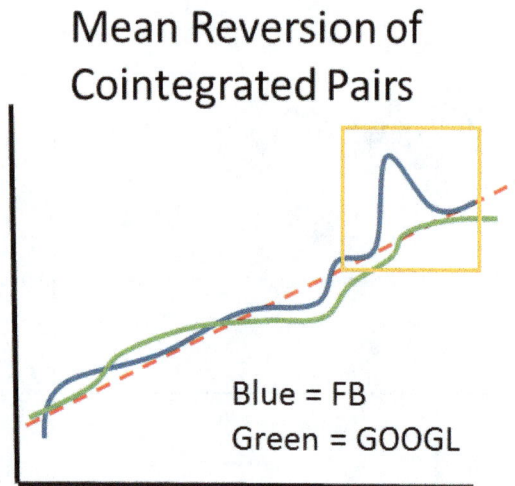

Mean Reversion of Cointegrated Pairs

Blue = FB
Green = GOOGL

Figure 25.20

Pairs trading can be extended to baskets of assets. All the stocks in a basket trend together, so they are said to be **cointegrated**. The idea behind using cointegrated asset baskets in a mean reversion strategy is that non-stationary time series (like stocks that follow random walks) are linearly combined to produce a stationary time series with a fixed mean and variance that can easily be predicted. The stationary time series is mean reverting by definition, meaning a long/short pairs trading strategy can be applied to the stocks in the basket when any of them deviates from the rest. The key to using this strategy is identifying which assets to include in the basket, which as we defined, must be cointegrated.

If two non-stationary time series form a stationary time series when linearly combined, they are said to be cointegrated. Recall that the Augmented Dickey-Fuller test is used to determine whether or not a time series is stationary. By forming a linear regression of two stocks, the coefficients will tell us which combination of the stocks to use for pairs trading. Then the Augmented Dickey-Fuller test applied to the regression will tell us whether we can rely on that strategy. Since it is never clear which stock to use as the dependent variable in the regression, it is best to try both and pick the dependent variable based on the regression that produces the smaller Augmented Dickey-Fuller test statistic.

Finding cointegrated pairs can be done by having prior knowledge or hunches as to which stocks could trend the same. By plotting two stocks over the same time period, it should be clear whether they trend the same. To be sure they are non-stationary, the Augmented Dickey-Fuller test can be used. When the residuals of the linear combination of the stocks are plotted, they should be almost perfectly linear.

Linear regression and the Augmented Dickey-Fuller test show the optimal hedge ratio for a pair of stocks in a mean reverting, pairs trading strategy. But for a basket of stocks, it is desirable to know the hedge ratios between several stocks, and not just two. The **Johansen test** combines several time series as predictors and produces coefficient matrices for each lag of the dependent variable. The test checks for the null hypothesis that there is no cointegration, meaning the rank of the coefficient matrix for the first lag = 0. If the rank of the coefficient matrix for the first lag > 0, then there is a cointegrating relationship between two or more of the time series. Eigendecomposition is used to determine the rank of the matrix. Whatever the rank is determined to be, that is how many of the time series should be combined. So if the rank of first lag's coefficient matrix is 3, then 3 of the stocks should be hedged for the strategy. The Johansen test checks for r = 0 as mentioned, but it also checks for r <= 1 vs r >1 and r <= 2 vs r > 2, meaning it can indicate whether a combination of 2 or 3 time series should be used. The largest eigenvalues of the decomposed matrix are the hedge ratios that should be used. We will go through an example of this in R and Python at the end of this chapter.

We have explored how cointegration can be used to determine the hedge ratio between stocks to execute a mean reversion strategy of an asset basket. While this approach works for a particular point in time, it assumes that the hedge ratio is stationary, and it almost never is: the hedge ratio should change over time. We could simply carry out this procedure on a daily basis to see what the best hedge ratio should be, but another approach is to use a Kalman filter (Halls-Moore, 2016). Recall that the Kalman filter can be used to estimate the value of a variable when it cannot be directly observed. So in this case, the hedge ratio cannot be directly observed for all points in time, but it can be estimated through the use of a Kalman filter.

Anomaly Detection

Time series anomalies can be additive or temporal. **Additive anomalies** are unexpected growths during a short time span. **Temporal anomalies** are periods of unexpected inactivity. Every anomaly and outlier detection method requires defining what is expected or normal, so that observations that defy this expectation can be tagged as abnormal. Defining what is normal or expected usually involves specifying an expected distribution or model. Therefore, observations that do not fit a model could be considered outliers or anomalies.

There are two extreme dataset types that anomaly and outlier detection methods must deal with: sparse data and high frequency data. It is hard to detect outliers in sparse data because there are few observations and it is hard to know which should be included in the definition of normal activity. Anomaly detection methods that rely on expected distributions for example are less robust when the data is sparse. On the other hand, when there are many observations or when the data is high frequency, anomalies can blend in.

One way to look for anomalies in time series data is to start by decomposing the time series using classical decomposition or STL. With seasonality and trend removed, spikes in the residuals are potential anomalies. An outlier detection test, such as one of the tests described in chapter 4, can be used to determine if a residual is an outlier.

Another way to look for anomalies in time series data is to use a model like ARIMA to predict the time series, and compare the predicted values to the true values. The model choice does not have to be a time series model like ARIMA, it could be a regression tree for example. Large differences between the predicted and observed value for a particular observation are an indicator that the observation could be an outlier.

Clustering methods are often used for time series anomaly detection. The idea behind using clustering to look for outliers is that observations that do not easily fit into clusters are outliers. There could also be a whole group of observations that form a cluster that differs greatly from all the other clusters.

Clustering is beyond the scope of this book, but it is worth mentioning that it can be used for time series anomaly detection.

Another anomaly detection technique that is common with business and clinical data is the use of control charts. A **control chart** (a.k.a. **individuals-moving range chart** or **I-MR chart**) shows how a process changes over time (Berardinelli, 2018). It is basically a sequence plot with upper and lower thresholds to define a range of normal behavior for the process. The upper and lower thresholds are often set to +/- 3 standard deviations from the mean. The mean +/- 1.5 times the interquartile range (IQR) is also commonly used. A set of heuristics is used to analyze the output of a control chart to look for anomalies in a process. The set of heuristics that is used is flexible, but here is a list of a few:

- One point beyond the control limit (outside 3 SD from the mean)
- Two of three consecutive points beyond 2 SD from the mean
- Four of five consecutive points beyond 1 SD from the mean
- Eight or more consecutive points on one side of the mean without crossing
- Six or more consecutive points that are all increasing or all decreasing

A process is said to be "in control" if there are no anomalies in the control chart and "out of control" when there are anomalies (Berardinelli, 2018). The anomalies in a control chart can be a sign of a meaningful change in the data. For example, a bank can plot a customer's transaction amounts over time in a control chart to look for changes in behavior. If the customer's transactional activity starts to go "out of control", then it may be a sign of something nefarious, like identity theft or money laundering. Points in a control chart that are outside 3 SD from the mean or at the end of a sequence of points that meet the criteria of one the heuristics listed above are said to be points of meaningful change. An example of a control chart is shown in figure 25.21.

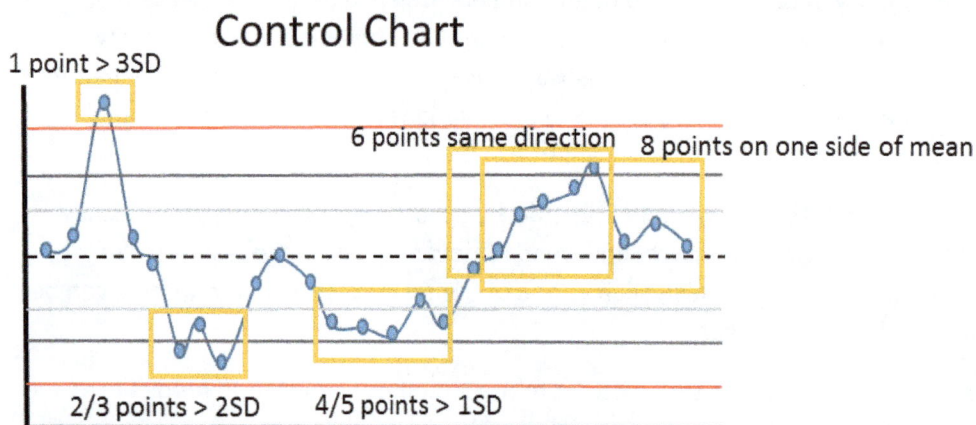

Figure 25.21

Control charts identify special cause variation. **Special cause variation** is defined as an unexpected real change that is not an essential part of a process, such as an unpredictable event or variation outside of the historical reference base. Special cause variation is the opposite of common cause variation. **Common cause variation** is random or expected variation that is not assignable to any specific cause and is a natural part of a process. Common cause variation does not change the predictability of a process. An example of common cause variation is normal wear and tear on an industrial machine that causes the machine to perform worse. An example of special cause variation is a faulty controller on an industrial machine that causes catastrophic failure.

Benford's Law

One anomaly detection method that is worth expounding upon is Benford's Law. Benford's Law is the assertion that the observed frequencies of digits within naturally occurring numbers should follow expected frequency distributions. Although it is being introduced in this chapter on time series, Benford's Law can be applied to any dataset. The law is named after physicist Frank Benford, who showed that naturally occurring numbers, like the surface areas of rivers, population sizes, physical constants, and molecular weights all have similar distributions of their digits (Nigrini, 2012). The first digit of a naturally occurring number, for example, is most frequently 1, followed by 2, and then 3, and so on, up to 9. The reason for this is that it is easy for small numbers to double in size, like doubling from 1 to 2. But each successive increase is harder. For the number 2 to double, 2 units are required. For 3 to double, 3 units are required. This holds true regardless of the order of magnitude. If this relationship were graphed, it would resemble a power law, and Benford's Law for the first digit of naturally occurring numbers is shown in figure 25.22, which resembles a discretized power law. The x-axis shows the first digit and the y-axis shows the probability of k appearing in naturally occurring numbers.

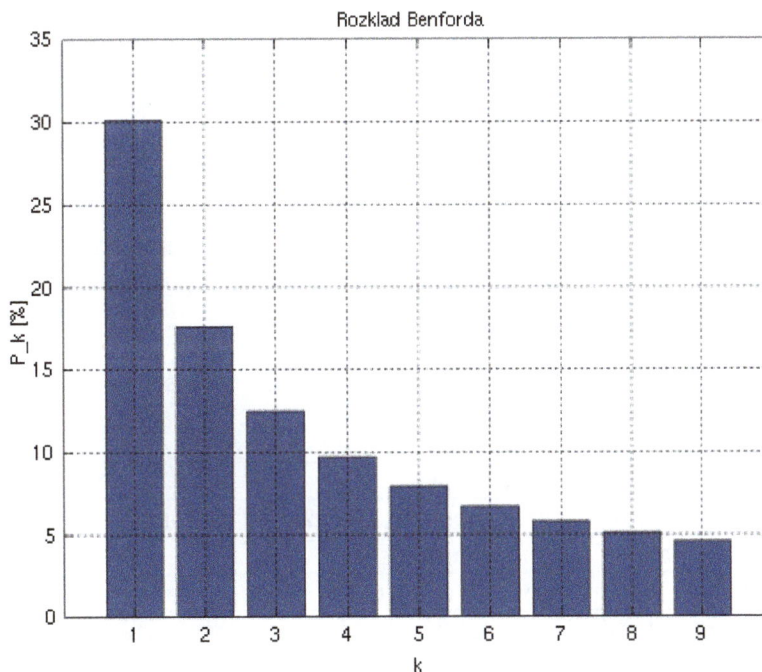

Figure 25.22

Since Benford's Law shows what the digit frequency distribution of naturally occurring numbers should look like, it is often used to detect manual manipulation of data or fraud. Benford's Law has been used to demonstrate election fraud in Iran, tax fraud, expense report fraud, and fraudulent financial statements (Nigrini, 2012). The reason it works is that people are bad at making up random numbers. It seems reasonable that when making up numbers, some of them should start with 8 or 9, but numbers starting with 8 or 9 occur only a fraction as often as numbers starting with 1 or 2. The digit distributions of manually created numbers therefore rarely resemble the digit distributions of naturally occurring numbers.

There are expected frequencies for the first digit, second digit, first two digits, and last two digits of numbers. Each of these has its own digit test. The formulas to calculate the expected proportions of each digit in a sample of numbers are shown in figure 25.23.

Test	Expected Ratio
First Digit	$$P(d_1) = \log_{10}\left(1 + \frac{1}{d_1}\right) \text{ for } d_1 \in \{1, \ldots, 9\}$$
Second Digit	$$P(d_2) = \sum_{d1=1}^{9} \log_{10}\left(1 + \frac{1}{d_1 d_2}\right) \text{ for } d_2 \in \{0, \ldots, 9\}$$
First Two Digits	$$P(d_1 d_2) = \log_{10}\left(1 + \frac{1}{(10 * d_1 + d_2)}\right) \text{ for } d_1 d_2 \in \{10, \ldots, 99\}$$
Last Two Digits	$$P(d_1 d_2) = 0.01$$

Figure 25.23

These formulas can be used to get the expected ratios of each digit. The expected ratio can be compared to the observed ratio using the z-test. The ratios can be converted to frequencies by multiplying by the sample size, and then the Chi-square test can be used instead. Since the digit frequencies are treated as categorical variables, either the Chi-square test or the mean or median absolute deviation (MAD) can be used to determine conformity to Benford's Law, given the expected and observed frequencies. These tests were described in chapter 4. The Chi-square test checks conformity for each digit test, whereas the MAD checks the overall conformity of all the digit tests.

Benford's Law relies on the assumption that there is at least one observation in each category. For example, there are 90 categories in the first two digit test, so the dataset should be large enough for each category to be adequately represented. Benford's Law expert Mark Nigrini recommends that each category should have at least 10 samples (Nigrini, 2012).

Each digit test is useful for targeting a particular type of anomaly. Figure 25.24 shows how the tests can be used.

Test	Use Case
First Digit	High level test for identifying obvious anomalies and overall conformity
Second Digit	High level test for identifying obvious anomalies
First Two Digits	Low level test that can identify individual samples for further review
Last Two Digits	Low level test that can identify individual samples for further review, and tests for rounding anomalies

Figure 25.24

In financial data, there are often normal patterns that might appear to make a dataset violate Benford's Law. For example, prices ending in 99 cents are very common in retail. Deposits or withdrawals from ATM's are always round dollar amounts of some multiple of 10, perhaps with an added service charge. These patterns would show spikes in the observed digit frequency distribution. One way to account for these explainable spikes would be to remove the observations from the dataset before plotting the digit frequency distribution. Another way would be to modify the equations in figure 25.23 to account for known fluctuations in the expected frequency distribution of the digits. This practice was introduced by forensic accountant Mark Nigrini as "My Law" (Nigrini, 2012). My Law is essentially a method for fitting the observed frequency distribution of digits to a custom expected frequency distribution. Like Benford's Law, a dataset's conformity to My Law can be tested with either the Chi-square test or MAD.

Time Series Segmentation: Change Point Detection and Regime Change

It is often desirable to partition a time series into discrete segments, where each segment has unique characteristics. For example, audio data may consist of different people holding a conversation, and the data could be segmented by speaker. Stock prices may consist of time periods with different attributes, like bear and bull markets. World events like terrorist attacks could shift the entire market, breaking stock price time series into time before the event and time after. There are two approaches to time series segmentation: change point detection and regime change.

Change point detection looks for break points in the time series between periods where some characteristic of the series changes. The simplest example would be a change point that causes the time series' moving average to shift, but other characteristics like the variance or spectral density can also be used to identify change points. Figure 25.25 shows an example of the results of running a change point detection algorithm to identify change points in a times series. The blue lines are the change points. The algorithm can be tuned to be more sensitive and find more change points, or fewer. It is critically important to note that change point detection does not give any insight into what caused a change – it only indicates that a change occurred. Looking at figure 25.25 for instance, we can only speculate why Apple's stock price had change points on those dates.

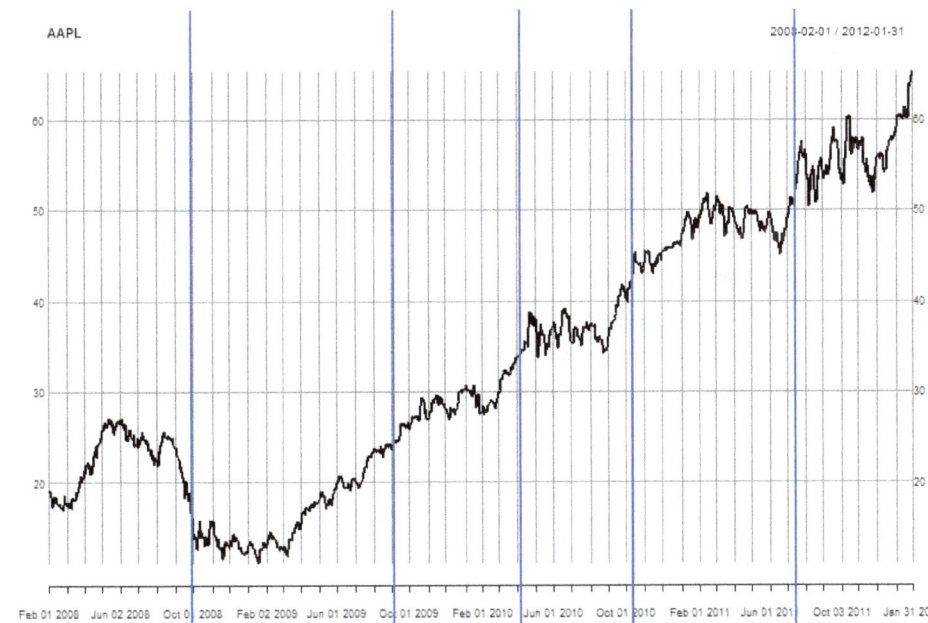

Figure 25.25

There are several approaches to change point detection. In this section, we will explore the e-divisive algorithm (ECP) and Bayesian change point detection (BCP).

The **e-divisive algorithm** uses the Euclidean distance to calculate a divergence measure between multivariate distributions. The equation, taken from the original paper (Matteson & James, 2013) is shown in figure 25.26.

$$\varepsilon(X, Y; \alpha) = 2E|X - Y|^{\alpha} - E|X - X'|^{\alpha} - E|Y - Y'|^{\alpha}$$

Figure 25.26

The equation in figure 25.26 leads to the lemma that for any pair of independent random vectors, X and Y, and for any alpha in the set (0, 2), if the expectation of $|X|^{\alpha} + |Y|^{\alpha} < \infty$, then the Euclidean distance between X and Y, given alpha, equals the nonparametric divergence measure D, the Euclidean distance is in the set [0, Inf), and if X and Y are identically distributed then the distance is 0. Matteson and James, the authors of the paper, go on to define an empirical divergence measure that is analogous to the equation above. If a scaled sample measure of divergence is calculated using the equation, a change point at location tau can be identified. The requirement for using the e-divisive algorithm is that there must be some pre-conceived notion of how many change points are expected to appear in the series.

Matteson and James wrote a follow up paper in which they introduced an R implementation of the e-divisive algorithm (James & Matteson, 2015) in the ECP package. The R implementation works by recursively partitioning a time series and using a permutation test to locate change points. Twitter developed its own version of the e-divisive algorithm using medians, so its implementation in the R package Breakout Detection is another popular alternative to the ECP package (Kipnis, 2015).

The key to remembering how the e-divisive algorithm works is knowing that the nonparametric divergence measure D, represents the **energy distance** between probability distributions (energy -> e-divisive).

Bayesian change point detection is carried out using Gibbs sampling, which was described in chapter 24. Suppose there is a sample of observations from a Poisson process (a counting process) where events occur randomly over time. The purpose of change point detection is to find when the occurrence rate of the events speeds up or slows down. For example, maybe the frequency of bank deposits is speeding up over a certain time interval. Change point detection seeks to determine when this speed up occurs and identify the occurrence rates before and after the shift to the time period of increased activity. If the observations are from a Poisson process, then they would be a sequence of counts having an average count from time step 1 to n, and a different average count from time step n+1 to N. The count at each time step can be modeled as a Poisson variable with a density function defined below (Yildirim, 2012).

$$Poisson(x; \lambda) = e^{x*\log(\lambda) - \lambda - \log(x!)}$$

Figure 25.27

In the equation in figure 25.27, lambda is the mean of the distribution. The mean is drawn from another distribution, called the Gamma distribution, with a density function defined below (Yildirim, 2012).

$$Gamma(\lambda; a, b) = e^{(a-1)*\log(\lambda) - b\lambda - \log(\Gamma(a)) + a*\log(b)}$$

Figure 25.28

Since the initial mean, λ_1, takes on a new value, λ_2, after time step n, a posterior distribution over λ_1, λ_2, and n can be inferred using Bayes theorem:

$$p(\lambda_1, \lambda_2, n | x_{1:N}) \propto p(x_{1:n}|\lambda_1)p(x_{n+1:N}|\lambda_2)p(\lambda_1)p(\lambda_2)p(n)$$

Figure 25.29

This equation shows the joint distribution of lambda sub 1, lambda sub 2, and n. Taking the log of this equation and substituting for each of the random variables (the Poisson and gamma distributions from figure 25.27 and 25.28) produces the full posterior.

$$\log\big(p(x_{1:n}|\lambda_1)p(x_{n+1:N}|\lambda_2)p(\lambda_1)p(\lambda_2)p(n)\big)$$
$$= \sum_{i=1}^{n}(x_i\log(\lambda_1) - \lambda_1 - \log(x_i!)) + \sum_{i=n+1}^{N}(x_i\log(\lambda_2) - \lambda_2 - \log(x_i!))$$
$$+ \Big((a-1)*\log(\lambda_1) - b\lambda_1 - \log(\Gamma(a)) + a*\log(b)\Big)$$
$$+ \Big((a-1)*\log(\lambda_2) - b\lambda_2 - \log(\Gamma(a)) + a*\log(b)\Big) - \log(N)$$

Figure 25.30

Now the posterior conditionals for lambda sub 1, lambda sub 2, and n can be found by extracting from the equation in figure 25.30 only the elements corresponding to each random variable. The simplified posterior conditionals are shown below.

$$\log(p(\lambda_1|n,\lambda_2,x_{1:N})) = \log\left(Gamma\left(a + \sum_{i=1}^{n}x_i, n+b\right)\right)$$

$$\log(p(\lambda_2|n,\lambda_1,x_{1:N})) = \log\left(Gamma\left(a + \sum_{i=n+1}^{N}x_i, N-n+b\right)\right)$$

$$\log(p(n|\lambda_1,\lambda_2,x_{1:N})) = \left(\sum_{i=1}^{n}x_i\right)*\log(\lambda_1) - n\lambda_1 + \left(\sum_{i=n+1}^{N}x_i\right)*\log(\lambda_2) - \lambda_2(N-n)$$

Figure 25.31

Now Gibbs sampling can be used to generate samples from the joint distribution in figure 25.29. The resulting sequence can be used to find the occurrence rates for each grouping of the observations between change points, as well as when the change points occurred.

Regime change is a segmentation method that assumes different segments of a time series are generated by a system with distinct parameters. These segments are called regimes. A hidden Markov model is used to infer the most probable segment locations within the series, because the regimes are hidden states. Refer back to the Python section of chapter 24 to see how a HMM can be used to detect regime change.

Time Series Analysis with R

The Air Passengers dataset contains monthly airline passenger numbers from 1949-1960 (Box, Jenkins, and Reinsel, 1976). If we load the dataset, perform some basic cleaning operations, and plot it, there is clear seasonality and a long term increasing trend.

```
library(tseries)
library(forecast)

d <- read.csv('AirPassengers.csv', header=T)
d$X <- NULL
plot(d$time, d$value, type='l')
```

If we plot the ACF and PACF, the first few lags appear to be correlated. We can test for stationarity by performing an Augmented Dickey-Fuller test. The ADF test p-value of 0.01 means that we can reject the null that the series is non-stationary. Be sure to note that the stationary may vary depending on the number of lags, k, used. By default, R uses the cube root of the number of observations minus one, which is 5 in this case.

```
#Look at ACF and PACF, and perform Augmented Dickey-Fuller test for stationarity
acf(d$value)
pacf(d$value)
adf.test(d$value)   #Stationary
```

The plot of the series shows both seasonality and a trend, so we may with to decompose the series to its residual before forecasting. If we do this, we would have to add the seasonality and trend back in to make any predictions though. This will be important to keep in mind. For now, let's just decompose the series. If we take the classical decomposition approach, we should start by determining if the trend is multiplicative or additive. Since the seasonal variation appears to increase as the time series increases, the data has a multiplicative trend. Therefore, when we fit a moving average to the trend, we will need to divide the series by the moving average to remove the trend. If it were additive, we would subtract instead. The window for the moving average could be anything, and we must choose wisely. Fortunately, it is pretty obvious from looking at a plot of the series that it has a monthly cycle, so a 12 month window for the moving average is appropriate.

```
#Smooth by using a centered moving average (data is monthly, so 12 month trend seems
appropriate)
trend <- ma(d$value, order=12, centre=T)
plot(d$time, d$value, type='l')
lines(d$time, trend, col='red')

#Remove the trend: remove multiplicative trends by dividing and additive trends by
subtracting
d_detrend <- d$value/trend
plot(d$time, d_detrend, type='l')
```

Averaging the seasonality over the 12 month period of the moving average and dividing the series by the result gets rid of the seasonality.

```
#Average the seasonality over the period of the MA trend (it was 12 for this dataset)
mseas <- t(matrix(d_detrend, nrow=12))  #Create a matrix of the time periods, where
columns are the periods
```

```
season <- colMeans(mseas, na.rm=T)         #Column means are the averaged seasonality
for each period
plot(as.ts(rep(season, 12)))

#Remove the seasonality: remove multiplicative seasonality by dividing and additive
seasonality by subtracting
d_random <- d$value / (trend*season)
plot(as.ts(d_random))
```

Note that everything we have done so far can be more easily accomplished through the decompose function:

```
#Do the same thing using the decompose function (requires a time series object) -
this is classical decompostion with MAs
?decompose
dts <- ts(d$value, frequency=12)   #Use 12 as frequency because data is measured
monthly so a 12 month period seems reasonable
d_decomp <- decompose(dts, "multiplicative")
d_decomp
plot(d_decomp)
```

No matter whether we do it manually or with the seasonal_decompose function, the results are the same. The residual series looks much more stationary than the original series. Let us now try LOESS decomposition using the stlm function.

```
#Do the same thing using LOESS decomposition
?stlm
dts <- ts(d$value, frequency=12)
d_stl_decomp <- stlm(dts, "periodic", allow.multiplicative.trend=T)
d_stl_decomp
d_stl_decomp_ts <- as.data.frame(d_stl_decomp$stl)
par(mfrow=c(4,1))
plot(d_stl_decomp_ts$Data, type='l', ylab='observed')
plot(d_stl_decomp_ts$Trend, type='l', ylab='trend')
plot(d_stl_decomp_ts$Seasonal12, type='l', ylab='seasonal')
plot(d_stl_decomp_ts$Remainder, type='l', ylab='random')
par(mfrow=c(1,1))

#Test for stationarity after seasonality and trend have been removed
adf.test(d_stl_decomp_ts$Remainder)   #Stationary
#Perform first differencing to see if that makes series stationary (don't need to
here, but it's good for example)
adf.test(diff(d_stl_decomp_ts$Remainder, differences=1))
```

A quick ADF test shows that the residual series is stationary. But if it were not, we may wish to try first differencing to see if that makes the series stationary. This can be performed using the diff function, as shown in the previous code block.

Now that we have seen how to decompose a time series, let's move on to ARIMA modeling. We will use the auto.arima function that tests every combination of orders (p, d, q) that are specified with the max.p, max.d, and max.q arguments, and returns the optimal ARIMA model using either the AIC or BIC as the evaluation metric. Let's try models up to (5, 5, 5) and use the BIC. Note that we are fitting the model to the residual series.

```
#Find best ARIMA model, using BIC, up to maximum orders (max.p, max.d, max.q for
ARIMA(p,d,q) models)
arima_model <- auto.arima(d_stl_decomp_ts$Remainder, max.p=5, max.d=5, max.q=5,
ic="bic")
plot(arima_model)
tsdisplay(residuals(arima_model), main="ARIMA(2,0,1) Residuals")
Box.test(residuals(arima_model), type='Ljung-Box')  #No evidence of autocorrelation
in the residuals
#Check residuals for normality
qqnorm(residuals(arima_model))
qqline(residuals(arima_model))
```

The Ljung-Box test looks for autocorrelation. If the model is a good fit, there should be no autocorrelation. We can use this model to forecast the residual series for the next 24 months. We can also rebuild the model, using a holdout test set to determine how accurate the ARIMA model is.

```
#Make future predictions for next 24 months
preds <- forecast(arima_model, h=24)
plot(preds)

#Do the same thing using a holdout test set to determine model accuracy
train <- ts(d[1:(144-24), 'value'], frequency=12)
test <- ts(d[(144-24):144, 'value'], frequency=12)
train_stl_decomp <- stlm(train, 'periodic', allow.multiplicative.trend=T)
test_stl_decomp <- stlm(test, 'periodic', allow.multiplicative.trend=T)
train_stl_decomp_ts <- as.data.frame(train_stl_decomp$stl)
test_stl_decomp_ts <- as.data.frame(test_stl_decomp$stl)
arima_model_train <- auto.arima(train_stl_decomp_ts$Remainder, ic='bic')
#arima_model_train <- arima(train_stl_decomp_ts$Remainder, order=c(2,0,1))
preds_test <- forecast(arima_model_train, h=24)
plot(preds_test, main="Forecasted (blue) vs Actual (red) for Last 24 Months")
lines(seq(120,144), test_stl_decomp_ts$Remainder, col='red')
```

Another common smoothing method is Holt-Winters exponential smoothing. We can try it here to compare it to MA and LOESS smoothing, and make smoothed forecasted predictions for the next 24 months.

```
#Holt-Winters exponential smoothing, similar to MA or LOESS smoothing, assumes trend
& seasonality are additive
hw_model <- HoltWinters(dts)
hw_model
plot(hw_model)
tsdisplay(residuals(hw_model), main="Holt-Winters Residuals")
Box.test(residuals(hw_model), type='Ljung-Box')  #There IS evidence of
autocorrelation in the residuals
#Check residuals for normality
qqnorm(residuals(hw_model))
qqline(residuals(hw_model))
#Plot forecasts
preds_hw <- forecast(hw_model, h=24)
plot(preds_hw)
```

All we have done so far is predict the future random values: the "core signal", or residual of the series. At this point, seasonality and trend would need to be added back to the model to transform the forecasts back to the original scale. Instead of doing all this work, we could have skipped decomposition altogether, since ARIMA models handle trends and seasonality with their AR, MA, and differencing components. So now we will just build an ARIMA model on the original time series and make forecasts at the proper scale.

```
arima_model_two <- auto.arima(dts, ic='bic')
preds_orig_scale <- forecast(arima_model_two, h=24)
plot(preds_orig_scale)
```

Let's now focus on modeling volatility with GARCH. To do this, we will download 4 years of S&P 500 daily closing prices.

```
library(quantmod)
library(timeSeries)
library(tseries)
library(forecast)
library(rugarch)
```

```
#Created differenced log returns of daily closing prices (Cl) for the S&P500 back to
2000
getSymbols("^GSPC", from="2000-01-01")
spret <- diff(log(Cl(GSPC)))
spret[as.character(head(index(Cl(GSPC)),1))] <- 0
```

First we will fit an ARIMA model and forecast the next 60 days of log returns.

```
#Fit ARIMA
arima_model <- auto.arima(spret, ic='bic')
arima_model
preds <- forecast(arima_model, h=60)   #Forecast next 60 days
plot(preds)
```

Now we will fit an ARIMA+GARCH model using the parameters of the best fitting ARIMA model and plot the predicted log returns vs actual, the volatility that was modeled, and the forecasted values.

```
#Fit ARIMA+GARCH
garch_model <- ugarchspec(mean.model=list(armaOrder=c(2, 0), include.mean=T),
                          variance.model=list(garchOrder=c(1,1)),
                          distribution.model='sged')
#Fit the model to all but the last 100 observations
garchfit <- ugarchfit(spec=garch_model, data=spret, solver='hybrid', out.sample=100)
garchfit@fit$coef   #Model coefficients
print(garchfit)
```

```
#Plot observed log returns vs predicted log returns - notice that the predicted
returns show what happends when volatility clustering is removed
plot(spret, main='Actual vs Fitted (red) Values')
lines(fitted(garchfit), col='red')
```

```
#Plot the volatility by itself
plot(garchfit@fit$sigma, type='l', ylab='Volatility')
```

```
#Plot the volatility with the observed log returns
par(mfrow=c(2,1))
plot(spret, type='l', main='S&P500 Daily Log Returns')
```

```
plot(garchfit, which=3)
par(mfrow=c(1,1))

#Plot forecasts
garch_preds <- ugarchforecast(garchfit, n.ahead=100, n.roll=100, out.sample=100)
plot(garch_preds)
```

With the last example, we have started getting into modeling financial time series. In this chapter, we explored a pair trading strategy that relies on mean reversion of cointegrated stocks. Now we will put this strategy into practice. We will cheat a little bit and pick 2 stocks that we know to be cointegrated. In the real world, an analyst would have to make a guess and then perform the cointegration test that we will define below, several times until a good pair is found.

```
getSymbols("GOOGL", from="2012-05-01", to="2018-05-01")
getSymbols("GOOG", from="2012-05-01", to="2018-05-01")
googlp <- unclass(GOOGL$GOOGL.Adjusted)
googp <- unclass(GOOG$GOOG.Adjusted)
plot(googlp, type='l', col='purple')
lines(googp, type='l', col='orange')
plot(googlp, googp)  #This should be generally linear
```

To test for cointegration, we need to try fitting 2 linear regressions. Each regression will have one of the stocks as the dependent variable and the other as the independent variable. If the coefficients of both regressions are equal, then the hedge ratio is simply the regression coefficient. This almost never happens though. If the coefficients are not equal (which is usually the case), then the hedge ratio will be the coefficient from the regression with the smaller ADF statistic. The logic is that the regression with the smaller ADF statistic is more stationary, and our goal is to use a mean reversion strategy, which relies on the assumption that the series is stationary.

```
#Test for cointegration by regressin each on the other
googlm <- lm(googlp~googp)
googm <- lm(googp~googlp)
adf.test(googlm$residuals, k=1)  #Cannot reject null for stationary series ->
nonstationary so cannot use cointegration
adf.test(googm$residuals, k=1)   #Cannot reject null for stationary series ->
nonstationary so cannot use cointegration
```

An interesting outcome has occurred: neither of the regressions produced stationary residuals! That means we cannot use these stocks for a pairs trading strategy, because we cannot rely on mean reversion. To run the test again, we will make a function to make the process faster, and we will use two stocks that have been shown in literature to be cointegrated, over a time period during which they are known to cointegrate nicely and be stationary.

```
#To try again for 2 other stocks, let's make a function
coint_test <- function(stock1, stock2) {
  stock1m <- lm(stock1~stock2)
  stock2m <- lm(stock2~stock1)
  adf1 <- adf.test(stock1m$residuals, k=1)
  adf2 <- adf.test(stock2m$residuals, k=1)
  if (adf1$p.value>=0.05 & adf2$p.value>=0.05) {
    return(print('Non-stationary, cannot do cointegration'))
  } else if (adf1$statistic < adf2$statistic) {
```

```
    return(print(paste0('Use ', colnames(stock1), ' as the dependent variable in the
linear combination.')))
  } else {
    return(print(paste0('Use ', colnames(stock2), ' as the dependent variable in the
linear combination.')))
  }
}

#Let's cheat and pick 2 that are known to be cointegrated to see what happens
#We'll also cherry pick a time frame during which these stocks were stationary
getSymbols("EWA", from="2008-02-01", to="2012-02-01")
getSymbols("EWC", from="2008-02-01", to="2012-02-01")
ewap <- unclass(EWA$EWA.Adjusted)
ewcp <- unclass(EWC$EWC.Adjusted)
par(mfrow=c(2,1))
plot(ewap, type='l', col='purple')
plot(ewcp, type='l', col='orange')
par(mfrow=c(1,1))
plot(ewap, ewcp)  #This should be generally linear
coint_test(ewap, ewcp)
ewa_hedge_model <- lm(ewap~ewcp)
summary(ewa_hedge_model)
```

The stocks we picked are two exchange traded funds (ETFs) that are known to be cointegrated. We checked for cointegration using linear regression and found the ideal hedge ratio for the 4 year time period between February 1, 2008 and Februrary 1, 2012. The regression of EWA on EWC resulted in a hedge ratio of 0.71, meaning that however EWA and EWC change, they should revert to a mean ratio of 0.71. This information can be used to long or short either stock if the observed ratio deviates too far from the mean. For example, suppose we construct a confidence interval for the hedge ratio and the observed ratio breaches the upper bound. This would mean either EWA is overpriced or EWC is underpriced. The two stocks can be traded according to the situation. The amounts bought or shorted depend on how much money is invested between the two stocks, but the idea is for the pair to revert to the mean hedge ratio.

Note that linear regression finds the mean hedge ratio between 2 stocks over a time period. If the hedge ratio changes over time, then it must be smoothed. One way to do this is to use a Kalman filter. Another thing to note about the example here is that we only tested for cointegrated pairs. If it is desirable to use a basket of stocks, then linear regression can again be used to find the hedge ratios, and the Kalman filter can again be used to smooth dynamic (changing) hedge ratios. The difference between hedging 2 stocks and a basket is the stationary test: for a basket, the Johannson test must be used, whereas the ADF test is used for 2 stocks. One final note about pairs trading is that the hedge ratio is found through linear regression, which means that it can be assumed to be drawn from a normal distribution. This means that a confidence interval can be constructed for the hedge ratio.

We have explored pairs trading in detail. Now let's move on to testing a dataset's conformity to Benford's Law to look for potential manual manipulation. For this exercise, we will use the air passenger dataset again. I have developed a package to quickly perform the first digit test, first two digit test, and last two digit test that can be downloaded from GitHub: https://github.com/nlinc1905/benfordsLaw. We will use a dataset of city and town populations from the US Census Bureau, because it conforms well to Benford's Law.

```
library(benfordsLaw)

census <-  read.csv("Census_Town&CityPopulations.csv", header=TRUE)
#Isolate town population and remove commas and NAs
x <- as.numeric(sub( ",", "", census$X7_2009))
x <- x[!is.na(x)]

first_digit <- benfordFirstDigit(x)
first_two_digits <- benfordFirstTwoDigit(x)
last_two_digits <- benfordLastTwoDigit(x)
plotFD(first_digit)
plotF2D(first_two_digits)
plotL2D(last_two_digits)
```

As the plots show, this dataset conforms closely to Benford's Law for all digit tests. The dataframes produced by the test functions contain the observed and expected frequencies and ratios, as well as the MAD value. The abnormal_flag shows the results of the z-test. If the flag is true, then the digit was statistically significant. Notice that even though the dataset conforms to Benford's Law overall, a few digit tests were abnormal. This is natural. With Benford's Law, it is up to the researcher to use best judgement in determining whether any abnormalities warrant further investigation. In this case, they are simply benign statistical anomalies.

We will now move on to time series segmentation with change point detection. For this exercise, we will explore Apple Inc's stock price from February 1, 2008 to February 1, 2012. Using the e-divisive with medians algorithm from the ecp package, we can calculate the change points in the series. The min.size argument alters the number of change points detected. The smaller the minimum size of the observations between change points, the more change points will be detected. We will use min.size=120, because this algorithm takes a long time for smaller values of min size.

```
library(ecp)

getSymbols("AAPL", from="2008-02-01", to="2012-02-01")
AAPL <- Cl(AAPL)

#AAPL test
aapl_ecp <- e.divisive(X=AAPL, min.size=120)
plot(AAPL)
abline(v=.index(AAPL)[aapl_ecp$estimates[-c(1, length(aapl_ecp$estimates))]],
col='blue', lwd=2)
```

The resulting plot shows where the change points in Apple's stock price occurred. Note that these changes are supposed to represent fundamental shifts in the price, not short term shocks. We can explore characteristics about the series between these change points to understand them better. For example, let's look at the median and mean closing prices for each segment, as well as the percent changes in the medians and means from one segment to the next.

```
#Get median and mean closing prices for each segment (area between change points)
aapl_segments <- AAPL
aapl_segments$AAPL.Segment <- aapl_ecp$cluster
aapl_seg_meds <- aggregate(aapl_segments, by=list(aapl_segments$AAPL.Segment),
FUN=median)
```

```
aapl_seg_means <- aggregate(aapl_segments, by=list(aapl_segments$AAPL.Segment),
FUN=mean)
#Find differences between segments as percent change from prior segment
diff(aapl_seg_meds)/aapl_seg_meds*100
diff(aapl_seg_means)/aapl_seg_means*100
```

Time Series Analysis with Python

The Air Passengers dataset contains monthly airline passenger numbers from 1949-1960 (Box, Jenkins, and Reinsel, 1976). If we load the dataset, perform some basic cleaning operations, and plot it, there is clear seasonality and a long term increasing trend.

```
import pandas as pd
import numpy as np
import matplotlib.pyplot as plt
from statsmodels.graphics.tsaplots import plot_acf, plot_pacf
from statsmodels.tsa.stattools import adfuller
from statsmodels.tsa.seasonal import seasonal_decompose
from statsmodels.stats.diagnostic import acorr_ljungbox


d = pd.read_csv('AirPassengers.csv')

#Fix the dates
months = ['Jan', 'Feb', 'Mar', 'Apr', 'May', 'Jun', 'Jul', 'Aug', 'Sep', 'Oct',
'Nov', 'Dec']
d['Month'] = months*12
d['time'] = d['time'].astype('str')
d['time'] = d['time'].apply(lambda x: x[0:4])
d['time'] = d['time']+'-'+d['Month']
d['time'] = d['time'].apply(lambda x: pd.datetime.strptime(x, '%Y-%b'))
d.index = d.time
del d['Month'], d['Unnamed: 0'], d['time']
print(d.head())
print(d.dtypes)

#Convert to series to simplify future commands
dts = pd.Series(d['value'])
#print(dts['1949'])
plt.plot(dts)
plt.show()
```

If we plot the ACF and PACF, the first few lags appear to be correlated. We can test for stationarity by performing an Augmented Dickey-Fuller test. The ADF test p-value of 0.99 means that we cannot reject the null that the series is non-stationary. Note that this is the exact opposite of the result of the ADF test performed with R. The reason is that statsmodels used 13 lags, as shown by the output, whereas R defaults to the cube root of the number of observations minus 1, which is 5 lags in this case. If we go back and change the number of lags to 13 with R by setting the k argument equal to 13, the test concurs with statsmodels that the series is non-stationary.

The plot of the series shows both seasonality and a trend though, so we may be able to convert the series to a stationary one through decomposition. The residual series might be stationary. If we do this, we would have to add the seasonality and trend back in to make any predictions though. This will be important to keep in mind. For now, let's just decompose the series.

```
#Look at ACF and PACF, and perform Augmented Dickey-Fuller test for stationarity
plot_acf(dts)
plt.show()
plot_acf(dts, lags=30)
plt.show()
plot_pacf(dts)
plt.show()
plot_pacf(dts, lags=30)
plt.show()
adftest = adfuller(dts, autolag='AIC')
print('Augmented Dickey-Fuller Test Results',
      '\nTest Statistic:', adftest[0],
      '\np-value:', adftest[1],
      '\nLags Used:', adftest[2],
      '\nNumber of Observations:', adftest[3])
#Cannot reject the null that series is non-stationary
```

If we take the classical decomposition approach, we should start by determining if the trend is multiplicative or additive. Since the seasonal variation appears to increase as the time series increases, the data has a multiplicative trend. Therefore, when we fit a moving average to the trend, we will need to divide the series by the moving average to remove the trend. If it were additive, we would subtract instead. The window for the moving average could be anything, and we must choose wisely. Fortunately, it is pretty obvious from looking at a plot of the series that it has a monthly cycle, so a 12 month window for the moving average is appropriate.

```
#Smooth by using a centered moving average (data is monthly, so 12 month trend seems
appropriate)
trend = dts.rolling(12).mean()
plt.plot(dts)
plt.plot(trend)
plt.show()
```

```
#Remove the trend: remove multiplicative trends by dividing and additive trends by
subtracting
dts_detrend = dts/trend
plt.plot(dts_detrend)
plt.show()
```

Averaging the seasonality over the 12 month period of the moving average and dividing the series by the result gets rid of the seasonality.

```
#Average the seasonality over the period of the MA trend (it was 12 for this dataset)
mseas = np.asarray(dts_detrend).reshape(12, 12).T  #Create a matrix of the time
periods, where columns are the periods
season = np.nanmean(mseas, axis=1)                     #Column means are the averaged
seasonality for each period
season = np.asarray(list(season)*12)
plt.plot(season)
plt.show()
```

```
#Remove the seasonality: remove multiplicative seasonality by dividing and additive
seasonality by subtracting
dts_random = dts / (trend*season)
plt.plot(dts_random)
```

```
plt.show()
```

Note that everything we have done so far can be more easily accomplished through the seasonal_decompose function:

```
#Do the same thing using the seasonal_decompose function (requires a pandas object) -
this is classical decompostion with MAs
help(seasonal_decompose)
dts_decomp = seasonal_decompose(dts, model='multiplicative')
dts_decomp.plot()
plt.show()
```

No matter whether we do it manually or with the seasonal_decompose function, the results are the same. The residual series looks much more stationary than the original series. Let us now try LOESS decomposition. We will need to define a custom function to perform LOESS decomposition. This function was adapted from GitHub ().

```
def stl_decompose(df, period=365, lo_frac=0.6, lo_delta=0.01,
allow_multiplicative_trend=False):
    '''Create a seasonal-trend (with Loess, aka "STL") decomposition of observed time
series data.
    This implementation is modeled after the ``statsmodels.tsa.seasonal_decompose``
method
    but substitutes a Lowess regression for a convolution in its trend estimation.
    Defaults to an additive model, Y[t] = T[t] + S[t] + e[t]
    For more details on lo_frac and lo_delta, see:
    `statsmodels.nonparametric.smoothers_lowess.lowess()`
    Args:
        df (pandas.Dataframe): Time series of observed counts. This DataFrame must be
continuous (no
            gaps or missing data), and include a ``pandas.DatetimeIndex``.
        period (int, optional): Most significant periodicity in the observed time
series, in units of
            1 observation. Ex: to accomodate strong annual periodicity within years
of daily
            observations, ``period=365``.
        lo_frac (float, optional): Fraction of data to use in fitting Lowess
regression.
        lo_delta (float, optional): Fractional distance within which to use linear-
interpolation
            instead of weighted regression. Using non-zero ``lo_delta`` significantly
decreases
            computation time.
    Returns:
        `statsmodels.tsa.seasonal.DecomposeResult`: An object with DataFrame
attributes for the
            seasonal, trend, and residual components, as well as the average seasonal
cycle.
    '''
    import numpy as np
    import pandas as pd
    from pandas.core.nanops import nanmean as pd_nanmean
    from statsmodels.tsa.seasonal import DecomposeResult
    from statsmodels.tsa.filters._utils import _maybe_get_pandas_wrapper_freq
    import statsmodels.api as sm
```

```
    # use some existing pieces of statsmodels
    lowess = sm.nonparametric.lowess
    _pandas_wrapper, _ = _maybe_get_pandas_wrapper_freq(df)

    # get plain np array
    observed = np.asanyarray(df).squeeze()

    # calc trend, remove from observation
    trend = lowess(observed, [x for x in range(len(observed))],
                   frac=lo_frac,
                   delta=lo_delta * len(observed),
                   return_sorted=False)
    detrended = observed - trend

    # calc one-period seasonality, remove tiled array from detrended
    period_averages = np.array([pd_nanmean(detrended[i::period]) for i in
range(period)])
    # 0-center the period avgs
    period_averages -= np.mean(period_averages)
    seasonal = np.tile(period_averages, len(observed) // period + 1)[:len(observed)]
    resid = detrended - seasonal

    # convert the arrays back to appropriate dataframes, stuff them back into
    #  the statsmodel object
    results = list(map(_pandas_wrapper, [seasonal, trend, resid, observed]))
    dr = DecomposeResult(seasonal=results[0],
                         trend=results[1],
                         resid=results[2],
                         observed=results[3],
                         period_averages=period_averages)
    return dr

stl = stl_decompose(dts, period=12, allow_multiplicative_trend=False)
stl.plot()
plt.show()
```

A quick test for stationarity reveals that the residual series is still non-stationary. So let's try first differencing and applying the ADF test again to see if the series is difference stationary.

```
#Test for stationarity after seasonality and trend have been removed
adftest = adfuller(stl.resid, autolag='AIC')
print('Augmented Dickey-Fuller Test Results',
      '\nTest Statistic:', adftest[0],
      '\np-value:', adftest[1],
      '\nLags Used:', adftest[2],
      '\nNumber of Observations:', adftest[3])
#Cannot reject the null that series is non-stationary

#Define differencing function
def diff(dataset, differences=1):
    diff = list()
    for i in range(differences, len(dataset)):
        value = dataset[i] - dataset[i - differences]
        diff.append(value)
```

```
    return pd.Series(diff)

#Perform first differencing to see if that makes series stationary
adftest = adfuller(diff(stl.resid, differences=1), autolag='AIC')
print('Augmented Dickey-Fuller Test Results',
      '\nTest Statistic:', adftest[0],
      '\np-value:', adftest[1],
      '\nLags Used:', adftest[2],
      '\nNumber of Observations:', adftest[3])
#Can reject the null that series is non-stationary, the series is difference
stationary
```

So it appears the series is difference stationary. Now that we have seen how to decompose a time series, let's move on to ARIMA modeling. Python does not have an auto ARIMA function like R, so let's make one. Then we can apply it to find the best fitting ARIMA model and plot the fitted values and residuals. Instead of fitting the decomposed series and adding trend and seasonality back in to make forecasts later, we will let the ARIMA model trend and seasonality with its AR and MA components. That will makes things much simpler.

```
#Replicate the auto.arima function from R - it uses grid search to optimize AIC or
BIC
def auto_arima(ts, max_p=5, max_d=5, max_q=5, ic='bic'):
    from statsmodels.tsa.arima_model import ARIMA, ARIMAResults
    import warnings
    with warnings.catch_warnings():
        warnings.simplefilter('ignore')  #Ignores all convergence warnings
        best_ic = 1e24
        best_params = (0,0,0)
        for p in range(max_p):
            for d in range(max_d):
                for q in range(max_q):
                    try:
                        model = ARIMA(ts.astype('float32'), order=(p+1, d+1, q+1))
                        fitted = model.fit()
                        if ic=='aic' and fitted.aic < best_ic:
                            best_ic = fitted.aic
                            best_params = (p+1, d+1, q+1)
                            best_model = fitted
                        elif ic=='bic' and fitted.bic < best_ic:
                            best_ic = fitted.bic
                            best_params = (p+1, d+1, q+1)
                            best_model = fitted
                    except:
                        continue
    return best_model, best_params, best_ic

arima_model, arima_params, arima_ic = auto_arima(dts, max_p=7, max_d=2, max_q=5,
ic='bic')
plt.plot(arima_model.fittedvalues)
plt.title('ARIMA Fitted Values')
plt.show()
plt.plot(arima_model.resid)
plt.title('ARIMA Redisuals')
plt.show()
```

```
print('Best ARIMA model had params', arima_params, 'and BIC', arima_ic)
#p-values are cumulative for this test, so restrict lags or get spurious significance
in distant lags
lag_test_stats, lag_p_vals = acorr_ljungbox(arima_model.resid, lags=10,
boxpierce=False)
```

Now let's forecast the number of airline passengers for the next 24 months.

```
#Make future predictions for next 24 months
preds = arima_model.forecast(steps=24) #Returns tuple of preds, stderr, conf_int
plt.plot(np.array(dts))
plt.plot(np.array(range(144, 144+24)), preds[0])  #Shift preds x-axis bc there are
144 obs in original data
plt.title('ARIMA Predictions for Next 24 Months')
plt.show()
```

Let's now focus on modeling volatility with GARCH. To do this, we will download 4 years of S&P 500 daily closing prices.

```
import datetime as dt
import numpy as np
import pandas as pd
import pandas_datareader.data as web
import matplotlib.pyplot as plt
import statsmodels.tsa.api as smt
from arch import arch_model

start_dt = dt.datetime(2012, 1, 1)
end_dt = dt.datetime(2016, 1, 1)

d = web.DataReader('SPY', 'morningstar', start_dt, end_dt)
dts = pd.Series(d['Close'])
dts.index= d.index.levels[1]
log_returns = pd.Series(np.log(dts/dts.shift(1)).dropna())
#Note that log returns = percent change
#print(d.pct_change().dropna())
log_returns.plot()
plt.show()
dts.plot()
plt.show()
```

First we will fit an ARIMA model and forecast the next 60 days of log returns.

```
def auto_arima(ts, max_p=5, max_d=5, max_q=5, ic='bic'):
    from statsmodels.tsa.arima_model import ARIMA, ARIMAResults
    import warnings
    with warnings.catch_warnings():
        warnings.simplefilter('ignore')  #Ignores all convergence warnings
        best_ic = 1e24
        best_params = (0,0,0)
        for p in range(max_p):
            for d in range(max_d):
                for q in range(max_q):
                    try:
                        model = ARIMA(ts.astype('float32'), order=(p+1, d+1, q+1))
```

```
                    fitted = model.fit()
                    if ic=='aic' and fitted.aic < best_ic:
                        best_ic = fitted.aic
                        best_params = (p+1, d+1, q+1)
                        best_model = fitted
                    elif ic=='bic' and fitted.bic < best_ic:
                        best_ic = fitted.bic
                        best_params = (p+1, d+1, q+1)
                        best_model = fitted
                except:
                    continue
    return best_model, best_params, best_ic

arima_model, arima_params, arima_ic = auto_arima(log_returns, max_p=7, max_d=2,
max_q=5, ic='bic')
plt.plot(arima_model.fittedvalues)
plt.title('ARIMA Fitted Values')
plt.show()
plt.plot(arima_model.resid)
plt.title('ARIMA Redisuals')
plt.show()
print('Best ARIMA model had params', arima_params, 'and BIC', arima_ic)
#p-values are cumulative for this test, so restrict lags or get spurious significance
in distant lags
lag_test_stats, lag_p_vals = acorr_ljungbox(arima_model.resid, lags=10,
boxpierce=False)

#Make future predictions for next 60 days
preds = arima_model.forecast(steps=60) #Returns tuple of preds, stderr, conf_int
plt.plot(np.array(log_returns))
plt.plot(np.array(range(len(dts), len(dts)+60)), preds[0])  #Shift preds x-axis
plt.title('ARIMA Predictions for Next 60 Days')
plt.show()
```

Now we will fit an ARIMA+GARCH model using the parameters of the best fitting ARIMA model and plot the predicted volatility for the next 60 days of log returns.

```
#Now fit a GARCH model, using the parameters of the best fitting ARIMA model
garch_model = arch_model(log_returns, p=arima_params[0], o=arima_params[1],
q=arima_params[2], dist='StudentsT')
garch_fitted = garch_model.fit()
print(garch_fitted.summary())
plt.plot(garch_fitted.resid)
plt.title('GARCH Redisuals')
plt.show()

plot_acf(garch_fitted.resid, lags=30)
plt.show()
plot_pacf(garch_fitted.resid, lags=30)
plt.show()

#View predicted volatility
garch_fitted.hedgehog_plot(horizon=60)
plt.show()
```

With the last example, we have started getting into modeling financial time series. In this chapter, we explored a pair trading strategy that relies on mean reversion of cointegrated stocks. Now we will put this strategy into practice. We will cheat a little bit and pick 2 stocks that we know to be cointegrated. In the real world, an analyst would have to make a guess and then perform the cointegration test that we will define below, several times until a good pair is found.

```
#Let's cheat and pick 2 that are known to be cointegrated to see what happens
#We'll also cherry pick a time frame during which these stocks were stationary
start_dt = dt.datetime(2008, 2, 1)
end_dt = dt.datetime(2012, 2, 1)

ewap = web.DataReader('EWA', 'morningstar', start_dt, end_dt)
ewcp = web.DataReader('EWC', 'morningstar', start_dt, end_dt)

ewapts = pd.Series(ewap['Close'])
ewcpts = pd.Series(ewcp['Close'])
ewapts.index = ewap.index.levels[1]
ewcpts.index = ewcp.index.levels[1]

plt.plot(ewapts)
plt.plot(ewcpts)
plt.show()

#Define function to check for cointegration using linear regression
def coint_test(stockA, stockB):
    from sklearn.linear_model import LinearRegression
    from statsmodels.tsa.stattools import adfuller
    stockA = stockA.values.reshape(-1, 1)
    stockB = stockB.values.reshape(-1, 1)
    stockAm = LinearRegression().fit(stockB, stockA)
    stockBm = LinearRegression().fit(stockA, stockB)
    stockAm_resid = stockA-stockAm.predict(stockB)
    stockBm_resid = stockB-stockBm.predict(stockA)
    adfA = adfuller(stockAm_resid.reshape(stockAm_resid.shape[0],), autolag='AIC')
    adfB = adfuller(stockBm_resid.reshape(stockBm_resid.shape[0],), autolag='AIC')
    if (adfA[1] >= 0.05 and adfB[1] >= 0.05):
        print('Non-stationary, cannot do cointegration')
    elif adfA[0] < adfB[0]:
        print('Use the first input as the dependent variable in the linear
combination.')
    else:
        print('Use the second input as the dependent variable in the linear
combination.')

coint_test(ewapts, ewcpts)
from sklearn.linear_model import LinearRegression
ewa_hedge_model = LinearRegression().fit(ewcpts.values.reshape(-1, 1),
ewapts.values.reshape(-1, 1))
print('Hedge ratio:', ewa_hedge_model.coef_)
```

The stocks we picked are two exchange traded funds (ETFs) that are known to be cointegrated. We checked for cointegration using linear regression and found the ideal hedge ratio for the 4 year time period between February 1, 2008 and Februrary 1, 2012. The regression of EWA on EWC resulted in a hedge ratio of 0.71, meaning that however EWA and EWC change, they should revert to a mean ratio of 0.71. This information can be used to long or short either stock if the observed ratio deviates too far from the mean. For example, suppose we construct a confidence interval for the hedge ratio and the observed ratio breaches the upper bound. This would mean either EWA is overpriced or EWC is underpriced. The two stocks can be traded according to the situation. The amounts bought or shorted depend on how much money is invested between the two stocks, but the idea is for the pair to revert to the mean hedge ratio.

Note that linear regression finds the mean hedge ratio between 2 stocks over a time period. If the hedge ratio changes over time, then it must be smoothed. One way to do this is to use a Kalman filter. Another thing to note about the example here is that we only tested for cointegrated pairs. If it is desirable to use a basket of stocks, then linear regression can again be used to find the hedge ratios, and the Kalman filter can again be used to smooth dynamic (changing) hedge ratios. The difference between hedging 2 stocks and a basket is the stationary test: for a basket, the Johannson test must be used, whereas the ADF test is used for 2 stocks. One final note about pairs trading is that the hedge ratio is found through linear regression, which means that it can be assumed to be drawn from a normal distribution. This means that a confidence interval can be constructed for the hedge ratio.

As mentioned, the Kalman filter that was explained in chapter 24 can be used to find a dynamic hedge ratio. Let's use the same 2 ETFs and find the dynamic hedge ratio with a Kalman filter. The trick here is using a Kalman filter to calculate the dynamic coefficient of a linear regression. We will plot the results to see the ideal hedge ratio for every point in time between Feburary 1, 2008 and Feburary 1, 2012.

```
#Use a Kalman filter to estimate the dynamic hedge ratio
from pykalman import KalmanFilter
def dynamic_regression_w_kalman(x, y):
    #w_t = covariance matrix of the noise term - decrease it to produce smoother
result (the smaller, the less sensitive KF is to true values)
    w_t = 1e-5
    #A_t or state transition matrix is the matrix of beta coefficients for the
regression - this will start off as a matrix of diagonal noise terms to simulate
random walk
    state_transition_matrix = w_t/(1-w_t)*np.eye(2)
    #x_t or the observed matrix is the matrix of values of the independent variable
on the left and the dependent variable on the right: [x, y] - initialize dependent
vars as ones
    observation_matrix = np.vstack([x, np.ones(y.shape)]).T[:, np.newaxis]

    kf = KalmanFilter(n_dim_obs=1, n_dim_state=2,
                      initial_state_mean=np.zeros(2),
                      initial_state_covariance=np.ones((2, 2)),
                      transition_matrices=np.eye(2),
                      transition_covariance=state_transition_matrix,
                      observation_matrices=observation_matrix,
                      observation_covariance=1.0)

    return kf.filter(y.values)

state_means, state_covariances = dynamic_regression_w_kalman(ewcpts, ewapts)
dynamic_reg_params = pd.DataFrame({
        'slope': state_means[:, 0],
```

```
        'intercept': state_means[:, 1]
        }, index=ewapts.index)
dynamic_reg_params.plot()
plt.title('Dynamic Hedge Ratio of EWA to EWC')
plt.show()
```

We have explored pairs trading in detail. Now let's move on to time series segmentation with change point detection. For this exercise, we will explore Apple Inc's stock price from February 1, 2008 to February 1, 2017. Using the Pelt function from the ruptures library, we can calculate the change points in the series. The penalty (pen) argument alters the number of change points detected. The smaller the penalty, the more change points. Note that the final change point needs to be dropped, as it is the end of the series.

```
import datetime as dt
import numpy as np
import pandas as pd
import pandas_datareader.data as web
import matplotlib.pyplot as plt
import ruptures as rpt

start_dt = dt.datetime(2008, 2, 1)
end_dt = dt.datetime(2017, 2, 1)

aapl = web.DataReader('AAPL', 'morningstar', start_dt, end_dt)

aaplpts = pd.Series(aapl['Close'])
aaplpts.index = aapl.index.levels[1]

plt.plot(aaplpts)
plt.show()

cp_model = rpt.Pelt(model='rbf', min_size=2).fit(aaplpts.values)
cp_preds = cp_model.predict(pen=3)  #Smaller penalty -> more change points
cp_preds = cp_preds[:-1]  #Drop the final change point, as it is at the end of the
series

plt.plot(aaplpts)
plt.title('Change Points in AAPL Price')
for cp in cp_preds:
    plt.axvline(x=aaplpts.index[cp], color='purple', linestyle='--')
plt.show()

print('Change points occurred at:')
for cp in cp_preds:
    print(aaplpts.index[cp])
```

The resulting plot shows where the change points in Apple's stock price occurred. Note that these changes are supposed to represent fundamental shifts in the price, not short term shocks.

26. Signal Processing

In this chapter, we will take an introductory look at signal processing. Signals are a synonym for time series data: they are measurements of a single continuous variable over time. Signals are usually distinguished from time series analysis though, because signal processing encompasses both time series and frequency analysis of signal data. Signal processing is where data science meets physics. Many of the signals we will explore in this chapter, like light and sound, are physical phenomenon.

Basics of Wave Physics

The difference between a signal and a wave is that a signal carries information, whereas a wave is a disturbance in a medium caused by energy transfer. Signals are functions of waves. So in order to understand signals, it is necessary to know a little bit about how waves work.

There are two types of waves: transverse and longitudinal. The difference between them is the direction of displacement of the medium through which the wave propagates. Transverse waves displace a medium perpendicular to the direction of the wave. Transverse waves are like ripples in a pond after a fish has jumped. Electromagnetic waves (light) are transverse waves. Longitudinal waves displace a medium parallel to the direction of the wave. Sound propagates as a longitudinal wave. Longitudinal waves can be thought of as Slinkys, as shown in figure 26.1.

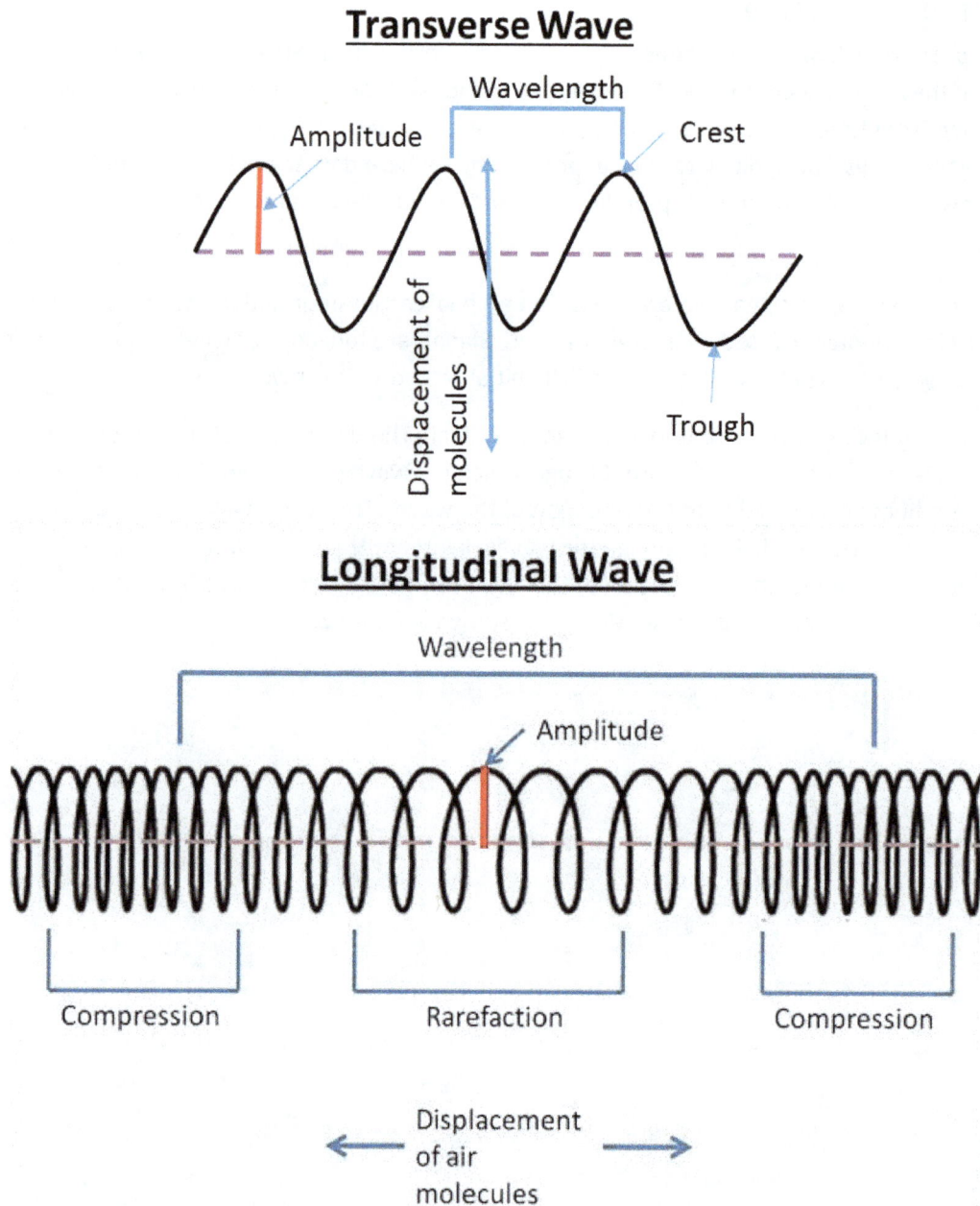

Transverse Wave

Longitudinal Wave

Figure 26.1

Amplitude is the magnitude or power of a wave. The wavelength of a transverse wave is the distance between two successive crests or two successive troughs. The wavelength of a longitudinal wave is the distance between two successive compressions or two successive rarefactions. Wavelength is also called the wave cycle. **Frequency** is the number of wave cycles per unit of measurement. If the unit of measurement is in seconds, then frequency is measured in **Hertz (Hz)**. Frequency measures oscillation for transverse waves and vibration longitudinal waves.

The frequency of a sound wave is the sound's pitch. Humans can hear sounds between 20 Hz and 20,000 Hz (20 kHz). Most adults can only hear up to around 16 kHz however, because the eardrum loses plasticity with age. Children can hear higher frequencies. The table below provides a reference point for sound frequencies.

Frequency (Hz)	Pitch	Sound Information
8.18	C-1	Lowest note that an organ can play
16.35	C0	Lowest note that a tuba can play
32.70	C1	Lowest note on a 88 key piano
261.63	C4	Middle C
1,046.50	C6	Highest note that a female voice can reach
4,186.00	C8	Highest note on a 88 key piano
16,744.00	C10	Near mosquito pitch (17 kHz)

Figure 26.2

Sound waves are longitudinal waves, so their amplitude is their volume or loudness. Sound volume is measured in decibels (dB). Decibels are a measure of sound pressure relative to a reference level of 0.0002 microbars, which is equal to 0 dB. The decibel is a logarithmic measurement. Therefore, a 100 watt speaker is able to produce only a slightly louder volume than a 50 watt speaker. Every 10 dB increase in sound volume requires increasing the wattage by 10 times, due to the logarithmic scale of sound volume. Humans perceive loudness differently than the true volume of a sound though. Lower frequencies are heard as being weaker even if they are at the same volume as higher frequency sounds. Due to the discrepancy between sound volume perception and real volume, doubling the perceived loudness only requires increasing the decibels by 3.

Frequency vs Time Domain
In the last chapter, time series models were described that analyze how signals change over time. All of the models explained in that chapter are considered part of the time domain of signals analysis. The analytic techniques in this chapter lie in the frequency domain. The frequency domain of a signal shows how much of the signal lies within each frequency band, over a range of frequencies.

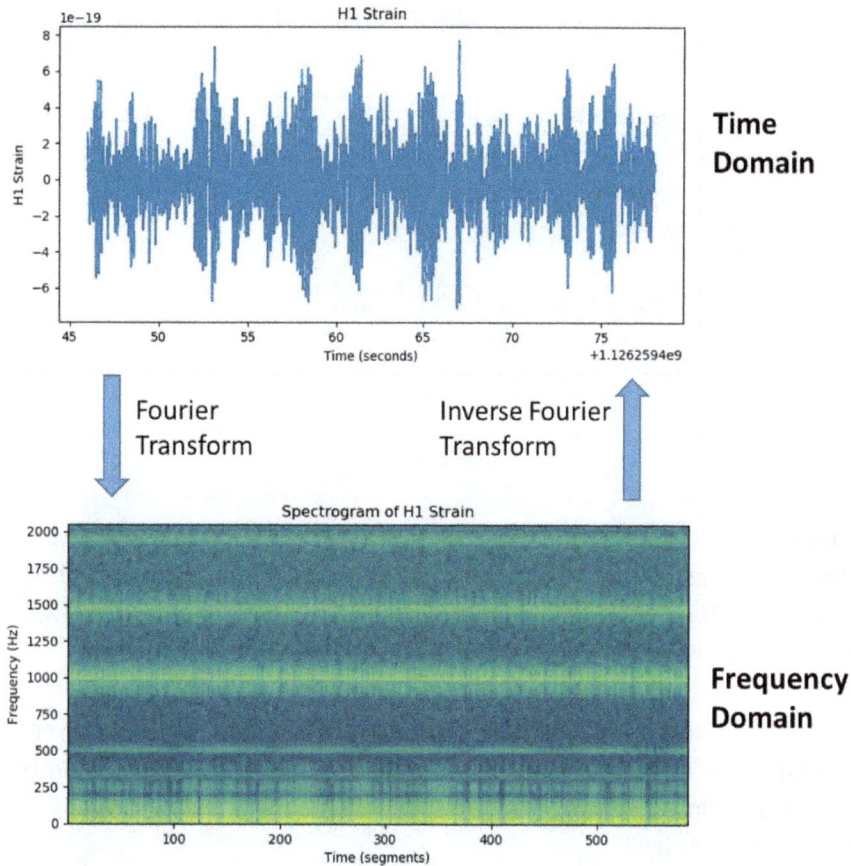

Figure 26.3

Converting a signal from one domain to the other requires a transform operation.

Fourier Transform

A signal can be converted from the time domain to the frequency domain using transform operations. The reverse can be done using inverse transform operations. The most common transform operation is the Fourier transform, which converts a signal from the time domain into the sum of sine and cosine waves of different frequencies (Pennsylvania State University, 2018). The range of all of these frequencies that the signal is converted to is called the spectrum. The easiest way to think of a Fourier transform is to imagine a ray of light passing through a glass prism, like on the cover of Pink Floyd's *Dark Side of the Moon* album. The prism splits white light into its constituent frequencies, causing a rainbow. Each color of light in the rainbow has a different wave frequency. In a way, a Fourier Transform is like PCA for signals or time series.

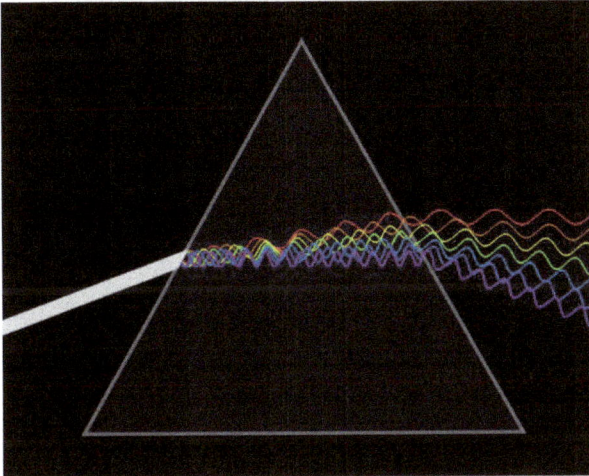

Figure 26.4

Just like how white light can be split into a rainbow, music can be split into different frequencies or pitches of tones. For example, a chord is made up of several tonal frequencies. But unless a musical tone is produced by a pure sine wave, it has **overtones** (tones at higher frequencies than the fundamental tone). These overtones give instruments their distinct sounds, resulting in a property known as timbre. **Timbre** is the reason the note C played by a violin sounds different from a C played by a guitar, even if the notes are played at the same loudness. The physical structure of an instrument causes different overtones to be generated. Similarly, different physical structures of human vocal chords produce different sounding voices. Timbre, or the pitch and volume (amplitude) of the overtones produced by a voice, acts as a fingerprint, allowing individual speakers to be identified.

Light and sounds are only two examples of signals that can be broken into sub-frequencies using a transform. Any type of signal can be broken down, unless the signal is only one frequency, such as a pure sine wave or pure blue light. The Fourier transform can be used to separate the frequencies of earthquake waves so that buildings can be constructed to avoid resonating with the strongest component frequencies of earthquake waves (Pennsylvania State University, 2018). **Resonance** is the tendency of a system's natural vibration frequency to be amplified when a wave vibrating at the same frequency interacts with it. Resonance is the reason why certain objects in a room will vibrate with certain frequencies of sounds that match the objects' natural vibration frequencies. It is also the reason why glass can be shattered when an intense sound vibrating at the glass' natural vibration frequency interacts with it. As with glass, resonance can damage any material if the amplitude of the interacting wave is strong enough, which is why it is important for buildings to be constructed so as not to interact with the vibration frequencies of earthquake waves.

The Fourier transformation converts a signal into the sum of sine and cosine waves at different frequencies. Sine and cosine waves are collectively called **sinusoidal waves**. If the conversion occurs over a continuous time interval, then the Fourier transform is an integral. The mathematical definition of the Fourier transform is shown in the equation on the lefts side of figure 26.5. The inverse Fourier transform is shown on the right.

Fourier Transform	Inverse Fourier Transform
$X(j\omega) = \displaystyle\int_{-\infty}^{\infty} x(t)e^{-j\omega t}\,dt$	$x(t) = \dfrac{1}{2\pi} \displaystyle\int_{-\infty}^{\infty} X(j\omega)e^{j\omega t}\,d\omega$

Figure 26.5

In the equations in figure 26.5, x(t) is the signal amplitude at time t, e is Euler's number, j is an imaginary number, and ω is frequency. Using this notation, X represents the whole signal. If time is measured in seconds, then the frequency is measured in Hz (cycles per second).

Astute readers will notice a quirk with the Fourier transform: it only applies to continuous functions. Signals are rarely ever continuous; they are usually sampled. Music for example, when it is heard in real life, is continuous. But when it is recorded and compressed into an audio file like an MP3, it must be sampled. If the samples are taken at extremely short intervals, the sampled signal sounds continuous. To grasp this idea, imagine a sequence of images with subtle changes, that when shown in quick succession, make the subjects in the image appear to move. This is the idea behind motion pictures. Movies are sampled, but because the image samples are taken at such short intervals, they appear continuous. The same concept applies to sound and other signals. As another example, daily stock prices are samples of stock prices taken once per day. When a signal is sampled, it is discrete, rather than continuous. The Fourier transform is called the **Discrete Time Fourier transform (DTFT)** when it is applied to a discrete or sampled signal. The equations in figure 26.6 define the discrete time Fourier transform and its inverse.

Discrete Time Fourier Transform	Inverse Discrete Time Fourier Transform
$X(e^{j\omega}) = \displaystyle\sum_{n=-\infty}^{\infty} x[n]e^{-j\omega n}$	$x[n] = \dfrac{1}{2\pi} \displaystyle\int_{2\pi} X(e^{j\omega})e^{-j\omega n}\,d\omega$

Figure 26.6

In the equations in figure 26.6, x[n] is the signal amplitude at sample n, e is Euler's number, 2*pi is the period of the signal measured in radians, j is an imaginary number, and ω is frequency. Using this notation, X represents the whole signal and N represents all samples.

Looking back at the equations in figures 26.5 and 26.6, the integral and summation measure the signal from negative infinity to infinity. This means that the frequency domain into which the signal is being decomposed is continuous. In the case of music for instance, a tone could be decomposed into tones that lie somewhere between two pitches. Musical scales are discrete though, rather than continuous. So it may be useful to decompose a signal into a discrete frequency domain. When a signal in continuous time is decomposed into discrete frequency ranges, the **Fourier series (FS)** is required. When a signal in discrete time is decomposed into discrete frequency ranges, the **Discrete Fourier transform (DFT)** is required. So there are four ways to decompose signals, based on the nature of the signal (continuous or discrete) and the nature of the frequency domain it is decomposed into (continuous or discrete). The equations for each type of transformation are shown in the matrix in figure 26.7.

	Discrete Frequency	Continuous Frequency
Continuous Time	Fourier Series $$X(\omega_k) = \frac{1}{T}\int_0^T x(t)e^{-jk\omega_0 t}\,dt$$ Inverse Fourier Series $$x(t) = \sum_{k=-\infty}^{\infty} X(\omega_k)e^{-jk\omega_0 t}$$	Fourier Transform $$X(j\omega) = \int_{-\infty}^{\infty} x(t)e^{-j\omega t}\,dt$$ Inverse Fourier Transform $$x(t) = \frac{1}{2\pi}\int_{-\infty}^{\infty} X(j\omega)e^{j\omega t}\,d\omega$$
Discrete Time	Discrete Time Fourier Series $$X[k] = \sum_{n=0}^{N-1} x[n]e^{-j\omega_0 kn}$$ Inverse Discrete Time Fourier Series $$x[n] = \frac{1}{N}\sum_{k=0}^{N-1} X[k]e^{j\omega_0 kn}$$	Discrete Time Fourier Transform $$X(e^{j\omega}) = \sum_{n=-\infty}^{\infty} x[n]e^{-j\omega n}$$ Inverse Discrete Time Fourier Transform $$x[n] = \frac{1}{2\pi}\int_{2\pi} X(e^{j\omega})e^{-j\omega n}\,d\omega$$

Figure 26.7

In the equations in figure 26.7, $x[n]$ is the signal amplitude at sample n and $x(t)$ is the signal amplitude at time t, e is Euler's number, 2*pi is the period of the signal measured in radians, j is an imaginary number, ω is frequency, k is a discrete frequency band, and $X(\omega_k)$ is the kth Fourier series coefficient. Using this notation, X represents the whole signal, T represents all times t, and N represents all samples n.

The period of a signal is also important. The Fourier transform, as it has been described so far, only applies to non-periodic signals (or one period, because the entire signal is assumed not to cycle). When the Fourier transform is applied to non-periodic signals, the signals are converted into a continuous frequency domain. When the Fourier transform is applied to periodic signals, the signals are converted into the frequency domain. When the Fourier series is applied to periodic signals, the signals are converted into discrete exponential functions, or sine and cosine functions. The Fourier series cannot be applied to non-periodic functions (Pennsylvania State University, 2018). Figure 26.8 shows how each type of transform applies to different types of signals.

Type of Fourier Transform	Signal Type
Fourier Transform	Continuous Frequency & Continuous Time, Aperiodic Signals
Discrete Time Fourier Transform	Continuous Frequency & Discrete Time, Aperiodic Signals
Fourier Series	Discrete Frequency & Continuous Time, Periodic Signals
Discrete Time Fourier Series	Discrete Frequency & Discrete Time, Periodic Signals

Figure 26.8

Since signals are usually sampled, the DTFT is more commonly used that the Fourier series. The DTFT is difficult to solve analytically, so algorithms have been developed to compute it. The **Fast Fourier transform (FFT)** is an algorithm for computing the DFT that uses matrix factorization (Amer, 2015). We will use the FFT at the end of this chapter.

Another common type of Fourier transform is the **Short Time Fourier transform (STFT)**. The Fourier transforms described so far operate over the entire duration of a signal. The STFT operates on a sliding window of fixed size over a signal. The sliding window is similar to the window used by moving averages for smoothing. When the transform is applied to a window of time, the frequency components are linked the time window. To understand why the STFT is useful, consider applying the Fourier transform to a song. A song typically has different parts so it would make no sense to look at the frequency components over the entire length of a song. A verse may have different chords than the chorus, for example. A song is a good example of a signal for which the STFT would be better than the standard Fourier transform, but the STFT is preferred for any type of non-stationary time series (Amer, 2015). The formula for the STFT is shown in figure 26.9.

$$X_m(\omega) = \sum_{n=-\infty}^{\infty} x(n)w(n-mR)e^{-j\omega n}$$

Figure 26.9

In the equation in figure 26.9, x(n) is the input signal at sample n, w(n-mR) is the window function that is moved over the time index m (as m increases, the window function w(n) moves to the right), R is the hop size in terms of the number of samples (R is often assumed to be 1 so it is not always written in the equation), j is an imaginary number, ω is frequency, and e is Euler's number. X is therefore a function of both time, t, and frequency, ω. Note that the window should have some overlap, usually 50%. The overlap is important because without it, information will be lost (DSP Related, 2018). Consider a sine wave component of the signal. Sine is zero at pi/2 radians. If the signal tapers to zero at the boundaries of the window, then there will be some area of the signal that does not have any values for the component frequencies. To avoid this, a slightly overlapping window ensures that there is never a point in the window with no values. Figure 26.10 shows the difference between a window with no overlap and a window with overlap (Jojek, 2014). Notice how information is lost as the signal tapers in amplitude when there is no overlap.

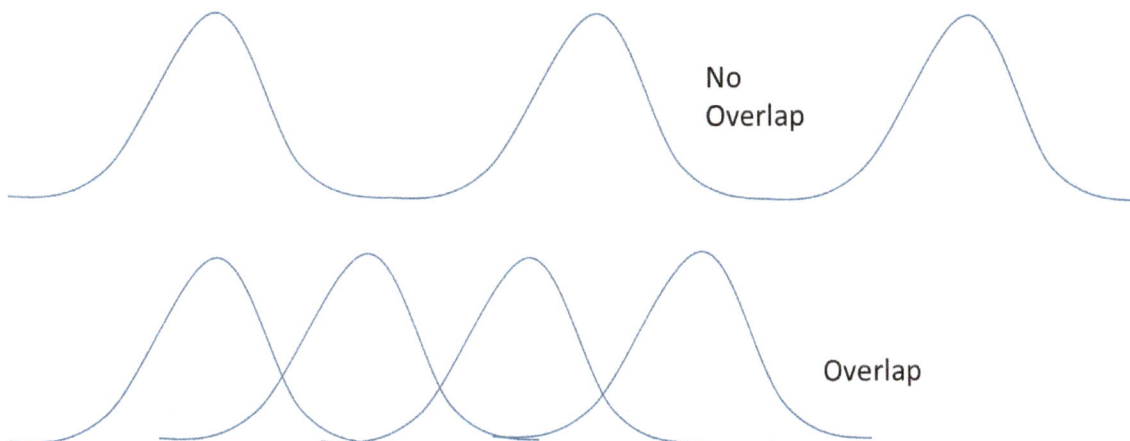

No Overlap

Overlap

Figure 26.10

A reasonable question at this point would be why the window could not be set as small as possible. The problem with using small windows is that there is a tradeoff between frequency resolution and time resolution. The smaller the window, the better the time resolution. The larger the window, the better the frequency resolution. This means that smaller windows are better able to represent when changes in the frequencies of a signal occur, but it is harder to know what those frequencies are. On the other hand, larger windows are better able to represent which frequencies are present in the signal, but it is harder to know when the frequencies change. The resolution tradeoff is a version of the Heisenberg Uncertainty Principle. The **Heisenberg Uncertainty Principle** asserts that there is a limit to the precision with which pairs of physical properties of a particle can be measured. The principle arose from the field of physics, where the position and velocity of particles was found to have a tradeoff: knowing the position limited the ability to know the velocity, and vice versa. The resolution tradeoff of the STFT spurred the creation of the wavelet transform (Dallas, 2018).

Fourier Transform in More Detail

It is hard to conceptualize how the Fourier transform converts a signal into its component frequencies. So in this section, we will walk through the logic step by step. This section can be skipped if the reader desires.

As previously described, the Fourier transform decomposes a signal into sinusoidal waves. These waves are continuous, but the signal has some starting point. The starting point is called the **phase** of the signal. Figure 26.11 shows an example of a wave with a phase of 0 (top) and a wave with a phase of 1 (bottom).

Waves with Different Phases

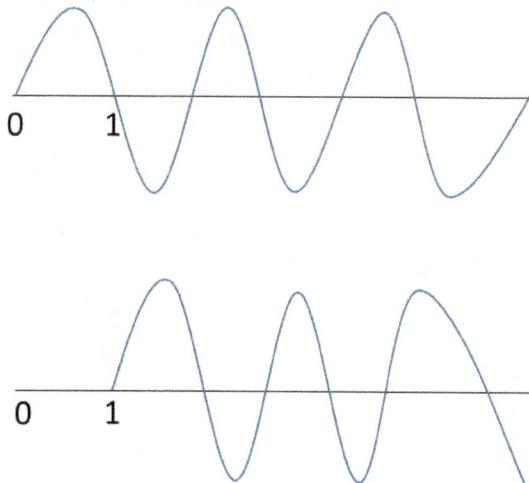

Figure 26.11

So every sinusoidal wave has a phase that defines where in the period of its cycle the wave begins. Every sinusoid also has an amplitude and a frequency, as defined in the first section of this chapter. Refer back to figure 26.1 to see how the amplitude and frequency are defined. The y-axis of the graph of a sinusoid is the real value of the wave at time t, where time is along the x-axis. It is important to distinguish between the real value and the imaginary value. A sinusoidal wave can be thought of as motion about a complex circle. A **complex circle** is a circle drawn on the complex plane, where the x-axis has real numbers and the y-axis has imaginary numbers. Numbers on the complex plane are complex numbers, such as a+bi, where i is the imaginary component, and a is the real component. Figure 26.12 represents a circle in the complex plane.

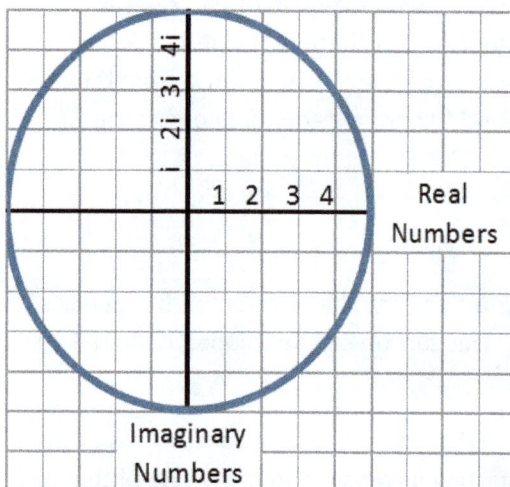

Figure 26.12

If we think of a wave as motion about a complex circle, then the starting point of the wave (the phase) is the starting angle of the motion about the circle. So if the phase of a sinusoidal wave is 0, then the first point of the wave is shown in figure 26.13.

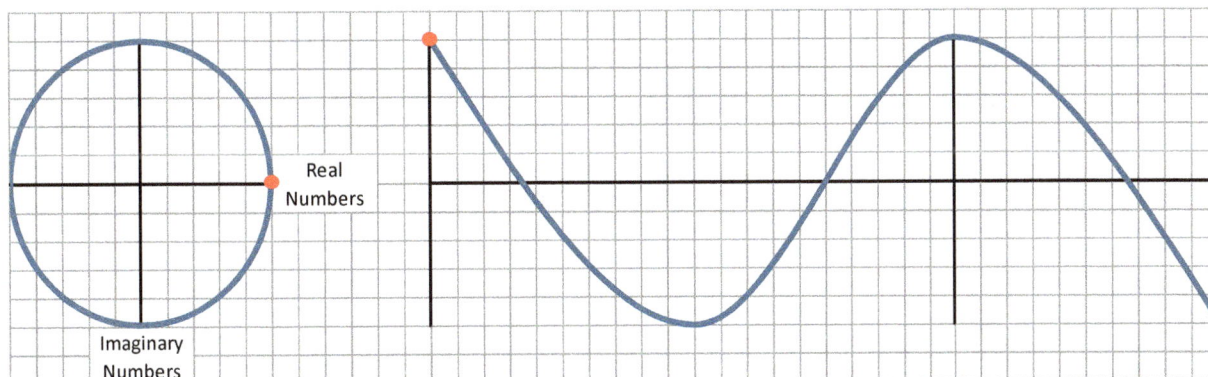

Figure 26.13

The starting angle 0 corresponds to a value r, the radius of the complex circle, which has an imaginary value of 0. In other words, for phase 0, the starting point (r, 0) only has a real component equal to r, which is equal to the amplitude of the sinusoidal wave. This reveals a useful identity: the radius, r, of the complex circle is the amplitude of the sinusoidal wave.

$$\text{Radius of complex circle} = \text{Amplitude of sinusoidal wave}$$

$$r = A$$

Figure 26.14

Now that we know the phase and amplitude of the sinusoid, all we need to know to completely describe the wave (motion about the circle) is the frequency. The frequency of a wave is how many peaks or troughs of the wave occur in a single unit of time. In terms of the complex circle, the frequency is the number of revolutions around the circle per unit time. For simplicity, we will assume a frequency of 1 (1 cycle per second would be 1 Hz). Figure 26.15 shows the movement about the circle corresponding to points along the sinusoid.

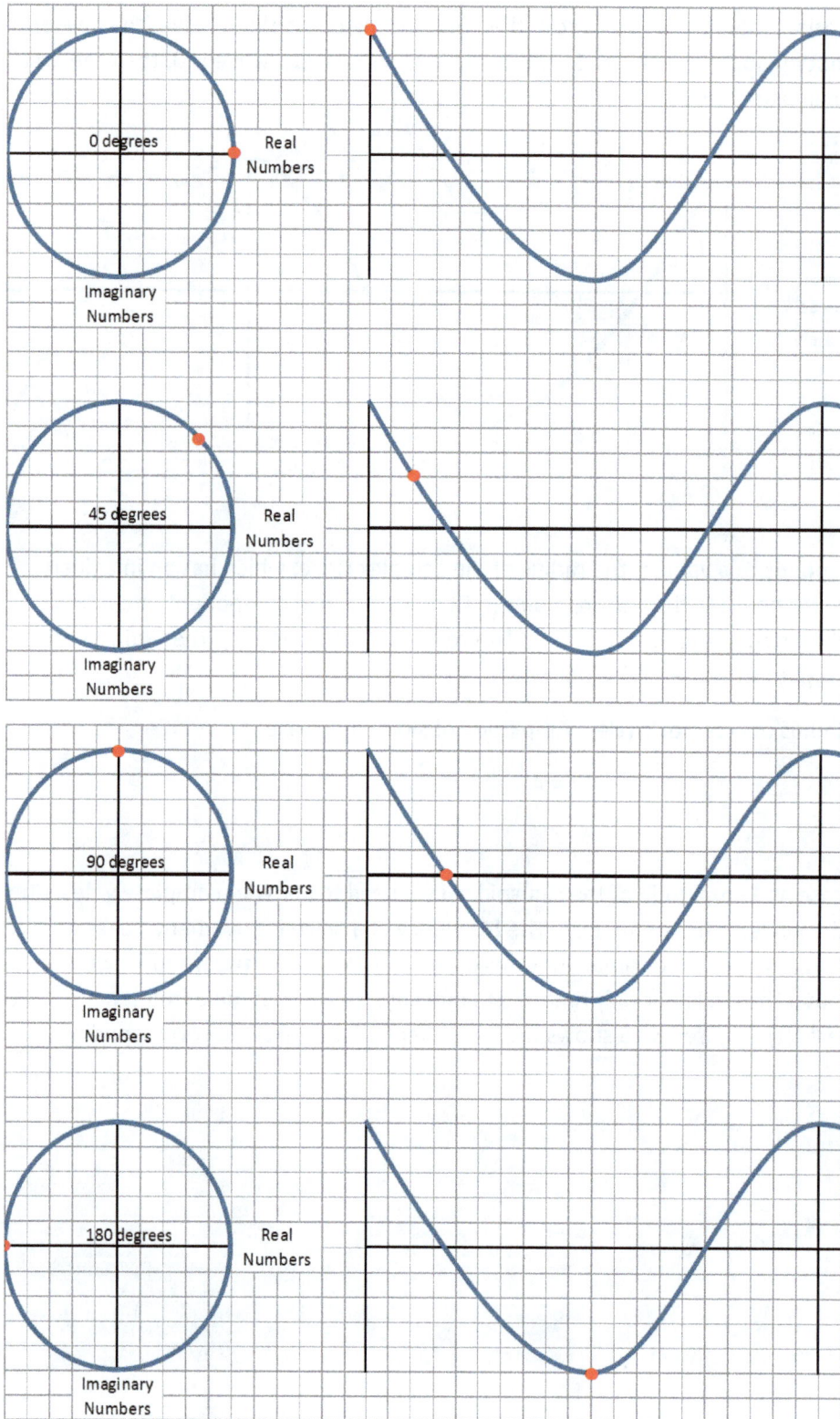

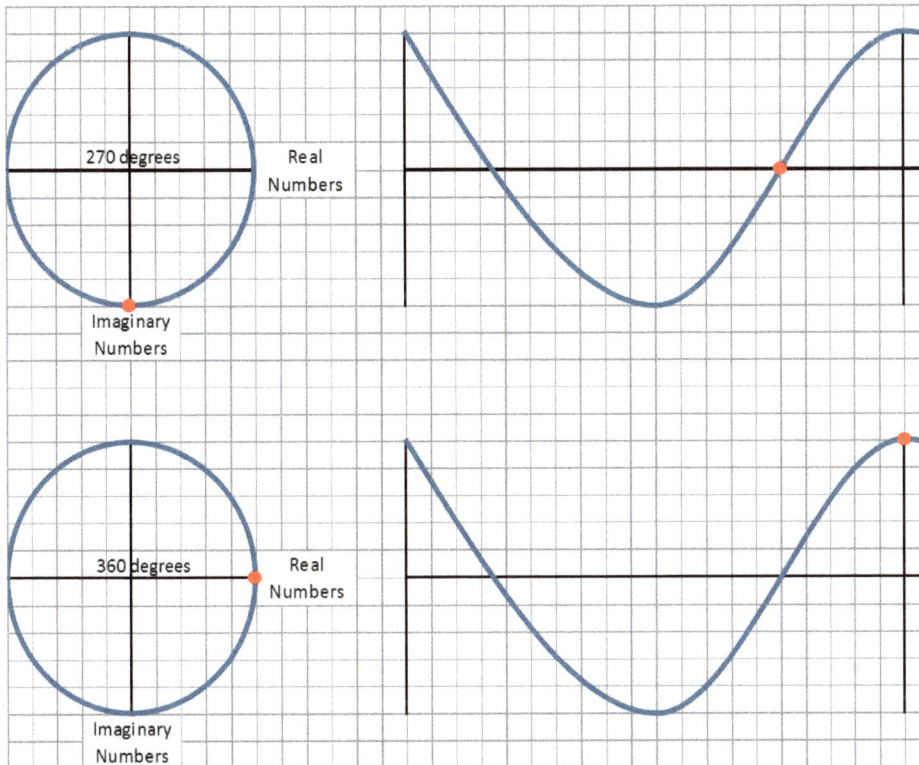

Figure 26.15

The complete period of this sinusoid is 2pi, because one revolution of the complex circle is 360 degrees or 2pi radians. Take a peek back at the equations for the Fourier transform to see the 2pi in the exponent. Now we know where it comes from. Motion about a complex circle with radius R, phase i, and angular frequency w is represented by the equation in figure 26.16.

$$z(t) = Re^{iwt}$$

Figure 26.16

The equation in figure 26.16 gives the value of the sinusoid as a position along the complex circle at time t. Now imagine two sinusoids. The equation to produce z(t) would be the sum of the motion about two complex circles:

$$z(t) = R_1 e^{iw_1 t} + R_2 e^{iw_2 t}$$

Figure 26.17

This equation is the same as combining 2 sinusoidal waves:

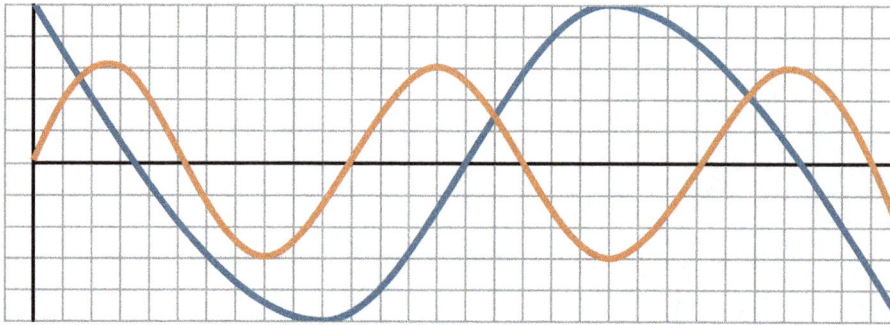

Figure 26.18

Now imagine stacking more and more sinusoids to figure, each representing a different frequency and having different amplitudes and phases. All of these sinusoids can be thought of as complex circles, and motion about all of these circles is represented by an integral. That is the logic behind the Fourier transform equations shown in figures 26.5, 26.6, and 26.7.

The great insight that was made when the Fourier transform was discovered was that if sinusoidal waves can be thought of as complex circles, and many sinusoidal waves add up to produce a signal, then any signal can be represented by a bunch of complex circles. More broadly, any time series or signal can be perfectly represented by the sum of infinitely many complex circles. Each complex circle represents one pure frequency. Therefore, by decomposing a signal using a Fourier transform, it is possible to acquire knowledge of the phase, amplitude of every frequency that makes up the signal.

Wavelet Transform

Recall that the Fourier transform has a resolution tradeoff. One way to reduce the magnitude of the tradeoff is to represent a signal using smaller waves instead of sinusoidal (sine and cosine) waves. Unlike sine and cosine waves that go on forever, wavelets are pulses, or mini-waves that burst quickly and then disappear. The graph of a wavelet can be envisioned as appearing similar to a heartbeat on an electrocardiogram. Wavelets generally have better time and frequency resolution than the Fourier transform. The reasons for this are that the wavelet transform allows windows to be of variable size, and instead of decomposing a signal into sine and cosine components like the Fourier transform, the wavelet transform decomposes a signal into functions of complex conjugates, called wavelets. Complex conjugates are numbers with equal parts real and imaginary. For example, the roots of the quadratic equation $x^2 + 4x + 5$ are $-2 + I$ and $-2 - I$. These roots are complex conjugates. The functions of complex conjugates (wavelets) are non-orthogonal for the continuous wavelet transform and orthogonal for the discrete wavelet transform. The equations for the continuous and discrete wavelet transform are shown to the left and right of figure 26.19, respectively.

Continuous Wavelet Transform	Discrete Wavelet Transform
$$X(a,b) = \frac{1}{\sqrt{|a|}} \int_{-\infty}^{\infty} x(t)\psi\left(\frac{x-b}{a}\right)(x)\,dx$$	$$X(a,b) = \sum_{k=-\infty}^{\infty} a_k \phi(Sx - k)$$

Figure 26.19

In the equations in figure 26.19, S is a scaling factor, a is the wavelet coefficient, b is a translational value, t is time, k is a sample that replaces time in the discrete transform, and phi (upper and lowercase) is some continuous function that can be chosen arbitrarily and is referred to as the mother wavelet. The mother wavelet serves as a function from which child wavelets are derived: the children are essentially scaled versions of the mother. Two types of mother wavelets are the Gabor and Morlet wavelets, as we will see shortly.

Notice that the discrete wavelet transform relies on a scaling factor S. Scaling is important when dealing with different frequencies, because higher frequency waves have shorter time lengths and lower frequency waves have longer time lengths. The scaling factor ensures that the different frequencies are transformed similarly (Dallas, 2018).

The most important concept to understand about the wavelet transform is that it improves upon the resolution tradeoff faced by the Fourier transform by transforming a signal into the wavelet domain, rather than the frequency domain.

As a consequence of decomposing a signal into non-orthogonal wavelets, the continuous wavelet transform produces an array that is one dimension larger than the input data. The wavelets are highly correlated because they are non-orthogonal. The mother wavelet function, phi, is important because it influences the time and frequency resolution of the result. The two most common mother wavelets are the **Morlet wavelet** and the **Gabor wavelet**. These wavelets are shown in the equations in figure 26.20.

Morlet Wavelet	$\Psi_\sigma(t) = -\sqrt{\left(1 + e^{-\sigma^2} - 2e^{-\frac{3}{4}\sigma^2}\right)} \pi^{-\frac{1}{4}} e^{-\frac{1}{2}t^2} \left(e^{i\sigma} - e^{-\frac{1}{2}\sigma^2}\right)$
Gabor Wavelet	$g_{\alpha,\xi,a,b}(x) = |a|^{-\frac{1}{2}} g_{\alpha,\xi}\left(\frac{x-b}{a}\right)$

Figure 26.20

Depending on the wavelet choice, the resolution may favor time or frequency. The Morlet wavelet favors frequency resolution, while the Gabor wavelet attempts to minimize the product of standard deviations in time and frequency, thereby attempting to optimize both (Dallas, 2018).

As a consequence of decomposing a signal into orthogonal wavelets, the discrete wavelet transform produces an array that is the same size as the input data. Since the wavelets are orthogonal and uncorrelated, the resulting array is sparse. This makes the discrete wavelet transform good for data compression, because there is no redundancy (Dallas, 2018).

Spectral Density
The frequency spectrum of a signal is the range of frequencies that make it up, which can be found by applying the Fourier transform to the signal. Spectral density is a measure of how much of something exists in each frequency per unit time. The "something" is usually energy or power, resulting in the **power spectral density (PSD)**. Spectral density is closely related to autocorrelation. Autocorrelation shows how the values of a signal relate to its previous values. Autocorrelation is a time domain representation. Spectral density shows how the frequencies of a signal relate to one another. It is a frequency domain representation of the same information that autocorrelation represents. The equations for autocorrelation and spectral density shown in figure 26.21 show how they are related (Ssk08, 2013 and Pennsylvania State University, 2018).

Autocorrelation (time domain)	$X(t) = \int_{-1/2}^{1/2} e^{2\pi j\omega t} x(\omega)d\omega$
Spectral density (frequency domain)	$x(\omega) = \sum_{t=-\infty}^{t=\infty} X(t)e^{-2\pi j\omega t}$

Figure 26.21

In the equations in figure 26.21, t is a time lag, j is an imaginary number, and ω is frequency. The spectral density, $x(\omega)$, is the sum of the autocorrelation, $X(t)$, times the complex frequency representation of the signal, $e^{-2\pi j\omega t}$. Note that the spectral density is defined for both positive and negative frequencies, but since the frequencies outside the range (-0.5, 0.5) are symmetrical, they are of no consequence (Pennsylvania State University, 2018). Since the total integrated spectral density equals the variance of the signal, the spectral density between frequencies A and B is the amount of variance in the signal that is explained by the frequencies between A and B.

A **periodogram** is a graph that provides a sample estimate of the spectral density of a stationary time series. It has frequency on the x-axis and power on the y-axis (power if often measured in amplitude). An example periodogram is shown in figure 26.22. It is a sample estimate of the spectral density because it only displays discrete fundamental frequencies, whereas the spectral density is continuous. Smoothing methods like moving averages can improve on the periodogram estimate, as can AR models. Note that if the signal is a non-stationary time series, spectral density is undefined.

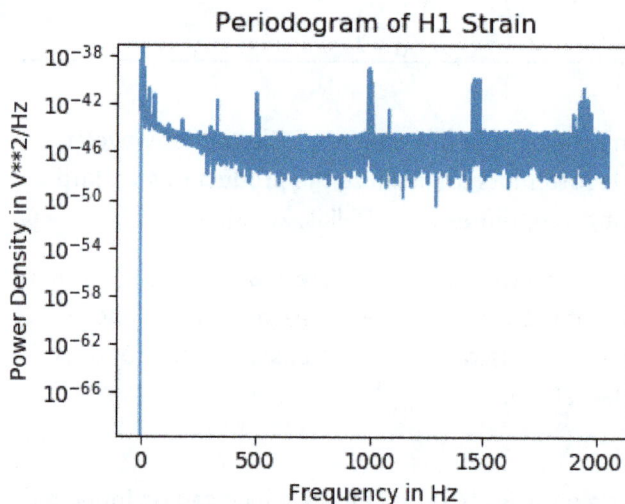

Figure 26.22

Spectrum versus Cepstrum

While the spectral density of a signal can be graphed with a periodogram, the frequency spectrum that is found by applying a Fourier transform can be graphed using a **spectrogram**. An example of a spectrogram is the graph on the left in figure 26.23. It is simply a colored representation of the frequencies present in a signal over time, where color intensity corresponds to the amplitude or power of the frequency. Spectrograms are useful for visualizing distinct sounds, because they easily visualize formants. A formant is the spectral maximum, or peak frequency of a sound. When people talk, each word that they speak naturally has a range of frequencies and a formant. Formants also occur at the sentence and phrase level. A smooth line can be formed by connecting the formants of a spoken sentence. This smooth line is called the **spectral envelope**. Figure 26.23 shows a spectrogram on the left and a spectral envelope on the right. The ability of spectrograms to capture unique characteristics about speech makes them good for speech recognition.

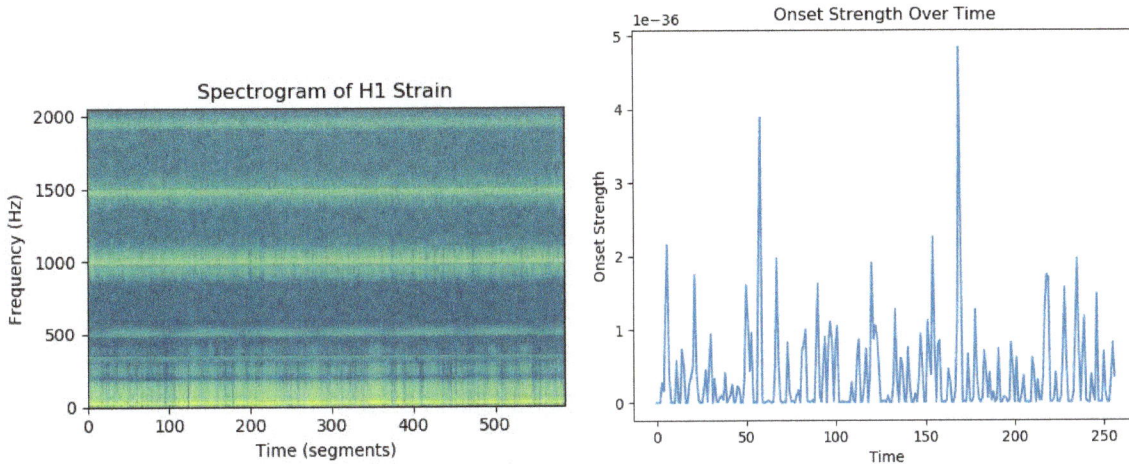

Figure 26.23

When the spectral envelope is removed from a signal's spectrum, only the spectral details remain. The spectral details are referred to as the **cepstrum**. Think of the removal of the spectral envelope as being similar to de-trending or removing seasonality from a time series. The spectral envelope can be removed from a spectrum by taking the **Inverse Fast Fourier transform (IFFT)** of the spectrum. Since the Fourier transform produces a spectrum, the cepstrum is often referred to as the spectrum of the spectrum. The amplitudes of the frequencies in the cepstrum are called the **cepstral coefficients**. The problem with using the IFFT to derive the cepstrum coefficients is that the IFFT produces complex coefficients. A simpler option is to use the **discrete cosine transform (DCT)**, which produces cepstral coefficients that are real numbers.

The cepstrum has linearly spaced frequency bands. For acoustic applications, or applications related to human hearing, frequencies should be scaled. The scaling is necessary because the human ear is most sensitive to frequencies between 2 and 5 kHz, and cannot hear frequencies above 20 kHz or below 20 Hz. Although the loudness of a frequency depends on its amplitude, the human ear perceives the loudness of different frequencies differently. A 100 Hz wave and 10 kHz wave with the same amplitude will be perceived as different volumes to a human. The variability of perceived loudness depending on frequency follows an **equal-loudness contour** or **Fletcher-Munson curve**, named after the physicists who discovered it, Harvey Fletcher and Wilden Munson. The equal-loudness contour shown in figure 26.24 shows that lower frequencies are generally perceived as quieter than higher frequencies. Guitars sound louder than bass guitars when they are playing at the same volume.

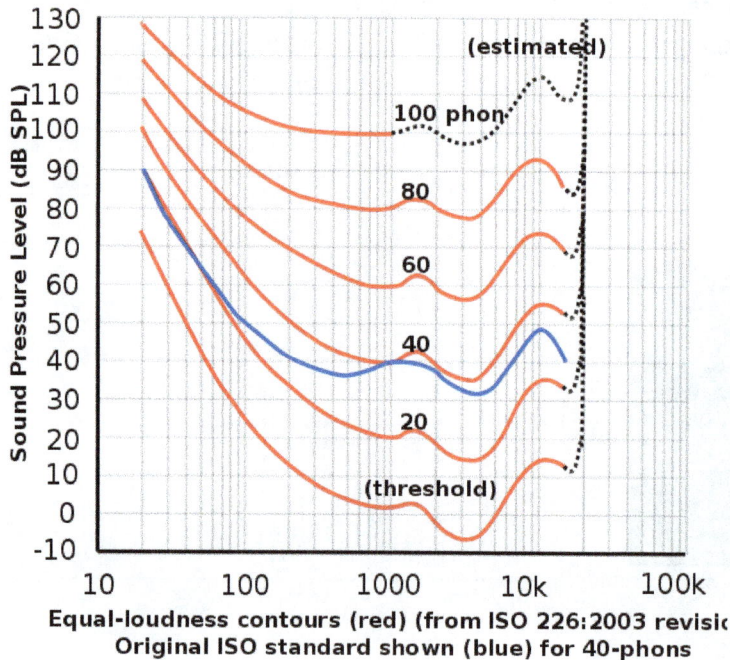

Equal-loudness contours (red) (from ISO 226:2003 revisic
Original ISO standard shown (blue) for 40-phons

Figure 26.24

Acoustic frequencies are most commonly scaled using the Mel scale. The **Mel scale** is named from the word "melody" to indicate that the scale is based on pitch comparisons, and it is based on a log scale. The equation to convert from the frequency in hertz to Mels is shown in figure 26.25.

$$M = 2{,}595 * \log_{10}\left(1 + \left(\frac{f}{700}\right)\right)$$

Figure 26.25

To account for acoustic loudness perception, the DCT is applied to the logs of the powers at each Mel frequency of the cepstrum of a sound signal. The result is called the **Mel-frequency cepstrum**, and the amplitudes of the resulting cepstrum are called the **Mel-frequency cepstral coefficients (MFCCs)**. MFCCs are commonly used as features for speech recognition, emotion detection, and voice identification models.

Filters

Filters simply remove certain components from signals. While there are numerous filters for signals, including the Kalman filter described in chapter 24, there are three basic filters: low pass, high pass, and band pass. In filter terminology, a frequency that is passed is kept or retained and a frequency that is attenuated is filtered or removed.

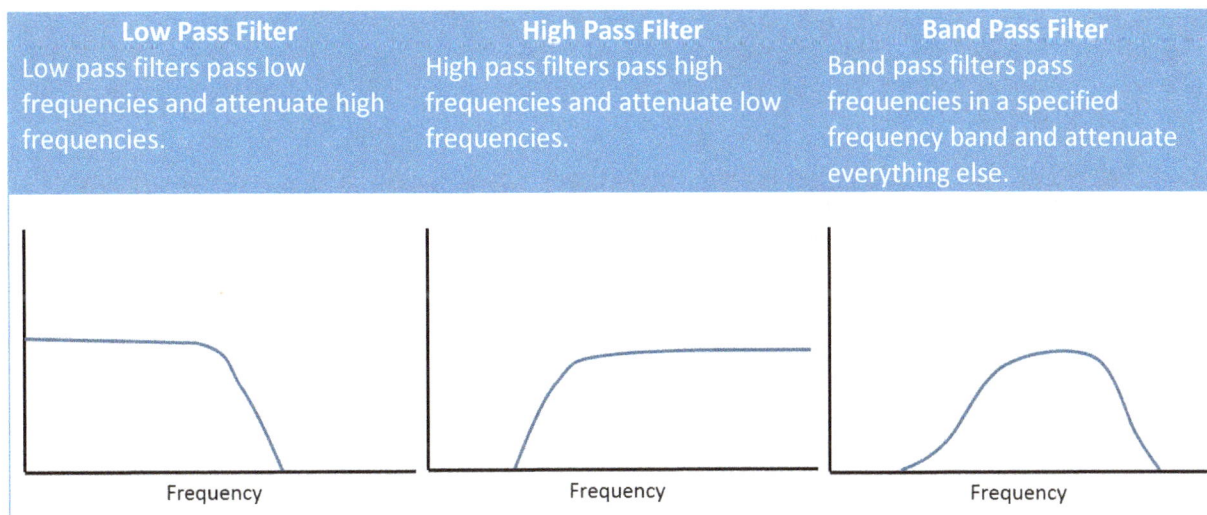

Low Pass Filter	High Pass Filter	Band Pass Filter
Low pass filters pass low frequencies and attenuate high frequencies.	High pass filters pass high frequencies and attenuate low frequencies.	Band pass filters pass frequencies in a specified frequency band and attenuate everything else.

Figure 26.26

Signal Processing with Python

In February 2016, the Laser Interferometer Gravitational-Wave Observatory (LIGO) announced that it had detected gravitational waves for the first time. The waves were caused by 2 black holes that collided and merged 1.3 billion years ago. At the moment the black holes merged, an amount of mass equivalent to 3 times the mass of our sun was converted into a burst of energy in the form of gravity. This energy burst was what was detected by LIGO. LIGO has released its signal data for this event through the LIGO Open Source Science Center, so we are going to apply signal processing techniques to see how gravitational waves were discovered, using Python. LIGO's data can be downloaded from its website: https://losc.ligo.org/events/GW150914/, and it is licensed under the Creative Commons Attribution 3 license. The files we will use are the 32 second HDF5 files for the Hanford and Livingston observatories. These signals are sampled at a rate of 4096 Hz.

For reference, LIGO is constantly "listening" for gravitational waves. The observatory is so sensitive that it can detect construction workers dropping a hammer to the ground more than a mile away. It has to be sensitive to be able to detect the subtle signal of gravity, but this means that the observatory records an overwhelming amount of noise or junk signal. In order to help find the true signal of a gravitational wave, LIGO researchers have provided a template for what a gravitational wave signal from colliding black holes is expected to look like. This template can be downloaded here: https://losc.ligo.org/s/events/GW150914/GW150914_4_NR_waveform.txt. Our goal will be to compare the time-series strain data to the template to detect gravitational waves from the colliding black holes. Note that this same general concept can be used to identify any type of signal hidden amongst noise, such as a certain person's voice hidden in a recording of conversations from a cocktail party, or an instrument playing in a band.

To read the HDF5 files, we need the h5py library. Before doing anything, let us explore the structure of the HDF5 data to see what we are dealing with.

```
import numpy as np
import matplotlib.pyplot as plt
import matplotlib.mlab as mlab
```

```
import h5py
from scipy.interpolate import interp1d
from scipy.signal import butter, filtfilt
from scipy.io import wavfile

#Read the file for the first detector to extract strain data and create a time array
fileName = 'H-H1_LOSC_4_V1-1126259446-32.hdf5'
dataFile = h5py.File(fileName, 'r')
print(list(dataFile.keys()))

#Explore each key's values
print(list(dataFile['meta'].keys()))
print(list(dataFile['quality'].keys()))
print(list(dataFile['strain'].keys()))
print(list(dataFile['strain']['Strain'].attrs))
```

It quickly becomes apparent that the file has a hierarchical structure with many levels. A HDF5 file can be viewed using a tool like PyHexad. But to save time, these are the parts we will need:

```
#Get the actual strain values and the time interval
hstrain = dataFile['strain']['Strain'].value
htime_interval = dataFile['strain']['Strain'].attrs['Xspacing']
```

We now know the time interval between samples. To get the actual time of each sample, we need to know the start time and duration of the signal. We know the duration is 32 seconds, but even if we did not know, these values can be found in the file's metadata.

```
#View the meta data to see what the attributes for start time and duration are
metaKeys = dataFile['meta'].keys()
meta = dataFile['meta']
for key in metaKeys:
    print (key, meta[key].value)
```

So the start time (UTC) is 2015-09-14T09:50:30 and the duration is 32 seconds, putting the end time at 2015-09-14T09:51:02. Instead of UTC, we will use the GPS time because it will be easier to work with. We can now create a numpy array that starts at the start time, ends at the end time, and increments at a value of htime_interval.

```
#Create time vector for the detector based on start time and duration
gpsStart = meta['GPSstart'].value
duration = meta['Duration'].value
gpsEnd = gpsStart + duration
htime = np.arange(gpsStart, gpsEnd, htime_interval)

dataFile.close()
```

Now we can repeat the process to read the second strain, and finally, we can read the template. LIGO's website shows how the template file can be read, so we know that it needs to be transposed, and that we will get both the reference time array (reftime) and reference strain values (ref_H1).

```
#Read the file for the second detector to extract strain data and create a time array
fileName = 'L-L1_LOSC_4_V1-1126259446-32.hdf5'
dataFile = h5py.File(fileName, 'r')
lstrain = dataFile['strain']['Strain'].value
```

```
ltime_interval = dataFile['strain']['Strain'].attrs['Xspacing']
ltime = np.arange(gpsStart, gpsEnd, ltime_interval)
dataFile.close()

#Read the template to serve as a reference
reftime, ref_H1 = np.genfromtxt('GW150914_4_NR_waveform_template.txt').transpose()
```

Let's plot the strains and the template over their respective time arrays to see what they all look like.

```
#Plot the detector strains and the template for comparison
fig = plt.figure()
numSamples = len(hstrain)
plth = fig.add_subplot(221)
plth.plot(htime[0:numSamples], hstrain[0:numSamples])
plth.set_xlabel('Time (seconds)')
plth.set_ylabel('H1 Strain')
plth.set_title('H1 Strain')
numSamples = len(lstrain)
pltl = fig.add_subplot(222)
pltl.plot(ltime[0:numSamples], lstrain[0:numSamples])
pltl.set_xlabel('Time (seconds)')
pltl.set_ylabel('L1 Strain')
pltl.set_title('L1 Strain')
pltref = fig.add_subplot(212)
pltref.plot(reftime, ref_H1)
pltref.set_xlabel('Time (seconds)')
pltref.set_ylabel('Template Strain')
pltref.set_title('Template')
fig.tight_layout()
plt.show()
plt.close(fig)
```

The H1 and L1 strains are quite noisy, as expected. Notice how clean the reference strain is – it is necessary for the reference template to be as pure as possible for any kind of signal processing problem where a signal is to be identified by matching it with a template. Also notice that the plot for the reference template has negative time. That is because the template shows how the strain should look leading up to the event (the collision of the black holes). The event itself occurs at time = 0.

At this point, it would normally be good to check the signal data for missing values. Fortunately, LIGO has already done the heavy lifting and there are no missing values during this 32 second time period for either detector. So the next step is to remove the noise to reveal the signal hidden in each strain. One step we can take to clean the data is compare the spectrum of the reference template to the spectrum of the strains.

```
#Specify the sampling frequency to use for Fourier transformation
samplefreq = int(1/htime_interval)   #4096

#Plot spectrogram for each strain and the template
h1samplefreqs, h1segtimes, h1sxx = signal.spectrogram(hstrain, fs=samplefreq)
l1samplefreqs, l1segtimes, l1sxx = signal.spectrogram(lstrain, fs=samplefreq)
refsamplefreqs, refsegtimes, refsxx = signal.spectrogram(ref_H1, fs=samplefreq)
fig = plt.figure()
plth = fig.add_subplot(221)
plth.pcolormesh((len(h1segtimes) * h1segtimes / h1segtimes[-1]),
```

```
                    h1samplefreqs,
                    10 * np.log10(h1sxx))
plth.set_xlabel('Time (segments)')
plth.set_ylabel('Frequency (Hz)')
plth.set_title('Spectrogram of H1 Strain')
pltl = fig.add_subplot(222)
pltl.pcolormesh((len(l1segtimes) * l1segtimes / l1segtimes[-1]),
                    l1samplefreqs,
                    10 * np.log10(l1sxx))
pltl.set_xlabel('Time (segments)')
pltl.set_ylabel('Frequency (Hz)')
pltl.set_title('Spectrogram of L1 Strain')
pltref = fig.add_subplot(212)
pltref.pcolormesh((len(refsegtimes) * refsegtimes / refsegtimes[-1]),
                    refsamplefreqs,
                    10 * np.log10(refsxx))
pltref.set_xlabel('Time (segments)')
pltref.set_ylabel('Frequency (Hz)')
pltref.set_title('Spectrogram of Reference Template')
fig.tight_layout()
plt.show()
plt.close(fig)
```

By plotting the spectrograms, we can see which frequencies are strongest in each signal. Right away it becomes obvious that the reference template has virtually no power in frequencies above 500 Hz. In fact, the expected frequency of gravity waves from colliding black holes is near 0. If we were to listen to it as an audio file, it would sound like a very low, bass heavy blip, almost like a heartbeat. We know it will be just a blip because the x-axis, the timespan of the reference template, is very short. Looking at the strain signals, we can see that they contain energy in higher frequency bands. The higher frequencies can be removed by a low pass filter.

The spectral envelope can be plotted using the Librosa library:

```
import librosa
def plotSpecEnvelope(wav, samplefreq):
    """
    The onset envelope, oenv, determines the start points for patterns.
    """
    mel = librosa.feature.melspectrogram(y=wav, sr=samplefreq, n_mels=128,
fmax=30000)
    oenv = librosa.onset.onset_strength(y=wav, sr=samplefreq, S=mel)
    plt.plot(oenv, label='Onset strength')
    plt.title('Onset Strength Over Time')
    plt.xlabel('Time')
    plt.ylabel('Onset Strength')
    plt.show()
    return oenv
plotSpecEnvelope(hstrain, samplefreq)
```

Recall that the power spectral density (PSD) shows how much energy or power lies within each range of frequencies that make up a signal. Also recall that signals can be analyzed from the time domain or frequency domain, and that the PSD is like the spectral version of autorrelation. We can plot the periodograms for the strain signals and the reference template to see how energy varies by frequency.

LIGO uses the amplitude spectral density (ASD) instead of the PSD. The ASD is simply the square root of the PSD. These plots will give us a much clearer picture of which frequencies are powerful for each signal. They are simply refinements of the spectrograms.

```
#Plot periodograms for each strain and the template
h1freq, h1power_density = signal.periodogram(hstrain, fs=samplefreq)
l1freq, l1power_density = signal.periodogram(lstrain, fs=samplefreq)
reffreq, refpower_density = signal.periodogram(ref_H1, fs=samplefreq)
fig = plt.figure()
plth = fig.add_subplot(221)
plth.semilogy(h1freq, h1power_density)
plth.set_ylim([np.min(h1power_density), np.max(h1power_density)])
plth.set_xlabel('Frequency in Hz')
plth.set_ylabel('Power Density in V**2/Hz')
plth.set_title('Periodogram of H1 Strain')
pltl = fig.add_subplot(222)
pltl.semilogy(l1freq, l1power_density)
pltl.set_ylim([np.min(l1power_density), np.max(l1power_density)])
pltl.set_xlabel('Frequency in Hz')
pltl.set_ylabel('Power Density in V**2/Hz')
pltl.set_title('Periodogram of L1 Strain')
pltref = fig.add_subplot(212)
pltref.semilogy(reffreq, refpower_density)
pltref.set_ylim([np.min(refpower_density), np.max(refpower_density)])
pltref.set_xlabel('Frequency in Hz')
pltref.set_ylabel('Power Density in V**2/Hz')
pltref.set_title('Periodogram of Reference Template')
fig.tight_layout()
plt.show()
plt.close(fig)

#Compute and plot the ASD for each detector strain
pxx_H1, freqs = mlab.psd(hstrain, Fs=samplefreq, NFFT=samplefreq)
pxx_L1, freqs = mlab.psd(lstrain, Fs=samplefreq, NFFT=samplefreq)
plt.figure()
plt.loglog(freqs, np.sqrt(pxx_H1), 'b', label='H1 Strain')
plt.loglog(freqs, np.sqrt(pxx_L1), 'r', label='L1 Strain')
plt.axis([10, 2000, 1e-24, 1e-18])
plt.legend(loc='upper center')
plt.xlabel('Frequency (Hz)')
plt.ylabel('ASD (strain/rtHz)')
plt.title('Strain ASDs')
plt.show()
plt.close()

#Store interpolations of the ASDs computed above to use later for whitening
psd_H1 = interp1d(freqs, pxx_H1)
psd_L1 = interp1d(freqs, pxx_L1)

#Plot the reference template ASD for comparison
pxx_ref, freqs = mlab.psd(ref_H1, Fs=samplefreq, NFFT=samplefreq)
plt.loglog(freqs, np.sqrt(pxx_ref), 'g', label='Template Strain')
plt.axis([10, 2000, 1e-28, 1e-21])
plt.xlabel('Frequency (Hz)')
```

```
plt.ylabel('ASD (strain/rtHz)')
plt.title('Template (strain/rtHz)')
plt.show()
plt.close()
```

It is clear from the reference template that gravitational waves mostly lie in the spectrum of 20 Hz to 300 Hz. So rather than a low pass filter like we originally thought, a band pass filter might be better, because we can set the lower limit to 20 Hz. However, even if we apply a band pass, we will still have white noise in the lower frequencies of the strain data. So before applying the band pass, we can use a technique called whitening. Whitening is a way to get rid of background noise/white noise by calculating the inverse of the signal and applying it in order to cancel it out through destructive interference. LIGO provides a handy whitening function in the tutorials section of their website.

```
#Use this whitening function provided by LIGO:
def whiten(strain, interp_psd, dt):
    Nt = len(strain)
    freqs = np.fft.rfftfreq(Nt, dt)

    # whitening: transform to freq domain, divide by asd, then transform back,
    # taking care to get normalization right.
    hf = np.fft.rfft(strain)
    white_hf = hf / (np.sqrt(interp_psd(freqs) /dt/2.))
    white_ht = np.fft.irfft(white_hf, n=Nt)
    return white_ht
```

```
#Whiten the data from H1, L1, and reference template
hstrain_whiten = whiten(hstrain, psd_H1, htime_interval)
lstrain_whiten = whiten(lstrain, psd_L1, ltime_interval)
ref_H1_whiten = whiten(ref_H1, psd_H1, htime_interval)
```

Looking at the whitening function, we can see that it first takes the DTFT of the continuous signal using numpy's fft.rfft function. This produces a one dimensional decomposition of the signal that can be thought of as a "principal component" of sorts. The decomposed signal is then scaled through division by the ASD. This division is the critical piece of the whitening function, as it scales the decomposed signal so that frequencies with more energy are weighted more heavily. The result is a "smoothed" signal that can be likened to a standardized or normalized numeric vector. A band pass to attenuate frequencies outside of the range of 20-300 Hz can then be applied to the whitened signals. Let's plot the whitened signals and the band passed signals to see how they make the true signal of the gravitational waves stand out from the background noise.

```
#Plot the cleaned detector strains
fig = plt.figure()
numSamples = len(hstrain_whiten)
plth = fig.add_subplot(221)
plth.plot(htime[0:numSamples], hstrain_whiten[0:numSamples])
plth.set_xlabel('Time (seconds)')
plth.set_ylabel('H1 Strain')
plth.set_title('Whitened H1 Strain')
numSamples = len(lstrain_whiten)
pltl = fig.add_subplot(222)
pltl.plot(ltime[0:numSamples], lstrain_whiten[0:numSamples])
pltl.set_xlabel('Time (seconds)')
```

```
pltl.set_ylabel('L1 Strain')
pltl.set_title('Whitened L1 Strain')
pltref = fig.add_subplot(212)
pltref.plot(reftime, ref_H1_whiten)
pltref.set_xlabel('Time (seconds)')
pltref.set_ylabel('Template Strain')
pltref.set_title('Whitened Template')
fig.tight_layout()
plt.show()
plt.close(fig)

#Apply band pass to remove everything outside of the desired spectrum
(b,a) = butter(4, [20/(samplefreq/2.0), 300/(samplefreq/2.0)], btype='pass')
hstrain_whitenbp = filtfilt(b, a, hstrain_whiten)
lstrain_whitenbp = filtfilt(b, a, lstrain_whiten)
ref_H1_whitenbp = filtfilt(b, a, ref_H1_whiten)

#Plot the cleaned detector strains
fig = plt.figure()
numSamples = len(hstrain_whitenbp)
plth = fig.add_subplot(221)
plth.plot(htime[0:numSamples], hstrain_whitenbp[0:numSamples])
plth.set_xlabel('Time (seconds)')
plth.set_ylabel('H1 Strain')
plth.set_title('Whitened and Band Passed H1 Strain')
numSamples = len(lstrain_whitenbp)
pltl = fig.add_subplot(222)
pltl.plot(ltime[0:numSamples], lstrain_whitenbp[0:numSamples])
pltl.set_xlabel('Time (seconds)')
pltl.set_ylabel('L1 Strain')
pltl.set_title('Whitened and Band Passed L1 Strain')
pltref = fig.add_subplot(212)
pltref.plot(reftime, ref_H1_whitenbp)
pltref.set_xlabel('Time (seconds)')
pltref.set_ylabel('Template Strain')
pltref.set_title('Whitened and Band Passed Template')
fig.tight_layout()
plt.show()
plt.close(fig)
```

We can now see a little spike near the middle of each strain signal. That is the signal we are looking for, but at this point we do not yet know this for sure. The next step is to compare the strains to the reference template. A cross correlation can be used to find any parts of the strain signals that match closely with the reference template. A **cross correlation** measures how similar two signals are by applying a sliding dot product along the length of the signals. Since the reference template has a different lengh than the strains, we set the mode to 'valid', meaning the shorter signal is move along the longer signal. When the signals overlap, their dot product will be maximized. If we plot the cross correlation, we will see spikes in the positive and negative directions when the template matches up to the strain signals. Note that this process can be applied to any kind of signal analysis where a template needs to be compared to a signal to determine if any part of the signal matches the template.

```
#Plot the cross correlation between each detection strain and the reference template
hcorr = np.correlate(hstrain_whitenbp, ref_H1_whitenbp, 'valid')
```

```
lcorr = np.correlate(lstrain_whitenbp, ref_H1_whitenbp, 'valid')
fig = plt.figure()
plthcorr = fig.add_subplot(211)
plthcorr.plot(hcorr)
pltlcorr = fig.add_subplot(212)
pltlcorr.plot(lcorr)
plthcorr.set_title('H1 and L1 Strain & Template Correlations')
fig.tight_layout()
plt.show()
plt.close(fig)
```

Looking at the plot of the cross correlation and the whitened strain plots, the event appears to have occurred between 20 and 25 seconds into the signal. Since the time period begins at 5 on the plot though, we must subtract 5 seconds from that range. We can plot the cross correlation of the H1 strain and reference template for that range, but adjusting the time axis a bit results in a better plot. After a few adjustments to zero in on the event, we can view the H1 and L1 strains compared to the reference template on the same plot.

```
#Plot the whitened H1 strain-template strain correlation between 15 and 17 seconds
startind = np.where(htime==(min(htime)+15))[0][0]
endind = np.where(htime==(min(htime)+17))[0][0]
hcorr = np.correlate(hstrain_whitenbp[startind:endind], ref_H1_whitenbp, 'valid')
plt.plot(hcorr)
plt.title('H1 & Template Correlation')
plt.show()


#Zero in on the event
startind = np.where(htime==(min(htime)+16.25))[0][0]
endind = np.where(htime==(min(htime)+16.5))[0][0]
plt.plot(htime[startind:endind], hstrain_whitenbp[startind:endind], 'b', label='H1
Strain')
plt.plot(htime[startind:endind], lstrain_whitenbp[startind:endind], 'r', label='L1
Strain')
plt.xlabel('Time (seconds)')
plt.ylabel('Strains')
plt.title('Whitened Strains')
plt.show()
```

The plots show that the H1 and L1 strains align very closely in this time interval. That means that whatever the event was, it was detected simultaneously by both observatories. This helps rule out the possibility that the event was a mistake, like a construction worker dropping a hammer nearby. Now that we have a good idea where the event occurred in the signal, we can cheat by looking at the actual time of the event provided by LIGO: Monday September 14, 2015 09:40:45 GMT. If we plot our cleaned strain signals and the reference template for a few time points around that event time, we can see that everything matches up closely. We can be confident that we have found a gravitational wave signal.

```
#Using the time of the event provided by LIGO, see if we found it
tevent = 1126259462.422     #Mon Sep 14 09:50:45 GMT 2015
deltat = 5.                 #Seconds around the event
indxt = np.where((htime >= tevent-deltat) & (htime < tevent+deltat))
plt.figure()
plt.plot(htime-tevent, hstrain_whitenbp,'r', label='H1 Strain')
plt.plot(htime-tevent, lstrain_whitenbp,'g', label='L1 Strain')
```

```
plt.plot(reftime+0.002, ref_H1_whitenbp,'k', label='Expected Strain')
plt.xlim([-0.1,0.05])
plt.ylim([-4,4])
plt.xlabel('Time (s) Since '+str(tevent))
plt.ylabel('Strain')
plt.legend(loc='lower left')
plt.title('Whitened Strain vs Expected Strain')
plt.show()
```

We have now found the gravitational wave signal. It would be interesting to hear the signal though, so let's export it to a wav file using scikit.io.

```
#Define a function to write to a wav file
def write_wav(data, samplefreq, filename):
    d = np.int16(data/np.max(np.abs(data)) * 32767 * 0.9)
    wavfile.write(filename, int(samplefreq), d)

#Write a wav file for each detector strain and the template
write_wav(hstrain_whitenbp[indxt], samplefreq, "H1_whitenbp.wav")
write_wav(lstrain_whitenbp[indxt], samplefreq, "L1_whitenbp.wav")
write_wav(ref_H1_whitenbp, samplefreq, "Ref_Template_whitenbp.wav")
```

While the whitening that was performed in Python was satisfactory, it is easier to hear the signal by processing it further in a digital audio workstation, like FL Studio. Using FL Studio and a plugin called Edison, the H1 strain audio is able to be de-noised even further. Equalization can also be applied to focus on the frequency range between 30 and 200 Hz. These actions make the gravity wave signal very clearly stand out in the audio file. It is also possible to view the frequency spectrum of the audio clip in FL Studio, using a plugin called Wave Candy. Turning the scale all the way up and setting the resolution to maximum results in the spectral graph in this video:

https://www.youtube.com/watch?v=IYq39kCjUns. The signal appears around 0:06 and sounds like a heartbeat.

27. The Data Analytics Process

In this book, we have explored data science through the lens of statistics. The first four chapters laid the foundation of statistics and data analysis. Then the rest of the book described a plethora of modeling techniques. It is important for data analytics professionals to know a variety of modeling techniques, but often there is too much emphasis placed on knowing modeling algorithms. The heart and soul of data science is data. Knowing how to use data to solve problems is more important than knowing every algorithm. So it seems fitting to end this book with a chapter that ties the whole process together, and lays out a framework that can be followed to approach any problem. The process we will follow is shown below. This process is called the **data analytics lifecycle**. Some sources may rename the steps, combine them, or break them into finer categories, but the general process is the same.

1. Defining the Problem
2. Gathering Data
3. Exploratory Data Analysis
4. Data Modeling
5. Interpreting and Communicating the Results
6. Putting Models into Production
7. Maintaining Models

Each of these steps will now be explained in detail.

Defining the Problem

Data science projects all start with a single problem statement. It is critical that the problem be as specific as possible, or else the entire data science project could be derailed before it can begin. All too often, customers will ask data scientists to provide them with wood, nails, and a hammer, when the customer's ultimate desire is a house. Data scientists are meant to build houses, but if the requirements are too loosely defined, or the problem is not completely identified, then neither the customer nor the data scientist is satisfied. The responsibility to obtain a clearly defined goal lies with the data scientist, and not the customer. The customer often may not even know what their ultimate goal is, so it is up to the data scientist to put themselves into their customer's shoes and figure out what the most useful product would be.

For a concrete example, consider the following problem: ABC Corp recently launched a social advertising campaign and wants to know whether the campaign has increased the sales of widgets. Suppose for the sake of argument that ABC Corp is able to find out where every customer heard about its widgets. ABC Corp hires a data scientist named Carol and asks her to determine how much sales has increased due to the advertising campaign.

The first thing Carol should do is reframe the question: has the campaign changed sales, and if so, in what way? Assuming that sales have increased due to the campaign is asking for trouble, as it paves the way for confirmation bias. The second task for Carol is to choose a target variable. In this case, sales is the clear target, but should the number of sales be used or the dollar amount of sales? If a customer buys 10 widgets, that would count as 1 sale. So instead of using the number of sales, or the number of transactions, the number of widgets sold should be used. Furthermore, it is important to look at widget sales for a constant price to ensure that the price did not play a factor in the number of sales. One more task for Carol is to determine whether the project will be an experiment or an observational study. In this case, there is no control group and Carol cannot directly manipulate any of the variables, so she must rely on data provided by ABC Corp on sales and source. Therefore, she will be performing an observational study and can only show correlation, not prove causation. In summary, Carol will perform a hypothesis test to see if the campaign has any effect on the number of widgets sold. Carol should inform ABC Corp about what she is able to do and why, so that their expectations from her are clear.

Another common scenario is for a company to recognize that it has a lot of data, but does not know how to use it to discover insights and trends. So the company hires a data scientist with the mandate: "tell us what our data reveals." In this case, there is no specific goal or expectation. The company expects the data scientist to be a wizard and find the answer to any question using the company's data. Once again, it is the data scientist's responsibility to put themselves into the shoes of their customer to figure out what would be beneficial.

Suppose that XYZ Corp hires a new Vice President of Human Resources. The new VP knows that data science is the hottest trend in HR analytics, so they hire Carol the data scientist to work her magic. Her only task is to "modernize XYZ Corp's HR analytics through the use of data science". Carol puts herself in the VP's shoes and decides that the biggest problem facing the HR department is the length of time it takes to onboard new employees. So she designs an observational study to determine which factors most significantly affect onboarding time. Her goal will be to determine how to modify those factors to reduce onboarding time. Carol informs the VP about her plan and her expected outcome (it will save the company $X), and everybody is happy.

The goal of defining the problem is to find a way to quantitatively solve the problem using data. This involves selecting a target and determining what data to collect and how. It also involves defining the type of problem: classification, regression, clustering, forecasting, or anomaly detection. The end product should be a specific outcome, such as determining whether x variables effected y, or producing a dashboard with descriptive statistics and a few plots. The outcome will set expectations and help define a time frame for the data science project.

Gathering Data

Once a problem has been defined and a quantitative target set, the next step is to gather data. Most of the time the data to be used is stored in some kind of database. Occasionally though, there is no existing data. When no data exists, it can either be collected through web scraping or by performing an experiment. Experimentation is referred to in the business world as A/B Testing. **A/B testing** is no different than the experimental designs described in chapter 1. Remember that only experimentation can identify causation. Observation studies where data is simply queried from a database or scraped from the internet can only identify correlation.

Data that has been collected needs to be cleaned, indexed, and stored. This process is called **data curation**, and people who curate data are called **data stewards**. Data stewards are usually exerts in enterprise data management. They can design and build databases using database management systems (DBMS). For most enterprise applications, the data must be stored in a relational database (a.k.a. tabular format, like a data frame in R or Pandas). Relational data is highly structured and easy to work with. To get data into structured form, data stewards often have to extract, transform, and load the data into several locations. This is called **ETL (extract, transform, and load)**. ETL tools can be as simple as structured query language (SQL) or highly specialized tools. It is beyond the scope of this book to describe the all the nuances of enterprise data management, but the reader should be aware that it is a field unto itself.

Many businesses employ people to query data stored in databases and produce dashboards and reports with plots and descriptive statistics. This is called **business intelligence**, and the people who carry out this function are called data miners or business intelligence analysts. Some businesses have gotten caught up in the trend of data science and inaccurately post jobs for data scientists or data analysts, when the job descriptions are purely business intelligence tasks. In part, this is due to confusion about what data science is, but it is also due to deceptive recruiting practices designed to ensnare new talent. Business intelligence can extend beyond reporting to include simple modeling tasks, but generally it is distinguished by reporting what has happened or what is currently happening in the business. Prediction and prescription are tasks that lie purely in the realm of data analytics. Business intelligence reports and dashboards are supposed to enable managers to quickly digest information through carefully designed data visualizations.

In the context of a data science project, the problem to be solved usually dictates which variables and types of data would be the most useful for analysis. For example, if the project goal is to measure sales performance, then the target variable is likely sales revenue, and all of the independent variables should be factors that are thought to impact sales revenue in some way. If the target variable is unavailable, then a proxy could be used.

When data is collected, it must be cleaned, parsed, and prepared for analysis. So the data gathering phase of the data analytics lifecycle includes preliminary data cleaning. At this point in the lifecycle, feature engineering can begin as well. It is generally a good idea to build a **data dictionary** for all of the variables. Documentation is important for any project, and in the case of data analysis, proper documentation can help explain anomalies in the data. For example, if one of the variables in a dataset is temperature, and half of the days have a value of 0, it is more likely that there was a reason the temperature was recorded as 0 than the actual temperature having a value of 0. Maybe many days had no temperature recorded at all, and instead of leaving the value blank, it was set to 0. All of the idiosyncrasies like this should be documented in a data dictionary. If a data model is used to gather the data, such as a data model from a relational database, the model should be documented as well. Suppose later in the project lifecycle it is discovered that there are 2 records per day when there should only be 1 record per day. This points to a bad join somewhere in the data model, and if the model is adequately documented, the bad join should be easy to spot and correct.

For a practical example of the data gathering phase, suppose ABC Corp wants to predict employees who will quit their jobs in the near future. Let us suppose that the target variable is a binary flag indicating whether the employee quit within 6 months of observation date. There are many variables that could be collected from a human resources database that could affect quit/not quit. Length of service, supervisor, salary, the average salary of peers under the same supervisor, last performance review rating, and the number of vacation days taken within the last year are all examples of variables that could be taken from a human resources database. These fields would come from a relational database that would require a data model to retrieve. Other variables that might serve as clues to job satisfaction include sentiment analysis of emails sent by the employee and hours worked. These variables are unlikely to be stored in a human resources database. The emails would have to be mined and parsed because they are unstructured data, and the hours worked would likely come from a time and expense database. The data models, code to retrieve unstructured email text data, and any feature engineering should all be documented. A data dictionary would define all of the variables.

Exploratory Data Analysis

Once the data has been collected and put into a format for analysis, it is time to do exploratory data analysis (EDA). EDA allows data scientists to get to know the data better. One of the first checks that should be performed is a search for null values by row and column. Another check should look for outliers. These steps were described in chapter 4. Once the missing values and outliers have been taken care of, univariate and bivariate plots and statistics can be examined. Correlation plots can quickly identify numeric variables that could be collinear, as well as variables that have linear relationships with the target variable. In lieu of correlation plots, cross tabulations or mosaic plots can be used to examine the relationships between categorical variables.

EDA can also lead to new features. Recall how we created a dummy variable in the Titanic dataset to indicate whether or not a passenger was a child, because we saw that children were more likely to survive. The dummy variable enhanced the explanatory power of subsequent models. Anyone familiar with the Titanic disaster probably had the pre-conceived notion that children were more likely to have survived, and this leads to another point about the EDA phase: the EDA phase of the data analytics lifecycle should include close interaction between data scientists and subject matter experts (SMEs). Data science projects usually include SMEs who know the data better than anyone else, and they can help explain outliers, anomalies, give ideas for feature engineering, and provide knowledge that can spur ideas for new features in the minds of a data scientist. Suppose for instance that we were completely unfamiliar with the Titanic disaster, but we were able to work with a SME who told us that children were given preference in life boats. After receiving this knowledge, it would have made sense to create the dummy variable and plot it against survival. This is part of the exploratory process.

For datasets with series of observations, time series analysis should be performed. Any autocorrelation should be identified, because it has implications for the assumptions of certain models. The same can be said for outliers: EDA informs the data modeling step of the data analytics lifecycle by providing ideas about which types of models may work best for the data.

Unsupervised learning should also be performed during the exploratory phase. Unsupervised learning includes the application of clustering algorithms and was not described in this book, but it should be noted that unsupervised learning is a useful step in the exploratory phase.

During EDA, the exploration of univariate, bivariate, and even multivariate analysis can uncover relationships that can be captured by the creation of new features. This is why it is important to leave no stone unturned when exploring data. Feature engineering is critical to building good models. In practice, EDA can take up nearly half of the total time of a data science project. If there is ever a question whether enough exploratory analysis has been done, the answer can be determined by asking if any of the relationships that could be found through modeling can be predicted. If it is possible to confidently guess what models will uncover, that is an indicator of thorough EDA. If it seems obvious which models will likely perform well with the data, that is another indicator of thorough EDA; with sufficient EDA, the right model choices will be evident.

Data Modeling

Exploratory data analysis usually informs data scientists which types of models are likely to be useful for a particular dataset. Using this information, models can be constructed. The data modeling phase is given the most attention in data science textbooks, but it is actually one of the shortest phases of the data analytics lifecycle. Models can often be fitted with a single line of code.

Data models should be built using resampling methods like cross validation. They should be compared using a common metric, such as log likelihood or mean squared error. The best models should be tuned if necessary, using the methods described in chapter 8.

Class imbalances should be accounted for when modeling, using the methods described in chapter 22. Data transformations, like standardization or normalization should be performed before models are applied, as described in chapter 4. Not all models require data transformation, but if even one of the models to be compared requires it, then all of the models should be fitted to the transformed data.

Feature selection, if it was not performed through dimension reduction in the EDA phase, should be performed in the data modeling phase. It is possible for some models to show certain variables as statistically significant while other models find the same variables to not be significant. The coefficients may vary between models too. Assuming overfitting and collinearity have been ruled out, it is not a problem if these things occur. A model is just one explanation for the relationships between variables. There could be several models with equally good fits that use different combinations of variables. When there are several good model candidates, it is up to the researcher to decide when to stop modeling. It may be the case that the target variable cannot be adequately explained by the variables present in the dataset though. So if there are several good models using different variables, there may still be exogenous variables that could produce even better models. Maybe the models are stuck in local minima in the optimization of the objective function (refer back to chapter 7). At the end of the day, a decision must be made: is it satisfactory to predict the target variable with a small error, even if the perfect model has not been found? It is almost never possible to find the perfect model for large datasets with many variables, so finding a model that is "good enough" is all that can be done.

Model selection can be a confusing process. To help with the process, a model picking assistant Python script is included with the scripts for this book. It is the file titled "model_picking_assistant.py" in this book's GitHub repository. Note that the models included in the script are only the ones that pertain to this book. Other models may be available.

An important note about model selection is that models are often dependent on the wishes of the customer of a data science project. Some customers do not care about the ability to explain the prediction process behind a model, so black box models, or highly complex models are acceptable. These types of customers care more about finding models that produce little error, regardless of how they work. Other customers prefer to be able to explain the models completely. For these types of customers, simpler models are usually preferable. For example, the decision process behind a generalized additive model is not nearly as clear as the process behind a decision tree regression. Customers who want to be able to explain their models should be given a decision tree instead of the GAM, even if it means slightly higher model error. Many customers have no choice in the matter, because they are obligated to explain their modeling process to auditors or government regulators.

One last note about black boxes is that they are misnomers, because there should be no black boxes for a data scientist. A data scientist should know the models they use well enough to explain every last detail, and be able to explain them in layman's terms. Black boxes are what business people call things that they do not understand, a.k.a. those things that data scientists pull rabbits out of. If a data science project has been carried out successfully, then even the customer should be able to explain what a model is doing at a high level. It is the data scientist's job to make sure that they can.

Interpreting and Communicating the Results

To begin this section, let us reiterate what was stated at the end of the last section: it is the data scientist's job to make sure the customer understands the models and results, at least at a high level. This requires interpretation of the model's output by looking at goodness of fit metrics like the RMSE, MAE, log-likelihood, ROC curve, confusion matrix, precision, recall, and R^2, measures of statistical significance like p-values, test statistics, confidence intervals, and credible intervals, and effect sizes of the predictors.

The effect sizes of the predictors and their coefficients are often the most important aspects of statistical models to interpret. The importance of predictors should also be assessed. Effect size and variable importance show how different features influence the target variable. The reason predictor assessment is so important is because it is one aspect of modeling where common sense should trump statistical analysis. For example, a model may have a near perfect fit, but if it relies nearly completely on one predictor, then is it really a strong model? Common sense checks should be performed during predictor assessment. Let us consider the data in figure 27.1 as a hypothetical example.

y	X1	X2	X3
1	Blue	16.1	3
0	Brown	37.58	3
0	Brown	11.09	3
0	Brown	45.3	0
0	Brown	20.0	2
1	Brown	24.98	2

Figure 27.1

Suppose we wanted to predict y using the predictors X1, X2, and X3. It is clear that X1 is by far the best variable for predicting y. If we rely on X1 alone, our model would be 83% accurate at classifying y, because it would likely learn to classify (if X1==Blue then y==1) and (if X1==Brown then y==0). This seems like a decent model, but suppose y is a flag indicating whether a student passed a test, and X1 is the student's eye color. The data may happen to show that a student's ability to pass the test is correlated with eye color, but that does not make any logical sense. So it would be imprudent to blindly operationalize this model because it is based on a faulty conclusion about the relationship between y and X1. That is why it is critical to assess the impact of the predictors on the target variable after models have been built.

Communicating the results of a data science project is usually best done through data visualization. The human brain is hardwired to understand visual cues with ease, so effective data visualization cannot be underemphasized. In fact, research by IBM has shown that data visualization "externalizes the data an enables people to think about and manipulate the data at a higher level", and furthermore, that it "enables humans to think more complex thoughts about larger amounts of information than would otherwise be possible." (Keahey, 2013).

Putting Models into Production

The last stage in the data analytics lifecycle is putting models into production. In many cases, models have lifespans. Even the best models can see performance degradation over time. Therefore, it is critical to monitor models when they have been put into production. Many businesses use several models, and adjust the weights of each depending on their performance and age. For example, in quantitative finance, a newly discovered trading strategy is relied on heavily because the trend is strong. As other quants catch on to the trend, it starts to fade, so the strategy is reallocated to have lower and lower weights, until it is finally decommissioned. In industries like finance, the key to a good data science department is not finding the silver bullet, but setting up a process to churn out many models like an assembly line.

Data scientists need to have the skills to operationalize models they develop. The skills required to operationalize models include job scheduling, code vectorization, multithreading, utilizing graphics processing units (GPUs), distributed computing in Hadoop/big data architectures, and slurm for high performance computing (HPC) in supercomputers. Describing these skills in detail is beyond the scope of this book.

There is a difference between training a model on a static dataset and deploying a live model. A **live model** is a model that makes live predictions for data that is streamed in real time. For example, when a poker match is shown in television, the live odds of winning a hand are shown. Whenever a new card is drawn, the odds are updated. This is an example of a live model. Live models must be extremely efficient to produce results in real time. Models that have been fitted to old data do not necessarily need to be tuned to be efficient, so there is often work required to rewrite the code to be more efficient.

One way to improve code efficiency is by vectorizing operations. **Vectorization** means using combining operations so that they happen concurrently. For example, if 2 columns of a data frame need to be added together, one way to do it would be to loop over each row and add the 2 numbers for each row. Another option would be to treat the columns as single column matrices and add them using linear algebra, which produces the sum of every element at once. A heuristic for vectorizing code is to replace loops with matrix operations wherever possible.

Another way to improve code efficiency is to parallelize it. If a computer has several processors, the code can be multithreaded so that each processor handles a share of the load. The same paradigm is behind using GPUs. GPUs can have hundreds or thousands of processing cores, whereas a standard desktop computer at the time this book was written has 4 cores. Interfaces like CUDA for NVidia GPUs are used to pass processes to GPU cores for computation.

Other ways to improve code efficiency are to use object-oriented programming and to use generator objects or lazy evaluation. **Lazy evaluation** delays the evaluation of an expression until the value is needed. Spark, a Hadoop based technology, uses lazy evaluation to perform operations quickly. For example, when Spark is asked to query data from a table, the data is not actually retrieved until it is called for use.

Maintaining Models

A model is simply one explanation for some phenomenon. Models are often incomplete and need to be updated to stay relevant and useful. In some cases, especially in areas where models are modeling human behavior, models lose their effectiveness over time until they are no longer useful. So part of a data scientist's job is to maintain models over time. This may mean retuning the parameters, retraining the model, adding or removing features, re-binning binned features, or adjusting the model for the bias-variance tradeoff or the type I/type II error tradeoff. Sometimes the model needs to be completely replaced.

The key to model maintenance is the ability to track model performance. So every model that is operationalized should be tracked with performance metrics that can serve as indicators for any loss in effectiveness. Usually, this is as simple as recalculating error as the true values for a model's predictions become available. The challenge is minimizing the feedback time. For example, if a model is predicting a risk score for bank customers asking for loans, then it may be weeks, months, or even years before the true risk of a new customer is completely known. Therefore, the feedback time for a risk scoring model could be lengthy. If a model becomes ineffective before its error can be updated, then it is making bad predictions for an unknown length of time. It is important that this time be minimized, and this can be accomplished by shortening the feedback loop.

Wrapping Up

Thank you for reading my book. By now, you should have the knowledge to go out and build models for just about any dataset. If you enjoyed this book, please visit the website to sign up for my data science blog and to get updates about my future publications: https://www.dsilt-stats.weebly.com. Any questions about the book or notifications about errors can be submitted through the contact form on the book's website. For questions or issues related specifically to the code, please see the book's GitHub repository: https://github.com/nlinc1905/dsilt-stats-code.

References and Bibliography

Chapter 1

Anderson, Edgar. (1936, September). The Species Problem in Iris. *Annals of the Missouri Botanical Garden, vol. 23(3)*. pp. 457-509. Retrieved from: https://www.jstor.org/stable/2394164?seq=1#page_scan_tab_contents

Kaggle Inc. (2018). *Titanic: Machine Learning from Disaster*. Kaggle Dataset. Retrieved from: https://www.kaggle.com/c/titanic#description

Chapter 2

Coladarci, T. (2010). Student's t statistic. In I.B. Weiner & W.E. Craighead (Eds.), *Corsini Encyclopedia of Pyshology (4th ed.)*. Hoboken, NJ: Wiley.

Glen, Stephanie. (2018). *Normal Distributions: Definition, Word Problems*. Statistics How To. Retrieved from: http://www.statisticshowto.com/probability-and-statistics/normal-distributions/

Lu, Ying & Belitskaya-Levy, Ilana. (2015, December 25.). The Debate about p-values. *Shanghai Archives of Psychiatry, vol. 27(6)*. pp. 381-385. Retrieved from: https://www.ncbi.nlm.nih.gov/pmc/articles/PMC4858512/

Mordkoff, J. Toby. (2000/2011/2016). The Assumption(s) of Normality [PDF Slides]. Retrieved from: http://www2.psychology.uiowa.edu/faculty/mordkoff/GradStats/part%201/I.07%20normal.pdf

Quinlan, Ross. (1993). Combining Instance-Based and Model-Based Learning. In Proceedings on the Tength International Conference of Machine Learning, 236-243. University of Massachusetts, Amherst. Morgan Kaufmann.

Chapter 3

Field, Andy, Miles, Jeremy, & Field, Zoe. (2012). *Discovering Statistics Using R*. London, UK: SAGE Publications Ltd.

Glen, Stephanie. (2018). *Hypothesis Testing*. Statistics How To. Retrieved from: http://www.statisticshowto.com/probability-and-statistics/hypothesis-testing/

McTague, Tom. (2015, June 25). *Britain's mid-life crisis: UK average age hits 40 for the first time as population jumps 500,000 to 64.6 million*. Daily Mail. Associated Newspapers Ltd. Retrieved from: http://www.dailymail.co.uk/news/article-3138853/Britain-s-mid-life-crisis-UK-average-age-hits-40-time-population-jumps-500-000-64-6-million.html

Chapter 4

Buuren, Stef van & Groothuis-Oudshoorn, Karin. (2010). MICE: Multivariate Imputation by Chained Equations in R. *Journal of Statistical Software, vol. VV(II)*. pp. 1-68. Retrieved from: https://www.jstatsoft.org/article/view/v045i03

Buuren, Stef van. (2018, May 26). *Package 'mice'*. R Package from CRAN. Retrieved from: https://cran.r-project.org/web/packages/mice/mice.pdf

Dorogush, Anna Veronika, Ershov, Vasily, & Gulin, Andrey. (2017). *CatBoost: gradient boosting with categorical features support*. Workshop on ML Systems at NIPS 2017. Yandex. Retrieved from: http://learningsys.org/nips17/assets/papers/paper_11.pdf

Glen, Stephanie. (2018a). *Interquartile Range (IQR): What it is and How to Find it*. Statistics How To. Retrieved from: http://www.statisticshowto.com/cooks-distance/

Glen, Stephanie. (2018b). *Cook's Distance/Cook's D: Definition, Interpretation*. Statistics How To. Retrieved from: http://www.statisticshowto.com/probability-and-statistics/interquartile-range/

Grubbs, Frank. (1969, Februaray). Procedures for Detecting Outlying Observations in Samples. *Technometrics, vol. 11(1)*. pp. 1-21.

Nigrini, Mark J. (2012). *Benford's Law: Applications for Forensic Accounting, Auditing, and Fraud Detection*. Hoboken, NJ: John Wiley & Sons, Inc.

NIST/SEMATECH e-Handbook of Statistical Methods. (2018a). 1.3.5.17. Detection of Outliers. Retrieved from: https://www.itl.nist.gov/div898/handbook/eda/section3/eda35h.htm

NIST/SEMATECH e-Handbook of Statistical Methods. (2018b). 1.3.5.17.3. Generalized ESD Test for Outliers. Retrieved from: https://www.itl.nist.gov/div898/handbook/eda/section3/eda35h3.htm

Press, Gil. (2016, March 23). *Cleaning Big Data: Most Time-Consuming, Least Enjoyable Data Science Task, Survey Says*. Forbes. Retrieved from: https://www.forbes.com/sites/gilpress/2016/03/23/data-preparation-most-time-consuming-least-enjoyable-data-science-task-survey-says/#635a3b476f63

White, IR, Royston, P, & Wood, AM. (2011, February 20). Multiple Imputation using Chained Equations: Issues and Guidance for Practice. *Statistics in Medicine, vol. 30(4)*. pp. 377-399. Retrieved from: https://www.ncbi.nlm.nih.gov/pubmed/21225900

Chapter 5

Field, Andy, Miles, Jeremy, & Field, Zoe. (2012). *Discovering Statistics Using R*. London, UK: SAGE Publications Ltd.

Galton, Francis. (1886). Regression towards mediocrity in hereditary stature. *Journal of the Anthropological Institute of Great Britain and Ireland, vol. 15*. pp. 246–263. Retrieved from: https://books.google.com/books?id=60aL0zlT-90C&pg=PA240#v=onepage&q&f=false

Glen, Stephanie. (2018). *Fisher Z-Transformation*. Statistics How To. Retrieved from: http://www.statisticshowto.com/fisher-z/

Pearson, Karl. (1895, June 20). Notes on regression and inheritance in the case of two parents. *Proceedings of the Royal Society of London, vol. 58*. pp. 240–242.

Chapter 6

Field, Andy, Miles, Jeremy, & Field, Zoe. (2012). *Discovering Statistics Using R*. London, UK: SAGE Publications Ltd.

Fitbit Inc. (2018). Retrieved from: https://www.fitbit.com/home

Hastie, Trevor, Tibshirani, Robert, & Friedman, Jerome. (2009). *The Elements of Statistical Learning: Data Mining, Inference, and Prediction (2nd ed.)*. New York, NY: Springer.

James, Gareth, Witten, Daniela, Hastie, Trevor, & Tibshirani, Robert. (2013). *An Introduction to Statistical Learning: with Applications in R (6th ed.)*. New York, NY: Springer.

Chapter 7

Boyd, Stephen, & Vandenberghe, Lieven. (2004). *Convex Optimization*. Cambridge, UK: Cambridge University Press.

Kennedy, James, & Eberhart, Russell. (2011). *Particle Swarm Optimization*. Sammut C., Webb G.I. (eds) Encyclopedia of Machine Learning. Boston, MA: Springer. Retrieved from: http://iranarze.ir/wp-content/uploads/2016/06/3226-English.pdf

McCall, John. (2004, February 27). Genetic algorithms for modelling and optimization. *Journal of Computational and Applied Mathematics, vol. 184*. pp. 205-222. Retrieved from: https://www.sciencedirect.com/science/article/pii/S0377042705000774

Ruder, Sebastian. (2016, January 19). An Overview of Gradient Descent Optimization Algorithms. Retrieved from: http://ruder.io/optimizing-gradient-descent/

Segaran, Toby. (2007). *Programming Collective Intelligence: Building Smart Web 2.0 Applications*. Sebastopol, CA: O'Reilly Media, Inc.

Chapter 8

Anna B. (2016). *Effectively running thousands of experiments: Hyperopt with Sacred*. Lab41. Retrieved from: https://gab41.lab41.org/effectively-running-thousands-of-experiments-hyperopt-with-sacred-dfa53b50f1ec

Bergstra, James, Bardenet, Remi, Bengio, Yoshua, & Kegl, Balazs. (2011). Algorithms for Hyper-Parameter Optimization. Part of: *Advances in Neural Information Processing Systems 24*. Retrieved from: http://papers.nips.cc/paper/4443-algorithms-for-hyper-parameter-optimization

Bergstra, J., Yamins, D., Cox, D. D. (2013). Making a Science of Model Search: Hyperparameter Optimization in Hundreds of Dimensions for Vision Architectures. To appear in Proc. of the 30th International Conference on Machine Learning (ICML 2013).

Brochu, Eric, Cora, Vlad M., & Nando de Freitas. (2010, December 12). A Tutorial on Bayesian Optimization of Expenseive Cost Functions, with Application to Active User Modeling and Hierarchical Reinforcement Learning. Eprint arXiv: 1012.2599. Retrieved from: https://arxiv.org/abs/1012.2599

Gonzalez, Javier. (2017, February 7). Introduction to Bayesian Optimization [PDF Slides]. Lancaster University. Retrieved from: http://gpss.cc/gpmc17/slides/LancasterMasterclass_1.pdf

Chapter 9

Field, Andy, Miles, Jeremy, & Field, Zoe. (2012). *Discovering Statistics Using R*. London, UK: SAGE Publications Ltd.

Hastie, Trevor, Tibshirani, Robert, & Friedman, Jerome. (2009). *The Elements of Statistical Learning: Data Mining, Inference, and Prediction (2nd ed.)*. New York, NY: Springer.

James, Gareth, Witten, Daniela, Hastie, Trevor, & Tibshirani, Robert. (2013). *An Introduction to Statistical Learning: with Applications in R (6th ed.)*. New York, NY: Springer.

Raschka, Sebastian. (2015). *Python Machine Learning: Unlock Deeper Insights into Machine Learning with this Vital Guide to Cutting-Edge Predictive Anlytics*. Burmingham, UK: Packt Publishing.

Chapter 10

Introduction to SAS. (2018a). *Negative Binomial Regression*. UCLA: Statistical Consulting Group. Retrieved from: https://stats.idre.ucla.edu/stata/dae/negative-binomial-regression/

Introduction to SAS. (2018b). *Zero-Inflated Poisson Regression*. UCLA: Statistical Consulting Group. Retrieved from: https://stats.idre.ucla.edu/r/dae/zip/

Kolassa, Stephan. (2010, September 23). *Why is Poisson Regression used for Count Data?* Cross Validated. Retrieved from: https://stats.stackexchange.com/questions/3024/why-is-poisson-regression-used-for-count-data

Pennsylvania State University. (2018). Stat 504: Lesson 9: Poisson Regression. Retrieved from: https://onlinecourses.science.psu.edu/stat504/node/168/

Chapter 11

Tobin, James. (1958, January). Estimation of Relationships for Limited Dependent Variables. *Econometrica vol. 26(1)*. Pp. 24-36.

Chapter 12

Chowdavarapu, Indra kiran, Shepherd, Dr. Scott, McGaugh, Dr. Miriam. (2017). *Analyzing the Effectiveness of COPD Drugs Through Statistical Tests and Sentiment Analysis*. SAS Paper 1031-2017. Retrieved from: http://support.sas.com/resources/papers/proceedings17/1031-2017.pdf

Field, Andy, Miles, Jeremy, & Field, Zoe. (2012). *Discovering Statistics Using R*. London, UK: SAGE Publications Ltd.

Laerd Statistics. (2018). *One-way ANOVA*. Lund Research Ltd. Retrieved from: https://statistics.laerd.com/statistical-guides/one-way-anova-statistical-guide-4.php

Chapter 13

Field, Andy, Miles, Jeremy, & Field, Zoe. (2012). *Discovering Statistics Using R*. London, UK: SAGE Publications Ltd.

Kohavi, Ron. (1996). Scaling Up the Accuracy of Naive-Bayes Classifiers: a Decision-Tree Hybrid. *Proceedings of the Second International Conference on Knowledge Discovery and Data Mining.* Retrieved from: https://archive.ics.uci.edu/ml/datasets/adult

Chapter 14

Field, Andy, Miles, Jeremy, & Field, Zoe. (2012). *Discovering Statistics Using R.* London, UK: SAGE Publications Ltd.

Glen, Stephanie. (2018). *Kruskal Wallis H Test: Definition, Examples & Assumptions.* Statistics How To. Retrieved from: http://www.statisticshowto.com/fisher-z/

Krzywinski, Martin, & Altman, Naomi. (2014, April 29). Points of significance: Nonparametric tests. *Nature Methods, vol. 11.* pp. 467-468. Retrieved from: https://www.nature.com/articles/nmeth.2937

Chapter 15

Deepajothi, S., and Selvarajan, S. (2012, October). A Comparative Study of Classification Techniques on Adult Data Set. *International Journal of Engineering Research and Technology, vol. 1(8).*

Field, Andy, Miles, Jeremy, & Field, Zoe. (2012). *Discovering Statistics Using R.* London, UK: SAGE Publications Ltd.

Hastie, Trevor, Tibshirani, Robert, & Friedman, Jerome. (2009). *The Elements of Statistical Learning: Data Mining, Inference, and Prediction (2nd ed.).* New York, NY: Springer.

James, Gareth, Witten, Daniela, Hastie, Trevor, & Tibshirani, Robert. (2013). *An Introduction to Statistical Learning: with Applications in R (6th ed.).* New York, NY: Springer.

Pennsylvania State University. (2018). Stat 505: Lesson 10: Discriminant Analysis. Retrieved from: https://newonlinecourses.science.psu.edu/stat505/node/89/

Poulsen, John & French, Aaron. *Discriminant Function Analysis (DA) [PDF Document].* San Francisco State University. Retrieved from: http://userwww.sfsu.edu/efc/classes/biol710/discrim/discrim.pdf

Ttnphns. (2013, October 1). *Three versions of discriminant analysis: differences and how to use them.* Cross Validated. Retrieved from: https://stats.stackexchange.com/questions/71489/three-versions-of-discriminant-analysis-differences-and-how-to-use-them

Chapter 16

Barber, David. (2012). *Bayesian Reasoning and Machine Learning.* Cambridge, UK: Cambridge University Press.

Field, Andy, Miles, Jeremy, & Field, Zoe. (2012). *Discovering Statistics Using R.* London, UK: SAGE Publications Ltd.

Hastie, Trevor, Tibshirani, Robert, & Friedman, Jerome. (2009). *The Elements of Statistical Learning: Data Mining, Inference, and Prediction (2nd ed.).* New York, NY: Springer.

Husson, Francois, Le, Sebastian, & Pages, Jerome. (2017, May 8). Exploratory Multivariate Analysis by Example Using R (2nd ed.). New York, NY: Chapman and Hall/CRC.

Powell, Victor, & Lehe, Lewis. (2018). *Principal Component Analysis.* Explained Visually. Retrieved from: http://setosa.io/ev/principal-component-analysis/

Raschka, Sebastian. (2015). *Python Machine Learning: Unlock Deeper Insights into Machine Learning with this Vital Guide to Cutting-Edge Predictive Anlytics.* Burmingham, UK: Packt Publishing.

Rashka, Sebastian. (2014, August 3). *Linear Discriminant Analysis: Bit by Bit.* Retrieved from: https://sebastianraschka.com/Articles/2014_python_lda.html

Chapter 17

Glen, Stephanie. (2018). Multidimensional Scaling: Definition, Overview, Examples. Statistics How To. Retrieved from: http://www.statisticshowto.com/multidimensional-scaling/

Hastie, Trevor, Tibshirani, Robert, & Friedman, Jerome. (2009). *The Elements of Statistical Learning: Data Mining, Inference, and Prediction (2nd ed.).* New York, NY: Springer.

Roweis, Sam, & Saul, Lawrence. (2000, December 22). Nonlinear dimensionality reduction by locally linear embedding. *Science, vol. 290(5500).* pp. 2323-2326. Retrieved from: https://cs.nyu.edu/~roweis/lle/

Sklearn User Guide. (2018). Retrieved from: http://scikit-learn.org/stable/modules/classes.html#module-sklearn.manifold

Tenenbaum, Joshua B., de Silva, Vin, & Langfor, John C. (2000, December 22). A Global Geometric Framework for Nonlinear Dimensionality Reduction. *Science, vol. 290(5500).* pp. 2319-2322. Retrieved from: http://web.mit.edu/cocosci/Papers/sci_reprint.pdf

Van der Maaten, L.J.P. & Hinton, G.E. (2008, November 9). Visualizing High-Dimensional Data Using t-SNW. *Journal of Machine Learning Research, vol. 9.* pp. 2579-2605. Retrieved from: https://lvdmaaten.github.io/tsne/

Wattenberg, Martin, Viegas, Fernanda, & Johnson, Ian. (2016, October 13). *How to Use t-SNE Effectively.* Retrieved from: https://distill.pub/2016/misread-tsne/

Chapter 18

Field, Andy, Miles, Jeremy, & Field, Zoe. (2012). *Discovering Statistics Using R.* London, UK: SAGE Publications Ltd.

Chapter 19

Hastie, Trevor, Tibshirani, Robert, & Friedman, Jerome. (2009). *The Elements of Statistical Learning: Data Mining, Inference, and Prediction (2nd ed.).* New York, NY: Springer.

James, Gareth, Witten, Daniela, Hastie, Trevor, & Tibshirani, Robert. (2013). *An Introduction to Statistical Learning: with Applications in R (6th ed.).* New York, NY: Springer.

Chapter 20

Aggarwal, Charu C., Hinneburg, Alexander, & Keim, Daniel A. (2001). *On the Surprising Behavior of Distance Metrics in High Dimensional Space*. Berlin, Germany: Springer-Verlag. Retrieved from: https://bib.dbvis.de/uploadedFiles/155.pdf

Hastie, Trevor, Tibshirani, Robert, & Friedman, Jerome. (2009). *The Elements of Statistical Learning: Data Mining, Inference, and Prediction (2nd ed.)*. New York, NY: Springer.

James, Gareth, Witten, Daniela, Hastie, Trevor, & Tibshirani, Robert. (2013). *An Introduction to Statistical Learning: with Applications in R (6th ed.)*. New York, NY: Springer.

Larsen, Kim. (2015, July 30). *GAM: The Preditive Modeling Silver Bullet*. Retrieved from: https://multithreaded.stitchfix.com/blog/2015/07/30/gam/

Raschka, Sebastian. (2015). *Python Machine Learning: Unlock Deeper Insights into Machine Learning with this Vital Guide to Cutting-Edge Predictive Anlytics*. Burmingham, UK: Packt Publishing.

Chapter 21

Field, Andy, Miles, Jeremy, & Field, Zoe. (2012). *Discovering Statistics Using R*. London, UK: SAGE Publications Ltd.

Goldstein, H., Rasbash, J., et. al. (1993). A multilevel analysis of school examination results. *Oxford Review of Education, vol. 19*. pp. 425-433.

Knowles, Jared. (2013, November 25). *Getting Started with Mixed Effects Models in R*. Retrieved from: https://www.jaredknowles.com/journal/2013/11/25/getting-started-with-mixed-effect-models-in-r

University of Bristol Centre for Multilevel Modeling. (2018). Exam Dataset. Retrieved from: http://www.bristol.ac.uk/cmm/learning/support/datasets/

Chapter 22

Blagus, Rok, & Lusa, Lara. (2013, March 22). SMOTE for high-dimensional class-imbalance. *BMC Bioinformatics, vol. 14*. Retrieved from: https://www.ncbi.nlm.nih.gov/pmc/articles/PMC3648438/

Brownlee, Jason. (2017, June 5). *How to Calculate Bootstrap Confidence Intervals for Machine Learning Results in Python*. Machine Learning Mastery. Retrieved from: https://machinelearningmastery.com/calculate-bootstrap-confidence-intervals-machine-learning-results-python/

Hastie, Trevor, Tibshirani, Robert, & Friedman, Jerome. (2009). *The Elements of Statistical Learning: Data Mining, Inference, and Prediction (2nd ed.)*. New York, NY: Springer.

James, Gareth, Witten, Daniela, Hastie, Trevor, & Tibshirani, Robert. (2013). *An Introduction to Statistical Learning: with Applications in R (6th ed.)*. New York, NY: Springer.

Raschka, Sebastian. (2015). *Python Machine Learning: Unlock Deeper Insights into Machine Learning with this Vital Guide to Cutting-Edge Predictive Anlytics*. Burmingham, UK: Packt Publishing.

Chapter 23

Barber, David. (2012). *Bayesian Reasoning and Machine Learning*. Cambridge, UK: Cambridge University Press.

Charniak, Eugene. (1991). Bayesian Networks without Tears. *AI Magazine, vol. 12(4)*. pp. 50-63.

Folkman, Tyler. (2015, February 27). *Introduction to Bayes Theorem with Python*. Dataconomy. Retrieved from: http://dataconomy.com/2015/02/introduction-to-bayes-theorem-with-python/

Ghahramani, Zoubin. (2001). An Introduction to Hidden Markov Models and Bayesian Networks. *International Journal of Pattern Recognition and Artificial Intelligence, vol. 15(1)*. pp. 9-42.

Glen, Stephanie. (2018). *Probability Introduction: Articles and Videos with Solutions!* Statistics How To. Retrieved from: http://www.statisticshowto.com/probability-and-statistics/probability-main-index/

Halls-Moore, Michael L. (2016). *Advanced Algorithmic Trading*. Retrieved from: https://www.quantstart.com/advanced-algorithmic-trading-ebook

Koehrsen, William. (2018, February 10). *Markov Chain Monte Carlo in Python: A Complete Real-World Implementation*. Towards Data Science. Retrieved from: https://towardsdatascience.com/markov-chain-monte-carlo-in-python-44f7e609be98

McDermott, Ryan. (2016, August 30). *Trump-speeches*. GitHub Repository. Retrieved from: https://github.com/ryanmcdermott/trump-speeches

Salvatier, John, Wiecki, Thomas V., & Fonnesbeck, Christopher. (2018). *Getting Started with PyMC3*. PyMC3 User Guide. Retrieved from: http://docs.pymc.io/notebooks/getting_started#Installation

Shaver, Ben. (2017, December 22). *Simulating Text with Markov Chains in Python*. Towards Data Science. Retrieved from: https://towardsdatascience.com/simulating-text-with-markov-chains-in-python-1a27e6d13fc6

Rice, Ken. (2014). *Bayesian Statistics (a very brief introduction) [Powerpoint Slides]*. Retrieved from: http://faculty.washington.edu/kenrice/BayesIntroClassEpi515kmr2016.pdf

Tim. (2016, December 20). *Bayes regression: how is it done in comparison to standard regression?* Cross Validated Stack Exchange. Retrieved from: https://stats.stackexchange.com/questions/252577/bayes-regression-how-is-it-done-in-comparison-to-standard-regression

Tsvetovat, Maksim, & Kouznestov, Alexander. (2011). *Social Network Analysis for Startups*. Sebastopol, CA: O'Reilly Media, Inc.

Chapter 24

Anton. (2018, April 5). *Pykalman: (default) handling of missing values*. Stack Overflow. Retrieved from: https://stackoverflow.com/questions/49662567/pykalman-default-handling-of-missing-values

Babb, Tim. (2015, August 11). *How a Kalman filter works, in pictures*. Bzarg. Retrieved from:
http://www.bzarg.com/p/how-a-kalman-filter-works-in-pictures/

Barber, David. (2012). *Bayesian Reasoning and Machine Learning*. Cambridge, UK: Cambridge
University Press.

Czerniak, Greg. (2018). *Kalman Filter Guide*. Retrieved from: http://greg.czerniak.info/guides/kalman1/

Emiswelt. (2016, June 20). *How do I choose the parameters of a Kalman filter?* Signal Processing Stack
Exchange. Retrieved from: https://dsp.stackexchange.com/questions/31632/how-do-i-choose-
the-parameters-of-a-kalman-filter

Esme, Bilgin. (2009, March). *Kalman Filter for Dummies*. Retrieved from:
http://bilgin.esme.org/BitsAndBytes/KalmanFilterforDummies

Grewal, Mohinder S., & Andrews, Angus P. (2001). *Kalman Filtering: Theory and Practice Using MATLAB
(2nd ed.)*. New York, NY: John Wiley & Sons, Inc.

Halls-Moore, Michael L. (2016). *Advanced Algorithmic Trading*. Retrieved from:
https://www.quantstart.com/advanced-algorithmic-trading-ebook

Hoover, Adam. (2017, November 21). *Introduction to Markov Models [PDF Slides]*. Clemson University.
Retrieved from: http://cecas.clemson.edu/~ahoover/ece854/refs/Ramos-Intro-HMM.pdf

Marcel. (2015, December 19). *Kalman Filter Transition Matrix*. Cross Validated Stack Exchange.
Retrieved from: https://stats.stackexchange.com/questions/107844/kalman-filter-transition-
matrix

Maucher, Johannes. (2013). *Tracking*. Computer Vision with Python. Retrieved from:
https://www.hdm-stuttgart.de/~maucher/Python/ComputerVision/html/Tracking.html

Protopapas, Pavlos. (2014, April 10). *Lecture 19: Hidden Markov Models*. Harvard University. Retrieved
from: http://iacs-courses.seas.harvard.edu/courses/am207/blog/lecture-18.html

Read, Jonathon. (2011, October 14). *Hidden Markov Models and Dynamic Programming*. Retrieved
from:
http://www.uio.no/studier/emner/matnat/ifi/INF4820/h11/undervisningsmateriale/20111014-
notes.pdf

Stamp, Mark. (2018, January 12). *A Revealing Introduction to Hidden Markov Models*. San Jose State
University. Retrieved from: https://www.cs.sjsu.edu/~stamp/RUA/HMM.pdf

Yildirim, Ilker. (2012a, August). Bayesian Inference: Metropolis-Hastings Sampling. University of
Rochester. Retrieved from:
http://www.mit.edu/~ilkery/papers/MetropolisHastingsSampling.pdf

Yildirim, Ilker. (2012b, August). Bayesian Inference: Gibbs Sampling. University of Rochester. Retrieved
from: http://www.mit.edu/~ilkery/papers/GibbsSampling.pdf

Chapter 25

Berardinelli, Carl. (2018). *A Guide to Control Charts*. i Six Sigma. Retrieved from: https://www.isixsigma.com/tools-templates/control-charts/a-guide-to-control-charts/

Box, G. E. P., Jenkins, G. M. and Reinsel, G. C. (1976) *Time Series Analysis, Forecasting and Control.* Third Edition. Holden-Day. Series G.

Christopher, Brian. (2016, November 8). *Time Series Analysis (TSA) in Python – Linear Models to GARCH.* Blackarbs LLC. Retrieved from: http://www.blackarbs.com/blog/time-series-analysis-in-python-linear-models-to-garch/11/1/2016#ARCH

Halls-Moore, Michael L. (2016). *Advanced Algorithmic Trading*. Retrieved from: https://www.quantstart.com/advanced-algorithmic-trading-ebook

James, Nicholas A., Kejariwal, Arun, Matteson, David S. (2014, November 28). Leveraging Cloud Data to Mitigate User Experience from 'Breaking Bad'. Cornell University and Twitter Inc. Retrieved from: https://arxiv.org/pdf/1411.7955.pdf

James, Nicholas A., & Matteson, David S. (2015, January 21). Ecp: An R Package for Nonparametric Multiple Change Point Analysis of Multivariate Data. *Journal of Statistical Software, vol. 62(7).* Retrieved from: https://www.jstatsoft.org/article/view/v062i07

Kipnis, Ilya. (2015, January 21). *An Introduction to Change Points (packages: ECP and BreakoutDetection).* R Bloggers. Retrieved from: https://www.r-bloggers.com/an-introduction-to-change-points-packages-ecp-and-breakoutdetection/

Matteson, David S. & James, Nicholas A. (2013, October 16). A Nonparametric Approach for Multiple Change Point Analysis of Multivariate Data. Cornell University. Retrieved from: https://arxiv.org/pdf/1306.4933v2.pdf

Nau, Robert. (2018). *Identifying the order of differencing in an ARIMA model*. Duke University. Retrieved from: https://people.duke.edu/~rnau/411arim2.htm

Nigrini, Mark J. (2012). *Benford's Law: Applications for Forensic Accounting, Auditing, and Fraud Detection.* Hoboken, NJ: John Wiley & Sons, Inc.

Nolan, Deborah, & Lang, Duncan Temple. (2015). *Data Science in R: A Case Studies Approach to Computational Reasoning and Problem Solving.* Boca Raton, FL: CRC Press.

Wooldridge, Jeffrey M. (2009). *Introductory Econometrics: A Modern Approach (4th ed.).* Mason, OH: South-Western Cengage Learning.

Yildirim, Ilker. (2012, August). Bayesian Inference: Gibbs Sampling. University of Rochester. Retrieved from: http://www.mit.edu/~ilkery/papers/GibbsSampling.pdf

Chapter 26

Amer, M.A. (2015). *DTFS, DTFT, DFS, DFT, FFT [PDF Slides]*. Concordia University. Retrieved from: http://users.encs.concordia.ca/home/a/amer/teach/elec442/notes/Ch8_DTFT_DFT_FS.pdf

Better Explained. (2018). *An Interactive Guide to the Fourier Transform*. Retrieved from: https://betterexplained.com/articles/an-interactive-guide-to-the-fourier-transform/

Dallas, George. (2018). *Wavelets 4 Dummies: Signal Processing, Fourier Transforms and Heisenberg*. Retrieved from: https://georgemdallas.wordpress.com/2014/05/14/wavelets-4-dummies-signal-processing-fourier-transforms-and-heisenberg/

DSP Related. (2018). *The Short-Time Fourier Transform*. Retrieved from: https://www.dsprelated.com/freebooks/sasp/Short_Time_Fourier_Transform.html

Eichenlaub, Mark. (2017, April 13). *Fourier transform for dummies*. Mathematics Stack Exchange. Retrieved from: https://math.stackexchange.com/questions/1002/fourier-transform-for-dummies

Jojek. (2014, November 25). *STFT: why overlapping the window?* Signal Processing Stack Exchange. Retrieved from: https://dsp.stackexchange.com/questions/19311/stft-why-overlapping-the-window

Pennsylvania State University. (2018). *Stat 510: Lesson 12: Spectral Analysis*. Retrieved from: https://onlinecourses.science.psu.edu/stat510/node/80/

Smith, David. (2014, January 3). *The Fourier Transform, explained in one sentence*. R Bloggers. Retrieved from: https://www.r-bloggers.com/the-fourier-transform-explained-in-one-sentence/

Ssk08. (2013, March 12). *PSD (Power Spectral Density) Explanation*. Signal Processing Stack Exchange. Retrieved from: https://dsp.stackexchange.com/questions/8133/psd-power-spectral-density-explanation

Chapter 27

Keahey, T. Allen. (2013, September). *Using Visualization to Understand Big Data*. IBM Software Business Analytics. Retrieved from: http://dataconomy.com/wp-content/uploads/2014/06/IBM-WP_Using-vis-to-understand-big-data.pdf

Acknowledgements

There are many people who I would like to thank, as I could not have written this book without them. Some of them are mentors and colleagues, others are friends and family, and others have just supported me or given me inspiration.

Bill Cormier	Kevin Vanhelden	Richard Steckel
Brandon Smith	Lenny Rodriguez	Ron Williams
Corey Lincoln	Lori Lincoln	Scott Shelters
Dan Anderson	Lou Stagner	Simran Ratra
Dan Coffman	Maryrose Edward Jayaraj	Steve Dayton
Jami Croteau	Matt Klotz	Wincent Yong
Jason Davis	Matt Lincoln	

Thank you all!

About the Author

Nicholas Lincoln is an interdisciplinary scientist and artificial intelligence researcher. He has patents pending in network science and artificial intelligence technologies, and has developed data science applications in multiple functional areas for fortune 500 companies.

Nicholas has a B.S. in Economics and a B.A. in International Studies, both from from The Ohio State University, and he is currently a M.S. student at Southern New Hampshire University in the field of Data Analytics. He is a computer scientist at the U.S. Army Research Laboratory, concentrating on high performance computing and scalable data science. He is also a co-founder of Cumulus Labs LLC, a data science and artificial intelligence consulting startup.

Future Works

Data Science in Layman's Terms: Machine Learning

Expected release in 2020